GOLD & GALENA

GOLD & GALENA

a

HISTORY OF THE MAYO DISTRICT

By

Mayo Historical Society

compiled by Linda E. T. MacDonald & Lynette R. Bleiler

with

ADDENDUM

MAYO, YUKON

THE MAYO HISTORICAL SOCIETY

1999

Second Printing (with Addendum) 1999

Canadian Cataloging in Publication Data

Main entry under title:
Gold & Galena

Includes bibliographical references.
ISBN 0-88925-934-8

1. Mayo Region (Yukon)—History. 2. Mayo Region (Yukon)—Biography. 3. Gold mines and mining—Yukon Territory—Mayo Region —History. 4. Silver mines and mining—Yukon Territory—Mayo region—History. I. Mayo Historical Society (Mayo, Yukon)
FC4049.M39G64 1990 971.9'1 C90-091404-1
F1095.5.M39G64 1990

Published by the
Mayo Historical Society
P.O. Box 32
Mayo,
Yukon Territory
Y0B 1M0

Design by
Aasman Design

Printed and bound in Canada by the
Friesen Corporation
Altona, Manitoba, R0G 0B0

Gold & Galena is dedicated to the pioneers and original people of the Mayo District.
We salute their courage, their fortitude and their ingenuity.

CONTENTS

Acknowledgements

THE MAYO HISTORICAL SOCIETY is indebted to Linda MacDonald, Lyn Bleiler, and Gordon McIntyre, who through their efforts have made this publication possible. We will remain ever grateful to Linda MacDonald for initiating the project in 1982, for her research, writing, interviewing, administration, and perseverance over the past seven years; to Lyn Bleiler for her work on our photograph collection, for her computer skills, travel, interviews and research; and to Gordon McIntyre for the wealth of firsthand information he was able to add to each topic, for his photographs, proofreading skills, and patience in answering our endless questions. We would very much like to thank Lowell Bleiler and Howard MacDonald for their patience, understanding, and support during the years their spouses worked on *Gold & Galena*, and for having their cheque books ready whenever a bill needed to be paid while the society's treasury was empty.

We are grateful to the late Dorrie Ewing for conducting taped interviews with Na-Cho Nyäk Dun elders. It was Dorrie's warm and friendly personality that made those interviews so successful. We are grateful for the firsthand information provided by Ed Bleiler, Ted Bleiler, Mary Rich, Betty and Charles Taylor, Annie and George Besner, Dorrie and Harry Ewing, Nora McLaren Swenson, Lillian Cheatham, Dr. Allan Duncan, Jean and Wilf Gordon, Doreen Doidge, Mary McDiarmid and Father Henk Huijbers.

A special thank you is extended to Helen Buyck for her assistance in preparing the Na-Cho Nyäk Dun history and for her counsel throughout the project. Thank you also to Nancy Sinnott, Catherine Germaine, Mary Hager, Lucy Peter, Sam Peter, and Paul Germaine for their contributions to the Na-Cho Nyäk Dun history.

We are grateful to our Treasurer Ina-Mae Klippert and and our Vice-President Bev Mason-Wood for their hard work and unwavering support. Thank you to Michael McGinnis and Frances Lew who assisted the book project in any way they could, and to Anne Meaney-Leckie for her work with Lyn on *The Mail Service & Radio Communications*. Thanks also to Bruce & Chris Moore, Sharalyn Brooks, Kim Amor, Jamshed and Shahaa Karkar who helped with photocopying and organizational tasks.

We thank Mabel McIntyre for her introductions to area pioneers and for her faith in our ability to complete this project.

We are grateful to the W.I.G.'s of Mayo for supplying operating funds during our lean times, and to the Mayo Community Club for granting money to buy a copy stand necessary for re-photographing.

We thank Roland Ronaghan, Jerald Graham, and Yolanda Burkhard at the Mayo Mining Recorder's office; Diane Chisholm, Lesley Buchan, and Doug Whyte at the Yukon Archives; Marilyn Kwock at the Alaska Historical Library; and Renne Blahuta at The Elmer E. Rasmuson Library, University of Alaska for their records, research, and interest in our project. We also thank Bruce MacGregor of the J. V. Clark School as well as Ron Hoeppner and Doug Harridge of the F.H. Collins Computer Lab.

We are grateful to the following people who contributed topical segments of information to *Gold & Galena*: Geordie Dobson, Margaret Wallingham, Flo Whyard, Alan Fry, Elsy Rowan, Jim Smith, Brian Denman, Tim Boyce, Lew Green, Jim McFaull and Pat Watson of U.K.H.M., Hans Barchen, John Ewing, Dorothy Howett at the Yukon Chamber of Mines, Chappy Chapman, Bessie Friesen, Mary Jean Morrison, Lowell Bleiler, Simon Mason-Wood, Suzanne Moses, Wendy Andre, Eileen and Emery Shilleto, Don Baker, Verna Klippert, Ted Wernecke, Alex McCarter, Harry Ewing, Dr. J. V. Clark, John Scott, Fred Terwilliger, Claire Briand, Cecile Curry, Pat Living of the *Yukon News*, the Village of Mayo, the Yukon Development Corporation, NorthwesTel, and the library staff of the Geological Survey in Vancouver.

Our heart-felt thanks to everyone who submitted stories for inclusion in *Gold & Galena*.

We are grateful to the following people for sharing their photograph collections with us: Gordon McIntyre, Ed Bleiler, Ted Bleiler, Ralph Rogers, Annie and George Besner, May Gillespie Fielder Morrell, Joe Long – N.C.P.C., Ken Schellinger, along with Joe & Nadine Schellinger – A. K. Schellinger Collection, Nancy Moulton – Theos Whitney Collection, Dorrie Ewing, Lucy Peter, Alla Blysak – David Hager Collection, Joanne Buyck – Buyck Collection, Fred Berger, Linda Brennan – J. Bonner Kunze Collection, Ed Kunze, Isabelle Poirier – Joe Clifton Collection, Rick Cocks – Charles Taylor Collection, Ruth MacDonald, Cecile Curry – Rich Collection, Brian Denman, Jean Gordon, Father Henk Huijbers – O'Neill Collection, Verna Klippert, Mrs. Sylvia Aho – Aho Collection, Ione Christensen, Nora Swenson (McLaren), Dorothy Walloch (McLaren), Lillian Cheatham (McLaren), Athol Retallack, and Alice and Jim Boyes.

We appreciate the formatting work done by Lowell Bleiler on the maps contained in this volume; and the editing, layout and design work by Margriet and Al Aasman of Aasman Design in Whitehorse.

We extend a special thank you to the following individuals, corporations, and agencies whose financial assistance helped to make the printing of this publication a reality: Hans & Regina Barchen, Bleiler Placers, C & I Construction Ltd., Canada Tungsten Mining Corp. Ltd., Canadian Imperial Bank of Commerce Whitehorse, Conde North Ventures Ltd., Mr. & Mrs. W. L. Drury, Duncan Creek Gold Dusters, Engelhard Industries of Canada Ltd., Finning Ltd., Gerald Mitchell Petroleum Ltd., Horwood's Office Supply, Hutton's Service, J & B Contracting, Lansing Enterprises, Lotteries Yukon, Mayo Caselots & Expediting Services, Mayo Recreation Board, Mayo Variety Store, Ruth & Gordon McIntyre, NorthwesTel, Betty & Charles Taylor, Total North Communications Ltd., White Pass & Yukon Corp. Ltd., Yukon Economic Development – Community Development Fund; The Yukon Foundation – Alec Berry Fund, and the Yukon Heritage Branch.

We will remain forever grateful to all those who realized the historical value of *Gold & Galena* and whose encouragement and faith have helped us to see this project to completion.

Preface

Richard Hartmier photo

THE RICH AND RELATIVELY UNDOCUMENTED history of the Mayo district sparked the Mayo history book project in 1982.

While parts of this district are steeped in aboriginal history, men came in from the "outside" and created history following the discovery of gold on the Stewart River, and later silver on Galena Creek and Keno Hill.

In 1883 fine placer gold was discovered, in paying quantities, on the bars of the Stewart River prompting the first gold rush in the Yukon Territory. These Stewart River miners prospected widely in the district and discovered gold on Johnson Creek in 1892, on Haggart Creek in 1895, and uncovered a major find on Duncan Creek in 1898. When the Duncan Creek discovery was made known, a gold rush occurred on that creek in 1901–02 that triggered the establishment of Mayo. Mayo grew as gold was found on other creeks in the district and boomed with the development of silver discoveries. The economic impact of these silver discoveries affected the entire Yukon Territory. Diane Chisolm Johnston wrote in *"Carrying the Territory, Silver-Lead Mining in the Mayo District, 1919–1942"*:

> Mayo quite literally carried the Yukon through the trough between the First World War and the revival of gold production by the Yukon Consolidated Gold Corporation in the middle 1930's. During this period Dawson was on a decline and many of the businesses established at the time of the [Klondike] Gold Rush had closed or were barely functioning. The economic activity associated with silver-lead production around Mayo offset this depression and provided enough stimulus to sustain the Yukon's faltering economy until the revival of gold production in the mid 1930's.

And so it was, with these tantalizing bits of historical information, that we began our pursuit of this untapped history, armed only with an intense interest and our own resources. As the years passed, the project gained a momentum and force all its own.

Our approach to this history has been "straight and true" and we have recorded it without speculation or embellishment. Mayo Historical Society volunteers carried the project alone during the seven years of research, photograph collecting, and manuscript preparation. Our limited budget restricted research that necessitated travel and only in rare instances were we able to pay for archival research. It was not until the manuscript was completed that canvassing began for funds to print the book.

Our intention was to record the history to the present time, but it soon became obvious that the quantity of material was simply too great and a decision was made to stop at the year 1955. In a few cases topical information has been recorded to the present where it seemed appropriate to do so. It will be apparent to readers that there is far more biographical information about men in *Gold & Galena* than about women. Historical records contain very little information about women and even then their names have often been obscured, i.e. married women were known only by their husband's surname. Most of the biographical information we have assembled about women has come from the recollections of pioneers.

There were once many old-timers in the Mayo district but now there are only a few. Those precious few have shared their memories with us. It is those memories that make *Gold & Galena* unique and priceless. The stories written by pioneers, or written on their behalf by others, appear as they were written. These stories reflect the individual writers' points of view and for that reason may be contradictory.

We have organized *Gold & Galena* chronologically by subject and sprinkled the text with historic photographs. These photographs are a tribute to those early photographers who, whether deliberately or accidentally, recorded history in the making. Many of these pioneer photographers developed their own films right in Mayo. Almost all of the photographs in this book have come from private collections, some as far flung as Australia and the Southern United States, which have been re-photographed by the Mayo Historical Society. It is the photographs that will remain in our memory long after the printed word has faded.

Gold & Galena is a long overdue tribute to the past and a gift to future generations.

Linda E. T. MacDonald,
Lynette R. Bleiler,
Beverley Mason-Wood,
Ina-Mae Klippert,
Mayo Historical Society

MAYO, 1990

Prologue

Above all else, the climate and the land itself have made the greatest impression on the people of Mayo district. Nature has made it a place of extremes. The landscape is both pastoral and magnificent and there is an intense contrast in colour and activity from season to season. Mayo is indeed the "hottest and the coldest" place in the Yukon. Temperatures can range 180 degrees Fahrenheit from winter to summer.

During the very cold winter, the land is still and isolated. Only the sky is alive with an occasional raven gliding overhead during the day and the flash and fall of stars and waves of northern lights during the night. Insignificant sounds are magnified. Simple breathing evokes the sights and sounds of a steam engine. This, combined with the creak of snow underfoot, makes it necessary to stop and hold your breath in order to hear other sounds. Logs burning in a wood stove sound from the roof chimney like corn popping in a kettle. The yapping of sleigh dogs, anticipating their evening meal, carries for miles over the vast, frozen land. As night falls and the temperature drops even further, nails loosen from wooden buildings like gun shots and by morning hoar frost droops from cabin ceilings, almost touching the faces of those sleeping within. All movement outside is slow. Machines clank and complain. Technology kneels in the face of the cold. Only the splitting of wood is easy, the rock-hard logs responding to the slightest swing of an axe and dropping in pieces from the chopping block with a tinkling sound.

The long hours of darkness are wearisome and when the sun shows itself in the late morning and early afternoon, human faces cleave to it like winter flowers. At night the winter-weary watch moonlight radiate through icicles above the window and dream of summer strawberries.

Then, quite suddenly, winter falters. Days grow longer and the snow begins to whisper and melt. Memories of sixty below flicker into the subconscious. The enchantment begins.

The earth shimmers to green. Grouse drum like heartbeats. The song of a hermit thrush pierces the evening, so singularly pure and perfect it touches even the most shrivelled soul. The fragrance of budding leaves and warm earth are intoxicating. Crocus, lupine and Jacob's Ladder splash purple and blue on green.

By the middle of May Nature's spell has completely won every heart. The creek gravels are laced with gold, the hills glitter with silver, the land is bountiful. People are alive and free in the heart of the Yukon and there is no better place to be.

Spring suspends time and then plunges into summer. Night disappears. Miners and merchants rush about, salmon thrash upstream and fish camps are reopened, gardens sprout, cabin walls are freshly chinked with mud and moss. The fireweed begins its radiant ascent, wild roses perfume the air and daisies circle in white and gold. Berry patches are buzzing with pickers. Now and then smoke or bear smell drifts on the wind. Friends visit long into those warm evenings so unique to the Stewart River valley. Mosquitoes and black flies are the only sting in paradise.

– by Linda E. T. MacDonald

GOLD & GALENA

Mayo from the air, circa 1960. Photographer, Gerry Podhora. *G. J. Gordon Collection*

Chapter 1

Geography

WRITING FOR THE GEOLOGICAL SURVEY OF CANADA, Memoir 284, Yukon Territory,[1] Hugh S. Bostock makes a full and accurate description of the extent and geography (*see maps on pages 18, 32 & 34*) of the Mayo district:

"The Mayo district includes the Stewart River and its tributaries from Mayo River eastward, the Mayo River and its tributaries and the north and south branches of the McQuesten River and their tributaries.

The Stewart River is one of the main tributaries of the Yukon River. It rises in the Pacific-Arctic watershed ranges lying between the heads of the Peel and Pelly rivers, and flows in a general westerly direction towards the Yukon valley. The Stewart enters the Yukon River from the east at a distance of 58 miles south of Dawson. From Fraser Falls to its mouth, a distance of nearly 200 miles, it is a large stream, seldom less than 150 yards in width and often more than double this breadth. At Fraser Falls, the Stewart flows for a third of a mile with great velocity through a narrow canyon bounded by vertical walls of (rock). The word falls is a misnomer, as the grade in the canyon is fairly uniform and the total descent was estimated to be only thirty feet. The principal tributaries of the Stewart below Fraser Falls are the McQuesten and Mayo Rivers, both fair sized streams, Clear Creek from the north, and Crooked River, Lake Creek and Scroggie Creek from the south.

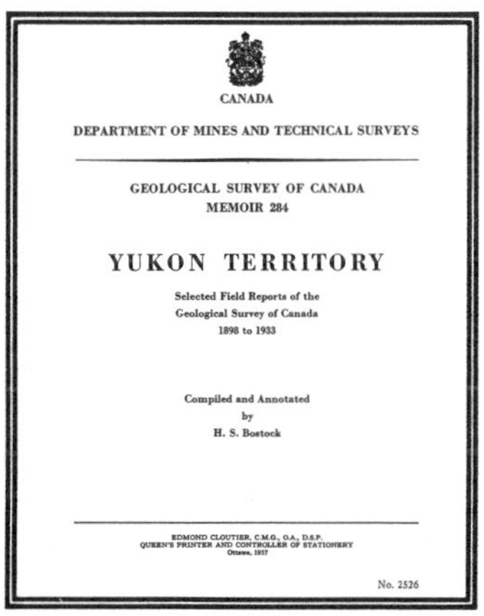
CANADA

DEPARTMENT OF MINES AND TECHNICAL SURVEYS

GEOLOGICAL SURVEY OF CANADA
MEMOIR 284

YUKON TERRITORY

Selected Field Reports of the
Geological Survey of Canada
1898 to 1933

Compiled and Annotated
by
H. S. Bostock

EDMOND CLOUTIER, C.M.G., O.A., D.S.P.
QUEEN'S PRINTER AND CONTROLLER OF STATIONERY
Ottawa, 1957

No. 2526

Memoir 284, Yukon Territory

Mayo Lake is the largest sheet of water in the district. The main body of the lake is twenty miles long and from one and a half to two and a half miles wide and lies in an east and west direction. A narrow arm of the lake, twelve miles long, extends to the southeast. The northern shores of the lake rise in gradual slopes to the Gustavus Mountains. Of the Gustavus Mountain group overlooking Mayo Lake, the highest point is Mount Albert, at 6,500 feet above sea level, while Mount Hinton, of the Duncan Creek group is the highest point of all, being 7,000 feet above sea level. [Mayo Lake is larger now due to the building of the N.C.P.C. dam. The construction of the dam also created Wareham Lake upstream of the Mayo River Canyon.]

Northeast of the Gustavus Mountains, but separated from them by the wide valley of the Ladue River, rise the Davidson Mountains, some of whose peaks are as high as Mount Hinton. This group is a spur from the Rocky Mountains, whose higher peaks appear in the distance in continuous array, sweeping in a great curve towards the northeast. Twenty-two miles west of Mount Hinton and rising from the valley of the McQuesten, is Mount Haldane (6,000 feet above sea level), a very prominent feature, visible from many points on account of the wide valleys which lead to and surround it. Locals also refer to Mount Haldane as Lookout Mountain because of these characteristics. [The Indian word for the mountain is Nátth'än which means lightning.]

"... whose higher peaks appear in the distance in continuous array, sweeping in a great curve towards the northeast."

The Stewart River above Fraser Falls drains an area of about 12,000 square miles. During its course through this region it receives four important tributaries, the principal one being the Hess River, or south branch of the Stewart, the Lansing River, Ladue River and Beaver River. The headwaters of the Stewart River and its branches have their source either in the Ogilvie range to the north or in the Selwyn range to the east. These two mountain chains form the watershed between the Yukon and Mackenzie drainage basins in this region."

Three Mile Rapids above Fraser Falls. *T. McKay Collection*

Chapter 2

The Na-Cho Nyäk Dun

THE STEWART RIVER, called Nahcho Nyäk – *Big River* – in Northern Tutchone, remained almost completely isolated until the discovery of gold there in 1883. Earlier explorers, therefore, wrote very little about the area and its people.

In *Through the Eyes of Strangers*, Yukon historian Julie Cruikshank writes, "Mayo Indian people originally came from nearby camps along the [Stewart] river, as well as from Fort Selkirk, Lansing, Fort Norman and even Fort McPherson."[2]

Robert Campbell of the Hudson's Bay Company, while at Fort Selkirk in 1849, sent his assistant clerk James Green Stewart to the as yet unnamed Stewart River in search of a group of Indian people who were believed to have a hunting camp along the river. Stewart explored the river further than any non-native had done before and as a result, Campbell named the river after him.

Chief Johnson in 1920
Ralph Rogers Collection

Explorers Schwatka (1883) and Dawson and Ogilvie (1887) were told by Indian people living at the mouth of the Stewart River that more of their people were camped along the river's drainage.

Mayo elder Sam Peter relates that Fort Selkirk Indians traditionally travelled the Stewart River country to trap and hunt. From Fort Selkirk they travelled up the McQuesten River (*Et'O Nyäk* — to paddle against the current) to the Beaver River, then down the Stewart to the McQuesten and back to Fort Selkirk. Peter went on to say that when gold was discovered on the tributaries of the McQuesten River in the late 1800's, a miners

The history of Mayo Indian people highlighted here is intended to complement the individual elders' stories in this book. All firsthand information was obtained from Mayo people. The remaining information was researched from books and records held in the Yukon Archives, Yukon libraries, and in private library collections. While the research is reliable, it should not be considered exhaustive. For those interested in pursuing this subject, there are several good books available today on Yukon Indian history and culture that include information on Stewart River people.

supply post was built at the mouth of that river and a group of Selkirk Indian people settled there.

By 1901, the N.W.M.P. referred to McQuesten village as the headquarters of Stewart River people with fishing/hunting camps located above Fraser Falls and at Ethel Lake. Sam Peter was born at the Ethel Lake camp. His grandfather, Chief Johnson, was a leader at McQuesten village.

Elder Paul Germaine (Jermie) tells us the history of Lansing people in this district:

> Indian people came from Fort Good Hope on the Mackenzie and stayed at Lansing. There were many tents there, more people than at the Village. People would dance far into the night, play drums and sing. They played drums and stick gambled.
>
> People travelled back and forth from Lansing to Fort Good Hope, Fort Good Hope to Lansing, until the border was drawn between the Yukon Territory and the Northwest Territories [1898]. The first time they travelled back to Fort Good Hope, all the people went.
>
> Most of the families who were there [in Lansing] the first time came back to Lansing again and some new families came with them. Again they lived at Lansing. Family groups went to various areas to make food, such as dry meat and fish, for winter. In winter they would leave to trap and come back to Lansing in the spring to trade fur. People travelled a wide area to trap and hunt. They went up the Stewart River, Rackla River, Nadaleen River, Beaver River, and Hart River.
>
> The third time, half of the Mackenzie people went back to Fort Good Hope and the rest stayed at Lansing. Then sickness fell upon the people of Lansing. Most of the people died. Only one or two families were left.
>
> St. Pierre (Izhihó) was head man of the Mackenzie people. His wife was Mezié. They had a son, Däboo, and two daughters Kätini (or Katrine) and Mary. Kätini married Philie Unëbel. Däboo was sent outside by Jim Mervyn and he died there of stomach disease.
>
> Grandma Tulä and her husband Jerroo also lived at Lansing.
>
> Some of the people's names I can remember who went back to Fort Good Hope are Pete So'-de'chili, Je begün, [and] Tsá Ta'tinì who I believe has a daughter still living in Yellowknife.

G. A. McIntyre's postcard – Yukon's oldest couple, Mr. & Mrs. Atsitsaw of Mayo. He died in 1959 at 105 years; both were born on Mackenzie River. *A. Aho Collection*

Et'tsì ts'ò and Mädlin [also known as Myr Lithon Atsitsaw and Taja] were from Aklavik. They adopted Söfiá Christine and raised her there. Et'tsì ts'ò left Aklavik when his younger brother died of starvation. The family moved to Fort Good Hope where

Söfiá Christine met and married Jermie. Söfiá Christine and Jermie were my mother and father. [See the Germaine Family story in Pioneer Accounts.] Jermie and Söfiá, Et'tsì ts'ò and Mädlin moved to Lansing with the rest of the Fort Good Hope people who went over.

... as told to his niece Catherine Germaine

Germaine's grandfather, Et'tsì ts'ò, died on May 20, 1959, at the remarkable age of 105.

In 1902, Frank Braine and Percival Nash asked a group of Fort Good Hope people to guide them to the Stewart River by way of the Peel River so that they could open a post at Lansing. Braine and Nash had previously rushed to the Klondike gold fields and then returned to the Mackenzie River. Approximately 15 families accompanied the traders. The group crossed the divide and settled at the confluence of the Stewart and Lansing Rivers. Braine and Nash's store was known as Lansing Post.

Approximately 90 miles southeast of Lansing Post, there apparently existed another Indian village circa 1918. In an article printed in *The Mayo Miner* on April 29, 1938, prospectors Dick McDiarmid and Robert Martin described the abandoned village they had discovered at Husky Dog Creek while on a trip to the McMillan River Country. Husky Dog Creek lies east of Fairweather Lake and flows into the North McMillan River. In the early days this stream was referred to as Husky Dog River rather than Husky Dog Creek.

> One of the most interesting parts of the trip was the finding of a deserted Indian village about two miles above the mouth of Husky Dog River. There were thirteen cabins, Dick related, one of these being a big hall not completed. The logs for finishing the building were lying on the ground close by, also a large pile of whipsawed lumber in a good state of preservation. Four wash tubs had been left by the Indians, before deserting the camp, a number of stoves and the frames of several sewing machines, as well as other paraphernalia which was apparently too heavy for the toboggans. McDiarmid believes the Indians were from the Mackenzie River side as the cabins and buildings were all built with dovetailed corners and joinings. Judging from what he saw Dick figured the camp must have been deserted about twenty years ago.

Lucy Peter, Persis Kendi, Mary (née Kendi) Hager, Ella Moses and a child cutting wood. *Dave Hager Collection*

Chief Alfred "Albert" Moses and his wife Alice in the 1930's. *William Hare Collection, Yukon Archives*

Sam Peter relates that this village was built by

Mackenzie people from Fort Norman. When he had trapped near Russell Post in 1916, he said, people at the post told him there were Fort Norman people trapping on Husky Dog River, and that they had trapped there every winter for quite some time. Peter said he believes there is a man named Andrews, now living in Ross River, who as a child, lived with his parents at Husky Dog River.

In *Part of the Land, Part of the Water*, anthropologist Catharine McClellan describes the seasonal activities of Stewart River people until the arrival of outsiders:

> People in the Mayo area could fish all year. In April, May and June they trapped grayling, especially at Fraser Falls, [and also] near the now abandoned trading post of Lansing. Families gathered there to catch and dry fish, and to hunt muskrats and beaver.
>
> Summer salmon fishing followed. In late July, the salmon arrived at Dawson, and a week later they reached Mayo. Traps were set on the Mayo River near the mouth of Mayo Lake and near the old village of McQuesten on the [Stewart] River. Salmon were also caught near Fraser Falls, in eddies in the Stewart River, and along the Hess and Beaver Rivers.
>
> The king salmon that come in early August are fat, and they were dried for people to eat. The dog salmon that come later are usually in poor condition, and they were dried only for winter dog food.
>
> In winter the people could still fish at Mayo Lake, Kathleen Lake, or Ethel Lake, taking grayling, trout, whitefish, and jackfish with spears or nets. They could also stay on at Fraser Falls, where they could combine hunting and trapping with fishing.
>
> Other families moved out in the fall and winter over a wider area to hunt and snare game. In the fall they snared sheep in the Ogilvie Mountains and around the Wind River or across the divide in the Mackenzie Mountains near the Arctic Red River. In winter, they built moose fences and caribou corrals in the mountains north of the Stewart River.[3]

Alice Hager with daughter Rosie on the left and Mary Hager with her daughter Bella. *Claude Tidd Collection, Yukon Archives*

Mayo Elder Helen Buyck remembers as a child in the 1930's, seeing thousands of caribou gathered at the mouth of the Mayo River and crossing on the ice of the Stewart River. *The Mayo Miner* reported caribou in the vicinity of Five Mile Lake as recently as 1937.

Stewart River people speak Northern Tutchone (which is an Athapaskan language) and some of the people speak Loucheux and Slavey. They identify themselves with the moieties Crow and Wolf. (The traditional social patterns of Indian people are very complex. The following is only a general description of the early Crow and Wolf tradition.) The relationship between persons of the same crest (Crow or Wolf) is considered to be nearer than that of the same tribe or geographical group. Members of the same tribe may, and do marry but those of the same crest are not allowed to do so. The child always takes the crest of the mother so that if the mother is a Wolf, all of her children will be Wolf. (In early times children received the name of their mother; so descent was traced from the mother, not from the father.) If a member of the Crow clan died, it was the responsibility of the Wolf clan to make the appropriate burial arrangements and the reverse if a member of the Wolf clan died.

Mayo elder Mary Hager sheds some light on Indian culture in earlier times:

> Women sewed pants and shirts made of moose skin. In winter caribou coats were worn. Caribou and moose skins were also used to make tents and boats. Stewart River women pierced their noses and ears [for adornment]. Lansing women did nice skin work and their clothing was different than that of Mayo women. The Lansing women also did porcupine quilling and embroidery work.
>
> Children were betrothed at an early age. Marriages were arranged by the parents and the family prepared a feast for everyone to celebrate the event.
>
> Fathers arranged a feast to celebrate the birth of a child and when a teenage boy shot his first moose or other large game animal.
>
> Boys were taught how to hunt by their fathers and as they approached manhood were not allowed to go out of the house for a time. Girls received instruction in food preparation and collection, housekeeping, sewing and tanning. During their transition from childhood to womanhood, girls lived three miles from the village in relative solitude, with only occasional visits from other women.
>
> Elders pass their knowledge of herbal remedies to someone within the family unit and in that way it is carried to future generations. Some examples of herbal remedies are: Indian root is a remedy for stomach ailments, skin rashes, itches, and acne. The correct name of this particular plant cannot be revealed because it looks very much like another plant known to be poisonous. Spruce gum maintained healthy teeth, eased toothaches, and was used for the prevention of scurvy in early times. [Spruce gum is an opaque lump of sap that appears on the trunks of spruce trees. The gum is broken off and chewed. Initially, it is very bitter and does not hold together while it is being chewed. The longer it is chewed, the more pleasant the taste becomes, and it acquires the consistency of modern day chewing gum.]
>
> *as told to her granddaughter, Nancy Sinnott*

ABOVE, Sam Peter and Dave Hager beaver hunt; drying beaver skins at Lake Creek, 1925. *D. Hager Collection*

BELOW, A group of Indian people travelling the river with canoes & dogs. *J. B. Kunze Collection*

In the early days, Stewart River people travelled to Fort Selkirk to trade and socialize. Mayo elders remember that Lucy Peter's mother and father, Eliza and Lucas, travelled to this meeting place.

When the town of Mayo was established in 1903, the people at McQuesten moved to Mayo and settled near the place where the Mayo River (*Tedze Nyäk* – muddy water) empties into the Stewart River. Their village at the mouth of the McQuesten River was abandoned.

In 1908 James E. Ferrell bought Lansing Post from Braine and Nash. The Lansing country continued to provide trappers with a rich supply of furs. Lansing people were excellent builders. They were hardy people, often travelling and hunting

Two views of the "Old Village" near Mayo. Above, *A.K. Schellinger Collection* Below, *D. Hager Collection*

for days in freezing weather with only one blanket. Church records show that Lansing people were Catholic and very much attached to their faith. Missionaries were not aware of their presence at Lansing for some years; therefore, it was not until the summer of 1908 that a priest arrived from Dawson to spend time with them.

In 1915, Ferrell sold Lansing Post to Jim Mervyn, a trader from Stratford, Ontario. Mervyn's wife, Julia, was a Lansing woman who had originally come from Fort Good Hope. They had a family of eleven children. Mervyn did not use cash for trade; instead he used tokens of aluminum and brass with his name stamped on them.

In the same year Julius Kendi arrived at Fraser Falls (*Tu Ninlin* – water flows over the rocks) where Mayo people were drying fish. Kendi was an Indian catechist of the Anglican faith, from the Peel River district. He was accompanied by Rev. Frank Buck and they held church services and baptized several people. Mayo people referred to the language spoken by Kendi as "Moosehide language," meaning Loucheux.

When Kendi came down from Fraser Falls to the Indian village at the mouth of the Mayo

Anglican Church and its congregation at the "Old Village" in 1936. *Claude Tidd Collection, Yukon Archives*

1. Paul Lucas
2. Charlie Jimmy
3. David "Little Dave" Moses
4. Ella Moses
5. Steven Louis
6. Jenny Jimmy ("Old Jenny")
7. Caroline Lucas
8. Sally Lucas
9. Mrs. Bud Fisher
10. Flora Harper
11. Billy Davis
12. Sophie Davis
13. Sarah Johnny
14. Lucy Peter
15. Rev. Bill Valentine
16. Arthur Davis
17. Mary Peter
18. Joseph Martin
19. Simon Lucas
20. Alice Jimmy
21. Rev. John Martin
22. Clifford "Kippy" Fisher
23. Alice Hager
24. Effie Martin
25. Edwin Hager
26. Charlie Profeit
27. Mary Hager
28. John Whitney
29. Henry Breaden
30. Alice Moses
31. Moses Lucas
32. Nancy Whitney
33. Jimmy Lucas
34. Mary Forbes
35. Chief "Tom" Alfred Moses
36. Arnold Theodore Hall (who at times filled in for minister)
37. Charlie Peter
38. Mary Moses
39. Doris Fisher
40. Johnson Peter
41. Mrs. Clifton
42. John Peter
43. Mrs. Theos Whitney
44. Phyllis Andison
45. George Andison
46. Mrs. Valentine
47. Sam Peter

River, he asked the people to select a site where they could establish a village of their own. After a lengthy discussion, the decision was made to locate two miles below the Village of Mayo on the banks of the Stewart River. "Tom" Moses was chief at that time. Once the site had been selected, people began the task of building the church and rectory, and homes for themselves. Kendi was ordained a Deacon in the Anglican Church in 1918 and went on to serve the Mayo people for many years. In her story, Helen Buyck, granddaughter of Julius Kendi, tells of the traditional lifestyle led by people at the village (see *Pioneer Accounts*). The village on the Stewart is now referred to as the "Old Village." People lived happily there, by all accounts, for 55 years.

In 1918 the village at Lansing was decimated by epidemics of influenza and tuberculosis, leaving only a few survivors. Some of these remained at Lansing, others returned to Fort Good Hope, and some journeyed down the river to settle in Mayo and other parts of the district.

Indian education, a federal responsibility, was in the hands of the church during the early days. Reverend Kendi realized the value of education and started a school in which Charlie Jimmy and Mary Moses taught. The school was eventually staffed by the Anglican Church and financed by the Department of Indian Affairs. This school was part of the Anglican mission. Classes were conducted in the church at the Old Village. The missionaries (usually stationed at St. Mary's Church in Mayo) were responsible for teaching the Indian children; in turn, the Diocese received payment for each day the "school" was open.

TOP, The Rev. Julius Kendi going down river with all of his family, including Mary & Edwin Hager. *Dave Hager Collection*

MIDDLE, An Indian fish trap below Mayo, circa 1920. *McLaren-Walloch Collection*

BOTTOM, Jenny Jimmy, "Old Jenny," in the final stage of tanning a caribou hide. *A. K. Schellinger Collection*

In the 1920's and 1930's, men began to work away from the village. They worked as market hunters and sold meat and fish in Mayo. It was legal to sell wild game at that time. Men also worked as deckhands on the river boats, cut firewood for the steamboat engines, worked as camp hunters and guides, and obtained work in the silver mines at Keno Hill. Colley Germaine, who hunted and trapped in the Ladue Lake district, was a licensed market hunter in 1935.

A tremendous flood occurred on the Stewart River in 1936. The floodwaters inundated the Old Village and destroyed the church, and thus the schoolhouse. Sadly, the beautiful costumes and ornaments used for traditional dancing at the Old Village were washed away and never replaced.

Lansing Post was also devastated and in the

flood's wake, Jim Mervyn closed his store.

After the flood, children from the village either attended school in Mayo, went to the Anglican residential school in Carcross (Chooutla), or the Catholic residential school at Lower Post. Residential school was an unpleasant experience for many Indian children. It meant long periods of separation from their families while trying to adapt to a completely different social environment.

The church at the Old Village was rebuilt a little further from the river bank, using the logs from the original structure.

Christmas was always a happy time at the Village and people who were away returned to celebrate with family and friends. An item in *The Mayo Miner* describes Christmas at the Old Village in 1937.

Christmas dance at the Old Village in 1935. *G. A. McIntyre Collection*

> The Christmas service and celebration of Holy Communion was conducted on Christmas morning at the Mayo Indian Village by Rev. Wm. Valentine. On Wednesday night the annual Christmas tree and concert was held at the village under the supervision of Rev. and Mrs. Valentine. A large number of villagers and their wives and families were in attendance to enjoy the yearly event. A dance, at which some 15 Mayoites were in attendance, was held after the Christmas tree party and feast. Every night this week a dance has been held at the village. All the residents of the village are

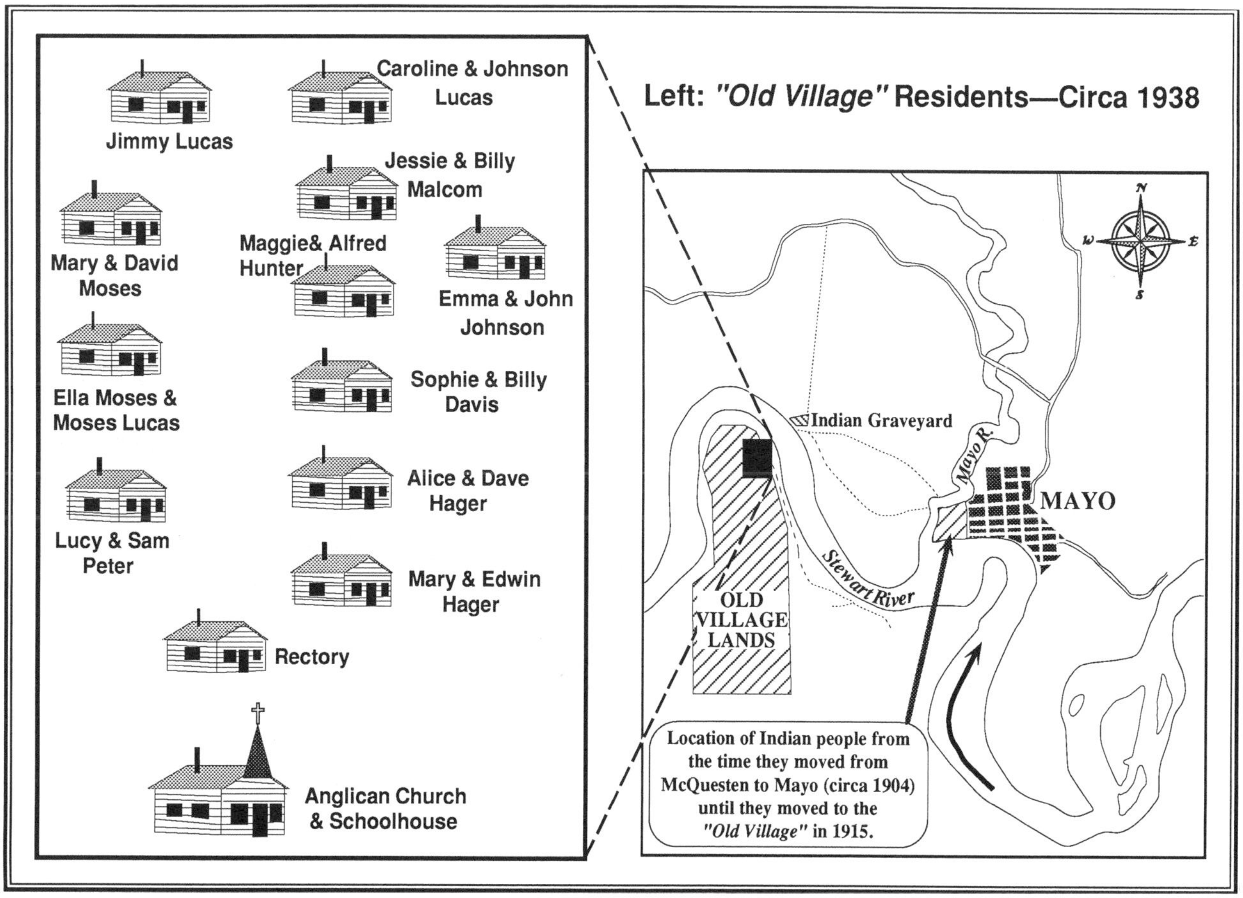

L. E. T. MacDonald & E. L. Bleiler

home for Christmas from many outlying parts and the Christmas festival is proving one of the merriest held in many years. (*The Mayo Miner*, December 31, 1937)

> **"... Old Village people had worked hard all of their lives to feed, clothe and provide shelter for their families..."**

By the 1940's, Indian people had moved out of Lansing entirely, leaving the place abandoned.

Around 1949, Indian Affairs altered the relationship between church and state by sending Sarah Anne "Sadie" Dougall, a certified teacher, to the Old Village. She taught there for two years and, by all accounts, was very successful. The church objected to this interference and arranged to have the department replace Miss Dougall with missionary teacher Lillian Connor. Miss Connor was replaced by Hilda Hellaby in 1952.

The Mayo Indian Band was organized in the early 1950's along with all other Yukon Indian Bands in the territory. The organization of Bands was directed by the Yukon Indian Agency and was controlled by the Federal Department of Indian Affairs. It introduced the concept of an elected chief and council. Until that time, the council of Elders had selected their leaders. A representative from each family unit constituted the general council.

In the early 1950's an R.C.M.P. officer visited the people at the Old Village and told them that the Indian Agent had arranged credit for them at the store in Mayo. Government assistance for Mayo Indian people was introduced in the 1920's but it was not fully understood by the people until the 1950's. The Old Village people had worked hard all of their lives to feed, clothe and provide shelter for their families; therefore, the concept of government assistance was very difficult for them to understand. Indian people referred to this assistance as "Jaw Bone."

In 1958 it was determined by local health authorities that the drinking water at the Old Village was polluted and Indian people were re-

Christmas celebrations at the Old Village in 1935. *G. A. McIntyre Collection*

quested to move back to the Town of Mayo. This began a new era for Stewart River people.

In March of 1987, Mayo Indian people chose the name "Na-Cho Nyäk Dun" which means "big river people" and in the summer of 1989 they returned to their traditional selection of chief and council.

Mayo Indian culture is rich in songs, stories and legends of varying themes.

The legend of bushmen is well known in the Mayo district. Bushmen are seldom seen but they have been encountered by people within recent memory. They are located mainly on the east side of the Stewart River and as recently as the early 1910's rained arrows and spears on people camped at Fraser Falls. Bushmen leave no tracks.

People of Lansing speak of "the place of shining rocks" in Bonnet Plume country which is guarded by huge birds with shaggy heads (like those of lions). Smaller birds ride on their shoulders to hunt for them. Any man who goes to that place must go in and come out very quickly.

Another legend, concerning the moon, is told by Mayo elder Mary Hager:

> An old woman raised one grandson. He wore an outfit made of two marten skins. The grandson was very special.
>
> The people were hungry. The grandson told the people that they would soon kill caribou. In a dream he saw the place where the caribou were. He told the people he would tell them where they could kill caribou but they must save one fat caribou for his grandmother.
>
> The people agreed and killed many caribou. After the hunt the grandson's uncle told the people to ignore the grandson's request and they gave the grandmother a caribou arm with no meat on it. When the grandson saw this he cried and cried – upset that the people did not listen.
>
> He told his grandmother to put the cari-

Dancers wearing traditional costumes during Christmas celebrations at the Old Village, 1935. *G. A. McIntyre Collection*

Jenny Jimmy, affectionately known as "Old Jenny," in the 1930's. *William Hare Collection, Yukon Archives*

bou arm in white skin after cooking it and the next morning there was meat on the caribou arm. Then he said, "I will be leaving you soon grandmother." When they went to bed the grandmother clung tightly to her little grandson but he disappeared. He told his grandmother, "You will see me again in the moon when it is full and bright as the sun, and when the people see me they will fall down as if they were hungry or carrying a heavy pack and they will be reminded of their selfishness. When they get up everything will be fine again."

Some people from the Mayo Band have seen this and believe it still.

... as told to her granddaughter, Nancy Sinnott August, 1987.

NA-CHO NYÄK DUN LEADERS to 1955[4]

CHIEF JOHNSON – *McQuesten Village era –1925*

CHIEF 'TOM' 'ALBERT' ALFRED MOSES – *1925–1947*

BILLY MALCOLM – *1947–1949*

DAVID MOSES – *circa 1950–1964* (traditional Chief)

ALBERT MOSES – *1951–1971* (elected Chief)

The Na-Cho Nyäk Dun prepared by Linda E. T. MacDonald

"Old Village" people, circa 1930. Children in the photograph are unknown. Front row from left: Sally Lucas, Ella Moses, Sophie Davis, Mary Hager, and Lucy Peter. Back row: Rev. Julius Kendi, "Big Dave" Jimmy Hager, and Edwin Hager. *William Hare Collection, Yukon Archives*

Chapter 3

The Stewart River

MCQUESTEN, MAYO & HARPER

MAYO LAKE, MAYO RIVER, AND THE TOWN OF MAYO are named for Captain Al Mayo while the McQuesten River is named for Leroy Napoleon "Jack" McQuesten. These place names, the only ones in Yukon to be named in their honour, pay tribute to the famous pioneers who were early explorers, prospectors, and later, traders in the firm Harper, McQuesten & Mayo.

Leroy Napoleon "Jack" McQuesten was born in 1836 at Litchfield, New Hampshire, U.S.A. As a young man he was attracted to life in the wilds and entered the employ of the Hudson's Bay Company at Fort Garry in what is now Manitoba.[5]

Al Mayo, 1895. *McKay Collection, Univ. of Alaska Fairbanks*

For approximately two years, he worked in the area between Minnesota and Hudson Bay before leaving the Company for the California gold fields. From there, McQuesten followed the gold trail north, working on the Fraser River in British Columbia from 1858 to 1863 and the Peace River to 1864. The end of that year found him trapping near Fort Vermilion, Alberta, where he took 200 marten and a few lynx over the course of the winter. It is not known if he was previously acquainted with Al Mayo or if he met Mayo that winter, but in the spring of 1865 the two men travelled together to Manitoba on Hudson's Bay Company boats.

At that time, Alfred H. "Al" Mayo was 18 years old. While it is not certain how he came to be in that part of the country, it is known that he was born in Bangor, Maine on February 7, 1847. It was said of him that he had been a circus acrobat in his younger days and later, in his northern travels, he often astounded the natives with displays of his agility.

McQuesten and Mayo traded and trapped in various parts of the country. In 1871 they worked the Great Slave Lake area and that fall they travelled 350 miles up the Hay River and wintered there. It was there

Prepared by Linda MacDonald from the George T. Snow Papers

they decided they would go to the Yukon and look for gold. The men in the employ of the Hudson's Bay Company (HBC) at Hay River told them a great deal about the Yukon River. James McKniff also wintered on the Hay River that year and he decided to accompany McQuesten and Mayo.

The trip would take them nineteen hundred miles through uncharted country. At that time there were no maps of either the Mackenzie or Yukon River valleys. The only information they could obtain would be gathered from Indians they met along the way.

In the spring of 1872, they crossed the divide to the Liard Valley and started down the Liard River. They travelled as far as the Nelson River (a tributary of the Liard) and wintered there. In November three men from the HBC arrived and wintered there as well in order to prevent McQuesten's group from trading furs. During the

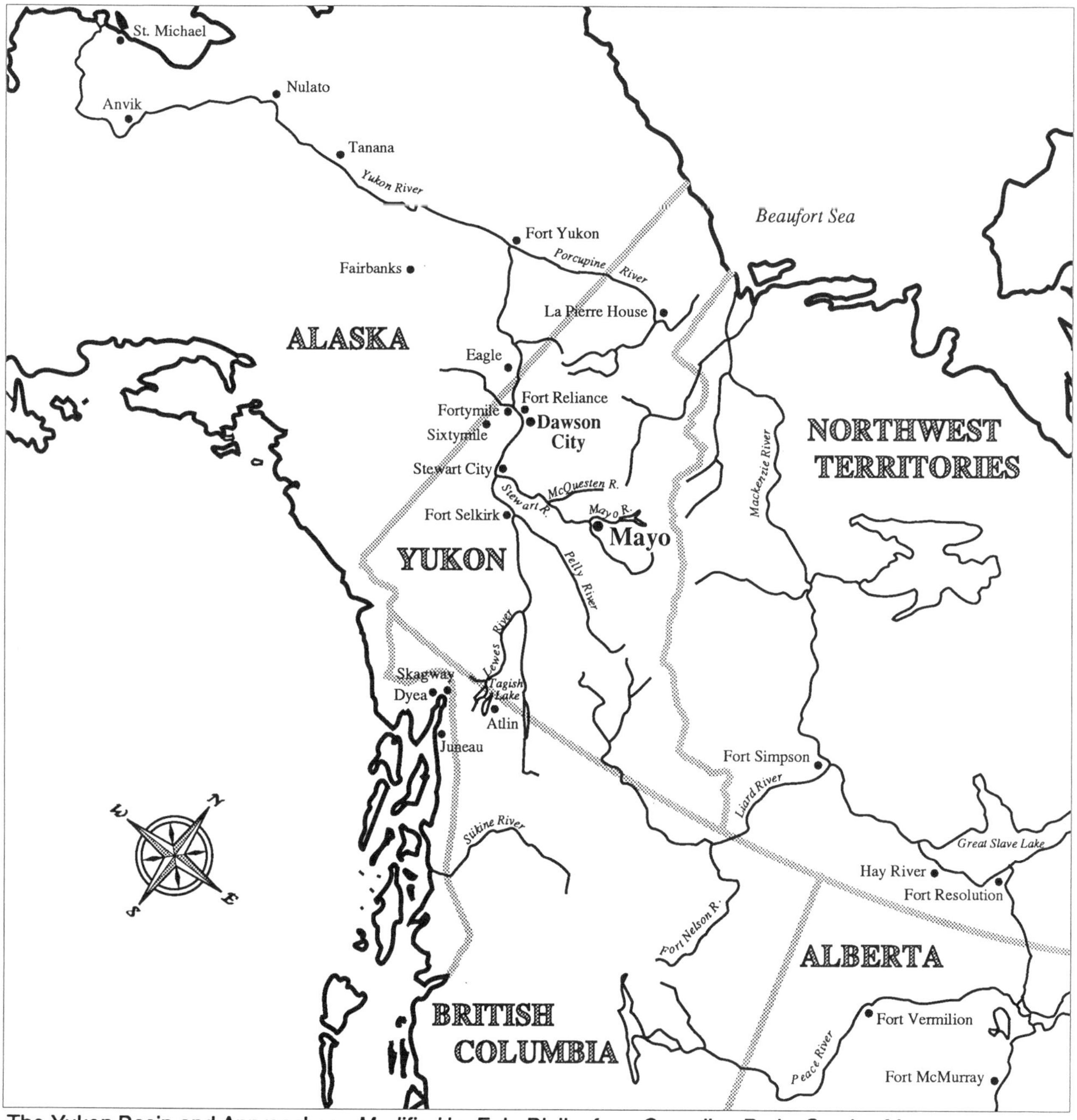

The Yukon Basin and Approaches. *Modified by E. L. Bleiler from Canadian Parks Service Map*

long winter evenings the HBC men told the trio more stories of gold on the Yukon River.

When the ice went out in the spring of 1873, five men came down the Nelson in a dugout canoe they had hewed from one of the huge cottonwood trees that grow along the Liard Valley. They were Arthur Harper, originally from County Antrim, Ireland; Frederick Hart, also from County Antrim; George Finch, a Canadian from Kingston, Ontario; Andrew Kansellar, a German; and Samuel Wilkinson, an Englishman. They had come from Fort St. John on the Peace River, having sledded their provisions across the mountains to the headwaters of the Nelson during the winter. Their plan was to prospect the Liard that summer and, in the long term, the Mackenzie and finally the Yukon. Now, however, their plans changed and they decided to accompany McQuesten, Mayo and McKniff into the Yukon. Only Samuel Wilkinson chose to remain and prospect the Liard.

The group of seven travelled together as far as Fort Simpson. McQuesten and Mayo had business to do in Fort Resolution on Slave Lake before they left, and so James McKniff went down the Mackenzie with the Harper group.

McQuesten and Mayo arrived back in Fort Simpson about a month later and proceeded down the Mackenzie, arriving at Peel River in July, 1873. Here George Nicholson joined them. Nicholson had been prospecting in the Peel River district for over two years and had found only traces of gold. The HBC provided the three with Indian packers and they packed across the mountains to the Porcupine, a distance of 75 miles. They rafted down the Porcupine to La Pierre House where McKniff was waiting for them with lumber to build a boat. They started for the Yukon River with 1,400 pounds of supplies and four dogs. On the trail they met Bishop Tachas. They arrived at Fort Yukon on August 14, 1873. Harper and party had gone up the Yukon two weeks before. Andrew Kansellar left the Harper group in Fort Yukon. Moses Mercier was the agent for the Alaska Commercial Company (ACC) at Fort Yukon and he gave them 50 pounds of flour. It was the first they had had for two years.

Mercier told them that there were three men camped about forty miles down the Yukon River on Beaver Slough, so the men decided to drift downstream for a visit, as one of the men was a cousin of McQuesten's. The three men, Sam Williams, Arthur Kensley, and George Finch (of Harper's party), were former Hudson's Bay Company employees. McQuesten, Mayo and Nicholson decided to spend the winter at the Beaver Slough camp. They built a cabin and obtained a good stock of meat — four moose and two bears. White fish were plentiful in a nearby lake. They cut two foot square pieces of ice from the lake to use as cabin windows and changed them once a week. McQuesten and George Nicholson made a trip to La Pierre House that winter to do some business with the HBC and arrived back at Fort Yukon on May 10, 1874, just as the ice went out on the Yukon River.

“During the long winter evenings the HBC men told the trio more stories of gold on the Yukon River.”

Moses Mercier took McQuesten and Mayo from Fort Yukon to Tanana in the ACC boat. They were in Tanana for about two weeks while the Indians traded their furs to Mercier. During this time Harper and party arrived. They had wintered at the mouth of the White River. The parties continued on to St. Michael with Mercier as there were no provisions in Tanana Station. Moses Mercier's brother Frank was the agent in St. Michael and he treated the group well.

In St. Michael, Jack McQuesten, Al Mayo, and Fred Hart were engaged by the Alaska Commercial Company. Frederick Hart was a life long friend of Arthur Harper. Hart, born in 1835, had graduated from the University of Dublin as a medical doctor. When an epidemic struck his father's cattle herd and destroyed it, the family moved to Canada. Hart never practised medicine after leaving the university. He was a school teacher for a time. In Canada he became a hunter, trapper and prospector.

Arthur Harper, George Finch and James

McKniff received supplies from the company and went to look for gold at Gold Mountain near Tanana. George Nicholson and Andrew Kansellar (also referred to as A. Gutter and Al Gesler) went to San Francisco.

McQuesten, Mayo and Hart left St. Michael on the steamer *Yukon* on July 7, 1874. They had four barges in tow loaded with ten tons each of goods for the Indian trade. They left one barge at Anvick, one at Nulato and one at Tanana. They arrived at Fort Yukon on August 10. On August 18 the *Yukon* took McQuesten, Mayo and Frank Bernfield up the Yukon River to establish a new post. They had four tons of merchandise and a whale boat in tow. It was slow going as the river was very low and it was the first time the boat had been above Fort Yukon. They selected a spot 350 miles from Fort Yukon and five miles below what later became Dawson City. The goods were landed and the steamer and Mayo returned to St. Michael. It was August 27, 1874. McQuesten and Bernfield were left to build winter quarters. The Indians were very friendly and helpful and supplied them with meat all winter. The trade in furs was very good and they named the station Reliance. Fort Reliance was the first trading post built in the Yukon Territory after the HBC's withdrawal.

It was the first winter in ten years that Jack McQuesten did not have to hunt for his living. Before freeze up he and Bernfield (also referred to as Bernstein or Banfield) had a house and store completed.

> Fort Reliance was the centre of trade on the upper Yukon River between 1874 and 1886. It was the Mile "0" of the Yukon River; all distances along the river were calculated from this point e.g. the Fortymile and Sixtymile Rivers.[6]

In the spring of 1875 they went to St. Michael. Arthur Harper and James McKniff had found nothing that would pay so they went to work for the ACC. Mr. Newman was the new agent and the stations were to be let out to the traders on a percentage basis. Mayo, Harper, and McQuesten had the area from Fort Yukon to the upper river. It was the first time the three could

Yukon Order of Pioneers at 40-Mile in 1895. Al Mayo is on the right at the top of the pyramid. Jack McQuesten is centre, front row. *McKay Collection, University of Alaska, Fairbanks*

be viewed collectively as a commercial concern, with Harper forming the third part of this pioneering trio.

> Arthur Harper was born in County Antrim, Ireland in 1835. He mined in California as a young man and then moved north in search of gold. From the Fraser River goldfields he went on to the Cariboo and then opened the Omineca diggings of northern B.C.[7]

Mayo and Harper left St. Michael to operate the post at Fort Reliance and McQuesten went to Fort Yukon. James McKniff stayed at Fort Yukon during 1876. Mayo and Harper remained at Fort Reliance from 1875 to 1878. They came down in a boat every spring and the little steamer *Yukon* would tow them back in the summer. Harper, who went prospecting each fall, found gold on the Sixtymile that he thought would pay.

“While one group of miners climbed the Chilkoot Pass, another came in by way of the Yukon River.”

By this time McQuesten had abandoned Fort Yukon and was in charge of Tanana Station. Mayo and Harper arrived at Tanana Station from Fort Reliance in the spring of 1877. They did not want to go back to Fort Reliance because they had had trouble with the Indians there. Another company arrived in Tanana that year and McQuesten decided to go to Fort Reliance because the competition's fur prices were so high that he could make little working on a percentage basis. In Fort Reliance he was able to assuage the Indians and to continue operating the post. The Indians had a large amount of furs to trade by this time. Mayo and Harper took charge of Tanana Station.

In the fall of 1877 McQuesten prospected the Sixtymile and then returned to Fort Reliance for the winter (1877–78). There were no other white men in that part of the country. Spring was late that year and the Yukon did not break up until May 22, 1878.

McQuesten was alone at Fort Reliance from 1877 to 1882. Every spring he would go down to St. Michael in a boat manned by ten Indians and would be towed back in the fall by the steamer. Furs were taken down in the spring and trade goods brought back in the fall.

Twelve miners came into the territory by way of the Chilkoot Pass on September 15, 1882. McQuesten heard through the Indians that white men had been prospecting on the Lewes and Pelly for two years. Among these men were Joseph Ladue, Charles Powell, J. Carr, George Spankizberg, Joseph Miller, J. Rogers, and Joseph Paris. They remained all winter and in the spring of 1883, they branched out to prospect. They found prospects on the Sixtymile and the Fortymile that would have paid, but as there were no provisions in the country, they all went out to San Francisco in the fall of 1883. Only Joseph Ladue and Charles Powell remained and obtained some supplies from McQuesten.

While one group of miners climbed the Chilkoot Pass, another came in by way of the Yukon River. Ed Scheiffelin, the discoverer of the *Tombstone Mine* in Arizona, arrived at St. Michael with a party of five men in that same fall of 1882. One of the party was Professor Jacobsen of the Royal Berlin Museum, another was Henry de Wolfe. In a report dated that year, de Wolfe said a party of three Arizona prospectors had worked their way 200 miles up the Stewart, which would have taken them to Fraser Falls. They reported the existence of gold, silver, nickel, copper, and coal to Edward Scheiffelin. (This account is doubtful, according to William Ogilvie, who thought that if the Arizona men had made their exploration at all it was in 1881.)

Scheiffelin brought along lumber and machinery for a small sternwheel steamer which he called the *New Racket*. He ascended the Yukon River with the steamer in 1883 as far as Tanana Station. Scheiffelin was interested in quartz lodes only. This party too, with the exception of a man named Phillip Fancio, left the country in the fall of 1883. McQuesten, Mayo and Harper bought the *New Racket* from them when they left.

Earlier that spring, an ice jam from the Yukon River had sunk the little steamer *Yukon* in the slough near Fort Yukon where it had been wintering.

EARLY BAR MINING ON THE STEWART RIVER

In the summer of 1883 Richard Poplin, Charles McCoskey, Benjamin Beach, and George Marks came over the divide from Juneau by way of the Dyea Pass and prospected the Stewart River from its mouth to the McQuesten River. They recovered $10 a day in gold using a rocker on the river bars. (Workings on river bars are traditionally named after the prospector or some incident connected with their discovery.)

They also found gold on most of the creeks they tested in the area. In the fall they returned to Fort Reliance where they learned that the steamer *New Racket* was broken down on the lower river. Since they would not be able to get supplies, they wintered at Tanana. In the spring, Marks left for the Kuskokwim and McQuesten took Poplin and McCoskey to Fort Reliance on the *New Racket*. From there, they made their way to the Stewart River where they spent the summer prospecting. In the fall, they left, again by way of McQuesten and the *New Racket*. On the way up the Yukon River, they met Thomas Boswell and Howard Franklin who were working the banks and bars of the upper river. The Poplin group told Boswell of the excellent prospects on the Stewart River. McQuesten took the Poplin party as far as Fort Reliance and from there they went out to Juneau.

“Called Wynn’s Bar, it paid Slim Jim $6,000 for less than 50 days work.”

Up to this time, there were no commercially available provisions or tools for miners in the territory except for flour, tea, and sugar, and no clothes or boots suitable for working men. With indications that there would be men coming into the country every year, McQuesten went “outside” that summer of 1884 to bring in miners’ supplies.

The winter of 1884–85 found the trio separated: Mayo was at Fort Reliance, Harper was at Eagle, and McQuesten was in San Francisco.

The following April, after wintering in Fort Reliance, Thomas Boswell, John Fraser (for whom Fraser Falls was named), and Chapman followed Poplin’s lead and trekked to the Stewart River. They pulled their own sleds loaded with supplies, as did most of the early prospectors when the waterways were not open. When the trail was heavy or a strong head wind was blowing, the men changed the sled ropes from across their chests and placed the loop under their arms and at the back of their necks.

By building fires on the bars, thawing the gravel and washing it, Boswell and party found several bars that produced 50 cents to the pan. They made rockers and went to work on what they called Chapman’s Bar located 90 miles up the river.

On June 7, Poplin returned, accompanied by Peter “Pete” Wibourg (sometimes written Wyberg), Francis “Frank” Morphat (sometimes written Moffat), and Jeremiah “Jerry” Bertrand (some accounts say Joe Macheal rather than Jerry Bertrand). They passed Boswell’s operation on Chapman’s Bar and continued on to what was later named Steamboat Bar (about seven miles further up the river).

Boswell and Fraser’s cleanup that summer was $6,000. McQuesten recorded the Poplin party cleanup as $3,500. Ogilvie wrote that it was $35,000. In the late summer “Slim Jim” Wynn, who had first come over the Chilkoot Pass in 1879, found a rich bar 100 miles upstream. Called Wynn’s Bar, it paid “Slim Jim” $6,000 for less than 50 days work.

In early August (1885) Boswell went down to Fort Reliance to get provisions for the parties working on the river. McQuesten landed approximately 30 tons of miners’ supplies at Fort Reliance on August 20. On his supply trip McQuesten had taken aboard Joe Ladue, Howard Franklin, H. Madison, Thomas Williams, and Mike Hess who were prospecting along the Yukon River. They had found nothing to speak of.

When Boswell left his party, it was understood that he was to tell no one how much they

Prepared by Linda MacDonald from the George T. Snow Papers

were making on the Stewart. Consequently, he said they had found nothing and were going to trap for the winter. He bought supplies for six men for one year.

McQuesten was going to Fort Selkirk to trade furs and he took Boswell back to the mouth of the Stewart. McQuesten was to take him up the Stewart, as far as the boat would go, when he returned from his trip to Fort Selkirk.

Thomas Williams, Joe Ladue, and Mike Hess went on to Fort Selkirk where they built a boat and went on up the Yukon to some bars that had been worked the summer before. They worked until late in the fall and then returned to winter at the mouth of the Stewart.

Meanwhile, when McQuesten returned to the Stewart, Boswell told him for the first time about the gold they had found and how much they had made that summer. McQuesten thought it wrong of him not to have told Williams, Ladue, and Hess about the Stewart prospects.

They started up the Stewart and had gone about 12 miles when they encountered six men rocking on a bar. The men had come in from Juneau that spring and arrived at the Stewart after Boswell had left to get supplies at Fort Reliance. The men were Frank Densmore, John Hughes, Ike Powers, Steven Custer, and the brothers Al and H. Day. They were making $5 a day. McQuesten and Boswell had dinner with them and Boswell told the men he had found nothing. McQuesten, however, told John Hughes that Boswell's party was making $20 a day. As they were short-handed on the *New Racket*, Boswell said the men could go up the Stewart with them. Densmore and Hughes wanted to leave the Stewart in a few days and so they decided not to accompany the others.

They travelled about fifteen miles more up the river but the water was so low that they could

A 1938 miner rocking on a river bar. *Claude Tidd Collection, Yukon Archives*

go no further with the steamer. Having stored the goods in a cache, Boswell and McQuesten continued on in a canoe while the steamer left. They found the men working near Steamboat Bar. The party had been up the river about 150 miles but found only one bar that they worked.

McQuesten worked five days with the group and made $250. The miners on the river that year gave McQuesten's name to the first north branch of the Stewart, thence the McQuesten River.

It was getting late in the season when they returned to the cache to build their winter quarters. Upon arriving, they were met by a man named J. Chapman, who had come in that spring with Charles Powell. Powell had gone down to Fort Reliance and Chapman had decided to remain by the cache in case someone struck it rich.

There were sixteen men wintering on the Stewart and/or in the vicinity of Fort Reliance that winter of 1884–85.[8] They were all expecting to strike it rich the following spring.

That winter was the coldest on record. On January 4, it was –80°F and the month's average was –56°F. Steven Custer froze his foot and had to go to St. Michael and have several of his toes amputated. Others suffered from scurvy that winter.

In April the men who had wintered over started up the Stewart with dogs. They went up the McQuesten River first but the stream was very rapid and shallow and they could not go very far with their boats. They returned and went up the Stewart above the falls. They continued to travel up until it was getting late in the season but found nothing. They returned and made grubstakes working on the bars.

News of gold on the Stewart River had reached the outside and, in 1885, approximately 75 men came in over the Dyea Pass to work the bars. McQuesten took most of the men in the steamboat as far as the McQuesten River. They worked the bars and did very well. The Stewart had now been prospected almost to its headwaters and miners were working on Wynn's, Dutch John's, Lac de Bar, Halfmoon, Joe Jay, Low Water, and Black Mike's Bar.

In 1886 McQuesten, Harper and Mayo established a post at the mouth of the Stewart to supply the miners gathering on the river.

Some of the miners called the post Fort Nelson after Pte. Edward William Nelson of the Signal Service of the US Army, who was said to have been there at the time collecting specimens of the flora and fauna of the region. (Nelson went on to make invaluable collections from Yukon and Alaska and later became one of America's most famous naturalists.)[9]

Al Mayo ran the store and a small town sprang up around his post. The town was built on the right bank (north) of the Stewart, at its intersection with the Yukon River.

There were 32 cabins, built in a half circle, following the shoreline, with the inverted side of the crescent facing the Yukon River. A slough ran through from the Stewart to the Yukon River,

The mouth of the Stewart River, 1898.

cutting the town in half. In the winter this backwater created an uninterrupted trail along the line of cabins. The cabins on the south, or up river side of the slough, were occupied by two or three men; while on the other side each man lived in a cabin by himself. The boys on the south side of town named the other side of the slough "Cranktown" and they responded by naming the south side "Blowville." During the winter a good-natured rivalry sprang up between the two towns and they would challenge each other to a whist tournament or a game of shin-maiming hockey on the ice of the slough. On occasion the losing side had to furnish the winners with doughnuts and coffee. When "hootch" was substituted for coffee, singing and laughter went on for many hours. However, not all was fun and games.

The first miners' meetings in the Canadian territories were held at the Stewart River post. Two men were sentenced to be exiled from the camp, one for the attempted poisoning of his partners with arsenic, and the other for stealing butter from Mayo's store at a time when the commodity was in very short supply—a very grave offence in a frontier community. On the verdict being given, each of the men "was furnished with a sled, provisions enough to get out if he could, and was ordered to move at least one hundred and fifty miles from the Stewart, and assured that if ever he was seen within that distance of it, anyone then present would be justified in shooting him on sight." Both of the exiles managed to make it to the coast. The would-be poisoner caused considerable alarm in Haines Mission by spreading a wild story of an Indian uprising on the Stewart River to cover up his unseasonable arrival. Not knowing that the story was a complete fabrication, it caused Ogilvie a great deal of unnecessary anxiety when he first ventured into the interior.[10]

Hegg Collection, Yukon Archives

Ogilvie, a surveyer with the Dominion Land Survey, recorded the Stewart River placer deposits as having been extraordinary. It was estimated that $300,000 was taken from the bars in 1885–86. On average a man could earn $30 per day and occasionally up to $100 a day.

In 1886 Howard Franklin and Henry Madison (two miners who had arrived with Joe Ladue in 1882) were among the men who prospected the Stewart to the falls and the McQuesten as far as they could ascend in their boats. They seemed not to like the rivers, as Franklin remarked to miners on the Stewart that he disliked the kind of trees that grew on the bars, and that gold was never found where wild onions and leeks grew. While Franklin and Madison left the Stewart for the Fortymile, the brothers Al and H. Day (from French Canada) made over $30 per day on one of the bars rejected by them. Later that fall, Franklin and Madison discovered coarse gold on the Fortymile.

Word of the discovery at Fortymile drifted back to the Stewart River post and the men who were at the post at the time as well as the ones who came in early that year bought all the supplies Mayo would give them and went to the Fortymile. Men arriving late from up river suffered in consequence. Due to the shortage of provisions there were 45 to 55 men on the verge of starvation that winter at Stewart River post. Many of the latecomers did not reach the post until well on into the winter that year. The situation was so desperate that when some men came in from Juneau that spring, Harper and Mayo had to buy flour for themselves from them.

McQuesten had gone to San Francisco that winter of 1886–87 to order supplies for the Alaska Commercial Company. The partners realized their post at the mouth of the Stewart would have to be moved to Fortymile. Harper, back at Stewart Post, recognized the full value of the Fortymile coarse gold discovery and knew that hundreds of prospectors would arrive in the spring. He feared that starvation might ensue if McQuesten did not bring in more stock than had been planned and asked if anyone at the post would be able to take a message for McQuesten to the coast. A steamboat man named Tom Williams and an Indian companion volunteered to take the message to John Healy's trading post on Dyea Inlet on the far side of the Chilkoot. It was a trip of over 500 miles which few men had attempted in winter and it exacted a terrible toll on the two men. At the summit of the pass they suffered frostbite and the Indian man carried the exhausted Williams part of the way. Williams died two days after they reached Healy's trading post.

In March of 1887, the men who had wintered at Mayo's post on the Stewart, prepared to leave for the Fortymile.

The following is derived from historian George Snow's colourful – though factually correct – account of the leave taking.

A meeting was held at the store that afternoon and it was decided to divide all the outfits equally among the stampeders as the distance to Fortymile was only about sixty miles and they estimated it would take about four days to get there. All the sleds were to be loaded and ready to start at five o'clock in the morning and because the

Five men and a dog outside a log cabin in winter, circa 1910. *Emmett Middlecoff Collection*

trail was good for two-thirds of the way, they proposed to make a twenty mile camp the first day.

In the morning everyone was ready for the start except Frank Densmore. His partner, John Hughes, came down to the river bank where the men were waiting with the loaded sleds and told them that Densmore was heating a coal-oil can full of water on the stove in order to take a bath and shave before he started. Hughes told him everyone was waiting and not to bother with the bath but Densmore told him to go ahead and that he would catch up with them in an hour. When Hughes told him the others wouldn't start without him, Densmore threw him out into the snow and bolted the cabin door. "French Joe" Beaudreau jokingly suggested that they smoke him out of the cabin. That comment gave Mayo an idea and he invited them all to come to the store. Once assembled, Mayo told them to go to Densmore's cabin and commence cussing and yelling for him to come out. That way he wouldn't hear Mayo climb up on the roof with a can of red pepper which he would dump down the chimney, which in turn would be covered with a box, forcing Densmore out of the cabin. The group left the store together. The cabin was the fourth from the store. Mayo went to the back of the cabin while the rest of the men gathered around the door. John Nelson, George Matlock, Frank Bowker, and Pete Wibourg were at the front of the lines. The plan was executed and the cabin door was soon flung open and Densmore charged out in his underclothes and stocking feet. The men immediately fell upon him and tried to hold him in the snow but Densmore was a powerful man and managed to free himself momentarily. Just as he started to run, Matlock grabbed him by one leg and down he went into the deep snow. The other men pinned his arms and legs and they showered him with snow. Being a determined man, it was several minutes before Densmore was cold enough to forget about the bath and agree to leave with the rest of the party. Densmore went to his cabin, changed his clothes, and then sprinted over to where his loaded sled was standing; throwing the sled rope across his shoulders, and taking the gee-pole in his hand he pulled the load over the bank and down onto the river where the rest of the sleds were waiting in line.

BUT THEY WERE GOOD DAYS

When asked by a writer if he could remember any humourous incidents from the early days, one of the old stampeders replied,

"Thirty-six years at the business end of a long waisted No. 2 and the long winters has worn most of the humour out of me. To recall anything really comical, especially from the Fortymile days, at present is impossible for me. But they were good days. The country is changing mightily, the population is getting smaller and there is nobody prospecting any more to amount to anything. Mostly everybody has either got a government job or is looking for one. The bean peddlers still carry a stock of shovels but they are poor sellers and if a man goes out prospecting all summer and comes in in the fall broke, everybody gives him the glassy eye. Then it takes him two or three years to get enough money, working for wages, to go prospecting again. Now no one has any time for the roughnecks and that's not the way it used to be." *George Matlock, veteran of the Stewart River, Fortymile, Circle, and Klondike gold rushes; November 7, 1923.*

The stampede to Fortymile was on.

When the ice went out, Mayo readied the *New Racket* for the trip to St. Michael, a distance of 1,800 miles, to bring up the much-needed provisions. On his return, work began on the new post at Fortymile and thus, in 1887, Stewart River was deserted.

Although other rushes were to follow the Stewart River Rush of 1884–85 — the Fortymile in 1886, the Sixtymile in 1891, Birch Creek (Circle, Alaska) in 1892, the Klondike in 1896, and others — the Stewart River strike had, for the first time, brought men into the Yukon in substantial numbers. From the Stewart, they fanned out to search other rivers and streams.

SOME OF THE PROSPECTORS AND MINERS ON THE STEWART RIVER, 1883–1888

BEACH, Benjamin
BEAUDREAU, Joe
BERNDT
BERTRAND, Jeremiah "Jerry"
BOSWELL, Thomas
BOWKER, Frank
CHAPMAN, J.
CUSTER, Steven
DAY, Al
DAY, H.
DENSMORE, Frank
FRANKLIN, Howard
FRASER, John
HARPER, Arthur "Cariboo "
HART, Frederick "Fred"
HESS, Mike
HUGHES, John
LADUE, Joseph "Joe"
MACHEAL, Joe
MADISON, Henry
MARKS, George
MATLOCK, George
MAYO, Alfred H. "Captain Al"
MCCOSKEY, Charles
MCDONALD, Alexander
MCQUESTEN, Leroy Napoleon "Jack"
MILLER, George F. "Buckskin"
MORPHAT, Francis "Frank"
NELSON, John
PIERCE, Walter H.
POPLIN, Richard "Dick"
POWELL, Charles
POWERS, Ike
SNOW, George T.
SONNIKSON, Chris
SULLIVAN, "Black Mike"
SUMMERS, Hank
WIBOURG (or Wyberg), Peter "Pete"
WILBORING, P .
WILLIAMS, Tom
WYNN, "Slim Jim"

MINING ON THE STEWART CONTINUES

Miners discovered that although the gold at Fortymile was coarse, it was not uniformly spread over the surface as it had been on the Stewart, and that it was much harder to mine. Some of the miners who were unsuccessful at Fortymile returned to the Stewart River.

> Miners referred to the Stewart as the 'Grubstake River' because enough gold could always be mined on it to provide money for supplies for another season of prospecting.
>
> William Ogilvie, who had been Commissioner of the Yukon from 1898–1901, became the President of the Yukon Basin Gold Dredging Co. Ltd., of Missouri, which held extensive leases on the Stewart and operated with limited success from 1908–1912.[11] *(See pp. 54-55)*

The record of bar mining on the Stewart continued over the years. In 1918 Robert Hager bar mined at Johnson's Point making from $6.50 to $9.60 per day. Also in 1918, Cornelius Driscoll rocked at Steamboat Bar, while Richard McDiarmid and his wife were on Joe Gee Bar (also written as Joe Jay Bar). Oscar Miller, Frank Fratas, and George Schweitzer worked Chapman's Bar (6 miles below Clear Creek) where they had an open cut about 200 feet long. They wheeled their dirt to the river to rock it in a grizzly.

In the summer of 1938 there was considerable activity along the Stewart River bars. There were boats of every description, even a small sternwheeler manned by the Rendell twins of Dawson, plying in and around the sandbars on the river, looking for places to land and do a little gold "sniping." In July of 1938, fourteen bar miners were washing bars on the Stewart River between Stewart City and Mayo. Among them were Hughes and Porsild who worked at Sniperville, just above Sterling Bend; Carl Burian and partner Brooks who made an ounce a day just below Independence Creek; Wilcox and Fresner who poled up the river; and Wootton and Hale, located at the 22-Mile below Mayo. Also working the river were P. Perrett, J. LeSaux, Clarence Currie, Ole Lunde, and H. Gibbs.

"Miners referred to the Stewart as the 'Grubstake River' because enough gold could always be mined on it to provide money for supplies for another season. . ."

In 1940 Gus Bergstrom and M. Gadda worked the bars 9 miles above Stewart City, Currie and Lunde were still at Chapman's Bar, Louis Lechner and H. Frieson were 50 miles below Mayo and Andy Anderson and Louis Hughes worked at a bar at the 26-Mile.

STEWART RIVER POST

Stewart River post, known as Fort Nelson Post in 1885, was later referred to as Stewart City or simply Stewart River settlement.

When George Snow and George Miller visited Stewart River post in 1888, only a year after it had been abandoned, it looked as though it had been deserted for ten years. Most of the cabins were destroyed; only the store and two of the cabins were in good repair. There was a note tacked on the door of the store from Al Mayo. It read, "Gone to St. Michael for load of grub. Will be back about the middle of September with the *New Racket*. Nothing but beans and flour in the grub line. If you are short, you'll find the key in the moss between the door casing and the fourth log from the bottom. Leave a memorandum of what you take." Snow and Miller worked up the Stewart that summer. When they saw Mayo a couple of months later, they asked him why the cabins at the post were in such poor condition. Mayo told them that the Indians had dismantled nearly all of them to get the iron nails.

In the Klondike days the old settlement was revived as a wood point for riverboats, a N.W.M.P. post, and a wintering place for miners unable to reach Dawson before freeze up. Nearly one hundred Klondike stampeders were trapped by the sudden onset of winter in 1897 at the mouth of the Stewart River. Among those who saw the winter out was Jack London. Many of these men prospected the tributaries of the river in 1897–98 and later.

In 1898 there was a N.W.M.P. post at Stewart City and in 1902 the townsite had a post office. Stewart City operated as a service and transfer point for steamboat traffic from the days of the Klondike gold rush until the 1950's.

Stewart City was the supply point for Barker, Black Hills, Donahue, Henderson, Scroggie, and Thistle Creeks. In the 1920's there were semi-weekly boat and mail connections with Dawson, Mayo and Whitehorse during the summer. It was also the government telegraph station on the line between Atlin, B.C., and Dawson City.

Margaret "Peg" and David C. Shand ran the Hotel Stewart on the old site from 1900 until it burned in 1918. They rebuilt it and operated the hotel until 1930. Another roadhouse was operated by Mrs. Emma Smythe. It is not known when the Smythe Hotel was originally built, but it was in operation in 1918 and was still there in the 1930's. Mrs. Smythe was also the postmistress. In 1918, she had a cow and calf, a large number of hens and, reputedly, a splendid garden. She made all of her own butter and had sufficient eggs and vegetables to supply her roadhouse. The Shands had a large garden as well. Also in 1918, J. C. Lawrence operated a store in Stewart City, with a "first class stock of goods." The population was estimated to be at 25 in 1922.

William Taylor owned and operated the Hotel Stewart in the 1930's and then sold it to

Smythe's Store and Hotel at Stewart City in the 1920's. *D. Walloch (McLaren) Collection*

W .E. "Bill" Bamford in 1937. Taylor returned to New Zealand. After purchasing the hotel, Bamford built a trading post of frame construction.

In July of 1937 Bamford sold the Hotel Stewart to Rudolph "Rudy" Burian and his sister Mrs. L. Lechner. Bamford continued to operate the store.

Mrs. Peggy Shand, left, at Stewart City in 1926. *Isabelle Poirier Collection*

In June, 1939, the Hudson's Bay Company bought Bamford's Stewart River post at the mouth of the Stewart River (Town Lots 5, 6, 7, 8 in Block S; 15, 16 Block P in Stewart River Townsite). W. H. "Jack" Houston arrived to organize the post for the Hudson's Bay Company and in 1940 built a store. The Hudson's Bay Company post managers were as follows:

1939–1943	W.H. "Jack" Houston
1943–1945	Thomas A. Retallack
1945–1949	George Gardner
1949–1950	F.J. Tapp
1950–1951	G.F. Ferrand

By 1951, river transportation had diminished while the population moved to points on the Whitehorse–Mayo Highway. On May 31, the Hudson's Bay Company closed its post at Stewart River.

In April, 1952, the HBC store, dwelling house, warehouse and power house were sold to Rudolph Burian.

There are about 20 graves in a small cemetery at Stewart River post dating from the days of McQuesten, Mayo, and Harper's Fort Nelson Post. Some of the graves and many of the old cabins on the site have washed into the river.

The Hotel Stewart at Stewart City in 1933. *G. A. McIntyre Collection*

Commencing a river trip down the Stewart from Mayo, 1920. *A. K. Schellinger Collection*

Three men pull away from Mayo on a raft loaded with their possessions. *A. Berger Collection*

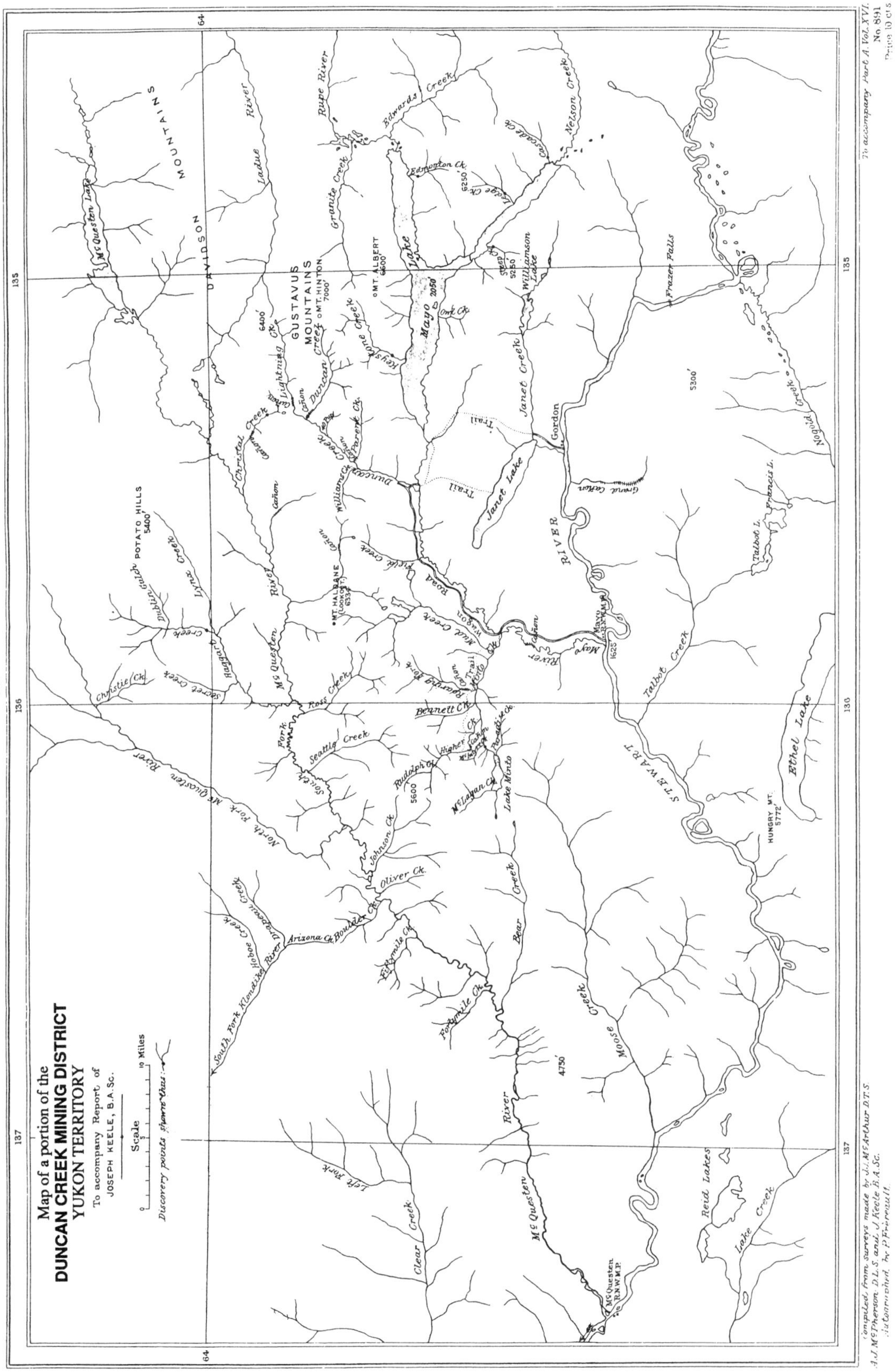
Map of a portion of the
DUNCAN CREEK MINING DISTRICT
YUKON TERRITORY
To accompany Report of
JOSEPH KEELE, B.A.Sc.
Scale
10 Miles
Discovery points shown thus:
Geological Survey of Canada
ROBERT BELL, D.Sc.,(Cantab), M.D., LL.D., F.R.S., I.S.O., ACTING DIRECTOR.
DAVIDSON MOUNTAINS
GUSTAVUS MOUNTAINS
POTATO HILLS
Mayo Lake
STEWART RIVER
Mc Questen River
Ethel Lake
Reid Lakes
Janet Lake
Duncan Creek
Lightning Ck.
Keystone Creek
Williamson Lake
Janet Creek
Gordon
Frazer Falls
Nogold Creek
Talbot Creek
HUNGRY MT. 5772'
Lake Creek
Clear Creek
Left Fork
South Fork Klondike River
North Fork Mc Questen River
Moose Creek
Bear Creek
Fortymile Ck.
Arizona Ck.
Boulder Ck.
Oliver Ck.
Johnson Ck.
Rudolph Ck.
Seattle Creek
Ross Creek
Bennett Ck.
Haggart Creek
Dublin Gulch
Lynx Creek
Christie Ck.
Secret Creek
Crystal Creek
Edwards Creek
Rupe River
Ladue River
Nelson Creek
Edmonton Ck.
Granite Creek
Wagon Road
Trail
To accompany Part A. Vol. XVI.
No. 891
Compiled from surveys made by J.J. McArthur D.T.S.
A.J. McPherson D.L.S. and J. Keele B.A.Sc.

Chapter 4

The Placer Gold Creeks

BY 1900, THE LUSTRE OF THE DAWSON GOLD FIELDS WAS DIMMING, gold production had peaked, claims were being bought up by large mining concessions, and dredges and large-scale hydraulic operations were proving to be the most economical means of mining the ground.

Unemployed men, eager for adventure, anxiously awaited rumours of new stampedes. Such was the feeling in Dawson City that fall when three Swedes brought in their gold, gold which eventually led to the Duncan Creek stampede of 1901.

This stampede brought a couple of hundred hopefulls into the Duncan Creek area, more people than were brought in by the earlier stampedes to Johnson Creek or Haggart Creek. By 1902 the Duncan Creek Mining District was established with an office at the mouth of Duncan Creek. The overflow of men spread to nearby creeks, staking tributaries of Duncan Creek and the creeks of the Mayo Lake area.

Gold Nuggets from the Duncan Creek region.
E. L. Bleiler Collection

In 1903, the search was expanded to include Minto Creek and its tributaries. Claims were staked and quickly sold to finance the next prospecting trip while the buyers recruited men to work their claims.

It was the placer gold deposits of Highet and Haggart Creeks that provided the stable economic environment for the development of roads, transportation, and communication for the Mayo area. The labour force stabilized and by 1904 there were about 80 miners in the district.[12]

In a 1904 report, Joseph Keele, a surveyor with the Geological Survey of Canada, warned that every prospector coming into the Duncan Creek Mining district should have one year's outfit of provisions and clothing. He also mentioned that there were stores at Mayo and Duncan Creek where clothing and provisions could be purchased. In the same report he stated that, "An excellent road with good grades, suitable for either summer or winter travel, was located and partly cut out this summer by Messrs. Gordon and Davidson from Gordon [Landing] to Duncan Creek a distance of eleven miles." Keele also mentioned that miners could supplement their provisions with fish, moose, caribou and even bear.[13]

The price of gold dictated the surges in activity somewhat. From 1879 to 1934 the gold price was $20.67 U.S. per ounce. In 1934 the U.S. Government banned the private ownership of gold and devalued the dollar by 41%. This revalued and fixed the price of gold at $35 an ounce where it stayed until about 1970. After 1970, the price was no longer fixed but was set by the free market and allowed to fluctuate.

The mining district was known by different names over the years. In 1901, it was known as the Clear Creek Mining District and from 1903 to 1919 it was renamed the Duncan Creek Mining District and finally in 1920 the name was again changed to the Mayo Mining District. Along with the name changes, mining methods also changed from the "gumboot miners" who sank shafts and drifted the rich pay streaks to hydraulic mining with nozzles, pipelines and sluice boxes which could economically mine the leaner ground. Each method was larger, more mechanized and less labour-intensive than the one before. Labour became especially scarce during World War II. However, bulldozer mining was becoming more common, making it possible to mine greater yardages with fewer men than ever before.

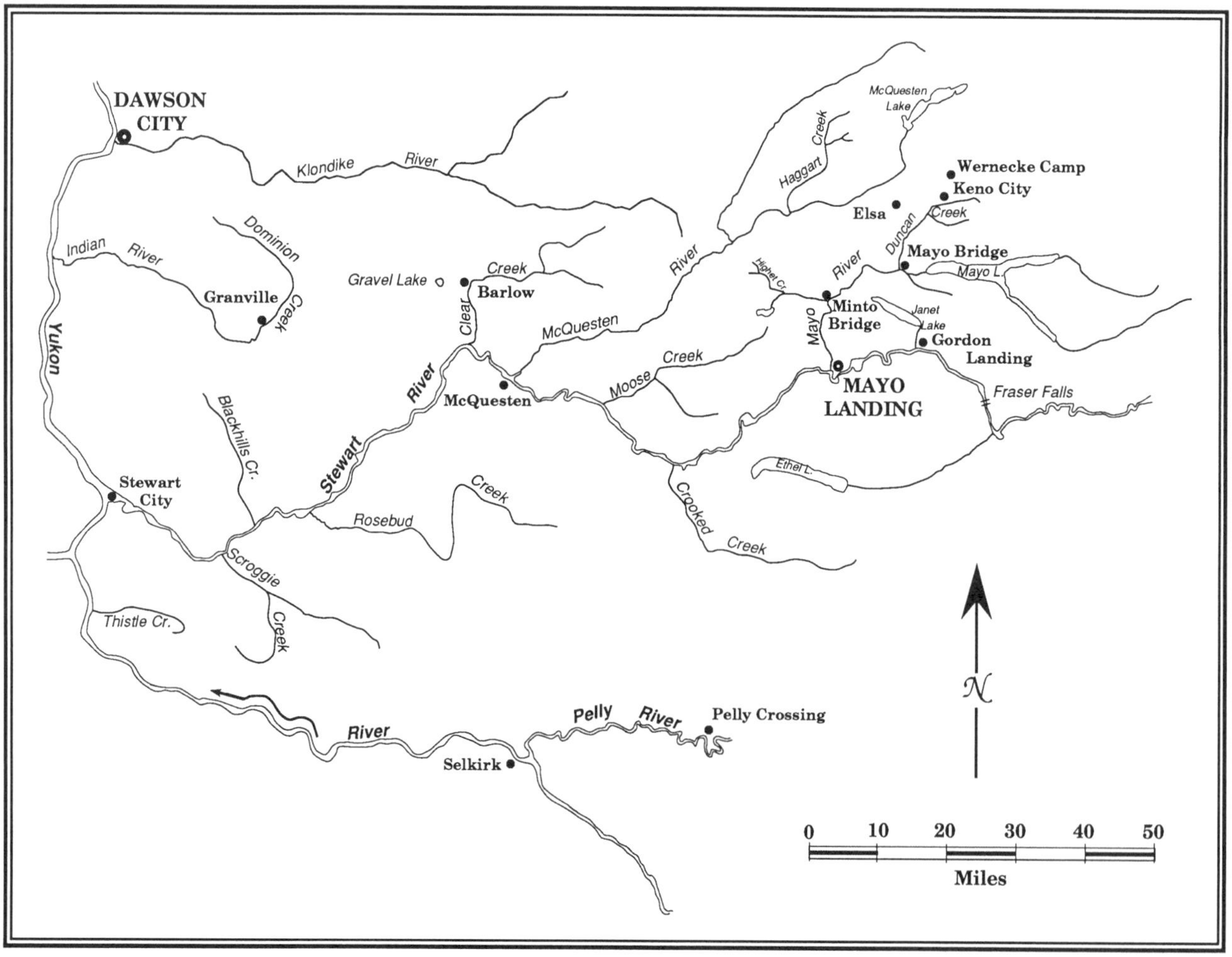

Map showing creeks and rivers in the Mayo area. *Map by E. L. Bleiler*

While the Mayo Mining District has experienced the boom and bust cycles common to mining communities, up to 1955 the gold and silver-lead mining intervals have worked together to smooth out the highs and lows of the cycle.

The economic backbone of the Mayo Region has always been the gold and silver mines. The same optimistic prospectors who staked the gold producing creeks in 1900 carried on their quest for new mines and were responsible for the silver discoveries in the 1920's. In fact, it was the revenue from the mining companies in the 1920's and 1930's that prevented the economic collapse of the Yukon.

JOHNSON CREEK *by Lynette Bleiler & Linda E. T. MacDonald*

Johnson Creek is a small stream about 6 miles long, whose drainage heads with Highet Creek but flows in the opposite direction into the McQuesten River. Due to its accessibility from the Stewart and McQuesten Rivers, prospectors found their way to Johnson Creek in their search for gold several years before the Klondike Gold Rush.

Early records are vague or nonexistent but it is known that Ray Stewart found pay on a bar of the McQuesten River a few miles below the mouth of Johnson Creek in 1892. He mined it with Ben Atwater and Charlie Dickson. He returned several years later and tried to locate the same bar, but the river had changed, making it impossible for him to find the area.[14]

In 1894 the two Garrison brothers found gold on Johnson Creek. They went to Fortymile to get an outfit but did not return. Later, they staked claims on Eldorado Creek, in the Dawson District, and died not long afterward from fever.

Fred Johnson prospected the creek in 1898 and staked a Discovery claim. That fall, thirty-eight men stampeded in and wintered there. Their exploration was unsuccessful because water came up in the shafts before pay was reached on bedrock in the thawed ground. Johnson was still working the creek in 1915 and it was reported that he recovered 47.5 ounces that year.[15]

Records show that there was very little activity on Johnson Creek from 1898 until the autumn of 1914 when Ogden Pickett Thompson, an old-timer who had been in the Yukon since 1898, built a cabin at the mouth of the creek and staked his Discovery claim, the only frozen ground encountered on the creek. He found pay gold and as a result the creek, together with a tributary, Sabbath Creek, was stampeded and staked during the latter part of January and in February of 1915.

Some of the men who staked Johnson Creek during the 1915 rush are: John B. Donley, David Connors, Mathias Veilhur, Ole Dahl, Malcolm McKinnon, Dan Vucer, Thomas Keating, Hector Morrison and David Morneau.

Thompson leased portions of his Discovery claim, where miners thawed the ground with steam points. They sank shafts and drifted the bedrock gravels, which were hoisted to the surface by windlass and washed. The results were encouraging, and up to September 1915 the total gold obtained from Johnson Creek amounted to about $800.[16] Thompson held his claims on Johnson Creek until 1922.

> ***"He returned several years later and tried to locate the same bar, but the river had changed, making it impossible for him to find the area."***

In later years, George Ortell, a former woodsman on Joseph Keele's 1905 G.S.C. expedition up the Stewart River, lived at the mouth of Johnson Creek. He and Norman Niddery staked a new Discovery on Johnson Creek on August 16, 1923. Ortell also trapped and prospected in the area for many years until his death in 1943. (See *Pioneer Accounts*.)

In July 1943, Dr. Hugh Bostock of the Geological Survey of Canada was on Johnson and Sabbath Creeks examining the mineralization of the area. He reports in his book *Pack Horse Tracks*, published in 1979, that amongst the abandoned cabins he came across a flume and large sluice boxes on trestles arranged for operation with a dragline. He was told by Joe Laska (who had been on the creek himself since 1901) that the workings belonged to "Pete" Minton who had mined there in 1916 and 1917. Laska reported that 400 ounces of gold had been recovered at an expense of $37,000.[17] Minton Creek another tributary of Johnson Creek, was named for Peter Minton by Bostock in 1957.[18]

According to information in *Yukon Mineral Industry 1941 to 1959* compiled and edited by R.L. Debicki, no active mining was done on Johnson Creek from 1941 to 1955 although some prospecting may have been carried out.[19]

However, in 1956, John Sandanger and Klaus Djukastein sank twenty-four prospect shafts on Johnson Creek. At first, their results were disappointing. The yield from the 1957 season's work, however, indicated the creek could be worked at a profit. In 1959 the partners incorporated under the name of Bardusan Placers Ltd. with the addition of another partner, Hans Barchen. By 1962, $150,000 had been mined from the creek.

While the partnership dissolved, Barchen retained the name Bardusan Placers Ltd., and continued to mine on Johnson Creek until 1966. The claims were later staked by Klippert Bros.

Another creek in the area is Savage Gulch, a tributary of Sabbath Creek which flows into Johnson Creek. It was named after Victor Savage who prospected and mined on it circa 1900. Savage also prospected on Duncan Creek.

HAGGART CREEK & DUBLIN GULCH *by Ed Bleiler*

Ed Barker's placer operation on Haggart Creek.
F. & T. Whitney Collection

Haggart Creek

Haggart Creek is one of the principal tributaries of the McQuesten River. It is over 20 miles long and joins the south fork of the McQuesten River about 13 miles above its confluence with the North Fork.

There does not seem to be much information on the early Haggart brothers' mining operation, but it is recorded in the Geological Survey of Canada, Memoir 284, that gold was first found on Haggart Creek in 1895. In 1896, the creek was known to have been prospected by Thomas Nelson and named Nelson Creek. Also in 1896 Thomas Haggart built two cabins on the creek and one on Dublin Gulch. However, that winter they left for the Klondike and it was not until 1898 that Haggart, Nelson, Peter Haggart and Warren Hiatt returned from Dawson. The party split up along the way with the result that Peter Haggart and Hiatt reached Nelson Creek first. They staked Discovery and renamed the creek Haggart Creek. While they did not work it much, Tom Haggart did prospect extensively in the area over the years.

At the confluence of Dublin Gulch and Haggart Creek and working downstream, one of the most extensive and successful operations was that of the Cantins: Frank Cantin and his cousins Joe, Phileas, and Louis Cantin. They also had a part-

"Cat" train bound for Haggart Creek. Left to right: Ed Barker, Irvin Ray, Con Lakness and Jim Gibson, circa 1940. *F. & T. Whitney Collection*

ner, Frank McKenna. The Cantins were French Canadians from Trois Rivières, Québec and they hydraulic-mined a large bench area on the right limit of Haggart Creek almost down to Jahnke Pup. From Frank's information they recovered over $50,000 at a time when gold was at $20.67 a fine ounce.

It was shallow and easy ground to mine, once a ditch was constructed to get water up on the hillside. They commenced mining in 1909 and held Discovery and three adjoining claims. While it is uncertain how long they worked their claims, Frank, Joe and Louis were still living in the Mayo area in the 1950's where they ended their days. Phileas returned to Québec as soon as the easily-mined areas were exhausted.

Downstream from the Cantins were three very hard working and resourceful miners, Albert

Having arrived in Mayo In June of 1933, . . .

Parts of this history were related to me by the pioneers whom I met in my early years in the Mayo area.

Having arrived in Mayo in June of 1933, I had the good fortune to meet many of the men and women who were very much involved in the actual mining of Haggart Creek shortly after its discovery.

My involvement in mining began on Dublin Gulch, a small tributary of Haggart Creek which enters on the left limit about 12 miles from the McQuesten River. In 1933 all of the old workings, log cabins, ditches, dumps, old shafts, tailing piles, wooden dump boxes, sluice boxes, wooden riffles and even an overshot water wheel, were very visible. I was, naturally, very curious about the people who had accomplished such prodigious works so long before my time, especially when one realizes that transportation was so difficult. Everything was hauled the 40–45 miles from Mayo during the winter months when the streams were frozen solid. The trail at that time, and even in my early years there, was over some very wet muskeg. The freighting of course, was done by horse and by dog teams.

These are the pioneers who were still alive in 1933 and to whom I spoke about the history of Haggart Creek and Dublin Gulch: Frank Cantin, Joe Cantin, Louis Cantin, William Walby, Andy Johnson, Bobbie Fisher, Robert Burns, Alex Nicol, Jack Maynard and Jack Alverson.

The latter, Jack Alverson, was a miner and prospector with extensive experience. He gained much of that experience in Colorado in the silver mining areas as well as in other areas of the U.S.A. It was Alverson who built the overshot water wheel used by Jahnke and Gill on Haggart Creek which both raised water to a dump box and sluice line and supplied the necessary power for hoisting gravel also. *Ed Bleiler, longtime miner in the Mayo district.*

Jahnke, Fred Gill and Jack Maynard who held claim Nos. 1, 2, and 3 from 1908. Maynard left after it was apparent he and Jahnke could not work together. Jahnke and Gill worked as a team for quite a number of years and had a very well located cabin and other buildings on a flat near where Jahnke Pup joins Haggart Creek. They were able to grow a nice garden in their spare time. They mined open cut mainly but also did some winter drifting on the right limit. As mentioned earlier, they used an overshot water wheel which was a unique and economical power supply. Both men were expert trappers and so lived on Haggart year round. When they left the country, Jahnke for Germany and Gill for the United States, they were reported to have sizeable fortunes.

Below Jahnke and Gill were three English men with their wives. William Portlock and Nathan and William Abbott together held about five claims which they worked from 1907 on. Each family had its own cabin and, having been gardeners in England, established quite large rhubarb patches which were enjoyed by the many miners who worked the area in later years. "Cat" operations, which began in the 1940's, eliminated the rhubarb patches. Portlock and the Abbotts mainly attempted drifting on the right limit and open-cutting in the creek bed during the summer months. They were not very successful in spite of their efficient mining plant which consisted of a boiler, engine and a self-dumping bucket. Portlock died in November 1924 and in later years his widow, Alice, married Frank Cantin. The Abbotts returned to England.

There was one other open cut operation below Portlock and Abbott circa 1910. It was the partnership of Clarence Kinsey, Charles Merriman and John Mawhinney. They had installed a boiler, engine and a self-dumping scraper, but were not very successful and folded after a year or two.

There was barely a mile of Haggart Creek suitable for mining from 15 Pup downstream. The depth to bedrock increased greatly. It is possible that some day the continuation of the pay streak will be found and mined.

In 1934 when the U.S. government set the price of gold at $35 per ounce, almost double the

Ed Barker's cabin on Haggart Creek. From left: Alec Berry, Ed Barker and Kate Broadfoot. *A. Aho Collection*

Ed Barker's mining operation on Haggart Creek. *A. Aho Collection*

existing price, Ed Barker and "Kit" Watters began prospecting in Jahnke and Gill's area near the mouth of Jahnke Pup after visiting an old-timer on Dublin Gulch. Barker looked into Abbot's abandoned cabin on Haggart Creek and on a pile of discarded papers found Abbott and Portlock's bank book which showed the returns from their clean-ups over their many seasons of mining. With this information and the encouraging gold price, they set to work.

They had fair encouragement both from underground and open pit mining. It was all pick and shovel work and, in winter, hoisting with a hand operated windlass. "Kit" Watters was quite elderly so Barker took in two younger men as partners, Roy McDermid and Norman Wightman along with Oscar Miller who was a silent partner. McDermid died of cancer in 1936 and Wightman later retired from mining.

At this point Barker decided the property would support a more mechanized operation, so in the late 1930's and with a new partner, Irvin Ray, who was quite knowledgeable about machinery and an expert welder, they bought a D-4 cat and formed the Haggart Mining Company. Together with a crew of 3 to 10 men, depending on the season, they carried out quite a successful operation and by 1940 held 22 claims and 12 miles of prospecting leases along Haggart Creek. Elsy Ray, Irvin's wife did the cooking assisted by Miss Kate Broadfoot.

During the war years, machinery was desperately needed at Whitehorse both in the construction of the airfield and the Alaska Highway and so, in 1942, they closed the mining operation on Haggart Creek and moved to Whitehorse.

Barker returned to the area and leased the lower part of his Haggart Creek ground, formerly Portlock, Abbott and Kinsey's area, to some very successful operators, some of whom had extensive experience in Alaska and Atlin. From 1953–1958 Waddco Placers Ltd. (Maynard Wilson, Jack Acheson, Bill Drury and Clyde Day were the partners) mined very successfully on Haggart Creek with very good heavy duty equipment, including a dragline, and a crew of 4 to 6 men. The ground was mined out in a few years, as it was, for the most part, very shallow. The Waddco partners moved to Dominion Creek in the Dawson district and Ed Barker again returned to Haggart and mined on a small scale near his cabin with a D-7 bulldozer. He alternated between hardrock and placer prospecting ventures until his death in 1961.

In later years Klaus Djukastein successfully mined some areas that had been left and presently there are two or three miners still finding areas that yield fair profits.

Dublin Gulch

Dublin Gulch has been a surprisingly productive stream, considering its small size. John J. Suttles bought claim No. 1 and mined successfully from approximately 1899 until the fall of 1915. It was reported at the time that he took out between $45,000 and $50,000 in gold before he lost the ground to the Cantins over a legal dispute concerning his tailings. His holdings on Dublin Gulch were acquired by the Cantin Bros. around 1916. They worked by ground sluicing and took out about $6,000 in gold. However, the Cantins did not have great success mining Dublin. It proved much more difficult to mine – due to greater depths, numerous, large, granite boulders and a shortage of water at times – than the shallow bench they had mined on Haggart Creek. They abandoned the creek in the 1920's.

Fred Taylor's operation on Dublin Gulch. *Fred Taylor Collection*

Several hardrock claims were staked on Dublin Gulch in 1907. Among them the Carscallen claim, and the Victoria claim which was the first claim staked on Dublin Gulch. In 1908 Robert "Bobbie" Fisher staked a hardrock deposit, the Olive claim named after his sister. It is located on the left side of Dublin Gulch, about two miles from Haggart Creek and is a crown grant.

In the early 1930's Theodore Bleiler and partner, Norval Lochore, staked part of Dublin Gulch and shortly afterward, Jim Gibson staked below them. Bleiler worked the claim a number of years, while Lochore left Dublin to return to B.C.

Fred Taylor's rock derrick on Dublin Gulch in the early 1940's. *Fred Taylor Collection*

In the late 1930's, Bleiler was joined by Fred Taylor who, after a year or two, bought his interest. Fred Taylor mined very successfully for a number of years using heavy duty equipment. At first he hired men to assist him until his two sons, Frank and Jim were old enough to help run the operation. It is interesting to note that during the early 1940's miners on Dublin Gulch received preferential treatment with regard to the availability of rationed heavy equipment parts and road maintenance, because the tungsten they recovered was in great demand for the manufacture of munitions.

Mayo was a centre of activity to recover scheelite (tungsten) for the war effort. All scheelite concentrate was shipped to the Mines Branch in Ottawa. In 1942, a special grant of $20,000 enabled the construction of a bridge over the McQuesten River as well as a road to Dublin Gulch.

Between 1941 and 1947 these miners worked either their own property or had lays: Vilhelm Lunde and Robert Swanson, Hugo Seaholm, and Cliff Greig.

In 1955 Cliff Greig purchased approximately half of Fred Taylor's Dublin Gulch property and continued to mine for several years.

Fred Taylor sold out to Ron Holway and his partner Darrell Duensing who operated under the name of Darron Placers and in 1978 they optioned the property to Canada Tungsten Mining Corp. Ltd.

Ground sluicing on Fred Taylor's property using a large automatic gate, circa 1955. *F. Taylor Collection*

Some of the other early stakers on Haggart Creek and Dublin Gulch are: Hector Morrison, Alex Nicol, Rodolph Rosmusen, Charles Turgeon, Frank Carscallen, Gervais P. Godbout, Louis Bouvette, Henri LeBlanc, Andrew Johnson, Elmer Middlecoff, William Hayden, "Jake" Davidson and John Stewart.

The Cantin cousins on Dublin Gulch. Frank Cantin is second from the left and the others, not in order, are Joseph, Louis, and Phil. *M. Rich Collection*

DUNCAN CREEK *by Lynette Bleiler & Linda E. T. MacDonald*

Duncan Creek is one of the larger streams in the Mayo area. It is about 14 miles long, and joins the Mayo River below Mayo Lake. About 12 miles from its mouth, Duncan Creek forks, with the larger branch becoming Lightning Creek. The smaller fork is known as Upper Duncan Creek.

Gold was first discovered on Duncan Creek in 1898 by three Swedes named Gustaveson, a father and his two sons. They were located on the canyon on Duncan Creek, about eight miles from the McQuesten River where they built their cabins and erected a water-powered sawmill. For over two years the trio mined undisturbed, making an occasional trip to Dawson for supplies.

One fall, the Swedes appeared in Dawson with thousands of dollars worth of coarse gold which they took to the bank. The appearance of the Swedes' gold was different in both colour and shape than the nuggets found in the Dawson district, and rumours of a rich new find began to spread. The Gustavesons had come from up the Yukon River but no one knew from where. In order to keep their find a secret, the Swedes had not recorded any claims. In the spring, the Gustavesons left Dawson with supplies and were followed to the Stewart River by hopeful stampeders, where they managed to elude their followers.[20]

In the fall of 1901, four young Dawson stampeders, Allan McIntosh, Colin Hamilton, Duncan Patterson and "Jake" Davidson, were camped on the bank of the McQuesten River. While there they saw the three Swedes heading down the river toward Dawson.

They decided to look for the Gustavesons' camp and poled a boat up the McQuesten looking for signs of a trail. Fifteen miles east of Haggart Creek they found a cache on the bank and indications that a boat had been launched there. From this point they crossed the flats of the McQuesten Valley. The trail led up near a creek later named Christal Creek after Charles Christal. After following the creek for three miles, they came to a box canyon nearly one mile long, and the Swedes' mine. The Swedes had not erected any claim posts.

On September 12, 1901 the four men each staked a claim up to the falls and named the creek Duncan Creek after Duncan Patterson who staked the Discovery claim. In order to stake another 500 foot Discovery claim, the head of the creek was named Lightning Creek and the south fork was referred to as the head of Duncan Creek or Upper Duncan Creek.

As soon as they recorded their claims in Dawson, word spread and the stampede to Duncan Creek began. This new strike proved to be the next most important discovery in the lower Yukon since the Klondike Gold Rush. Some of the names associated with the Duncan Creek Rush are: Albert Desautels, Victor Savage, Matthew Levi Foley, Elphege Desautels, Joel M. Zane, Francis W. Moore, Ambrose Corbeil, Fred H. Wade, C.M. Comstock, William "Billy" A. Burrell, John Durand, J. Scrivener, Alex Nicol, Peter O'Boyle, H.W. McWhorter, James Mervyn,

Minto Bridge, 1922. *Middlecoff-Bleiler Collection*

Joesph Cantin, Frank Cantin, Louis Cantin, Mary Cotter and Triffle Paris.

The Swedes left the country and never returned. However, it was reported by Joseph Keele in his G.S.C. Summary Report for 1904 that the Swedes took out gold to the value of $30,000 or more.[21]

A settlement was established at Gordon Landing in 1902 on the Stewart River near the mouth of Janet Creek to serve the miners on Duncan Creek. Rather than taking the all water route, stampeders began to travel up the Stewart to Gordon Landing, hike the two miles from there to Janet Lake, and the twelve miles north to Duncan Creek. As word began to spread about the new gold strike, gold seekers arrived from all over and by 1902 Duncan Creek was staked from its head-waters to the Mayo River. Cabins were built on almost every claim and active preparations were made to develop the ground. A good wagon road was constructed by the government from the mouth of the Mayo River to Duncan Creek, a distance of approximately 24 miles. Roadhouses were also established at several points along the route. Gordon Landing was abandoned in 1903 when the townsite of Mayo Landing on the Stewart River was surveyed and established as the main transportation and supply link to Duncan Creek.

The miners used quite primitive methods to work the shallow gravels. Water, picks, shovels and wooden sluice boxes with pole-riffles were the implements used to mine the richest and most accessible ground. Dr. A. Aho states that McIntosh, Hamilton, Patterson and Davidson took out approximately $45,000 in coarse gold in 1902.[22] However, in 1903, the year of greatest activity, Joseph Keele reports that $30,000 was recovered from the canyon claims, declining to $15,000 in 1904.[23]

It was a different story on lower Duncan Creek where in 1903 shafts were sunk to depths of over 100 feet. They were soon abandoned when water filtered through the thawed gravels and filled the shafts faster than the pumps could drain them. The total value of the miners' clean-ups was reported as between $800 and $1,200.[24]

Williams Creek, a tributary of Duncan Creek, was staked on December 2, 1901 by F. Williams, F. Cantin, J. Lelievre, D. Parent and O. Letourneau.

During the first week in March, 1903, the miners on Duncan and Williams Creeks contested the baseline survey done on those creeks by Raoul Rinfret of the Dominion Land Survey. The Duncan Creek petitioners wrote in their complaint that, "According to said survey, many of us are thrown out of ground which we legitimately staked and recorded according to the mining regulations of the territory." Rinfret's survey was indeed in error and A.J. McPherson, of the Dominion Land Survey in Dawson, who wrote a three-page report outlining the deficiencies, had to redo the survey.

Miners continued to do enough work to hold their ground although little mining was done on an economic scale for many years below the canyon.

Joe Walsh "walked" a Keystone drill from Dawson to Duncan Creek in 1912. He drilled the ground in 1913 and 1914 and worked it with William L. Bramley and John "Jack" Adair. However, even though the prospecting results were encouraging, the actual mining operation was not successful.

By 1915, only eleven men were working on Duncan Creek and its tributaries. Many stampeders left the area for World War I. George MacKenzie, Gold Commissioner of the Yukon Territory, estimated that by the close of 1915, Duncan Creek had produced $55,000. Oldtimers, however, estimated that the production was probably closer to $75,000. This figure does not include amounts taken out by the Swedes, which totalled $30,000 or more.

By 1920, a number of the cabins and buildings on Duncan Creek were dismantled and moved to build the mining camp of Keno as interest in the placer deposits dwindled in favour of the nearby and newly discovered silver deposits.

According to the Mayo Mining Recorder's reports from 1920 on, placer mining on both Upper Duncan and lower Duncan Creek and its tributaries was intermittent with only the following operations appearing in the records over the next 39 years.

Mrs. Joe Walsh outside her cabin on Duncan Creek.
T. McKay Collection

During the 1921–22 season David Sparks and partners prospected on 13 Pup on Duncan Creek, but the next burst of activity was ten years later when, in 1932, Andy Johnson put in a mile long ditch above the first roadhouse below Keno. This ditch was extended to three miles in 1933 and Andy Johnson mined with some success.

That same year, Meredith and Henderson made wages working previously mined bedrock on Upper Duncan Creek.

Arthur Chambers and William Bramley had a favourable year in 1936 and Matt Butyer made wages during the summers of 1938 and 1939 working his ground on Duncan Creek.

From 1940 to 1941, C.E. "Bud" Fisher prospected a mile lease located at the bridge on Duncan Creek. There was no production or further activity reported until 1958 when several employees of United Keno Hill Mines bitten by the gold "bug" recovered several ounces of gold prospecting on weekends on Duncan Creek.

Fred Taylor, who had an active operation on Dublin Gulch, staked a prospecting lease on Duncan Creek in 1958. He staked his lease into claims which he kept in good standing until he eventually sold to his son, Frank, in 1975. Frank, inspired by the increase in the gold price, decided to follow in his father's footsteps with a mining career. Frank Taylor with his wife, Bonnie, and family continue to mine on Duncan Creek today (1989).

Duncan Creek has seen renewed activity with increases in the price of gold but this activity occurred outside the parameters of this publication.

After the most accessible areas on Duncan Creek were staked the miners spread out to its tributaries in search of the precious metal.

In 1901, Denis Parent staked Discovery on Parent Creek, a tributary of Duncan Creek, and some time later Jim Mervyn also mined on Parent Creek. Initially he was unsuccessful, but after moving his operation further up the creek, he made enough to help start his trading business at Lansing. In 1907, Peter Divine mined a $14 nugget on Parent Creek but nothing more.[25]

Miners sank a shaft on Lightning Creek just above the forks reaching bedrock at 105 feet with disappointing results. Nearby Thunder Gulch which joins Lightning Creek proved more promising but was filled with huge boulders. Martin Malesich held claims on Thunder Gulch

Fritz "Fred" Fluhmann and crew at Fluhmann's claim on Duncan Creek, circa 1910. *R.M. Clark Collection*

from approximately 1910 until 1940 and mined enough gold to make a living. In 1941 John Backe purchased three claims formerly owned by Malesich and intended to mine his claims with the help of two other men.

Some mining was done on Field Creek with little success, but it did become the location of the Field Creek roadhouse after the construction of the road between Mayo and Keno. During the late 1920's the Treadwell Yukon Company Ltd. kept a garage at Field Creek to house their 10-ton tractors. The garage burnt down in March 1928 and was not rebuilt.[26]

Mining activity reached a peak on Duncan Creek in 1903. As the news spread of the "new Bonanza" miners flocked to the Duncan Creek gold fields only to find all the ground staked. These men, along with miners discouraged with their ground on Duncan Creek, formed the group responsible for the 1903 stampede which led to the major discoveries of Highet Creek, Minto Creek and Bennett Creek which flow into Minto Creek and then the Mayo River as well as Ledge Creek, Steep Creek and Cascade Creek in the Mayo Lake region, which were also staked in 1903.

MAYO LAKE REGION *by Lynette Bleiler & Linda E. T. MacDonald*

Alexander MacDonald of New Brunswick was the first man to prospect the Mayo Lake region. He discovered the lake in 1887 and named it after Captain Alfred H. Mayo.

Several of the creeks draining into Mayo Lake yielded discoveries of coarse nugget gold in 1903, and each caused a small stampede. The workable ground on Ledge, Cascade and Steep Creek was shallow and was mined with very primitive methods by today's standards. The miners ground-sluiced off the top dirt, constructed a dam across the creek and built a board flume which connected to wooden sluice boxes. Miners would then shovel the pay gravels into the sluice boxes fitted with pole riffles which would catch the gold.

By the 1924–25 season R.L. Gillespie, the Mayo Mining Recorder, reported that most of the gold produced in the district came from the claims on Mayo Lake. He also stated that during the 1925–26 seasons, gold activity was at a low ebb but silver production was booming.

However, in 1933 with the increase in the price of gold, there was a flourish of activity in the Mayo Lake area.

Steep Creek

Steep Creek, which flows into the south arm of Mayo Lake about eight miles from its southern end, was staked in September 1903 by Thomas Mikesich who staked the Discovery claim. Others who also staked in 1903 were Hector Morrison, Frank Chasni, James Butler, Thomas Maher, Matt Gaspar who held his ground until 1905, Mike Pavisich, Andrew Johnson, James Mervyn, Pedro Longa, Walter Harris, John Mawhinney, Paul Lester and Pierre David. This group of miners did not hold their ground long.

The next flurry of activity on Steep Creek was in 1905 when Charles Augustus Anderson staked 500 feet of Upper Discovery and in 1908 and 1909 when John August Kinman staked Discovery and a Bench Discovery. Norman Wightman recorded the next activity in 1931. Steep Creek was abandoned for other creeks in the Mayo Lake area until staked by Pete Nord in 1953.

Ledge Creek

Ledge Creek, which flows into Mayo Lake, has seen almost continuous activity since it was staked in 1903. The Upper 500 feet Discovery claim on Ledge Creek was staked by Mike Pavisich on August 18, 1903, then sold to Matt Gaspar that October. The Lower 1000 feet Discovery on Ledge Creek was staked by Gaspar on September 16, 1903. He held the ground until 1906.

Claim No. 1 was staked by Thomas Mikesich on August 18, 1903. Many more claims were staked in 1903 by: Thomas Curry, Andrew Johnson, Frank Ayerde, John Carlin, John Mawhinney, Walter Harris, Hector Morrison, Hugh Carlin, Frank B. Williams, John Hamilton Baker, Frank Chasni, Joseph Rivard, Peter O'Boyle, Gustav Schogren, Victor Butler, Adrieu Cadieux, Peter Falcouer, John Charles Brown, Paul Lester, Louis Bouvette, Samuel Blackmore, George Keyes, Lena Potter, George F. Potter, William Edward Potter, James Wood, George Cunningham, Frank Cantin and Albert Jahnke.

John Kinman staked on Ledge Creek in 1910 and in 1922 Hans R. Johnson was prospecting at the mouth of Ledge Creek. Johnson staked 1 Below Discovery in 1926 and was the only man working that creek in 1926, 1927 and 1928. In 1931 he entered into an agreement with George Reynolds and William Jeffrey Sr. Two years later in 1933, Reynolds and David Fotheringham worked together just covering wages with their production. By 1938, Reynolds was the only producing miner in the Mayo Lake area. He continued on alone through the 1939 season, making little more than wages. He was, however, hampered by inordinately high water that year. Reynolds continued to hold the ground until 1949 when William McCoomb optioned his ground and did some preparatory work for the 1950 season.

There was no gold production reported through the Mayo Mining Recorder's office from 1949 to 1952 even though 79 claims were held in good standing as at March 31st, 1952. According to R.L. Debicki, in *Yukon Mineral Industry 1941 to 1959*, there was no activity on Ledge Creek during this period.

> ***"By 1926, there were few miners left in the Mayo Lake Region. . ."***

Other creeks in the Mayo Lake region also attracted the miners' attention. On Cascade Creek, which enters the south arm of Mayo Lake about two miles south of Ledge Creek, Jacob Furrer staked the Upper Discovery on October 3, 1903, and Thomas Edward Heney staked the Lower Discovery on October 7, 1903. On Edmonton Creek, Ernest Beliveau staked the Discovery on June 13, 1904, and was joined by other stampeders such as A.E. Desautels, Arsene Tessier, Ambrose Corbeil, Walter Oliver Moore, Francis William Moore and Nicholas Hammon Johnson.

R. L. Gillespie, the Mayo Mining Recorder, noted in his report of 1923 that, "Arthur Hester, who owns Discovery claim, on Gull Creek, has sunk a number of holes on that creek during the winter, with fairly good results." In 1924, Gillespie again reported that Hester had uncovered some good pay late in the season which resulted in over twenty claims being staked.

By 1926, few miners were left in the Mayo

Lake Region apart from Hester who was still mining on Gull Creek, E. Burnell on Cliff Creek, H. Johnson on Ledge Creek and McMaster and Flynnen on Anderson Creek. Malcolm McCown also prospected Anderson Creek in 1933 while Arthur Chambers and Matt Butyer were on Gull Creek that same year.

Davidson Creek

Davidson Creek flows into the Mayo River not far from Mayo Lake. Discovery claim was staked by Jacob A. Davidson on August 5, 1907. Davidson then entered into a partnership with Herbert G. Wilson in 1908 and later sold out to Wilson. Some miners who followed were Frank German in 1913, David Sparks, Frank Chasni and J.T. Scrivener, the last two also selling to Wilson.

Gillespie's report of 1934 stated that the partnership of Rich, Hester and Ross had found "good prospects" and they were preparing to mine on a large scale in 1934.

He also reported that, "Fred Lefebvre and associates have put 8,000 feet of lumber on their Davidson creek property where they plan on working their prospecting lease with water power this coming summer [1934]." Lefebvre worked the 1936 and 1937 season, but there are no further details available either on him or Davidson Creek until 1954, when R.L. Debicki reports in *Yukon Mineral Industry 1941 to 1959* that Ellis Johnson reworked some old placer workings on Davidson Creek, taking out five ounces of gold.

MINTO LAKE REGION *by Lynette Bleiler & Linda E. T. MacDonald*

Once all the easily mined ground had been staked on Duncan Creek, the prospectors crossed from the Duncan Creek Valley to explore the creeks lying between the Mayo and McQuesten Valleys.

Highet Creek, circa 1910. *E. Middlecoff Collection*

Bennett Creek, which flows into Minto Creek, was staked on May 25, 1903, by Charles Edward Bennett. This creek has seen little activity over the years.

It was on Minto Creek, which flows out of Minto Lake and enters the Mayo River about ten miles from the Stewart River, that F. Leresque staked claim No. 1 on June 16, 1903, and E. Proulx staked claim No. 2 on the same day. Although shafts were sunk below Discovery, some to depths of 100 feet, bedrock was not reached. There was a flurry of activity over the next number of years with Louis Bouvette staking a Discovery on the left limit of Minto Lake on May 4, 1908.[27] James A. Scougale, a Dawson merchant and Otto F. Kastner bought some claims at Minto Lake from "Jimmy" Scrivener and proceeded to develop the ground. They hired engineer John A. "Jack" Ross to manage the operation. In 1913, Ross brought his wife and family from Dawson to live at Minto Lake where he was constructing ditches, later referred to as the Kastner Ditch, to bring water from nearby McIntyre, McLagan and Turnip Creeks to run the monitors and sluice boxes. An estimated $60,000 was spent trying to mine the Minto Lake claims, without success.[28]

Harvey J. Ray staked a Discovery Bench on the left limit of Minto Creek on July 18, 1916, but it wasn't until 1934 and 1935 when Frank Gillespie reconstructed and repaired the old Kastner Ditch in preparation for mining the lower benches adjoining the lower end of Minto Lake[29] that there was any activity in the area. Gillespie toiled for two summers cleaning out the six miles of ditch and building wooden flumes over the rocky areas to get water to his sluice box. He managed to get some sluicing done before his ditch was wiped out by a raging forest fire. Disillusioned, he did not rebuild his flumes and ditch, but continued to live in his cabin at Minto Lake and pursue his love of fishing.

James Thomas Scrivener staked claim No. 1 on Highet Creek on June 3, 1903. The Mining Recorder's Office would not allow a Discovery claim to be registered due to the proximity of Discovery on Minto Creek. Claims numbered from the mouth of the creek with none being more than 250 feet long.[30]

Highet Creek has had a long and prosperous history. From 1903 on, Missouri-born Elmer Middlecoff acquired most of the claims which, upon his death in 1951, fell to his heirs. One of his daughters, Flora, married Ed Bleiler, a young miner who had been working on Dublin Gulch and Highet Creek since 1933. After the tragic death of Middlecoff's son, Elmer Jr., on Highet in 1936, Ed and Flora took over the family business which, at that stage, was going through some lean times. The history of Highet Creek comes to us from Ed Bleiler, who has mined on Highet Creek for over fifty years.

Highet Creek. *Middlecoff-Bleiler Collection*

Highet Creek

Highet Creek is a small stream about 8 miles in length. It joins Minto Creek about 2.5 miles below Minto Lake which flows into the Mayo River.

From information related by some of the old-timers involved in placer mining in the Mayo area, notably Elmer Middlecoff Sr., Alex Nicol, Frank Cantin, Rodolph Rosmusen and others, gold was first discovered on Highet Creek by Warren Hiatt on one of the benches well up the creek, possibly even above Rudolph Gulch,[31] or near the junction of Rudolph and the main Highet Creek. Exactly which year is uncertain, but it was prior to 1903. Hiatt may have come over the divide from Johnson Creek because, well before the Klondike Gold Rush, quite a few people were mining on Johnson Creek. Hiatt did not even bother to stake a claim, his decision being influenced by very low water in the creek. He did not think there was enough water to carry out mining on a moderate scale.

As a result of the stampede on Duncan Creek and an overflow of stampeders from Dawson, some of the prospectors began looking for new areas nearby to prospect. According to Dr. A. Aho,[32] Jimmy Scrivener and Peter Haggart staked on Minto Creek where prospects were found in the surface gravels and according to records in Yukon Archives James "Jimmy" Thomas Scrivener, Archie McIntyre, Peter Findlay Haggart, Raleigh Lawson Bernard, George Henry Miller, George Gordon, Phidime Cantin, Peter O'Boyle, George Keyes, Louis Bouvette, Jacob "Jake" Davidson, Albert Jahnke, Alex Nicol, Warren Elmer Hiatt, James Mervyn, Grant Huffman and George Potter were among the first stakers on Highet Creek.[33] Fred Wade also staked some ground and found bench gold which he mined later.

Once on a walk up Highet Creek with Middlecoff Sr. he pointed out Wade's workings, still quite visible due to stacked boulder piles, even though completely overgrown with second growth.

In the succeeding years, an extensive bench area was mined almost as far downstream as McRae Pup and half a mile above Rudolph Pup.

The richest and most extensive was in the Rudolph Pup area, which was worked by Rodolph Rosmusen and partners. The ground here yielded gold to a value of between $100,000 and $140,000 according to a 1915 report by D.D. Cairnes.[34] They did not bother with the main creek valley because of depth, boulders, drainage and tailings disposal problems. This area of the creek was worked in later years by Middlecoff.

There were many miners up and down the valley at first. The most notable successes were some of the drift miners and the bench miners, all of whom came and went early in the life of Highet Creek.

Some of the early miners, notably Hector Morrison, found very rich pay in the main valley, above and below the mouth of Morrison Pup on the right limit. The area was mined by underground drift mining. As the ground was thawed it required much support by timbering. Apparently, it was Morrison who, naming Highet Creek, spelled it as it has always been shown on maps and records. Morrison did name the stream after Warren Hiatt, who made the original discovery, but was not aware that his spelling was wrong.

Middlecoff acquired some claims in 1903 but did not mine them until 1906. At first, his crew consisted of one old-timer who taught him some of the rudiments of open cut mining. Up to that time, Middlecoff's experience had been underground mining on Lovett Gulch, a tributary of Bonanza Creek in the Dawson area, where he had made a rich discovery. His luck was good on Highet Creek as the ground was rich and at first, only 10 feet to bedrock. The 1915 Geological survey reports that in his first seven years he obtained

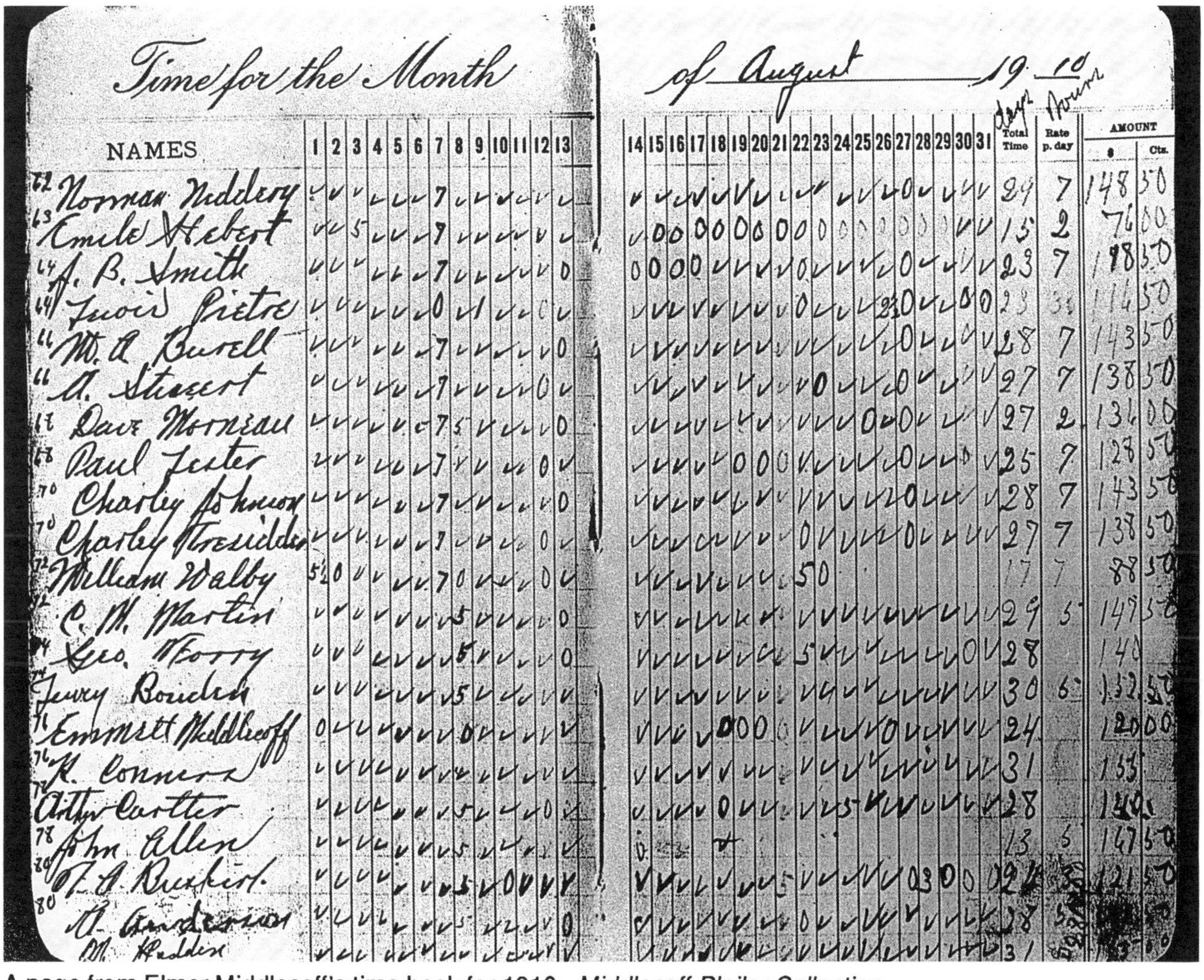

A page from Elmer Middlecoff's time book for 1910. *Middlecoff-Bleiler Collection*

gold to a value of almost $250,000.

His first camp was at the mouth of Morrison Pup which he staked in April, 1909. As the years went by, he experimented with different mining methods until he settled on a system of nozzling the banks down with monitors and sluicing everything through a dump box and sluice line. In the beginning, his riffles in the bottom of the sluice boxes consisted of flat stones[35] selected from the tailing piles. He had one steam boiler rigged to hoist the boulders out of the way and a second larger boiler was used to stack tailings with a clam-shell bucket. He had a large dams made of logs, moss and earth for water storage. At these dams he had automatic gates, 6 feet and 12 feet wide, with which he "boomed" off the top dirt.

The Middlecoff cabin on Highet Creek. Ethel and Elmer Middlecoff Jr. are on the right and the crew is on the left, circa 1910. *Middlecoff-Bleiler Collection*

This was a very expensive way to mine. The two boilers gobbled up great quantities of wood. At first, this was not a problem as the Highet Creek Valley and adjoining hills were heavily wooded. However, Middlecoff was not the only one using wood. The drift miners and the dredge further down the creek, and all of the other miners were denuding the hillsides until the shortage of timber became a great problem.

The need to have so many people, as well as

Elmer Middlecoff's placer operation on Highet Creek, 1916. *Middlecoff-Bleiler Collection*

horses, involved in acquiring wood, digging and attending ditches, building dams and working around the mining operation (it operated night and day) as well as hunters to feed the crew, which numbered up to 30 men at any one time, made for much overhead. This could be borne only while the rich pay lasted.

About 1914, when the First World War broke out and young men were no longer available, Middlecoff abandoned this method of mining. He began experimenting with a monitor and grizzly method which proved quite successful and was very economical, requiring few men. He used this system until World War II and a shortage of labour forced him to close down. He died in Mayo in 1951.

In 1957, Flora, Middlecoff's youngest daughter, and her husband Ed Bleiler took over the operation. By this time roads and communications had been greatly improved. It was the era of bulldozers, backhoes, rubber-tired and track type loaders and 4-wheel drive vehicles. The Bleilers continued using the machinery-intensive Middlecoff system with the help of son, Lowell. Fair production was attained very economically with one man on each of the two daily ten-hour shifts.

As bigger machinery was acquired, the monitor and grizzly were abandoned and a one-shift operation was used, pushing materials into a large dump box and washing plant with a D8 bulldozer

Two Middlecoff crew members doing maintenance work on a ditch carrying water to the Highet Creek mining operation. *Middlecoff-Bleiler Collection*

Middlecoff's cabin at Highet Creek. *Middlecoff-Bleiler Collection*

and stacking tailings with various sized rubber-tired loaders.

Lowell took over when his father retired from full-time mining in 1984. Larger equipment was acquired in order to increase the yardage handled each day. This was necessary since the ground just above the canyon upstream almost to the dredge was covered with as much as 65 feet of overburden which had to be stripped off, hauled and piled a long way from the area being mined. Working two 10-hour shifts with four men and using a grizzly and washing plant designed by Lowell, much larger yardages have been obtained with less effort. The values have remained fairly consistent over the years.

Ballarat Mines, Ltd., Highet Creek, 1952, employed five men and produced 600 fine ounces of gold valued at $20,406.88. It was the only operation in Mayo district that year. *Middlecoff-Bleiler Collection*

One other operation of historical importance, though short lived, was a partnership, humorously referred to as the "Little Guggs" after the Guggenheims who had huge dredging operations in the Dawson area. George Miller, Gervais P. Godbout, M. P. Lindquist, Rodolph Rosmusen and Charles Rockney put their claims together and, using a large dam for ground sluicing when the water was high, they removed much of the overburden. Men with picks and shovels and wheelbarrows loaded the pay gravel into a self-dumping bucket which was hoisted up to a washing plant where the gravels were sluiced and the gold recovered. They were one of the larger producers around 1910–1911. This group of partners leased their ground to the Highet Creek Dredging Company in 1919.

"Claim No. 67 Highet Creek June 27, 1914, Wolfe Photo Dawson Y.T." This mine was known locally as the "Little Guggs." *Dora Nicol Collection*

GOLD DREDGES

Designed and Built to Suit all Conditions Improved All-Steel Prospecting Dredge.

LIGHT, STRONG AND CHEAP

A. W. ROBINSON

14 Phillips Square Montreal, Can.

25 YEARS' EXPERIENCE IN MORE THAN 125 DREDGES

Cable Address "Dredge Montreal"
Code A. B. C. Fourth Edition

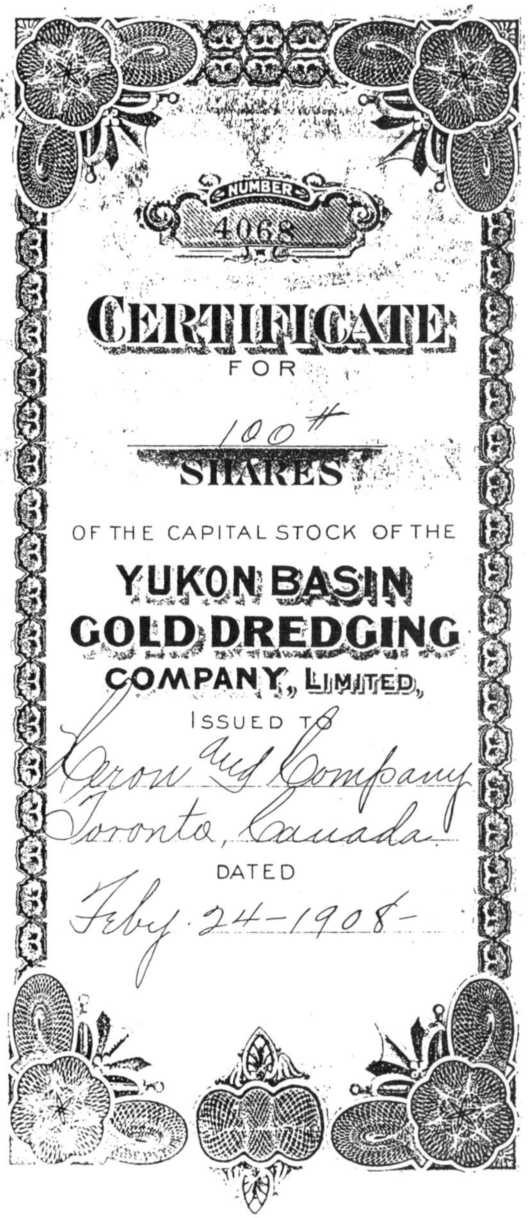

Artifacts. ABOVE, Item on the Ogilvie Dredge from the "Journal" of May 2, 1903. RIGHT, Dredging company stock certificate. *Brian Denman Collection*

One other miner, Harvey Ray, was mining on Highet Creek in the late Thirties and early Forties. He changed from underground methods to a hydraulic method of mining when his brother Irvin with his wife Elsy joined him for a year or two. The small piece of bench ground was soon mined out and Harvey Ray left for greener pastures.

Presently, aside from Lowell Bleiler's operation and Ed Bleiler's small sniping operation, two others, Wilf and G. Jean Gordon and Frank Erl, have been mining towards the Rudolph pay area using cats, loaders and various types of washing plants. These are one or two man operations.

Highet Creek Dredges

Unlike the Klondike, gold dredges in the Mayo area were an uncommon sight. The fact that Highet Creek was the site for the only two dredges to be constructed in the area warrants further attention.

The first dredge belonged to the Klasco Dredging Company. According to records in the Mayo Mining Recorder's Office, in February 1917 a power-of-attorney was registered between M.P. Lindquist and the Klasco Dredging Co. of Seattle as well as agreements between the dredging company and Joseph. J. Langram and his wife, Lydia, R. Rosmusen, E. Middlecoff, C.L. Rockney, G.H. Miller and Alex Nicol to mine their claims on the upper area of Highet Creek.

The hull of the dredge was constructed near

the mouth of 80 Pup and a building housing the pipe boiler and electric generator was built below the Highet Creek canyon, near the mouth of the creek. A pole line was built as the steam boiler would provide the electricity needed to power the dredge further up the creek.

The project seemed doomed from the start after a newly designed steam tractor, brought in to haul equipment, broke down in Mayo, never to reach the creek. Teamsters Alex Nicol, Joe and Louis Cantin were hired to haul the materials needed to build the dredge. However, with only the hull completed, the Klasco Dredging Co. encountered financial difficulties and the project was abandoned, leaving parts of this dredge scattered from Highet Creek to San Francisco.

During the months of July and August, 1917, liens were recorded against the Klasco Dredging Co. et al, by George Shimada, Mrs. J.W. Nicholas, H.B. Sugbus, Alex Nicol, J.A. Connon, Wigo Anderson, M.P. Lindquist et al.

Dredge on Highet Creek, 1921. *E. Barker photo, A. Aho Collection*

The second dredge, of Scottish design, was built in the British Yukon Navigation shipyards in Whitehorse in 1908–1909 for the Yukon Basin Gold Dredging Company (Ltd.), a public company with former Commissioner of the Yukon, William Ogilvie, as its president.

The company's office was in Kansas City, Missouri and the list of Directors included: William Ogilvie, who was cited as Ex-Governor Yukon Territory; Jas. P. Batchelor, President of Cotton Belt Lumber Co., Kansas City; Charlie H. Clouston, General Southwest Agent for W.H. Baker Cocoa and Chocolate, St. Louis Mo.; and Thomas B. Flint, Clerk, House of Commons, Ottawa.

The prospectus, a forty page document, extols the virtue of dredges and tells of the great wealth to be uncovered along the Stewart River, "A River of Gold." The cover page makes the claim, "We hold exclusive title by concession from the Canadian Government to the largest and richest area of dredgable gold-bearing gravel, controlled by one management on this continent, extending for one hundred and five miles along one of the richest gold producing rivers on earth."

They described their concessions as beginning at a point 15 miles above the confluence of the Stewart and Yukon Rivers, to the mouth of the McQuesten River, and on up to and beyond Fraser Falls. The prospectus also points out that, as Fraser Falls was in the centre of their concessions, it could be used to generate power to run all the dredges the Yukon Basin Gold Dredging Company envisioned building for the Stewart River. They calculated their reserves as lasting 130 years with one dredge or ten years with twenty dredges.

The prospectus also contained endorsements by well known Yukoners, Robert Henderson, Asst. Government Mining Engineer; Dan Matheson, engineer; John Grant and Sam Moreau among others. The company offered twenty million shares with a par value of one dollar. The shares were offered to the public at 25 cents per share.

The company's dredge was steam operated and was not self-propelled. It was towed by sternwheeler from the Whitehorse shipyards to the Stewart River where it began to dig the bars. The dredge would quickly mine the bar and then lay idle as it waited for the next sternwheeler on its way from Stewart City to Mayo to push or tow it to the next bar. It got as far as the mouth of the McQuesten River when, presumably, financial difficulties forced it to quit.

Records in the Mayo Mining Recorder's Office show that from July 23, 1909, to September 1909, the Yukon Basin Gold Dredging Co. Ltd. bought claims on the right limit of the Stewart River from John B. McLeod, and Joe Longtin, also the Gatensbury claim, the Ralph claim, the Nita claim, the Geo. Clawson claim and the Gilbert Skelly claim.

A company called either the Titus Dredge Co. or the Highet Creek Dredging Co. made up of a mixture of Canadian and American entrepreneurs, McVicar, Edward O. Finlaison, (Manager of the Bank of Montreal in Dawson City), Harry Jones and "Lou" Titus was the next group to try their hand at dredging in the Mayo area. These local men were so impressed by the dredges in the Klondike Goldfields that they called in Bucyrus and the Marion Steam Shovel Company. After reviewing the site conditions, the two companies refused to recommend a dredge as they felt that Highet Creek had too many boulders for dredging.

Dredge on Highet Creek, circa 1941, built circa 1920. *Christena Matheson Andison photo, G. Jean Gordon Collection*

However, the determined group acquired the old harbour mud dredge abandoned by the Yukon Basin Gold Dredging Co. Ltd. and the news spread as shown by A.K. Schellinger's report to E.E. McCarthy, Resident Manager of the Yukon Gold Co.in Dawson City, dated April 7th, 1919.

> The dredge which has stood idle for many years at the mouth of McQuesten River is to be moved this spring [1919] onto Highet Creek near Minto Lake. Mr. Titus is one of the promoters and is in charge of the work.

In June 1919, the dredge was towed from the mouth of the McQuesten River to Mayo by the Str. *Dawson*[36] where it was disassembled and transported overland piece by piece by horse power to Highet Creek.

"Eddie" Kimbel and his brother were brought in to build the hull in 1920 and upon completion Ed Barker, who had had experience dredging in Siberia before the Russian Revolution, was hired as dredge master, Joe Walsh and John A. Ross the engineers and Rupert "Rupe" Steeves the winchman.

"Lou" Titus took out lays with the "Little Guggs" on Highet Creek and held them between March 1919 and August 1922. The dredge began operation on claim No. 66 below Lindquist's cabin. However, it was forced to close down in the early part of the 1922 season after one and a half years of effort.

Ed Bleiler, a long-time miner on Highet Creek, had this to say about the dredge:

Highet Creek Dredge, circa 1921. *D. Walloch (McLaren) Collection*

Ed Barker informed me it was broken down more hours than it was in operation due to the flimsy material in the bucket line. Actually, on inspection, the bucket line was quite massive and looked like it could dig any ground; however, it was the huge boulders that it couldn't cope with. The sudden impact with a boulder would make the whole dredge shake and shudder – something had to give.

The company paid a percentage to the claim owners and according to Evelyn Lee (Middlecoff) they each got $5,000 during the time of operation. However, on the last shipment of gold to Dawson the bank manager, E.O. Finlaison recovered his total investment, leaving nothing to pay the crew of 20 men or expenses of the dredge. Liens were put against claims held by "Lou" Titus by S. Blackmore, Joe Hoder and John A. Ross.

Once again the dredge was abandoned, this time for good. The partially dismantled skeleton of the dredge is all that remains on Highet Creek.

MAYO LANDING

With the numerous gold discoveries and increasing numbers of miners gathered in the district, the town site of Mayo Landing was surveyed in 1903 by Raoul Rinfret, Joseph Edward Beliveau, and John Dease Bell. The site, encompassing 185 acres at the confluence of the Mayo and Stewart Rivers, was selected for its proximity to the Duncan Creek trail and because the flat river banks at that point on the Stewart were suitable for steamboat docking. Mayo was formally referred to as Mayo Landing until 1958.

Alex Nicol, hearing during the winter of 1902–03 that a town site would be established at Mayo Landing, built the first log cabin on the site in March and was there when the surveyors arrived.

Gene Binet also heard rumours that a town was to be surveyed and roads built and he promptly bought the entire townsite. Before the ice went out on the Stewart that spring, he had erected the outer walls and roof of his log hotel. By late spring, lots were sold and cabins (complete with pole caches and outdoor privies) began to spring up, connected only by wagon trails and foot paths. Water was drawn in buckets from the Stewart River.

The arrival of the first boat, spring, 1922. Binet Bros. store and hotel is in the background. *Ralph Rogers Collection*

Binet's hotel was completed by July. It housed a dining room and the entire complex was equipped with Delco lighting. He soon added a store onto the hotel in order to sell much-needed miners' supplies.

In 1904, a N.W.M.P. detachment was established in Mayo. Also in 1904 and 1905 the government built an overland trail from Dawson to Duncan Creek and a good wagon road from Duncan Creek to Mayo.

“Gene Binet also heard rumours that a town was to be surveyed . . . and he promptly bought the entire townsite.”

By 1914, men were working on Duncan, Haggart, Highet, Johnson, Minto and Secret Creeks, as well as Dublin and Thunder gulches. Huffman and Alverson's shipment of silver ore from the Silver King and Thomas Aitken's subsequent development of the mine brought more men into the district. Also in 1914, the Mining Recorder's Office was moved from Minto Bridge to Mayo and the Mayo Landing Post Office was established. The Mayo school opened and Alex Profeit taught the first classes. Jim Ferrell moved from Lansing Post to Mayo and opened a store. John Davis opened a blacksmith shop and Isaac Mallette and Fred Lefebvre began operation of their sawmill. In addition to these services, there was a livery stable and liquor store.

Steamships with good passenger and freight accommodation were making weekly trips from Dawson to Mayo. During the winter there was a monthly and, for part of the season, a bi-monthly stage service between Dawson and Minto Bridge. Allan McIntosh carried mail to Mayo, Minto Bridge, Duncan Creek and Haggart Creek.

Activity in the district began to wane with the outbreak of World War I. John Suttles, who operated the largest placer outfit on Dublin Gulch, went overseas and did not return to the Mayo district after the war was over.

Cunningham Hotel and Store, Mayo, circa 1910. *Middlecoff-Bleiler Collection*

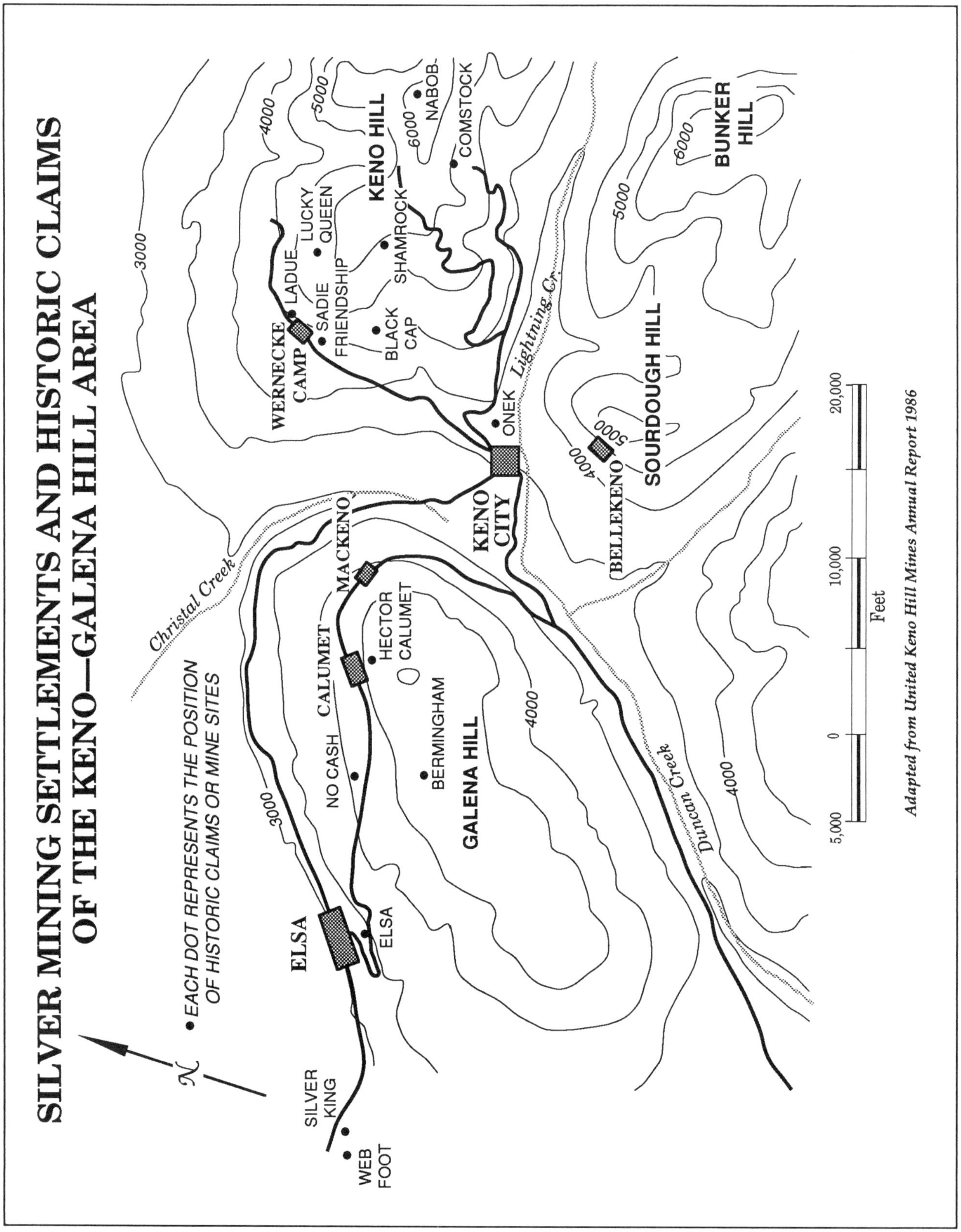
SILVER MINING SETTLEMENTS AND HISTORIC CLAIMS OF THE KENO—GALENA HILL AREA
N
• EACH DOT REPRESENTS THE POSITION OF HISTORIC CLAIMS OR MINE SITES
Christal Creek
WERNECKE CAMP
LADUE
SADIE
FRIENDSHIP
LUCKY QUEEN
BLACK CAP
SHAMROCK
KENO HILL
NABOB
COMSTOCK
BUNKER HILL
Lightning Cr.
ONEK
KENO CITY
SOURDOUGH HILL
BELLEKENO
MACKENO
HECTOR CALUMET
CALUMET
BERMINGHAM
NO CASH
GALENA HILL
ELSA
SILVER KING
WEB FOOT
Duncan Creek
3000
4000
5000
6000
5,000
0
10,000
20,000
Feet
Adapted from United Keno Hill Mines Annual Report 1986

E. L. Bleiler

Chapter 5

Silver Mining

DAVIDSON

JACOB A. "JAKE" DAVIDSON, JAMES CORKERY, WILLIAM HODGE WILLIAMSON, and Harry McWhorter prospected widely in the Duncan Creek area during the years following the rush to that creek. In 1903 Davidson found an exceptional silver-rich galena specimen on the trail to Duncan Creek.

His find was located in the canyon of what was later known as Galena Creek, three miles from the mouth of the creek. (Galena Creek joins the south fork of the McQuesten River approximately 25 miles above the river's confluence with its North Fork.) On July 20, 1903, Davidson staked the Hell's Gate claim, recording it on August 3, 1903, at the Duncan Creek Mining Recorder's Office, and became the first man to find silver-lead ore and stake a silver claim in this district.

Davidson later moved on to prospect the Beaver River area, to the north, and let the claim lapse. In 1905 he heard of the rich silver discovery in Cobalt, Ontario, and decided to go there. Before he left, he gave the galena samples from Hell's Gate to Henry William "Harry" McWhorter. He told McWhorter about his find and asked him to have the samples assayed and let him know if they were any good. McWhorter was leaving for the Alaska gold fields but did get the samples assayed.

Jack Alverson, circa 1920. *M. Rich Collection*

The assay results were 300 ounces of silver to the ton. Davidson wrote back to McWhorter, inquiring about the assay results, but received no reply. McWhorter did not consider the find worthwhile because the vein was small and not gold-bearing.

McWHORTER, ALVERSON, HUFFMAN & EVANS

McWhorter eventually returned to the district and in 1912, he and Jack Alverson leased the Midas and Ophir lode gold claims on Dublin Gulch. Finding nothing worthwhile, they moved from Dublin down Haggart Creek toward the McQuesten valley where Davidson had found the galena. At the McQuesten River they met Grant Huffman and Mark Evans who decided to accompany them to the find.

They arrived at Davidson's Hell's Gate claim and on February 23, 1913, McWhorter restaked it as the Silver King claim. Alverson staked the Webfoot, Huffman staked the Mabel, and Evans staked the Adam.

Jack Beckman and Dan Ray's first prospect at the Silver King claim. *M. Rich Collection*

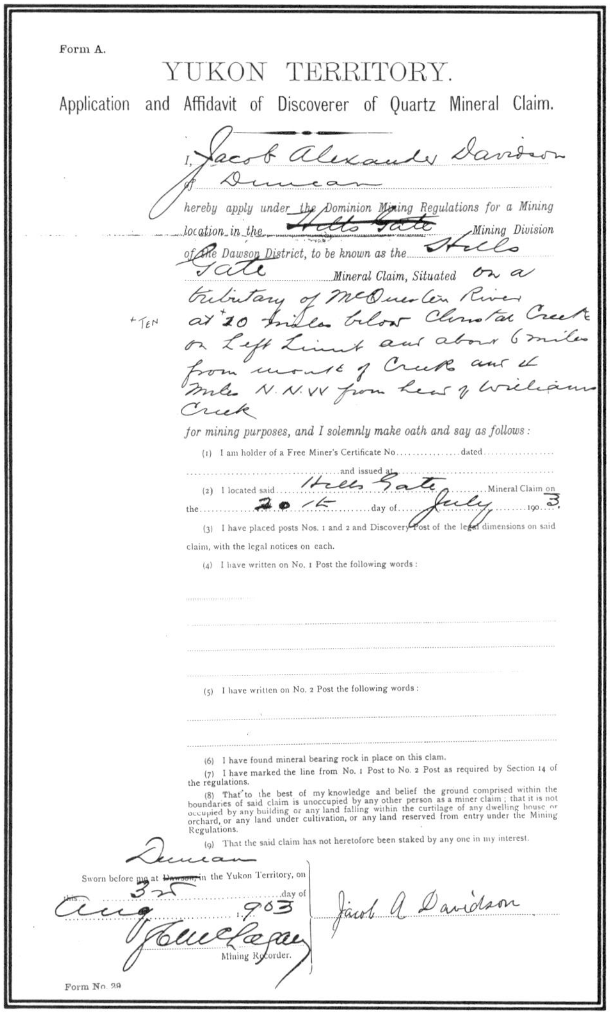

Form A.

YUKON TERRITORY.

Application and Affidavit of Discoverer of Quartz Mineral Claim.

I, Jacob Alexander Davidson of Duncan hereby apply under the Dominion Mining Regulations for a Mining location in the ~~Hells Gate~~ Mining Division of the Dawson District, to be known as the Hells Gate Mineral Claim, Situated on a tributary of McQuesten River at 20 miles below Christal Creek on Left Limit and about 6 miles from mouth of Creek and 4 miles N.N.W from head of Williams Creek for mining purposes, and I solemnly make oath and say as follows:

(1) I am holder of a Free Miner's Certificate No .... dated .... and issued at ....

(2) I located said Hells Gate Mineral Claim on the 20th day of July 1903.

(3) I have placed posts Nos. 1 and 2 and Discovery Post of the legal dimensions on said claim, with the legal notices on each.

(4) I have written on No. 1 Post the following words:

(5) I have written on No. 2 Post the following words:

(6) I have found mineral bearing rock in place on this clam.

(7) I have marked the line from No. 1 Post to No. 2 Post as required by Section 14 of the regulations.

(8) That to the best of my knowledge and belief the ground comprised within the boundaries of said claim is unoccupied by any other person as a miner claim; that it is not occupied by any building or any land falling within the curtilage of any dwelling house or orchard, or any land under cultivation, or any land reserved from entry under the Mining Regulations.

(9) That the said claim has not heretofore been staked by any one in my interest.

Sworn before me at Duncan ~~Dawson~~, in the Yukon Territory, on this 3rd day of Aug 1903

[signature] Mining Recorder.

Jacob A Davidson

Form No. 29

"Application Form A for a Full Claim" for Jake Davidson's Hell's Gate (or Hill's Gate) claim, staked July 20, 1903. *Mayo Mining Recorder's Records*

Alverson and McWhorter worked the Silver King that winter. By spring McWhorter had tired of the project and granted Alverson and Huffman a 100% lay on the claim, for one year, in exchange for the construction of a cabin and development work on the vein. Gene Binet, owner of the Chateau Mayo Hotel, grubstaked Alverson and Huffman and the pair assembled a small crew. Gilbert Rich, Huffman's nephew, was hired to pack supplies; George Crisfield, Jim Christie's former partner, was hired as cook; and Evans helped Rich and Huffman cut a ten mile trail to Field Creek on the Duncan Creek road to provide transportation access to Mayo. Huffman and Alverson named Galena Creek and Galena Hill.

A deal was struck with Alex Nicol to haul

ABOVE, Alex Nicol handling ore in 1913–14. *A. Aho Collection.*
RIGHT, Men gathered in front of Grant Huffman's first cabin at the Silver King. *M. Rich Collection*

sacked ore to Mayo for $50 per ton and Cora Rich, Gilbert Rich's mother, was contracted to sew ore sacks from doubled grain bags. Alverson and Huffman began to high-grade the property.

By June of 1914, 59 tons of ore was stacked at the Mayo dock awaiting shipment to the Selby smelter in San Francisco. Sid Barrington, Captain of the *Vidette*, picked up the ore and the first shipment of high-grade silver-lead ore in the district

Silver King mine, circa 1914. *M. Rich Collection*

ABOVE, Alex Nicol. *McLaren-Swensen Collection*
BELOW, Silver King mine shaft, circa 1914. *M. Rich Collection*

began its journey to market.

In September the smelter returns netted Alverson and Huffman $269 per ton for a total of $15,871 or $5,000 apiece, after expenses. The news caused an immediate stampede of stakers to the Silver King area. Approximately 40 men began prospecting in the vicinity and 95 claims were staked.

McWHORTER & AITKEN

One result of Alverson and Huffman's success was that McWhorter immediately cancelled their lease. He contacted Thomas P. Aitken of Fairbanks, a successful Alaskan gold miner, who agreed to finance the project with an option to buy the claim if it proved profitable.

McWhorter hired a crew and contracted Greenfield and Pickering to haul ore from the Silver King to Mayo for $19 a ton. By the spring of 1915, 1,180 tons of ore was stacked at the Mayo dock, ready for shipment. McWhorter continued to mine the Silver King profitably for another year. The Silver King mine produced major quantities of ruby silver (pyrargyrite).

AITKEN

In September of 1916, Aitken exercised his option and bought the Silver King from Henry McWhorter for $75,000. The sale was formalized on September 13, 1916. With a large crew, Aitken mined a further 1,386 tons of high-grade ore during the winter of 1916–17. In the summer of 1917, Aitken optioned the property to his Alaskan gold dredging partners J.E. Ives and F.G. Manley and their partner J.L. McGinn. During his involvement with the Silver King, Aitken earned approximately half a million dollars. He had removed all of the visible high-grade ore, leaving Ives, Manley, and McGinn to mine only 80 tons. They spent a year searching for more ore, then abandoned the project when the money ran out.

SCOUGALE

James A. Scougale was a successful Dawson City businessman. He leased the Silver King and, with government assistance, had the property explored by diamond drill. No ore was found.

Map of the Silver King area claims. Part of G.S.C. claim map, N.T.S. No. 105-M13

YUKON SILVER-LEAD MINING COMPANY

On September 26, 1916, Andrew Johnson staked five lode claims on Lookout Mountain (Mount Haldane). The claims were located on the western side of Bighorn Gulch. In 1918 the claims were organized into a company called Yukon Silver-Lead Mining Company by John Ernest "Jack" Pickering, James Greenfield, James A. Anderson, and Dr. Randy Maclennan. The company made every effort to locate an ore body and even had the property diamond drilled. In 1919 they contacted Fred Bradley, President of Alaska Treadwell, and invited him to examine the property. Bradley visited the site but was not interested in the prospect. With no substantial ore discovery, their exploration funds used up, and Alaska Treadwell's lack of interest, Yukon Silver-Lead Mining Company became inactive.

BOUVETTE

Louis Bouvette arrived in this district during the Duncan Creek stampede. In the fall of 1918 he found silver-rich galena float while hunting sheep on what was known as Sheep Mountain (later named Keno Hill). The following summer

Old workings of the Silver King Mine. *Ed Kunze Collection*

he returned and traced the float up the hill. On July 10, 1919 Bouvette staked the Roulette claim. When he heard that Fred Bradley, President of Alaska Treadwell, was in Mayo examining Andrew Johnson's Lookout Mountain property, Louis rode into Mayo with samples from his find but Bradley had already gone back to Dawson. He discussed the situation with Jack Pickering, and Pickering said he would take the samples to Dawson and catch Bradley there.

Yukon Silver-Lead Mining Co. Ltd., Lookout Mountain, February 1919. From left: Billy Burrell, Axel Erickson, Jack Adair, Jim Anderson, George Cunningham, Jack Pickering, Randy Maclennan, and John Kinman. *A. K. Schellinger Collection*

"On a stampede to Keno." From left: Axel Erickson, Dorothy McLaren (rear), Lillian McLaren, (front), Louis Bouvette, and a man known as Joe, April 1920. *McLaren-Walloch Collection*

Bouvette returned to the hill with James A. Anderson, one of the partners in the Yukon Silver-Lead Mining Company, and together they re-examined Louis' discovery. On July 21, 1919, Anderson staked the Rico.

When Pickering arrived in Dawson he learned that Bradley had gone up river with his engineer, Livingston Wernecke, to examine a property. Thus, Alaska Treadwell missed the opportunity to investigate this initial discovery on Keno Hill.

The Yukon Gold Company owned by the Guggenheim brothers, operated a successful gold dredging operation on the creeks around Dawson. Pickering took Bouvette's samples to the Yukon Gold Company assayer, Alfred K. Schellinger, who agreed to assay the samples. When the assay indicated 200 to 300 ounces of silver per ton, Schellinger immediately contacted E.E. McCarthy, General Manager of the Yukon Gold Company. McCarthy asked Schellinger to return with Pickering to Mayo, on the *Vidette*, and examine Bouvette's ground.

From left: Jim Greenfield, A. K. Schellinger and E. E. McCarthy at Duncan Creek, on their way to Keno Hill, 1919. *A. K. Schellinger Collection*

YUKON GOLD COMPANY & KENO HILL, LIMITED

Schellinger and Pickering arrived in Mayo and met Bouvette and Anderson who had come to town to record Anderson's Rico claim. They hired teamster Scotty Mowatt to take them to Keno Hill. On July 29, 1919, Schellinger staked the Keno claim (thereafter Sheep Mountain became known as Keno Hill) and Mowatt staked the Scotty. On July 30, 1919, Pickering staked the Pinochle.

Schellinger returned to Dawson and McCarthy drew up an agreement between the Yukon Gold Company and Bouvette and Pickering.

From left: Emil Forrest, Louis Bouvette, E. E. McCarthy, Jim Greenfield and Jack Pickering on the cruise of the *Tango*, August 1919. *A. K. Schellinger Collection*

In addition to the Roulette and the Keno, the Yukon Gold Company's original claim acquisitions included the Scotty, Pinochle, Rico, Solo No. 2 staked by James Greenfield on August 15, 1919, and the Wolverine staked by Schellinger on October 3, 1919.

A staking rush ensued. Tom McKay staked the Nabob and the Shamrock; Axel Erickson staked Solo No. 1 and the Reno; Dr. Randy Maclennan staked the Frog and the Faro; A.R. "Deep Hole" Thomson staked the Gambler, on the location where Louis Bouvette found the first float, and the Lakeview; Joel Sunderland staked the Union and the Maiden's Hope; Louis Bouvette staked the Minto; Emil Forrest staked the Tango; E. E. McCarthy staked the Porcupine Fraction (on which the Porcupine vein was later found); Rodolph Rosmusen staked the Silver Basin; Joe Walsh staked the Caribou; Murdock Mitchie staked the Gold Hill. Other early stakers included Oscar Le-

From left: Jim Greenfield, Frank ?, Axel Erickson, Louis Bouvette, Harvey Ray, Hugh Monohan, Isaac Lusk, Jack Pickering, Stanley White and Tom Collins going down the Stewart River, May 1920. *A. K. Schellinger Collection*

tourneau, Ed Burnell, John Kinman, Thomas Hinton, Oscar Miller, Alex Nicol, Robert "Bobbie" Fisher, Anthony "Tony" Hollenbeck, Bert Lamb, Matt Butijer (Butyer), Hector Morrison, and others.

Over 500 claims were located in the months that followed. During the winter of 1919–20 the Yukon Gold Company prospected their holdings as well as the Caribou, the Gold Hill, and the Lucky Queen, and secured options on a number of other claims.

In 1920, a subsidiary company, Keno Hill Limited, (KHL) was formed to direct the Yukon Gold Company's operations in the Mayo district and the entire holdings of the parent company were turned over to this subsidiary.

Simon R. Guggenheim, who had become President of the American Smelting and Refining Company in 1919, sent engineer R.H. Humphrey from New York to examine their Keno Hill holdings and make recommendations for its development. Guggenheim placed Frank Short in charge of the exploration program, J.H. Farrell was sent to map the top of Keno Hill; Schellinger was made engineer, surveyor, and assayer; and Jack

Tom McKay, left, and Alex Nicol, in the Keno Hill area. *D. Walloch (McLaren) Collection*

Hauling ore from Keno. *L. Cheatham (McLaren) Collection*

Keno Hill and Bunker Hill claims. Part of G.S.C. claim map, N.T.S. No. 105-M14

Stewart bossed the 20–25 man operation. E.E. McCarthy, who had been in charge of the Yukon Gold Company in Dawson, was moved to the Guggenheims' New York office; and Col. O.B. Perry succeeded McCarthy as General Manager of Yukon Gold Company in Dawson City. Fraser Falls was studied as a possible source of hydroelectric power.

In 1920, the Territorial Gold Commissioner, George P. MacKenzie visited various properties on Keno Hill to determine transportation and other service requirements. Paul S. Hogan, Yukon Councillor for the Klondike, and Dr. Alfred Thompson, M.P. for the Yukon, visited Keno Hill. Col. Perry, General Manager of Yukon Gold Company in Dawson, and McCarthy of their New York office also visited the property to determine production plans. W.E. Cockfield with the Geological Survey of Canada arrived to map the district. Among other dignitaries to visit Keno Hill in the early 1920's was Lord Byng, Governor General of Canada.

The Yukon Gold Company's subsidiary, KHL, shipped the first high-grade ore from Keno Hill in the winter of 1920–21.

In 1921, the company acquired the Sadie, staked by James R. Clark on March 18, 1920; and the Friendship, originally staked by Joel R. Sunderland on May 7, 1920. Sunderland had given the Friendship to Raoul Binet in exchange for the $10 recording fee, and so it was Binet and his partner Dick Mercure who received the $40,000 option on the Friendship. The Sadie was optioned for $30,000.

ABOVE, Keno Hill Ltd., "the richest silver mine in Canada," 1920. *A. K. Schellinger Collection*
BELOW, from left, Jack Stewart, Frank Short and R.H. Humphrey, Keno Hill, 1920. *A. K. Schellinger Collection*

The company immediately began to sink a shaft on what became known as the Sadie-Friendship vein. Working conditions on Keno Hill were very primitive in the early days of the Yukon Gold Company's operation. In March of 1921, the men went on strike for better working conditions and an eight hour day. Improvements were desperately needed at the "Top of the Hill" tent camp. The camp lacked proper facilities and the miners were forced to work, live, sleep, and eat, in the pervasive clouds of lead-silica dust which hung over the operation. Many of the crew fell ill or died from lead poisoning and silicosis. These men were completely capable of building their own cabins and living off the land, and could easily have endured a long strike. Realizing this, the Yukon Gold Company agreed to the workers' conditions. The tent camp at the "Top of the Hill" mine was replaced by a permanent camp of wooden buildings which included a lumber bunkhouse.

In 1921, a $13,000 warehouse was built in Mayo and 325 feet of waterfront was leased for

ABOVE, "Rawhiding" – hauling silver ore down Keno Hill with sleigh and rawhide. *M. Rich Collection*
LEFT, The Keno Hill Limited office. *T. McKay Collection*

stockpiling ore. In 1922, Yukon Gold brought the first Jeffery "Quad" vehicle, a solid-rubber-tired 2-ton affair, to Keno Hill to be used for hauling purposes, as well as the first 5-ton Holt gasoline cat to haul wood for the power-house boiler.

The company supplied itself with electrical power from a power plant on Duncan Creek near its junction with Lightning Creek. The plant was a steam generator fired by wood, which it consumed at the alarming rate of 9 cords every 24

Assay office at the Sadie, 1924. *A. K. Schellinger Collection*

ABOVE, From left: Warren McFarland, Charles McLeod, Mrs. Hellman, Frank Short and Fred Hellman in a Model T Ford, on a visit to Keno Hill, 1922. *A. K. Schellinger Collection*
RIGHT, Keno Hill, 1920. From left: Dr. Alfred Thompson MP, E. E. McCarthy, J. H. Farrell, Jack Stewart, O. B. Perry, Thomas McKay, and Frank Short. *A. K. Schellinger Collection*

hours. The first plant opened December 9, 1920, but exploded due to a flaw in a steam cylinder. A replacement boiler was obtained from the Highet Creek Dredge (also known as the Titus Dredge

Early development of Wernecke Camp, circa 1921. *A. K. Schellinger Collection*

Co.) after the dredge ceased operation circa 1921. The power was used mainly for the shaft hoists.

In 1921, the company acquired prospect options on the Caribou, the Gold Hill, the Lucky Queen and the adjoining Mathole claim. Hector Morrison received $5,000 for the Lucky Queen option and was hired by Yukon Gold to prospect the claim.

KENO CITY

In 1919, John Kinman found some galena float on the trail just above Lightning Creek, near Malesich's placer mining cabin, and staked it with partner "Rocky Mountain Bill" Elliott. Kinman built a cabin on the claim. The site was soon used as a staging area where loads were assembled for the pull up Keno Hill. Alex Nicol and others began pitching tents nearby or stopping off at Kinman's cabin on their way up the hill. As the settlement grew, it became known as Keno City and John Kinman was referred to as "The Mayor," since he owned the claims the village stood on. By 1920 cabins, shacks, stables, and a hotel had been built. The old placer cabins on Duncan Creek, including the "Swedes" cabin, were torn down and the cabin logs used to build the new town of Keno.

Three hotels sprang up in Keno over the years. The Keno Hotel built by Bobby Greaves, the Northern Lights built by Mike Tomoff, and Yamasaki's Hotel (roadhouse and laundry) owned by Harry Yamasaki and later operated by Jimmy Sugiyama.

Prostitutes arrived and sales of moonshine were brisk. Major stills were located at Hanson Lakes and Mayo Lake. Typically, miners worked hard and played hard. Fist fights were considered a form of recreation.

A liquor store was opened in Mayo in 1921 and that same year a new liquor store was established at Keno City. For approximately a decade, the profits from the Mayo and Keno liquor stores amounted to 26 percent of the profits generated by liquor sales in the entire territory and averaged $15,000 a year.[37]

Robert Greaves opened a post office in Keno City on July 1, 1921.

In April 1922, acting on a request from the Mayo Board of Trade, O. S. Finnie, Director of the N.W.T. and Y.T. Branch of the Department of the Interior, recommended that the government sell 100 acres for the purpose of creating a town site at Keno City.[38]

The population of Keno City during the initial silver boom has often been incorrectly reported as 20,000. Although there are no official statistics for Keno, the entire population of the Yukon Territory was 4,157 at the time. The population of the Mayo District has been estimated at just under 1,000 in the 1920's.

Keno City, 1927. *Ione Christensen Collection*

The freighting company of Greenfield and Pickering had the contract to haul ore by horse-drawn sleighs from Keno Hill to the Mayo waterfront; a distance of approximately 40 miles. This was done during the winter months when the road surfaces were at their best. When the rivers opened, sternwheelers took the ore barges to Stewart City where they were transferred to the Dawson sternwheeler for the trip to Whitehorse. It was then moved over the White Pass Railway to Skagway and down the coast by ship to the smelters in San Francisco or Kellogg, Idaho. Alternatively, ore took the river route from Mayo to Nenana, Alaska, where it was transferred to the Alaska Railroad for shipment to the ocean. Ocean steamers of the Pacific Coast Co., and the Alaska Steamship Co., took it to the smelter.

At this time, the cost of the ore haul from Keno Hill to Mayo was $30 a ton. The cost to

Lord Byng of Vimy visits Keno Hill August 7, 1922. From left: Jim Fairborn, Jim Ferrell (Member of Yukon Council), the Police Commissioner, George Black MP, unidentified aide, John Fawcett, Frank Short, and Lord Byng, the Governor General of Canada. *A. K. Schellinger Collection*

Rough-locking down Glacier Hill, "Let her ramble," 1922. *Ralph Rogers Collection*

Reloading ore in Keno City, 1922–23. *A. K. Schellinger Collection*

move the ore from Mayo to San Francisco was also $30 a ton. The price of silver at the time was 65 cents an ounce and KHL's cut-off grade was 200 ounces of silver per ton.

During the 1922 season Guggenheim sent Fred Hellman, a senior mining engineer, to the district to evaluate the Yukon Gold Company's overall position on ore reserves and the production potential of the district. Hellman's conclusions were cautious regarding ore depth and continuity of high-grade material at depth on the No. 9 vein.

The Yukon Gold Company's subsidiary, KHL, was strictly a high-grade operation. By September, 1923, production of ore was dropping so the No. 3 and No. 9 workings at the Top of the Hill were cleaned up and all crews and equipment were transferred to the Sadie-Friendship property. The following year, the company ran out of high-grade ore and on May 31, 1924, Keno Hill, Limited closed down its operations.

McKAY

Thomas "Tommy" McKay staked the Shamrock, near the summit of Keno Hill on August 4, 1919. The Shamrock was rich and McKay and his lessees mined the property independently. Initially, McKay and partner, Axel Erickson, worked the claim. McKay then leased the Shamrock to Axel Erickson, Pat McIvor, and Hans Formo. It was later leased and worked by others but McKay always earned 15% of the smelter returns. He held the claim for the rest of his life.

TREADWELL YUKON COMPANY, LTD.

The Treadwell Yukon Company Limited established itself in the district in 1921, three years before Keno Hill Limited ceased operations. During that period, these two major companies worked the district simultaneously. Treadwell Yukon was to remain for 20 years.

Fred Bradley, President of the Treadwell corporation in San Francisco, had followed the developments in this silver district since Bouvette's strike in 1919. In 1921, he sent geologist and mining engineer Livingston Wernecke to Mayo to examine the district and gave him complete authority to negotiate deals on behalf of Treadwell. Wernecke arrived on the first boat of the season on June 5, 1921, and immediately examined the Shamrock and Guggenheim workings. Working out of the Keno Hotel, he began to contact the various claim holders in the area who were not bonded to the Yukon Gold Company's subsidiary, Keno Hill, Limited.

The Treadwell acquisitions included, among others, the Bluestone, which had been staked by Joel R. Sunderland on May 7, 1920; the Ladue staked by David A. Cunningham on September 11, 1920 (Cunningham had turned over 50% of the Ladue ownership to H.A. "Ray" Stewart on June 21, 1921); the Lotus staked by William L. Bramley on February 22, 1920; the Bluebell staked by Frank McNeill on May 7, 1920 and the Mary, also staked by McNeill on May 9, 1921; the Lansing staked by Thomas McKay on June 16, 1920; the Poca Plata staked by Stewart on May 7, 1920; the Travice staked by John E. Hollenbeck on March 12, 1921; and the Silver Bell staked by Bramley on February 20, 1920.

This is the waste and drainage tunnel Treadwell Yukon put in at the 600-foot level of the Ladue. All waste from the mine shaft was sent to the tunnel at the 600-foot level where it was loaded on a horse-drawn ore car, pulled out of the tunnel, and dumped. *Besner Collection*

Bradley financed the operation through a group of investors called Associated Companies. The investors were Bunker Hill and Sullivan Mining and Concentrating Company 50%, Alaska Treadwell 30%, Alaska Mexican 10%, and Alaska United 10%. By the end of the 1921 season, Associated Companies were said to have committed a total of $469,800 for claim options and the Treadwell Yukon Company, Limited was formed to manage their interests in the district.

The Ladue property was looked upon as a potential mine and William B. Hargraves, engineer, and Wernecke, geologist, were put in charge of development work on the property.

During the fall and winter of 1921–22, Treadwell Yukon opened the Ladue vein with several pits, trenches, and shafts, and started drifting on the vein. Prolems arose when excessive water flooded the workings. In 1923, a 2,600-foot drainage tunnel was driven on the 600 Level to drain the mine.

By the fall of 1923, 1,000 claims had been staked in the vicinity of Keno Hill, while on Galena Hill on the Arctic and Mastiff claims, an ore discovery was made by Charles R. Settlemier and Claude H. Bermingham.

Also in the fall of 1923, Treadwell brought in two 10–ton Holt tractors to take over their ore haul from Keno Hill to Mayo. Ed Barker and Danny McDonald brought the first cat load into Mayo, marking the end of horse drawn ore. That winter, the Holt hauled as much ore as had the 90–100 head of horses in the previous winter.

A considerable tonnage of ore was outlined on the Ladue vein by 1924 and in August of that year, Treadwell Yukon negotiated a ten-year lease on Keno Hill Limited's (Yukon Gold Company's) Sadie and Friendship claims. The ore was not of sufficiently high grade to be shipped raw, but enough ore reserves had been developed on the Ladue, Sadie and Friendship to justify Treadwell's decision to build a milling and concentrating plant. Mill construction began on the Sadie-Friendship in the fall of 1924, near the former Keno Hill camp and workings, and the town site which became known as Wernecke Camp began to develop near the mill site.

WERNECKE CAMP

Wernecke Camp evolved from the camp that KHL had established on the Sadie-Friendship. Wernecke was on the sunny, protected, lower slope of the hill, whereas the Ladue workings had been on the north-facing side. Every effort was made to develop the town as a pleasant place to live and work. Wernecke overcame the dust clouds from the mine by using water in the operation, thereby reducing silicosis and lead-poisoning.

Men from the Treadwell Yukon Company, Ltd.'s Wernecke Camp. *J. B. Kunze Collection*

The town boasted a recreation hall with poolroom, a bowling alley, library and radio, outdoor skating and curling rink, silent movies shown by Emil Forrest, and a laundry service. In addition to these facilities, there was a warehouse, powerhouse, assay office and various residences, cabins and shacks. Both Wernecke and Hargraves had homes in Wernecke Camp.

Treadwell's Ladue mine drainage tunnel at the 600-foot level, 1923. From left: David W. Ballentine, unknown, Black Mike Berovitch, Jimmy Ballentine, Jack McLean, unknown. *Ione Christensen Collection*

TOP, The ball mill at Treadwell Yukon's Wernecke Camp. *T. McKay Collection*
MIDDLE, The company's sawmill on the Mayo River, July, 1926. *J. B. Kunze Collection*
BOTTOM, Wernecke Camp, circa 1925. *Ed Kunze Collection*

The mill (125 tons per day capacity) was completed in 1925 and milling commenced on January 6, 1925, processing 120 tons of ore per day. A.K. Schellinger (who had been kept on a retainer by the Yukon Gold Company) was hired by Wernecke as assistant engineer. Alec F. Berry was hired as mill superintendent and assayer. The mill, the first ever constructed in the district, operated at Wernecke for seven years.

On September 13, 1924, Charlie Brefalt had staked the Elsa claim on the strength of a single surface showing of vein material which assayed a meagre 1.5 ounces of silver per ton. It was necessary to obtain financing in order to carry out exploration on the property and Dan Tolmie, another Keno prospector, agreed to share 50% of the work and expenses. In July of 1925, Brefalt turned over 50% of the Elsa ownership to Tolmie and the partners began to prospect the claim in earnest. While hunting for grouse, near the Elsa, Brefalt stumbled onto vein material which assayed at an incredible 3,000 ounces of silver per ton. This find became his Lucky Strike claim, which adjoins the Elsa. His next, and most impressive, discovery was located about 500 feet from the first, back on the Elsa claim, and was under only three feet of overburden. Brefalt was asking $250,000 for the property and Wernecke had just committed Treadwell Yukon to heavy development on the Sadie and Ladue veins and could not make a deal on the Elsa at that time. Instead, Wernecke used his influence with the government to open a winter road and build a bridge over Galena Creek so that Brefalt could haul his ore to Mayo. In October, 1925, Brefalt and Tolmie hired a crew of 12 men and soon had 50 tons of high grade ore sacked and shipped to Mayo. In spite of this success, expenses were high and Brefalt needed further financing. Taking his assay slips to Wernecke, Brefalt asked for an advance of $5,000 and received it without hesitation. By 1928 it appeared that they had mined out all of the high grade ore and Treadwell Yukon optioned the property that

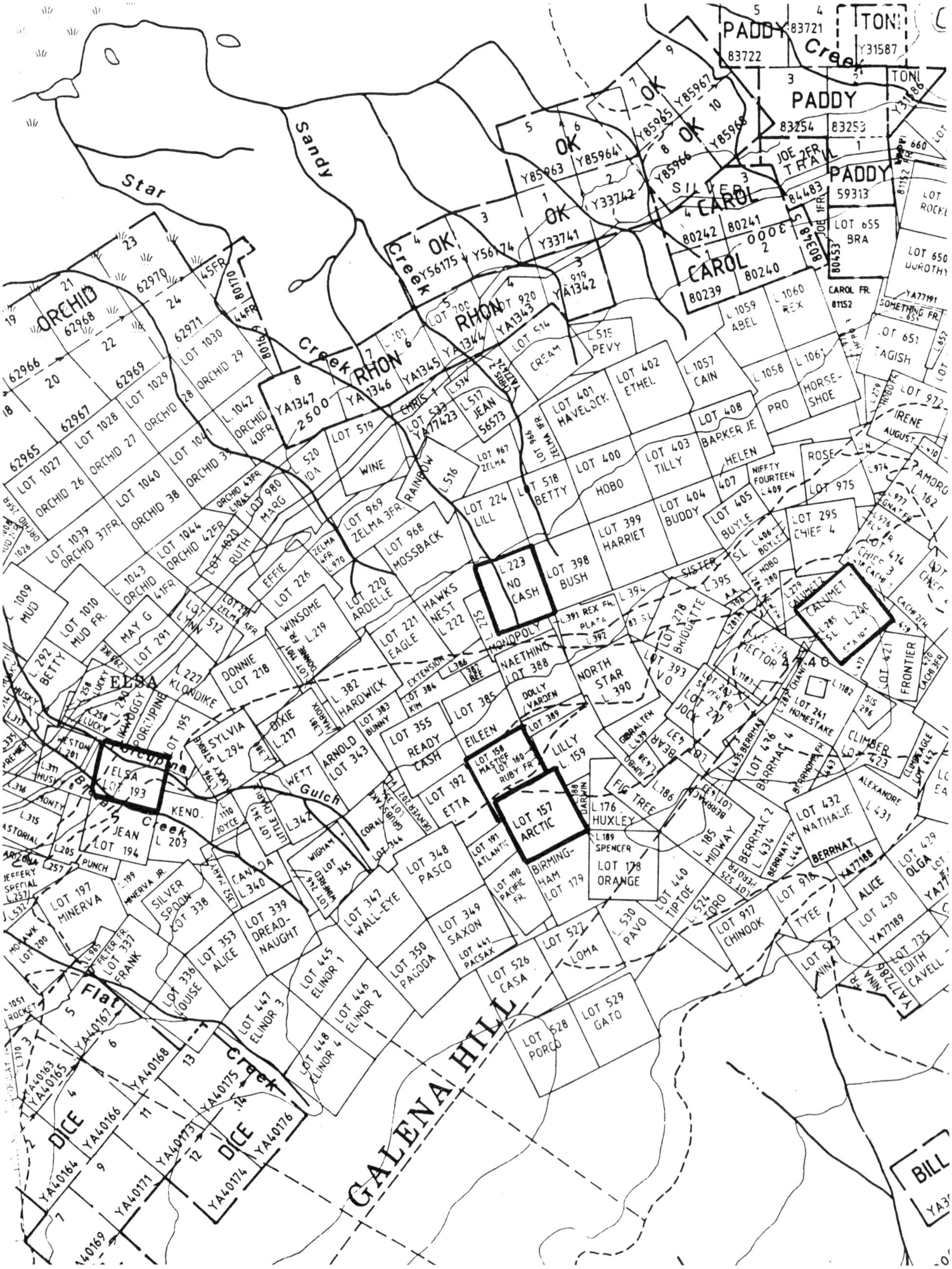

Galena Hill claims. Part of G.S.C. claim map, N.T.S. No. 105-M14 v.

year for $150,000. The partners Brefalt and Tolmie had mined 500 to 600 tons from 1925 to 1928 netting nearly $70,000. Next to the Lucky Queen, the Elsa became Treadwell Yukon's richest mine. Treadwell established a 30 man camp at the Elsa complete with mess hall and bunkhouse.

Mining continued on the Sadie-Friendship until 1929, when known ore reserves became exhausted and silver prices declined.

In 1929, Jack Hawthorne and Amos Noyd, who was working for Hawthorne at the time, struck an ore shoot on Hawthorne and Ferguson's Webfoot claim which was located next to the Silver King. The Webfoot was originally staked by Jack Alverson on February 13, 1913, and, over time, Hawthorne and Ruth Ferguson acquired total interest in the claim. Treadwell Yukon optioned the Webfoot following Hawthorne and Noyd's ore strike.

In March, 1928, Sandy McPherson, Jack McLean and Jack McHugh obtained an option on the Silver King mine from Thomas Aitken. The trio were known as the Rolled Oats Syndicate. After a dispute over the possible location of an ore vein, the Silver King option was split between McPherson (50%) and McLean and McHugh (50%). McPherson's former partner, Horace Glendenning, returned to Mayo to assist McPherson on his half of the claim. In March of 1929 McPherson and Glendenning struck a solid ore vein and sold the option to Treadwell Yukon for a reported $125,000.

Also in 1929, Treadwell Yukon acquired the rich Lucky Queen which had been staked by Hector Morrison on February 18, 1920, optioned to KHL and returned to Morrison when KHL ceased operations in 1924. Treadwell transferred its operations from the Sadie-Friendship to the Lucky Queen. An aerial tramway was constructed to move ore from the Lucky Queen mine site down the slope to the mill site at Wernecke Camp. By the end of 1929 Treadwell Yukon's acquisitions included

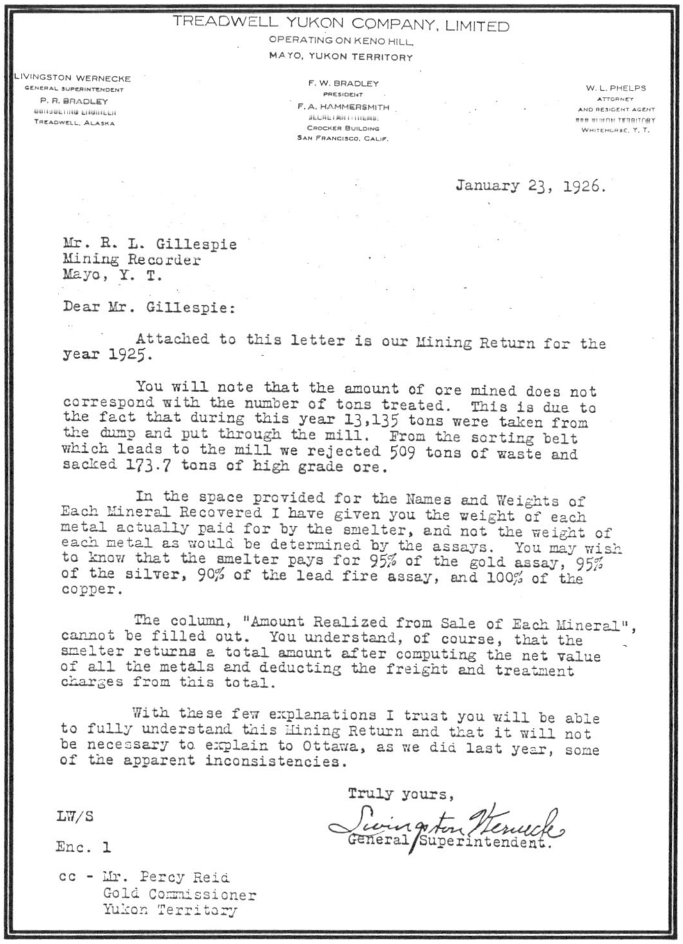

TREADWELL YUKON COMPANY, LIMITED
OPERATING ON KENO HILL,
MAYO, YUKON TERRITORY

LIVINGSTON WERNECKE
GENERAL SUPERINTENDENT
P. R. BRADLEY
[illegible]
TREADWELL, ALASKA

F. W. BRADLEY
PRESIDENT
F. A. HAMMERSMITH
[illegible]
CROCKER BUILDING
SAN FRANCISCO, CALIF.

W. L. PHELPS
ATTORNEY
AND RESIDENT AGENT
[illegible]
WHITEHORSE, Y. T.

January 23, 1926.

Mr. R. L. Gillespie
Mining Recorder
Mayo, Y. T.

Dear Mr. Gillespie:

Attached to this letter is our Mining Return for the year 1925.

You will note that the amount of ore mined does not correspond with the number of tons treated. This is due to the fact that during this year 13,135 tons were taken from the dump and put through the mill. From the sorting belt which leads to the mill we rejected 509 tons of waste and sacked 173.7 tons of high grade ore.

In the space provided for the Names and Weights of Each Mineral Recovered I have given you the weight of each metal actually paid for by the smelter, and not the weight of each metal as would be determined by the assays. You may wish to know that the smelter pays for 95% of the gold assay, 95% of the silver, 90% of the lead fire assay, and 100% of the copper.

The column, "Amount Realized from Sale of Each Mineral", cannot be filled out. You understand, of course, that the smelter returns a total amount after computing the net value of all the metals and deducting the freight and treatment charges from this total.

With these few explanations I trust you will be able to fully understand this Mining Return and that it will not be necessary to explain to Ottawa, as we did last year, some of the apparent inconsistencies.

Truly yours,

Livingston Wernecke
General Superintendent.

LW/S

Enc. 1

cc - Mr. Percy Reid
Gold Commissioner
Yukon Territory

TOP, Letter to R. L. Gillespie, Mayo Mining Recorder, from Livingston Wernecke, General Superintendent, Treadwell Yukon Company Limited, dated January 23, 1926. *Mayo Mining Recorder's Records*

BOTTOM, Left to right: Emil Forrest, Herb McAdam, Don Morrison, Sandy McPherson, and Amos Noyd in Mayo. *F. & T. Whitney Collection*

Wernecke Camp, looking down the McQuesten Valley. *Ed Kunze Collection*

the Silver King, Arctic, Lucky Queen and Elsa Groups.

In 1928 Wernecke began a Yukon-wide air reconnaissance prospecting program and the following year six prospecting parties were sent out. By 1931 there were ten parties in the field prospecting the Hyland River, Frances Lake, Hoole River, Pelly River, Morley River, Ruby Range, Kathleen Lake, Pelly River, MacMillan River, White River, Old Crow and Blow River, Bonnet Plume Lake, and Britannia Creek. Wernecke's pilots included Everett Wasson, Goldsborough, Tom Stephens, and Chuck Gropstis. His prospectors were Jack Carpenter, Jack McLean, Bill Forbes, "Hard Rock" MacDonald, Emil Anderson, and Joe Walsh, among others. The Depression brought an end to Treadwell Yukon's airborne prospecting.

Some of Wernecke's prospectors were still

Power plant at Wernecke Mine, 1925. *Ione Christensen Collection*

alive during the 1950's boom and they claimed to have made significant discoveries during Treadwell's air reconnaissance era. Unfortunately, when they were taken back to the various lakes they had ostensibly prospected, all of the prospectors said they had never been there before. It would appear that Treadwell Yukon was not specific about reconnaissance locations, to protect possible discoveries. All the company records of those explorations have been lost.

The rich ore discoveries made on the Webfoot, Silver King, and Lucky Queen were made in the year that the stock market crashed.

By 1931 silver had dropped to 26 cents an ounce. The last known ore was worked out of the Lucky Queen and the mill at Wernecke Camp was shut down on November 16, 1932.

During the years 1931 and 1932 no development work was done by Treadwell Yukon. Individual prospectors continued their search for ore but times were tough and financing was impossible. Wernecke did whatever he could to assist the independent prospectors by supplying them with tools, grub, and professional advice. Merchants in Keno and Mayo either extended credit or grubstaked prospectors in the hard times as they had in the good times. Bob Palmer in Keno and Gene Binet, James Mervyn, and T.C. Richards for Pat Burns in Mayo all financed prospectors.

When the price of silver rose to 64 cents in the spring of 1935, Treadwell management began the process of reopening the Silver King and the Elsa mines. The mill was moved from Wernecke Camp and the 150 ton per day Elsa mill began operation on March 5, 1936.

In 1935, Treadwell Yukon optioned the Hector claim originally staked by Clem Sinyard on July 21, 1920 and the Calumet staked by Harry Colley, Joe Walsh, and Harry McCrimmon on April 27, 1920. Sinyard, with financial backing from "Soapy" McCown, had prospected his claim for

ABOVE, "March 29, 1934. Don Morrison viewing the prospect hole where he is taking out galena only an hour after discovery. I [A. K. Schellinger] took this picture on a trip over to Clem Sinyard's hunting tent with Kenneth Schellinger. This hole started all the excitement on Galena Hill. Very good of Don Morrison rolling a smoke. First discovery of Calumet." *A. Aho Collection*

BELOW, Keno Hill Ltd. ore piled at Mayo dock in 1923. The Northern Commercial store is on the left and Binet's hotel and store is on the right. Ralph Rogers is standing on the extreme left. *William Hare photographer, Ralph Rogers Collection*

Wernecke Camp, looking north, circa 1925. *Ed Kunze Collection*

many years and finally struck an ore vein on the property. Little work was done on the Calumet until 1932 when Colley and Tom McKay began to prospect it further. In the spring of 1934 Don Morrison was added to the partnership and he struck a rich ore shoot that the partners mined out before optioning the Calumet to Treadwell.

Charlie Brefalt's "No Cash" camp, circa 1934–35. From left, Elmer Gustavson, Bob Miller, Charlie Brefalt, and Charlie Taylor. *Donated by Rick Cocks, Charles Taylor Collection*

In 1936, a 14,200 foot aerial tramline was built to carry ore from Calumet to the mill in Elsa. The tramline, supported by some 42 towers, was designed and constructed by A.K. Schellinger and John Scott.

ELSA

In 1928, Treadwell Yukon established a 30 man camp on the Elsa claim and constructed a mess hall and bunkhouse.

The town of Elsa developed in 1935 when the mill was built on the Elsa claim. It was initially referred to as Elsa Camp. In addition to the mill, there was a machine shop, bunkhouses, framing shed, staff residences and one or two other structures. Dick Major ran a store and poolroom. The Elsa school first opened on September 6, 1938, and the first teacher was Arthur Lanyon. The first post office in Elsa opened on May 26, 1949 with Henry Brodie Hicks as postmaster.

J. B. "Bon" Kunze standing beside Bosch-Sulzer diesel generator at the Werneke Camp, circa 1925. *Ed Kunze Collection*

Cats doubled up for the Glacier Hill pull, circa 1925–26. *William Hare photographer, Ralph Rogers Collection*

Ten-ton Holt Caterpillar taken between 1925 and 1930. *Ed Kunze Collection*

Clem Sinyard's first shaft on the Hector claim, 1935. From left: Charlie Brefalt, Malcolm McCown, "Little Nick" Milosevich, Elmer Gustaveson, and Clem Sinyard. *Schellinger Collection, Yukon Archives*

In 1939 the Silver King was closed, five years after its revival.

With the outbreak of World War II and the U.S. government's announcement in 1940 that they would no longer buy foreign silver, Treadwell management decided to close their Mayo district operations. In February, 1941 Wernecke was instructed by the company's board of directors to sell all assets in the district if a buyer could be found.

On October 21, 1941, the dynamic Livingston Wernecke was killed in a plane crash off the coast of British Columbia. So ended Treadwell Yukon's 20-year operation in this silver district. The company was dissolved on May 4, 1942.

By August of 1942, the only activity in the district came from small placer operations, trapping, and the dismantling and shipping of equipment from the Treadwell Yukon mines. Three stores remained open in Mayo but did very little business. The bank was closed. The United States Public Roads Administration in Whitehorse purchased most of the diesel engines from Treadwell Yukon, as well as the complete machine shop with all supplies, tools and equipment from the mines, and even the corrugated iron from the buildings. Only the mill, tramline, buildings, and mineral claims remained in place.

During its operations the Treadwell Yukon Company Limited, produced 625,000 tons of ore which yielded 44,000,000 ounces of silver and 96,000,000 pounds of lead.

UNITED KENO HILL MINES LIMITED

Assets of the Treadwell Yukon Company remained idle through most of World War II. In 1943, Thayer Lindsley,[39] of Ventures Limited, became interested in the Mayo silver district and involved Fred M. Connell (Metals Comptroller for Canada at the time) and his companies, Conwest Exploration Ltd., and Frobisher Ltd., in the prospect. They sent John Reid and Frank Buckle to the district in the spring of 1945 to examine Treadwell Yukon's properties. Upon assessment of Buckle's positive report, a 3 million share operating company called Keno Hill Mining Company Limited was formed with $100,000 in the treasury. Treadwell Yukon's assets were purchased by the new company, the shareholders receiving $75,000 in cash and 750,000 shares of non-par-value, non-assessable capital stock in Keno Hill Mining Company Limited. (This company is not to be confused with the earlier, and totally unrelated, Keno Hill, Limited, a subsidiary of Yukon Gold Company.)

Elsa townsite with the road to Mayo in the background.
George Hunter photographer, UKHM Collection

Buckle returned to implement the recommendations he had made and set about rehabilitating the mill, tramline, and buildings. His task was a difficult one.

Due to the tremendous outflow of capital during the start up phase of the project, the company was reorganized in 1947 under the name United Keno Hill Mines Limited (UKHM) to increase cash flow. Following the reorganization, H. Brodie Hicks was hired as manager, Buckle remained as overall consultant, W.O. McBride was chief geologist, Bob Segsworth was mine superintendent and C.D.N. "Red" Taylor was chief engineer. Jack Meyers was foreman at the Calumet.

Elsa tramway terminal. *UKHM Collection*

At this time, the assets of the company included 323 mineral claims, including those held under option. Among them were the Silver King, Elsa, No Cash, Hector-Calumet, Sadie-Ladue, Lucky Queen and "Top of the Hill" mines.

UKHM's Elsa mill began production on April 1, 1947, milling ore from the Hector-Calumet. All other mines were dormant.

By this time other companies had become interested in the district. Northwest Exploration Ltd. was prospecting on the Blue Rock claims and the McLeod property, Hoyle Mining Company was in the Beaver River district, and Noranda was exploring claims acquired from John R. O'Neill. Individuals continued to prospect.

UKHM bought Kimbel's sawmill, which supplied its mine timbers until Leo Proctor set up a portable sawmill in the McQuesten Valley.

In 1948, UKHM optioned the Bermingham

Calumet townsite. *UKHM Collection*

mine which later became the first open pit mine in the silver district. Other open pit mines were developed later.

The Hector-Calumet continued to produce good ore throughout the post-war period. The Elsa was reopened in 1948, as was the No Cash. Exploration work was done on the Onek, Flame, Moth, and the old Silver King workings. The company acquired the Shamrock from the Thomas McKay estate in 1949.

On June 11, 1949, fire destroyed the mill and assay office. Craigie Hood, of Kilborn Engineering in Toronto, was dispatched to design and erect a new 250 ton per day capacity mill with the assistance of the UKHM staff. The work went well and the new mill at Elsa was in operation by October 24, 1949. The capacity of this mill was increased to 500 tons per day in 1951.

When the company's success was revealed at its annual meeting in 1951, an incredible boom occurred in the Mayo silver district. UKHM announced cash and concentrates on hand worth over two million dollars, mounting production, and a rich vein discovery on the Hector-Calumet. These announcements coupled with the development of hydroelectric power and the completion of the highway from Elsa to Whitehorse drew companies and speculators to the district. At the peak of the 1950's boom, some 30 companies acquired interests in the district.

ABOVE, Calumet boiler house and filtration plant. BELOW, United Keno Hill Elsa Mill fire, June 1949. *Both photos, UKHM Collection*

Ore cars at the Elsa Camp, 1938. *T.R. Bleiler Collection*

Most of these were gone from the area by the middle 1950's. United Keno Hill Mines Limited continued to operate and gradually acquired additional ground in the area. By 1958, it was the only significant operator remaining in the district.

UKHM played a major role in the Yukon's economy in the 1950's. In addition to the growth of the Elsa and Calumet townsites, the company was involved in the Tantalus Butte coal mine and owned a greenhouse in Carmacks. It operated sawmills and lumber yards, maintained the road to Whitehorse and operated the ferries. UKHM established a garage and transport section in Whitehorse for its ore trucks that transported ore not only from the Mayo district but from Cassiar as well. The company was involved in docks in Skagway as well as in shipping.

In 1953, Clare White became general manager; Red Taylor was assistant manager; Al Pike was mine manager; Walter Clarke, general superintendent; Bruce Lang, mine superintendent; A.C. Carmichael, chief geologist; and Nick Gritzuk, transportation superintendent for UKHM.

The company reopened the Shamrock vein by a new adit in 1954–55 and exploration of the "Top of the Hill" workings, referred to by UKHM as the Keno 700 camp, began in 1956.

In 1957, Clare White left for Labrador and Albert Pike became general manager. Bruce Lang was production manager, Don R. Delaporte was mine superintendent, and A.H. "Moose" Manifold was chief geologist.

A signpost was erected adjacent to the No. 9 vein at the head of Faro Gulch in 1957–58, when Canada hosted the "International Geophysical Year." A part of that convention came to the Yukon and were hosted by UKHM. The company erected the signpost for this occasion, with signs pointing to the various countries represented at the convention.

In 1962, Falconbridge Nickel Mines acquired 48.42% of UKHM and assumed management control.

In 1964, the company's exploration geologist, R.E. "Dutch" Van Tassell, located the first new major vein discovery in the district in over 20

Main buildings at Elsa in the early 1950's. *UKHM Collection*

years. This discovery became known as the Husky mine. Van Tassell's discovery occurred at a time when UKHM was faced with a critical shortage of ore in its existing mines and was very close to shutting down its entire operation in the Keno Hill area. The discovery of the Husky mine came just in time to prevent this from occurring.

From its inception, the company grew as a world producer of silver and eventually became Canada's second largest producer and the fourth largest in the world. Since 1947, it has produced revenue shipments of over 148 million ounces of silver, 488 million pounds of lead, 331 million pounds of zinc and 4 million pounds of cadmium.

The total minehead silver and lead production by all operators in the area since 1913 is over 213 million ounces of silver, over 710 million pounds of lead and over 436 million pounds of zinc.

THE 1950's BOOM

John R. O'Neill from Almonte, Ontario, and former Federal Conservative Member of Parliament, played a major role in promoting the district. Archie Martin, former Yukon Councillor and originally from Aylmer, Quebec, sparked O'Neill's interest in the district. O'Neill arrived in Mayo, shortly after United Keno Hill Mines Limited began mining on Galena Hill, and optioned almost every property not controlled by UKHM.

William "Billy" Sime, left, and Miles Flynn, 1948.
Herb Foster photograph, O'Neill Collection

Some of the companies that were involved in the 1950's boom are as follows:

❖ *Yukeno Lead and Silver Mines Limited.* This company was organized in 1948 by John R. O'Neill with Billy Sime as president and Sandy McPherson director. The company held McPherson's portion of the Silver King, as well as the Highlander and Bunny claims, and eight claims on Cobalt Hill, northeast of Keno.

❖ *Yukon Galena Hill Mines.* Yukon Galena Hill Mines was a 5-million share company organized by John R. O'Neill, with Billy Sime as president. The company held the McLeod and adjoining Sime property.

❖ *Silver Basin Mines.* This company was formed by John R. O'Neill, who was also the president. The company acquired the Gold Queen, the Rasmusen (or Rosmusen), and the Formo claims.

❖ *Consolidated Yukeno.* Consolidated Yukeno was incorporated in 1949 when John R. O'Neill consolidated his three companies: Yukeno Lead and Silver Mines Limited, Yukon Galena Hill Mines, and Silver Basin Mines. The consolidated holdings comprised over 150 claims on various parts of Keno and Galena Hills. Consolidated Yukeno was a 10 million share company. The company optioned the Formo vein in 1950 and explored the vein by two adits and drifts on two levels during 1952–53. The company was renamed Yukeno Mines Ltd. in 1951. In 1952 the company carried out some underground development on the Silver Basin veins but the results were equally disappointing and the option dropped.

This company also held a 33% interest in Mackeno Mines Ltd., which acquired the Galkeno property from Yukeno Mines. After 1953, Yukeno Mines lay dormant until its properties (except for Galkeno) were sold to Rio Plata Silver Mines Ltd., in 1962.

❖ *Mackeno Mines.* Mackeno Mines was formed

when the Bryan W. Newkirk brokerage firm of Toronto purchased the vendor's interest and control of Consolidated Yukeno. Consolidated Yukeno received $57,000 in cash and 650,000 escrowed vendors shares of Mackeno for the McLeod and Sime property. Consolidated Yukeno retained its remaining 130 claims. The company proceeded with extensive underground exploration and development on the Sime, McLeod and Sugiyama veins. They produced a sizeable tonnage of ore from this ground. The company changed its name to Galkeno Mines in 1957, and in 1958, after suffering a drop in production, sold out to UKHM in the fall.

❖ *Yukeno Mines Limited*. Yukeno Mines Limited formed when shareholders of Consolidated Yukeno voted to reorganize the company with one new share for two old shares.

Yukon Galena Hill Mines Ltd., north slope of Galena Hill. Miles Flynn, Terry O'Neill and Jim Cox, 1948. *Herb Foster photograph, O'Neill Collection*

❖ *Bibis Yukon Mines*. Bibis Yukon Mines was formed with the same sponsorship and management as Mackeno Mines and Yukeno Mines Limited to make an agreement with Yukeno Mines Limited to develop the Highlander-Bunny property. Mackeno, Yukeno, and Bibis Yukon raised enough money to build a 150-ton mill on the east side of Christal Lake, which began production in June, 1953.

❖ *No Cash Keno Lead and Silver Mines Limited*. This company was formed by Keyboycon Mines and held eight claims north of the No Cash mine. After mapping and prospecting the claims, they ceased activity in 1952.

❖ *Mayo Mines Limited*. Mayo Mines Limited was incorporated in 1947 by J.R. O'Neill. The company held 11 claims in three groups in the area. Included in these were the Ram claims on Sourdough Hill. Promoter M.J. Boylen bought shares in the company. Mayo Mines carried out surface work on their claims plus underground development on the Ram group on Sourdough Hill. A small tonnage of high-grade was removed from the Ram group. In 1953, the company changed its name to Maybrun Mines Ltd. By 1966, Maybrun held only two claims in the district, the Tundra and Ram, which were leased to private operators.

❖ *Brewis Red Lake Mines*. Brewis Red Lake Mines was an Ontario based mining company formed in 1945. The company acquired nine claims on Keno Hill in the early 1950's. Their best option was a showing on the Black Cap claim, which lies on the southwest extension of the Lucky Queen vein. On the strength of the showings in surface pits, trenches and shafts, the company drove an adit to the veins. Bill Laberge was in charge of their Keno office. Some drifting was carried out on the vein but the results were poor and the operation was closed down in 1951.

❖ *Ankeno Mines*. This company was originally named Consolidated Ansley Gold Mines Ltd., of Toronto. The name was changed to Ankeno Mines Ltd., in December, 1951, after acquiring 12 claims in the Keno district, including part of the Hogan vein. The claims included those of Corp and Ryan and the Iron Clad claim from Barry O'Neill. From 1952 to 1954 the company drove an adit, cross-cut and drifted on the Hogan vein without success. In March 1955, the company changed its name to Bankeno Mines Ltd. Very little work was done on the property until 1965, when UKHM optioned it and carried out minor reconnaissance surface prospecting. The option was dropped and the Bankeno Mines property has been dormant since then.

❖ *Klondyke Keno Mines Ltd.* Klondyke Keno Mines Ltd. was incorporated in 1951 to carry out exploration work on the Klondyke Keno vein on the Blue Rock property. The company carried out surface diamond drilling and some underground development under the supervision of Jack Walli and consulting engineer, Murray Watts. Development on the property ceased in 1953. In 1955, the company changed its name to Jaye Explorations Ltd. and in 1957 sold its Klondyke Keno property to UKHM.

From left: Peter Orlovitch, Terry O'Neill and John R. O'Neill, 1948. *Herb Foster photograph, O'Neill Collection*

❖ *Kenmayo Yukon Mines.* This company was formed on the Macade group of 16 claims below the Sadie-Ladue. Nothing worthwhile was found on the property.

❖ *Packeno Yukon Mines Ltd.* This company acquired claims north and west of the Silver King in an area of heavy overburden. The company's directors were linked to Kenmayo Mines.

❖ *Yukore Mines.* Yukore Mines was originally named Lac D'Or Mines. When the company acquired the eight-claim Regent Group on Galena Hill, its name was changed to Yukore Mines.

❖ *Mount Keno Mines Ltd.* Mount Keno Mines was chartered in 1950, in Toronto, and acquired 35 claims on the south slope of Keno Hill, including those covering the Hogan and Runer veins, as

Stoneboat (called a "Go-Devil") and Caterpillar on Keno Hill. Jim Cox, Spud McMurphy, John R. O'Neill, and Terry O'Neill, 1948. *Herb Foster photograph, O'Neill Collection*

well as an option on the Vanguard. Under the supervision of Jack Hogan, the Hogan vein was opened up by an adit, cross-cuts and drifts. No ore bodies were found, although the vein was fairly strong. The Runer vein was explored and the company drove two adits known as the 3800 and the 3900 adits. Numerous surface pits, bulldozer cuts, and shafts also opened the vein for investigation. Two ore shoots were encountered on the Runer vein and Mount Keno entered into an option with American Metals Corporation. A total of 80 tons of ore was shipped. When the ore ran out in 1953, operations were suspended. These properties were leased to Silden Mines Ltd. in 1954, who mined out some small high-grade pods.

J.B. O'Neill leased the Mount Keno property in 1957 and high-graded the claims off and on. He made a discovery on the south west extension of the Runer vein in 1978. O'Neill high-graded the new discovery through 1980 and then sold his interest in the property to Canada Tungsten Mines Ltd. Martin Swizinski optioned Mount Keno in 1981 and he and his company, Springmount Operating, continues to mine the property today [1989].

Garry Gauvreau and Aurele St. Louis sacking high-grade silver-lead ore. *Herb Foster photographer, O'Neill Collection*

❖ *Jersey Yukon Mines*. This company was incorporated in 1951 and 40 claims and fractions in the district were held under option by J.R. O'Neill of Almonte, Ontario. J.J. Colt, of New York, was president of the company and Miles Flynn was vice-president and secretary-treasurer. The company had almost a half a million dollars in the treasury. Jersey Yukon Mines conducted the first geophysical surveys in this district. The company optioned the Stone claim from Matt Butijer (also spelled Butyer) and drove an adit which failed to locate the vein that was exposed on the surface. The company also acquired Angus McLeod's Eagle claims and sank a small shaft and cut numerous trenches on the western extension of the Eagle vein on Galena Hill.

The company changed its name to Jersey Consolidated Mines Ltd., in May of 1964 and to Holofile Technology Inc. in 1976. J.B. O'Neill subsequently optioned the property and turned it over to the present owner, Teck Corporation Ltd.

❖ *Lustre Yukon Mines*. This company was formed by Dan Cushing of Toronto. The company explored the 16 claim Homestake property on Bunker Hill, which was originally staked by Andrew Johnson. However, Cushing acquired the claims from Bon Kunze and Gordon Lee who held the major interest in the property.

❖ *Lookout Mountain Mines*. This company was organized on the 20 claim Lookout Mountain property owned by M.H. Ewing and E.C. Bleiler.

❖ *McQuesten Slope Mines Ltd.* McQuesten Slope Mines was formed in 1951 by W.S. Kennedy. The company acquired the Raven claim and property near Galena Hill.

❖ *Comstock Keno Mines Ltd.* Comstock Keno Mines was chartered in 1950, and acquired the

Gambler, Nabob and Comstock-Keno veins from J.R. O'Neill. The company president was George Boyd Webster, who was also President of Kerr-Addison Mines Ltd. Minor work was done on the Nabob vein with no significant results. The Comstock-Keno vein, an extension of the Porcupine vein, was investigated in the 1920's and ore had been mined. Comstock outlined further ore in the early 1950's and a joint venture with UKHM was undertaken to mine this out. The Gambler vein was also examined underground from 1951–53, but nothing significant was found and operations ceased. In 1957, J.B. O'Neill leased and high-graded the Gambler vein and shipped more than 40 tons of ore that ran 207 ounces of silver to the ton.

"Bellekeno carried out underground work and developed two ore shoots on one of the strongest veins in the area in 1953"

❖ *Bellekeno Mines Ltd.* Bellekeno Mines was a 3.5 million share company formed in 1950 during the Keno Hill boom. The company's president was S.A. Percy. In 1953, the company acquired the principal vein showings on Sourdough Hill. These were properties that had been originally organized by John R. O'Neill. Bellekeno carried out underground work and developed two ore shoots on one of the strongest veins in the area in 1953. It produced about 11,000 tons before the shoots were exhausted. In March of 1955 the company changed its name to Consolidated Bellekeno Mines Ltd. The company sold its Sourdough Hill property to UKHM in 1965, and ceased operations in the Keno Hill area.

❖ *Galkeno.* Galkeno was a new company created in 1957 by the reorganization of Mackeno, based on one share of Galkeno for three shares of Mackeno. The company name was later changed to Canadian Northwest Mines & Oils, who sold its Keno properties and assets to UKHM, in 1958.

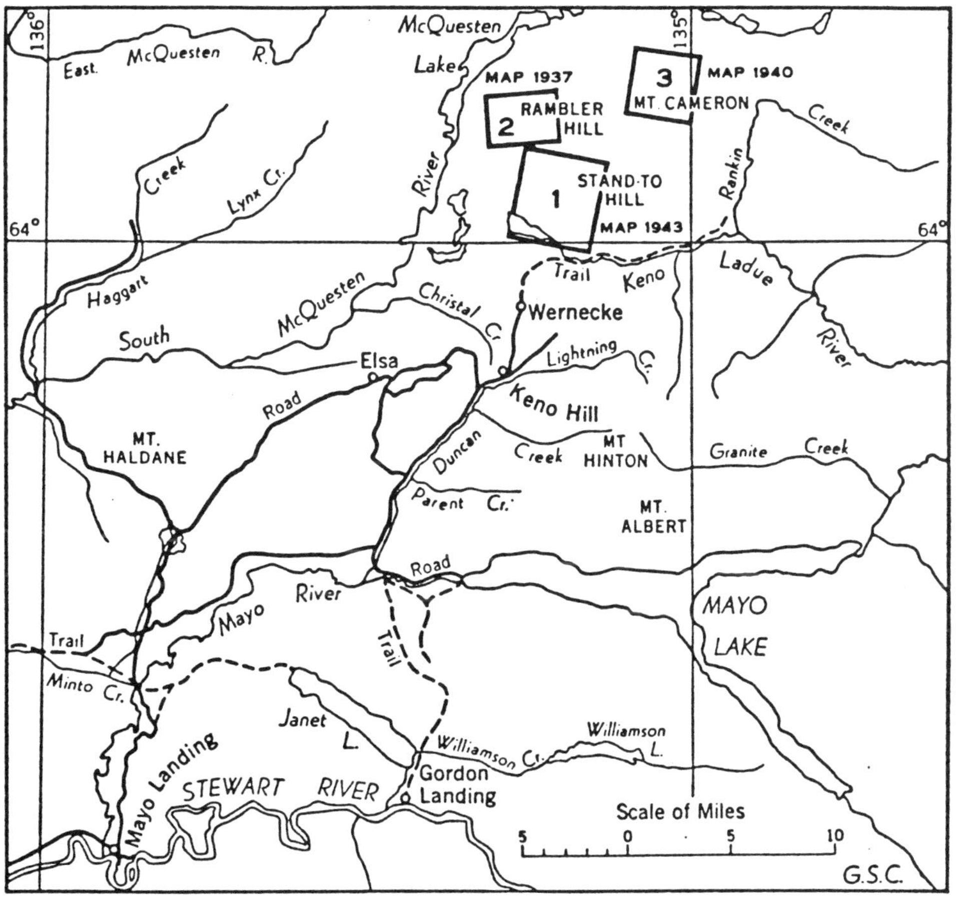

Index map showing location of mineralized areas of the Mayo district. *G.S.C.* circa 1921

EARLY SILVER CLAIMS

Claim Name	Original Staker & Date Staked
Adam	Mark Evans February 23, 1913
Ajax Fraction	John Fawcett September 7, 1919
Ala	Frank Cantin July 6, 1923
Apex	Thomas McKay February 3, 1920
Arctic	Robert Fisher May 3, 1921 (Later acquired by Charles R. Settlemier & Claude H. Bermingham)
Argentum	Thomas McKay May 3, 1920
Bell York	Al Gammon June 29, 1922
Black Hawk	Charles Beck June 29, 1922
Bluebell	Frank McNeill May 7, 1920
Bluestone	Joel R. Sunderland May 7, 1920
Bunny	Dorothy McLaren April 3, 1920
Calumet	Harry Colley, Joe Walsh & Harry McCrimmon April 27, 1920 (to Harry Colley, Thomas McKay and Don Morrison in 1934)
Caribou	Joseph A. Walsh September 6, 1919
Coral	Robert Fisher May 4, 1921
Croesus	Joseph A. Walsh February 20, 1920
Darling	Walter Johnston November 23, 1925 (Sold to Clem Sinyard and renamed the Jock in 1926)
Dixie	John V. O'Sullivan October 31, 1923
Dorothy	George Gray October 19, 1921
Dragon	Oscar Miller September 26, 1921
Dublin King	Hector Morrison August 11, 1909
Eagle	Miss Jessie Stewart June 13, 1926 (transferred to Angus McLeod)
Elsa	Charlie Brefalt September 13, 1924
Faro	Dr. Randy Maclennan August 5, 1919
Forest	Robert Greaves July 24, 1920
Friendship	Joel R. Sunderland May 7, 1920
Frog	Dr. Randy Maclennan August 5, 1919
Galena	Alex Nicol September 10, 1919
Gambler	A.R. "Deep Hole" Thomson August 13, 1919
Gold Hill	Murdock Mitchie August 26, 1919
Greenstone Fraction	Livingston Wernecke August 24, 1921
Grey Copper King	Robert Fisher October 15, 1923 (Acquired by James McCloskey)
Happy claim	James "Jimmy" Sugiyama July 5, 1936

Claim Name	Original Staker & Date Staked
Happy Jack	John Jackson Suttles October 6, 1908 (Sold to J. S. Stewart)
Hector	Clem Sinyard July 21, 1920
Hell's Gate	Jacob Alexander Davidson August 3, 1903
Highlander	Thomas McKay April 3, 1920
Hobo	Oscar Miller January 19, 1927
Homestake No. 2	Lenore Walsh April 13, 1920 (Sold 1/5 interest to Thomas H. Hinton on August 31, 1922)
Homestake No. 3 & No. 8	Thomas H. Hinton April 13, 1920
Homestake No. 5	Thomas H. Hinton April 17, 1920
Ike	Thomas A. Dickson July 5, 1938 (Sold to Treadwell Yukon on July 9, 1938)
Jessie	John H. Carpenter April 7, 1924
Jock (formerly the Darling)	Bought by Clem Sinyard on February 16, 1926
Jupiter	Robert Fisher June 7, 1925
Keno	Alfred K. Schellinger July 29, 1919
Kid	Louis Bouvette August 17, 1919
Ladue	David A. Cunningham September 11, 1920
Lake No. 1	Anthony A. "Tony" Hollenbeck June 8, 1920
Lakeview	A.R. "Deep Hole" Thomson August 13, 1919
Lansing	Thomas McKay June 16, 1920
Lion	Thomas McKay August 27, 1919
Lone Star	David Sparks September 2, 1919
Lotus	William L. Bramley February, 22, 1920
Luna	Horatio A. "Ray" Stewart September 9, 1920
Lucky Queen	Hector Morrison February 8, 1920
Lucky Strike	Charlie Brefalt May 22, 1925
Mabel	Grant Huffman February 23, 1913
Maiden's Hope	Joel R. Sunderland August 13, 1919
Mary	Frank McNeill May 9, 1921
Mastiff	Edward K. Jones July 10, 1921 (Acquired by Charles R. Settlemier and Claude H. Bermingham)
Mathole	Garnet A. Watt October 12, 1919
Minto	Louis Bouvette August 16, 1919
Mowhawk	Isaac N. Paynter May 16, 1928 (Sold to Treadwell Aug. 20, 1928)
Nabob	Thomas McKay August 4, 1919
No Cash	Charlie Brefalt August 9, 1924
Onek	A.G. Mensforth May 15, 1951

Claim Name	Original Staker & Date Staked
Pinochle	Jack Pickering July 30, 1919
Poca Plata	Horatio A. "Ray" Stewart May 7, 1920
Porcupine Fraction	Edward E. McCarthy October 3, 1919
Punch	John "Jack" Adair May 28, 1925
Roy	Matthew Butyer April 12, 1921
Reno	Axel Erickson August, 1919
Rico	James A. Anderson July 21, 1919
Rocket	Arthur Stewart June 14, 1925
Roulette	Louis Bouvette June 10, 1919
Ruby Fraction	Ellef Bjonnes June 9, 1924
Sadie	James R. Clark March 18, 1920
Scot	Thomas McKay January 6, 1921
Scotty	Scotty Mowatt July 29, 1919
Shamrock	Thomas McKay August 4, 1919
Shepherd	Andrew Johnson October 15, 1919
Silver Basin	Rodolph Rosmusen August 25, 1919
Silver Bell	William L. Bramley February 20, 1920
Silver King (Formerly Hell's Gate)	Henry William McWhorter February 23, 1913
Solo No. 1	Axel Erickson August 5, 1919
Solo No. 2	James E. Greenfield August 15, 1919
Stone	Mathew Butyer May 18, 1920
Sylvia	Angus A. McLeod July 20, 1921
Tango	Emil J. Forrest August 16, 1919
Tick	William H. Jeffrey Jr. March 31, 1928
Tiger	Axel E. Erickson August 27, 1919
Tin Can	William L. Bramley March 7, 1920
Travice	John E. Hollenbeck March 12, 1921
Tunnel Fraction	William B. Hargraves June 4, 1923
Tyee	Hans E. Formo August 10, 1925
Tyee	Colin Chisholm October 18, 1929
Twin Sisters	Samuel J. Thurber June 8, 1920
Union	Joel R. Sunderland August 13, 1919
Upton	Archie Martin May 7, 1921
Vanguard	Alex Nicol June 19, 1920
Webfoot	Jack Alverson February 23, 1913
Wigwam	Edward K. Jones May 4, 1921
Wolverine	Alfred K. Schellinger October 3, 1919
X Fraction	Howard "Harry" Colley May 9, 1929

(Information from Roland Ronaghan, Mayo Mining Recorder)

In addition to the claims staked on Galena Hill, Keno Hill, and Mount Haldane, there were also claims staked in the 1920's on a number of hills in the district including hills in the Davidson Mountains. Ore discovered in the Davidson Mountains was found to have a lower silver content and less mineralization than those on Keno and Galena Hills. The following information has been compiled from reports by W.E. Cockfield, *Geological Survey of Canada Memoir 284*, compiled and annotated by H.S. Bostock:

❖ *Stand-To Hill (Davidson Mountains)*. Seven claims, the Elsie, Dorothy Brown, Janet Agnes, Two Donalds, Victoria, Mary Bell, and Glengarry were staked on August 15 and 16, 1920 on Stand-to Hill by Joe Zahn, William Forbes, D. Forks, D. MacDonald, John Faulkner, and J.A. MacDonald.

Three other claims on the hill, the Surprise, the Enterprise, and the Hillside, were owned by Charles Coutts and J. McKinnon.

Charlie Brefalt and Joe Michaud also prospected Stand-to Hill.

❖ *Mount Cameron (Davidson Mountains)*. Jack Alverson, James Scougale, and J. Philip staked Cameron No. 1, Cameron No. 2, and Cameron No. 3 on Mount Cameron in 1918.

❖ *Rambler Hill (Davidson Mountains)*. Eight claims were staked on Rambler Hill by Archie Martin, Albert Lamb, A.R. "Deep Hole" Thomson who staked the Lucknow, Harry Colley, Jack Alverson, George Forry, John Lake, and J. Robertson.

Archie Martin, "Rocky Mountain Bill" William Elliott, Fred Arnold and B. Verschoyle were involved with the Homestake group on Cache Creek, Rambler Hill.

❖ *Sourdough Hill (south of Keno Hill across Lightning Creek)*. M. Mellish staked the Thunder claim and leased it to J. Gillis, J. McHugh and J. Curley in 1923.

❖ *McKay Hill (vicinity of Upper Beaver River)*.

> Silver–lead deposits were discovered in 1922 on McKay Hill in the vicinity of Upper Beaver River, and as the discovery followed so closely that of Keno Hill, a stampede to the area occurred, and numerous claims were staked. Subsequent prospecting however, showed that the deposits were low grade, and many of the claims were abandoned. In 1923, prospecting of the area continued and resulted in the discovery of deposits on Silver Hill, and also of high-grade float on Grey Copper Hill. These discoveries did not become known until midwinter when a second stampede to the area took place and some fifty claims were staked around the discovery claims of Grey Copper Hill. As the staking took place in the winter it was not until the spring of 1924 that any development could be undertaken, and on account of the retarded summer very little effective work could be done on most of the claims until June or July. Many of the claims, therefore, had no showings of mineral in place on them.[40]

Chief claims in 1923 were the Carrie and Whiterock owned by L.B. Erickson, the Snowdrift owned by W.F. McKay and the Black Hawk owned by Charlie Beck. Adjoining claims included the Snowball, Big Windy, Wild Goose, Eagle, Wild Duck, Bessie, Tiger, and Red Rock.

The Yellow Rock Group of six claims were staked by A.N. Martin, O. Dahl, E. Anderson, and C. Williamsen.

❖ *Silver Hill (near the head of Ervin Creek, a tributary of Carpenter Creek)*. Eight claims and two fractions were staked on Silver Hill by Jack Carpenter, E. Ervin, and J. McLean.

❖ *Grey Copper Hill (four miles north of the mouth of Carpenter Creek)*. In 1923 Robert Fisher staked the discovery claim Grey Copper King on Grey Copper Hill. This led to a stampede to Grey Copper Hill in the winter of 1923–24 and about 50 claims were staked. Fisher also staked the King Tut claim. L.B. Erickson staked the Silver Queen.

❖ *Horseshoe Hill (east of McKay Hill)*. The four claim Independence Group was staked on Horseshoe Hill by Archie Martin, O. Dahl, E. Anderson and C. Williamsen.

❖ *Elliott Hill*. The Apex claim was staked on Elliott Hill by J. McCloskey in 1922.

Silver Mining prepared by Linda E. T. MacDonald

ABOVE, Mayo crowd gathered on the waterfront for the arrival of a steamboat, 1923. The White Pass & Yukon Route warehouse is on the top left and their conveyor is in the centre, Chateau Mayo hotel is on the right. Ralph & Ruth Rogers are standing in the bottom left corner. George Black is in the lower centre of the photo, removing his hat and shaking hands. Gus Besner and Jimmy Ballentine are in the lower right corner. *Ralph Rogers Collection*

BELOW, Group of Mayo residents in 1922. Front row (left to right): Irvin Ray and Ted Sunderland. Second row: Jim Fairborn, Eddie McKenzie, Theos Whitney, unknown. Third row: Jack Pickering, unknown, unknown, Eva Ladd, Ruth Rogers, and Dorothy Hoggan. Fourth row: Gordon Lee, unknown, Kippy Boerner, Sgt. Dempster, and Walter Murphy. Back row: Frank Whitney, Ralph Rogers, unknown. *Ralph Rogers Collection*

Chapter 6

Mayo

THE TOWN

THE DISCOVERY AND DEVELOPMENT OF THE SILVER KING MINE from 1913 to 1917 increased activity in the Mayo district, but Louis Bouvette's Keno Hill discovery in 1919 sparked a boom that lasted for almost 20 years.

Both the development and atmosphere in Mayo Landing, as a result of the "silver tide," is reflected in an article that appeared in the *Dawson Daily News – Yukon and Gold Fields Edition*, in 1922:

Chateau Mayo with approximately 12,000 tons of silver-lead ore, 1938. *G. A. McIntyre Collection*

> The music of the carpenter's hammer was heard throughout Mayo all the spring, summer and fall of this year. Large and small buildings, all of significance in the growth of the place were erected and all are now occupied. There are no vacant houses in the town. Notable among the structures which went up during the season were: a fine two-storey hospital and a new building for the Mayo fire department headquarters, both attractive frame buildings and both owned and conducted under government; attractive and modern frame church building and manse, erected on the high ground facing the waterfront in the east end of town by the Church of England, under direction of Bishop I.O. Stringer, and in charge of Rev. F.H. Buck, M.C., as rector; a large and substantially built waterfront warehouse, at the foot of Center Street, owned by the White Pass Company, for accommodation of its general river transportation;

A view looking north on Front Street in Mayo. The buildings identifiable are Oscar Letourneau's Hotel Royal Alexandra, the Bake Shop and Coffee House, Gene Binet's Chateau Mayo Hotel, and the Northern Commercial Co. store. The photograph was taken at midnight on June 22, 1923. *Ralph Rogers Collection*

business block and warehouses erected by the Northern Commercial Company, Limited, and occupied with a large general merchandise stock; a fine two-storey business block, at corner of First and Center Streets, erected and owned by Joseph Beaupre, formerly of Dawson; a large new two-storey hotel, the Broadway, on First Avenue and Montreal Street, erected, owned and conducted by S. Tunnecliffe, formerly of the R.C.M.P. at Carmacks; large two-storey business block, close to the Broadway, on First Avenue erected and conducted by G.S. Churchward, pioneer tinsmith from Dawson; neat one-storey building erected by Mrs. Fanny Mitchell, formerly of Black Hills, located on First Avenue, near Center, where she has a bakery; fine large one-storey office and warehouse building on First Avenue, Mayo headquarters of the Treadwell Yukon Company, which also has other quarters in the city; neat one-storey building on Center Street, owned by Mrs. Agnes Kinsey, where she has the Silver Star Bakery; one-storey building on Second Avenue, owned and occupied by Olaf Anam as harness and shoe shop; fine barn, one of the largest and most commodious and costly in Yukon, located on Third Avenue, and owned by Greenfield & Pickering, freight and mail contractors; large modern two-storey frame residence on Second Avenue, erected and owned by John A. McDonald; fine large two-storey residence, erected near the new Church of England, by Mr. and Mrs.

Binet & Lefebvre's sawmill in Mayo. *Ed Kunze Collection*

Front Street in Mayo, October 5, 1923. Roy McDermid is driving a government Holt tractor which is pulling an early model scraper controlled by Charlie Rankin. *W. S. Hare. photograph, F. & T. Whitney Collection*

Malcolm McLaren, formerly of Dawson; and a number of other cottages and smaller business improvements.

In addition to the public and business institutions mentioned in the foregoing, Mayo City also has a school, where instruction is given in grades from primary to high school subjects with about twenty children attend-

A view of stacks of milled wood and saw shed at Kimbel's sawmill in Mayo during the 1930's. *Hare Collection, Yukon Archives*

ABOVE,The Bank of Montreal in Mayo in 1937. Left to right: teller Chuck Harrison, old-timer Jack Smith, and bank manager Arnold Hall. *G. A. McIntyre Collection*

BELOW, Business license issued by the Town of Mayo for Churchward's Sheet-Metal Works, April 18, 1933, signed by Joseph S. McIntosh collector of business license fees. *Verna Klippert Collection, Mayo Historical Society*

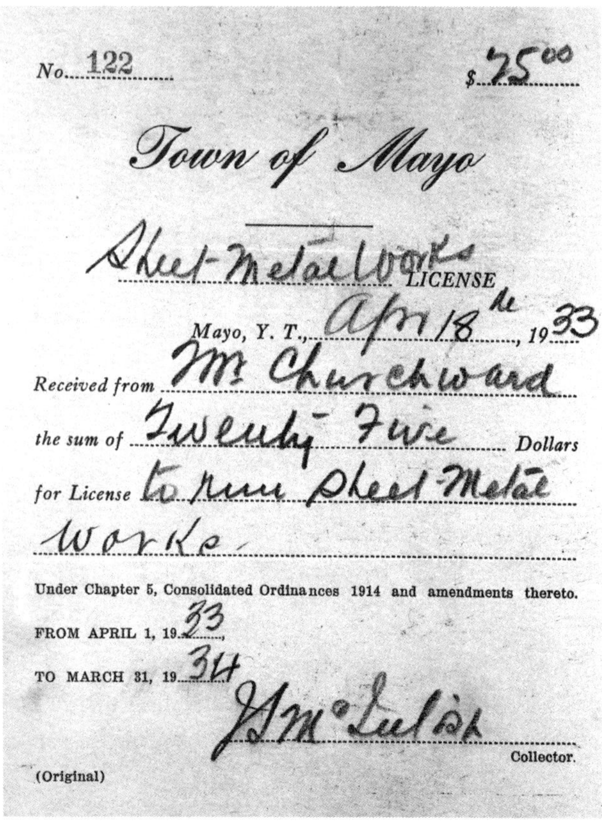

No. 122 $25.00

Town of Mayo

Sheet-Metal Works LICENSE

Mayo, Y. T., Apr 18th, 1933

Received from Mr Churchward

the sum of Twenty Five Dollars

for License to run Sheet-Metal Works.

Under Chapter 5, Consolidated Ordinances 1914 and amendments thereto.

FROM APRIL 1, 1933,

TO MARCH 31, 1934

J.S. McIntosh Collector.

(Original)

ing; barracks of the R.C.M.P., and home for the officer in command, having jurisdiction in the entire Mayo District, namely, Sergeant W.J.D. Dempster, who has authority as a magistrate and is ex-officio tax collector; a mining recorder's office, in charge of R.L. Gillespie, formerly in the gold commissioner's office at Dawson as recorder; a fine large hall, built and owned by Mayo lodge of the Yukon Order of Pioneers; large business block of Binet Brothers, comprising hotel and mercantile establishment, both two stories, at First Avenue and Center Street, with large basement, electric light plant and other auxiliaries, together with several large warehouses and barns belonging to the same concern and located on the waterfront and elsewhere in town; the Taylor & Drury Company, Limited's, large new business block, on First Avenue, and waterfront and other warehouses, carrying an extensive general merchandise stock; Oscar Letourneau's large business block, at First Avenue and Laurier Street, comprising a three-story building with many rooms and cafe, barber shop and merchandise store in connection; Lefebvre & Binet sawmill, at west end of First Avenue, with large lumber yard in connection; Mayo Utilities Company's large business building, with telephone headquarters and merchandise store in connection, in charge of D.A. Matheson and Mrs. Matheson; Waechter Brothers Company's meat shop, with floating refrigerator plant on waterfront; P. Burns & Company's meat market; Dr. W.W. Chipman, physician; Cantin Brothers and Alex Nicol, barns and freighting headquarters; large waterfront warehouse, garage and other warehouses of the Keno Hill Limited, and the Treadwell Yukon, Limited, with large ore-piling yards on waterfront for both companies; warehouse and garage of O.S. Laning; barns and warehouse of Joseph Longtin, wood and water dealer; blacksmith shop of Greenfield & Pickering; drug store of D.R. Maclennan, M.D.; lodging house and undertaking establishment of Alex Lesperance; business office of J.E. Ferrell, Yukon Councillor; merchan-

Front Street, Mayo, outside Binet Bros. Hotel and Store. *F. & T. Whitney Collection*

dise store and auto headquarters of R.H. "Bob" Palmer; merchandise store of E.J. "Eddie" McKenzie, and headquarters for McKenzie & McCown's auto stage line; Joseph Dubois' barber shop, and the government liquor store.

In various parts of the town also are many fine cottage homes, some erected this year, including those of Tom Chernuta, Bob Henderson, Alex Nicol, Helen Marlowe,

"Home Bakery," Mayo 1920's. Sam Blackmore's cabin. Betsy Cunningham ran the bakery and in 1932 Mrs. Parker operated it. *M. Rich Collection*

TOP, Hotel Royal Alexandra. *L. Cheatham (McLaren) Collection*
MIDDLE, Silver-lead ore on the river bank at Mayo, 1923. *Ione Christensen Collection*
BOTTOM, Government liquor store, middle 1930's. Vendor Jack Gallagher is standing at the door. *G. A. McIntyre Collection*

Mr. and Mrs. Isaac Taylor, Ed Kimbel, Mr. and Mrs. J.E. Pickering, Mr. and Mrs. James Greenfield, Mr. and Mrs. J.E. Ferrell, Cantin Brothers, Isaac Mallette, Mr. and Mrs. T.H. Ladd, Mr. and Mrs. Portlock, Mr. and Mrs. Parker, Mr. and Mrs. Rogers, Gideon Letourneau (brother of Oscar), Thomas McKay, Jack Turner, Grant Huffman, Mrs. Cora Rich, Jack Smith, George Miller, Mr. and Mrs. H.W. Reinhardt, James A. Fairborn, Mr. and Mrs. James Breaden, Jeff Wynen and mother, Mr. and Mrs. Rupert Steeves, Mr. and Mrs. Richard O'Loane, Mr. and Mrs. Frank Whitney, Mr. and Mrs. R.L. Gillespie, Mr. and Mrs. A.K. Schellinger, Mr. and Mrs. Robert Douglas, Wm. Douglas, Mr. and Mrs. Alguire, Joseph Hoyne, George Cordery, Mr. and Mrs. George Wright, Wilfred Guilbault, Stanley White, Rodolph Rosmusen, C.L. Snell, James Anderson, Charles Turgeon, Alex Profeit, Frank Carscallen, Mr. and Mrs. Elmer Middlecoff, Henry Ehlen, Tom Hibbert and quite a number of others.

The town has spread over quite an area, far in all directions from the centre, and several new cottages are being erected this winter. The Catholic Church secured lots this summer through efforts of Rev. Father Allard for erection of a church, and arrangements were made and a good deal of money subscribed for the proposed building, which likely will be erected next summer. Other new buildings also are planned.

Situated on the Stewart River, a fine large stream navigable for large steamers throughout the open season, Mayo City is favourably located. It is 180 miles up the Stewart from its mouth, where it flows into the main Yukon River, and 250 miles from Dawson, by water, and 570 miles from Whitehorse, by water. From Mayo to

Keno City, at the foot of Keno Hill, is forty miles over a road which, since the strike on Keno Hill, has been greatly improved and is now travelled extensively by motor cars and other conveyances.

Mayo City townsite, comprising 185 acres, originally was secured by J.D. Bell and Ernest Belliveau [J.E. Beliveau], and a little later taken over by J.E. Binet, who went there twenty years ago, from Dawson, hewed a place in the heavy forest at which to erect a cabin, now the site of his extensive business block in the town. Mr. Binet also connected Mayo with Duncan Creek, then thriving with a population of about 800 miners. Mr. Binet cut out the trail to Duncan, a stream about twenty miles in length, and thus diverted the traffic of that locality to the site of the present town. The townsite is on an extensive flat, originally heavily wooded, and now largely cleared in the process of town building.

Diane Johnston Chisholm's paper *Carrying the Territory, Silver-Lead Mining in the Mayo District 1919–1942* describes public services in the Mayo district which were supplied by the territorial government.

Territorial funding for public services in the Mayo district supported schools, a library, a hospital, medical health officer, municipal improvements for Mayo, and a territorial agent. As the population of Mayo and the surrounding area expanded with the mining boom, public pressure was brought to bear on the territorial government to provide certain services. The Mayo Board of Trade was particularly vocal and shortly after its formation in June 1921 requested, among other things, that Mayo be provided with a properly equipped school. By May of 1923 it had also sought provision for fire fighting

OFFICE OF

LICENSE AND TAX COLLECTOR

TAX NOTICE FOR THE YEAR ENDING MARCH 31ST, 19........

Mayo, Y. T., April 1, 19 35

MR. Olaf Annin

Take notice that you are assessed on the following lots in the unincorporated town of Mayo for the year ending March 31st, 19 36

		Arrears to March 31, 193....	$............
Unoccupied Lot No............in Block	No..............at $ 5.00		....
Residential Lot No. 25 in Block	No. 12 at 10.00		10.00
Business Lot No. ....in Block	No..............at 25.00		....
	Total		$ 10.00

The above amount is now due and payable at my office, and is a first lien on real estate.

J S McIntosh

Collector of Taxes and Licenses for the Unincorporated Town of Mayo, Yukon Territory.

Bring this notice with you.

☞ A discount of 10 per cent. will be allowed on all taxes paid on or before November 15th, and from that date to December 31st at par, and from that a penalty of 4 per cent. per month for ten successive months will be charged if taxes and penalty be not sooner paid.

TOP, Bags of coal from Carmacks stacked at the Mayo waterfront, 1948. Terry O'Neill in centre of photograph. *Herb Foster photograph, O'Neill Collection*

BOTTOM, Olaf Anam's Tax Notice from the Unincorporated Town of Mayo, signed by Joseph S. McIntosh, Collector of Taxes and Licenses, April 1, 1935. *Verna Klippert Collection, Mayo Historical Society*

Joe Longtin, his horse Eli and the water wagon used to deliver water to the homes in Mayo, 1935. *Besner Collection*

equipment, a hospital, and improvements to the town.

Grants for the maintenance of a circulating library and reading room were made in 1923 at the insistence of people in Mayo and were continued through the forties. The grants ranged from $200 a year to a peak of $500 in the late thirties. In 1926, additional funding was requested so that the Mayo public library could service Keno and the surrounding camps. [This library was run by volunteers from the I.O.D.E.]

In 1922, Mayo was brought under the provisions of the "Assessment Ordinance" whereby taxes could be collected on lands and licences in the area. [The Town of Mayo collected business license fees. J.E. Philip was the license collector in 1932.] At the same time, it was provided with about $4,500 for the development of fire fighting

Mayo residents, 1923. *Ralph Rogers Collection*

Mayo residents in front of Mrs. Kinsey's Bakery in 1923. *Ralph Rogers Collection*

services in the community. In succeeding years, funds were allotted in two separate accounts, one for fire fighting services and the other for streets and sidewalks. After a disastrous flood in 1936, additional monies were provided for repairs. The amount for municipal improvements averaged about $6,000 per year.

Between 1921 and 1942 the territorial government spent, on average, over $14,000 a year on public services in the Mayo district.

Dr. Aaro Aho has written a vivid account of the devastating flood that struck Mayo in 1936. (Yukon Archives, Aaro Aho, *Notes on the Keno Hill Area*, unpublished manuscript, p. 16–9, 16–10.)

The fire hall and tower during the flood in 1936. *G. A. McIntyre Collection*

The Kimbel Brothers' Ford truck with a load of ore got stuck in the high water outside town but did make it to the higher land on Front Street without its fan belt. "Spot Cash" Breaden went by singing in a kayak; George Nagano, the Japanese cook in his G.N. Cafe continued serving meals in hip waders until the water was three feet deep; the Burns & Co. butcher rushed his meat up to camp; the Bank of Montreal manager moved his safe into the church; and the R.C.M.P. and mining recorder moved some of their records. Tents were pitched outside town. Dr. H.W. "Herk" Herkenrath, "the whistling dentist" was playing his ragtime piano as it was being floated away on Kimbel's boat, and St. Mary's Rectory and the Catholic church became the main local refuge with cooking being done in the Rectory. The sternwheeler *Aksala* churned her way inland up Main Street and moored at the front door of the Anglican church to make her galley and accommodation available.

Friday morning, June 6, the water was still rising and was five feet deep in the lobby of the Chateau Mayo Hotel. The two plate glass windows were stoved in by the current and driftwood. The White Pass freight shed, Chateau Mayo Hotel, N.C., and Taylor and Drury store, and other buildings on Front Street would, in all probability, have been swept away by the full current of the Stewart if it had not been for the but-

ABOVE, A view of Mayo's Front Street completely covered with flood water. The W.P. & Y.R. freight shed (long building, left), the N.C. Co. (log building, right) and Lefebvre's sawmill (end of the street) are inundated with water, June 1936. *W. S. Hare Collection*
BELOW, Dr. H. W. "Herk" Herkenrath playing his ragtime piano as Kimble's boat is poled along a Mayo street in the June 1936 flood. *G. A. McIntyre Collection*

tress of the modest pile of ore and concentrates and the lower river barges. Only the Catholic and Anglican Churches, the school, the library and the government liquor store remained unflooded. Patients in the hospital were cared for in the top floor; the ground floor had been flooded and the X-ray machine ruined.

Several of the local characters sat on roof tops with cans of King Oscar sardines and crackers, cooling their beer on strings in the water and watching the ceaseless spectacle of evacuation which looked like a stampede.

Saturday morning the water ceased to rise and by Sunday it was receding. It was reported to have risen 16 feet above the usual high water mark. The town was left in a complete mess with several buildings having floated away; some shifted off foundations; the fire hall hose tower at a slant; and cordwood, lumber, oil drums, tires, barrels, garbage, sheds, outhouses and varied flotsam lodged in windows against buildings and fences, all sitting in a slimy mud up to a foot thick. The mining recorder's office was a terrible mess: five feet of water lay inside; five outhouses and other people's cordwood floated around in the yard and records of the quantity and value of shipping ore from Keno Hill were destroyed.

Taylor and Drury and the N.C. Company store managers told the populace to go in and help themselves to the label-less cans on the shelves. Nearly all other items were ruined and the bins were split by swelling of dried foods. A large supply of spoiling fish, thrown out of the meat market, floated around town on the last of the flood waters and added to the stench.

However, no disease epidemic resulted and the only outside help was chloride of lime shipped in for the stagnant pools of water. Within a few weeks the community was back to normal, with the S.S. *Keno* gingerly unloading the first powder shipment and taking out the first barge load of ore.

The flood of 1936. TOP, The Bank of Montreal building. MIDDLE, Burn's Butcher Shop. BOTTOM, Mayo residents' belongings piled beside the Catholic Church. *All photographs, G. A. McIntyre Collection*

ABOVE, Mayo's Front Street during the flood in 1936. BELOW, White Pass freight shed on the left and the Northern Commercial store during the flood in 1936. *Both photographs, G. A. McIntyre Collection*

A view of wood strewn about the front yard of the Gillespie's residence in Mayo, deposited by the spring flood. The woman in the photograph is indicating the high water mark on the house. Summer 1936. *W. S. Hare photograph, E. C. Bleiler Collection*

DISTRICT MINING RECORDER'S OFFICES *by Linda E. T. MacDonald*

Before a Mining Recorder's office was established in this district, claims were either worked and not staked, or miners travelled to Dawson City to record their claims.

In 1901 a Mining Recorder's Office was opened at Clear Creek and the district was referred to as the Clear Creek Mining District. This office remained open for the 1901 season only. In 1902, recording offices were opened at Gordon Landing and at the mouth of Duncan Creek (referred to at the time as Mayo Bridge settlement). The Gordon Landing office was closed by 1903, and the district was renamed the Duncan Creek Mining District with a sub-office at Dublin Gulch. This office was open for the 1903 season only. In 1904, the Mining Recorder's Office was moved to Minto Bridge and again, in 1914, to Mayo Landing. Although located in Mayo, it was referred to as the Minto Bridge Mining Recorder's Office until 1918, after which it became the Mayo Landing Mining Recorder's Office.

This office in turn had just been closed and all records moved to the Mining Recorder's office in Dawson City, when Louis Bouvette struck high-grade silver ore on Keno Hill, in 1919. J.E. Ferrell, who operated a store in Mayo, doubled as Mining Recorder and he recorded Bouvette's Roulette claim.

With the stampede to Keno Hill, it was necessary to reopen the office in Mayo. R.L. "Dick" Gillespie was appointed Mining Recorder and began recording claims in September, 1919. Harry Colley copied all the Mayo records he could find in the Dawson Mining Recorder's Office so that copies could be returned to Mayo. F.G. Berton, bringing along the record copies, officially reopened the Mayo Mining Recorder's Office in 1920, changing the name of the district to the Mayo Mining District. It remains under that name today.

In some instances, Mining Recorder's Agents were appointed to assist the Mining Recorder and Acting Mining Recorders performed his duties if he was absent from his post. The practice of appointing Mining Recorder's Agents was prevalent from 1901–1920 and, therefore, several names may appear in the Mining Recorder's capacity during a single mining season. Following is a list of Mining Recorders in the Mayo district from 1901 to 1965:

1901–Sept. 1902	W.C. Young *Clear Creek* *Clear Creek Mining District*
September 1902	John Rowley *Gordon Landing* *Clear Creek Mining District*
1902	Isaac Burper W.C. Young James W. Davies *Mouth of Duncan Creek* *Clear Creek Mining District*
March, 1903–September 1903	J.C. McLagan *Mouth of Duncan Creek* *Duncan Creek Mining District*
March 1903–September 1903	A. McDougall *Dublin Gulch* *Duncan Creek Mining District*
August, 1904–1911	Thomas H. Hinton *Minto Bridge* *Duncan Creek Mining District*
December, 1909	D.R. "Randy" Maclennan *Acting Mining Recorder* *Minto Bridge* *Duncan Creek Mining District*
April, 1912	James C. Philip *Acting Mining Recorder* *Minto Bridge* *Duncan Creek Mining District*
June, 1912	J.M.E. Henry *Mayo Landing* *Agent for Minto Bridge Mining Recorder* *Duncan Creek Mining District*
1913–1915	Phillip M. Kelly *Minto Bridge* *Minto Bridge Mining Recorder* *Duncan Creek Mining District* *(In 1914 the Minto Bridge Mining Recorder's Office was moved to Mayo Landing)*
1915–1918	D.R. "Randy" Maclennan *Mayo Landing* *Minto Bridge Mining Recorder* *Duncan Creek Mining District*
August, 1918	J.E. Ferrell *Mayo Landing* *Mayo Landing Mining Recorder* *Duncan Creek Mining District*
September, 1919–1937	R.L. Gillespie *Mayo Landing* *Mayo Landing Mining Recorder* *Mayo Mining District*
1920–1927	Howard "Harry" Colley *Acting Mining Recorder* *Mayo Mining District*
1937–1947	Sam Wood *Mayo Landing Mining Recorder* *Mayo Mining District*
1947–1965	Gordon McIntyre *Mayo Mining Recorder* *Mayo Mining District*

[Information supplied by Roland Ronaghan, Mayo Mining Recorder]

The cabin known as "Mabel McIntyre's Cabin," built in 1921 by Samuel Blackmore, was the first Mining Recorder's Office in Mayo. In 1933, Blackmore built a frame building (currently Joan and Dick Ewing's shed) in front of this cabin and the Mining Recorder's Office moved into it.

During World War II, in order to save the Territorial Government much needed money, the Mining Recorder, Stanley Simon Mason-Wood (known as Sam Wood), was appointed Liquor Vendor. His office was moved to the Liquor Store on Front Street. However, circa 1949, the Federal Government bought the Besner house and moved the Mining Recorder's office into it. At the same time, the first Park Warden was stationed in Mayo and he (Larry Requa) occupied the upper floor of the building.

Subsequently, both the Mining Recorder and the Park Warden were moved to the Panabode building which is the present location of the Mining Recorder's Office. The Panabode was built in 1954 and is solely occupied by the Mining Recorder's office [1989].

Three places were named for Mining Recorders in the Mayo district: McLagan Creek which flows into Minto Lake, Mount Hinton, six miles north of Mayo Lake in the Gustavus Range, and Mount Colley, west of the head of the Snake River.

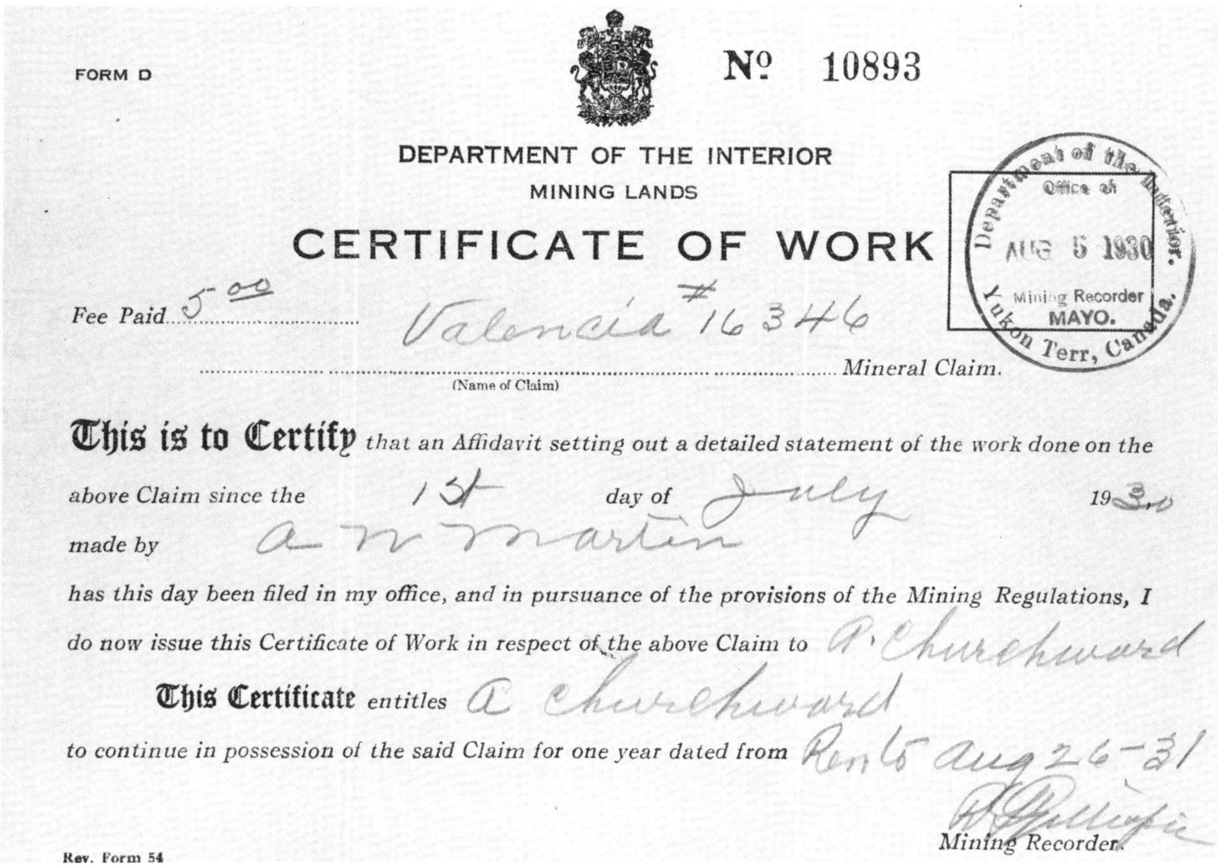

FORM D

No. 10893

DEPARTMENT OF THE INTERIOR

MINING LANDS

CERTIFICATE OF WORK

Department of the Interior, Office of, AUG 5 1930, Mining Recorder MAYO. Yukon Terr., Canada.

Fee Paid 5 00 Valencia #16346 *Mineral Claim.*

(Name of Claim)

This is to Certify *that an Affidavit setting out a detailed statement of the work done on the above Claim since the* 1st *day of* July *19*30 *made by* a W Martin *has this day been filed in my office, and in pursuance of the provisions of the Mining Regulations, I do now issue this Certificate of Work in respect of the above Claim to* R Churchward

This Certificate *entitles* a Churchward *to continue in possession of the said Claim for one year dated from* Ren'l to Aug 26-31

R Gillespie
Mining Recorder.

Rev. Form 54

Certificate of Work for the Valencia claim, signed by Mining Recorder R. L. Gillespie, July 1, 1930.
Verna Klippert Collection, Mayo Historical Society

TAYLOR & DRURY LTD *by Charles D. Taylor*

Taylor & Drury delivery truck, circa 1954.
J. & A. Boyes Collection

Mining activity in the Mayo area provided an attractive prospect for a new Taylor & Drury general merchandise store. This required the purchase of a larger steamer to provide necessary shipping. The Steamer *Kluhane* was replaced by the Steamer *Thistle*, formerly a B.Y.N. ship.

Isaac Taylor accompanied the first shipment of freight to Mayo in 1919. On their arrival in June, they docked at a bend in the Stewart River, just below where the Anglican Church now stands. The demand for supplies was urgent, and Taylor soon had a sales area on the deck. The sale of case lots of canned goods, large gunny sacks of flour, cereals and sugar at reduced prices for cash, was brisk. By the end of the first day, there was a scarcity of merchandise to store in the tent warehouse, which had been hastily constructed on the river bank.

Dick O'Loane was Taylor and Drury's first manager. His primary project was the construction of two buildings, a store and a warehouse, on property he leased on the waterfront further south along the Stewart River. This spot was very convenient for off-loading from the river steamers. Unfortunately, after construction, it was learned that the Federal Government would not allow the retail area to be on the waterfront lease. O'Loane then purchased the ground across the road from Angus McIntyre and had a frame building hastily erected.

Within the month, the *Thistle* returned fully loaded, having pushed a barge crammed with supplies to meet the increasing needs of the growing community. As more staff was required, Joe Clifton and several part time employees were hired. Keno, some 40 miles of wagon road north, became the centre of activity. Clifton was transferred to open and manage a store there. In 1932 George Capell replaced Clifton. In 1938 Charlie Rosenberg took over, and was joined by his wife, Laura.

Taylor & Drury soon built up a stable family trade. To achieve good service, large tonnages of supplies had to be stockpiled to carry the community over the freeze up period following the closure of navigation. At first, large amounts of hay and oats were required for the horses. Later, Caterpillar tractors took over and pulled long sleigh trains of sacked concentrates to the river bank in Mayo to be stored until navigation reopened.

Taylor & Drury often extended credit to prospectors or miners who were in the process of opening their claims. This policy was followed through the years with few losses sustained. The company felt that it was assisting in the development of the country and found the long-term results rewarding. Taylor & Drury never engaged in grubstaking, or taking shares.

ABOVE, Two dog teams on Front St. outside the Taylor and Drury Ltd. store in Mayo built in 1921. Frank Whitney on left in fur coat.
BELOW, S.S. *Thistle*, bought by Taylor & Drury Ltd. from White Pass in 1921 for the run to their store in Mayo. It sank in Lake Laberge (July 1928) with a full load for Mayo on board. *Both photographs, F. & T. Whitney Collection*

In 1928, roads improved and trucking became the chief mode of transportation. Taylor & Drury opened a General Motors agency with its headquarters in Whitehorse. The sale of Chevrolet cars and trucks was a major step forward for the company and soon entailed running a gasoline filling station.

W.H. Jeffrey, assisted by Mr. Richie, the accountant, took over from O'Loane. Richie was followed by Billy Cathro. In 1930 Bill Taylor, the eldest son of founder Isaac Taylor, took over the books and became manager of the operation, assisted by Jeffrey. In 1932, Charlie Taylor succeeded his brother Bill, who was transferred to the company's head office in Whitehorse. Due to the Depression, the owner's sons were forced to forfeit higher education to join the business. A low-salaried, hard working team ensued that managed to stave off the bankruptcy suffered by many similar operations.

In 1936, Charles Taylor married Betty Maclennan, the eldest of three girls of a pioneer family, and remained in charge of the store until 1942. As Treadwell Yukon closed its operations and thousands of workers poured into Southern Yukon to build the Alaska Highway, Charles was transferred to Whitehorse to help cope with the escalation of sales there.

Lack of customers meant closing the Keno store. Jeffrey and his daughter, Nancy, were able to handle the reduced business in Mayo. Following Jeffrey's death, George Andison took over the store and was assisted by his step-daughter, G. Jean Gordon. The new, all-season highway to Mayo from railhead at Whitehorse, enabled supplies and ore products to be moved daily. United Keno Hill Mines bought out Treadwell Yukon after Wernecke's unfortunate death. Prior to the sale, Treadwell reopened a mine at the rich Elsa deposit with a new mill. This new operation revitalized the area's economy.

The Taylor & Drury Store was then managed by the Earl Graham family, who bought it in 1970.

N.W.M.P. – R.N.W.M.P. – R.C.M.P. *by Linda E. T. MacDonald*

There was not a police force in the territory when the Stewart River was mined in the 1880's, and disputes were settled at miner's meetings. It was not until after the Matlock-Washburn incident at Fortymile (see Matlock, George H., *The Pioneer List*) that the North West Mounted Police were established in the Yukon. They were there when the Klondike was struck. During the rush to the Klondike, a N.W.M.P. post was established in Stewart City at the mouth of the Stewart River.

In 1901 a N.W.M.P. post was established at the mouth of the McQuesten River due to the number of miners gathering on that river and its tributaries. There was also a detachment at Gordon Landing. The N.W.M.P., posted at McQuesten, visited Mayo on a regular basis until the Mayo detachment was established in 1904.

The following men served at McQuesten post from 1901 until it was closed in 1906:

Sgt. J.W. Davis	June 1901–June 1903
Const. G.B. Joy	June 1901–July 1902
Special Const. D.R. Shaw	June 1901–January 1902
Special Const. A. Linklater	June 1901 only
Special Const. T. Shinoda	March 1902–July 1902
Const. T. Matheson	August 1902 only
Const. J. Rowley	August 1902 only
Special Const. E. Borrow	August 1902–May 1903

Const. J. Jones, Keno, 1927. *Ione Christensen Collection*

TOP, Keno R.C.M.P. Detachment, 1925. From left: Cpl. Thurgood, Cpl. Stephens, and Const. Schouard. *Ione Christensen Collection*
MIDDLE, R.C.M.P. dogs, Keno City 1927. *Ione Christensen Collection*
BOTTOM, From right: Gilbert Rich and Sgt. W. J. D. Dempster at the Minto Bridge Roadhouse. The two gentlemen on the left are probably roadhouse owners G. Schogren and F. Chasni. *C. A. Boerner Collection*

Const. A.E. Forrest	September 1902–November 1905
Const. S. Carter	November 1902 only
Staff Sgt. C.A. Corneil	August 1903–October 1905
Const. J.F. (J.B.) McLennan	September 1903–May 1906
Special Const. K.C. Takeda	September 1903–May 1904
Const. G.C. Morton	June 1904–August 1904
Const. J.A. Perry	June 1904–August 1904
Special Const. R. Hager	November 1904–February 1905
Const. W.J.D. Dempster	November 1905–May 1906
Const. J.A. Millard	November 1905–May 1906

(Special constables were local people who lived off the land and were aware of local customs. In those days, they were usually employed as guides, interpreters and dog team handlers.)

The Mayo detachment was opened in 1904, as part of Dawson's "B" Division, with a sergeant and two constables being stationed here. The Gordon Landing detachment was transferred to Mayo when Gordon Landing was abandoned as a point of departure from the Stewart River to Duncan Creek and the road from Mayo to Duncan Creek was built. The Mayo detachment was closed in October, 1906 due to the reduction in strength of "B" Division. The detachment reopened in the winter of 1910 as a result of the numbers of miners and prospectors coming to the Minto, Highet Creek and Dublin Gulch discoveries. The Mayo detachment handled post office duties until a post office was established in the district in 1910.

The men serving the Mayo detachment from 1904 to 1906 were:

Sgt. W.J. Beyts	January 1904–March 1905 and November 1905–March 1906
Const. B. Jackson	January 1904–March 1904

Const. J. Rowley	January 1904–May 1905
Const. C.R. Barker	March 1904–May 1905
Const. F. Corby	December 1904–March 1906
Corp. H.G. Mapley	June 1905–November 1905 and April 1906–July 1906
Const. W.F. Richards	September 1905–May 1906

One case mentioned in the N.W.M.P. 1912 Annual Report, concerns the accidental shooting of Albert E. McKay on October 2, 1912 at Gordon Landing. McKay and Alex Nicol arrived at Gordon Landing after dark intending to remain there for the night. As McKay moved up the bank of the river toward a cabin, a man named Henry McWhorter mistook him for a wolf and shot him. McWhorter had seen several wolves around the cabin earlier. When the dogs started barking, McWhorter called out but McKay did not reply. McWhorter shot into the bushes, hitting him. Nicol and McWhorter took McKay to Mayo but he died three hours after the shooting. McKay and McWhorter had once been partners.

The Mayo detachment remained open through World War I and through the 1920's as part of "B" Division. [In 1914 the North West Mounted Police became the Royal North West Mounted Police and in 1920 the Royal North West Mounted Police were renamed the Royal Canadian Mounted Police.]

In 1922, Mayo became headquarters for a subdistrict. The Mayo Subdistrict was commanded for ten years by Staff Sergeant and later Inspector W.J.D. "Jack" Dempster. Dempster had served at McQuesten Post as a constable. (While a corporal, Dempster made a name for himself when he searched for, and found, the lost Fitzgerald Patrol who perished on the MacPherson-Dawson Patrol

Keno City, 1925. *Ione Christensen Collection*

in 1911. The Dempster Highway was named for him in 1963. See *Pioneer Accounts*) Mayo Subdistrict was in charge of the Keno detachment.

Law enforcement was sometimes difficult in those days of miners, adventurers, and frontiersmen, especially during boom periods. Moonshine and homebrew were readily available, while fist fights and prostitutes supplied the entertainment. In 1925, Constable Gordon Cameron was sent from Vancouver to investigate the alleged bootlegging and prostitution in the district. The miners worked hard and played hard. Others searched for an escape from the long winter nights, as did the man who was arrested for embezzling money from his place of work to cover his panguingue (a card game commonly known as "pan") debts.

Officers were often required to settle disputes over mining claims and property. They patrolled miles in the dead of winter by dog team. They checked on the well-being of isolated prospectors, miners, and trappers, and often assisted with transportation during medical emergencies.

The R.C.M.P. post in Keno was closed in October, 1937. Constable W.E. McElhone, who was stationed there at the time, moved to Mayo to assist Lance Corporal A. Mason-Rooke who was the officer in charge of the Mayo post.

Prior to 1937, the Mayo detachment occupied a building on Centre Street on what is today the location of Stewart Valley Enterprises. Directly across the street was the commanding officer's residence (today, NorthwesTel). In 1936 the detachment quarters were in the frame dwelling known as the Mervyn residence, owned by James H. Mervyn. This house was later sold to Martha Noel Mast and continued to be used by the Force until the early 1950's. Today the house is owned by Margaret and Dick Wallingham.

During the silver boom of the 1950's, the Mayo detachment was increased to five men.

Following is a list of officers who served in Mayo from 1911 to 1912 and from 1918 to 1957:

January 1911–December 1911	Corp. F.H. Thompson
January 1912–March 1912 and September 1912	Const. W. Hocking
April 1912–August 1912	Corp. F.W. Schutz
January 1919	Const. W.C. Waters
December 1919	Corp. W.C. Waters
January 1920	Corp. W.C. Waters
January 1921	Corp. W.C. Waters Const. J.S. Cooley

Mayo Subdistrict created in December of 1921

January 1922	Staff Sgt. W.J.D. Dempster Sgt. C.S. Harper Const. G. Waston
January 1923	Staff Sgt. W.J.D.Dempster Const. H. Dean Const. R.J. Scott
January 1925	Staff Sgt. W.J.D. Dempster Corp. C. Stephens

Silver King claim, 1913. Mark Evans and Ed Burnell, R.C.M.P. Mayo. *M. Rich Collection*

January 1928	Staff Sgt. W.J.D. Dempster
	Const. P.B. Smith
January 1929	Staff Sgt. W.J.D. Dempster
	Const. D. Perks
January 1931	Staff Sgt. W.J.D. Dempster
	Const. D. Perks
March 1931	Inspector W.J.D. Dempster
	Const. D. Perks
June 1931	Inspector W.J.D. Dempster
	Corp. T.G. Coleman
January 1932	Inspector T.B. Caulkin
	Corp. M.P. Delaney
January 1933	Sgt. C.B. Tidd
	Const. L.C. Thomas
January 1934	Sgt. C.B. Tidd
	Corp. D.A. McAskill
	Const. L.C. Thomas
January 1935	Corp. L.F. Fielder
	Const. P.P. Lees
January 1936	Corp. L.F. Fielder
January 1937	Corp. L.F. Fielder
	Const. W.E. McElhone
October 1938	Corp. A. Mason-Rooke
January 1939	Corp. A. Mason-Rooke
January 1941	Corp. W.W. Sutherland
January 1943	Corp. B. de R. D'Easum
January 1945	Const. I.A. Mast
January 1947	Const. I.A. Mast
January 1950	Const. R.R. Johnson
January 1953	Corp. R.R. Johnson
	Const. C.W. Smith
	Const. J.A.L. Foster
	Const. R.D.H. Morrice
	Const. I.R. Smith
January 1957 to	Corp. H.D. Ferguson
January 1960	Const. R.K. Pollard
	Const. W.F. Semenoff
	Const. J.T.A. Stewart
	Const. J.C. Morrison

Shooting at the Chateau Mayo Hotel *by Linda E. T. MacDonald*

In 1946, a dispute over some mining claims ended in tragedy. George Ferguson Potter was shot in his room at Jim Mervyn's Chateau Mayo Hotel, at approximately 11:30 p.m., April 11, by Peter Leonard Nord.

The shooting stemmed from a misunderstanding over some hardrock claims on Johnson Creek which were actually worthless. Pete Nord made the discovery and offered to supply old-timer, and veteran prospector, George Potter information about the prospect for a fee. A deal was struck and Nord and Potter went to Johnson Creek where they sampled the quartz veins and each of the men staked a claim. Nord also blanketed claims for his brother, Carl. Potter had taken an unspecified poisonous substance along on the trip and, while they were at Johnson Creek, Nord's small black dog ate some of it and began a slow painful death. Nord was very upset about his pet and asked Potter to shoot the dog but was told that, if he wanted the dog shot, he had to do it himself.

They brought their samples as far as Highet Creek where they found an old retort and crushed some of the samples in it. They took the samples to assayer Billy Sime in Keno City and Sime assayed the crushed samples. Unfortunately, the retort had been lined with gold from many years of use on Highet Creek and these remnants caused the assay results to be high. Sime discovered the error when he assayed the uncrushed samples but the men had already gone to Mayo. As the telephone lines were not working, Sime wrote Potter to tell him the samples were worthless. Potter did not believe Sime and left Mayo in the middle of the night to stake more claims. The news leaked out and a staking rush ensued before Nord could stake the claims he had blanketed. "Dinky" Mervyn made money hauling people back and forth from Mayo to Johnson Creek with his dogteam.

Both Potter and Nord were seasoned miners and prospectors and yet they overlooked what seemed obvious to others. A number of experienced prospectors, who went to Johnson Creek to stake claims, sampled the Discovery and realized almost immediately that there was no visible gold in the samples. When Ed and Clarence Bleiler and Ed Kunze crushed the samples they had taken from the Discovery, and found not a trace of gold, they knew something was amiss and didn't bother to register the claims they had staked. Other stakers used the retort to crush samples and were excited by the gold they found. On his return, Ed Bleiler retrieved the Middlecoff retort from Ed Kimbel's cabin in Mayo where it had ended up. Unfortunately, the Bleilers and Kunze did not get back to Mayo until late in the evening of April 11.

Potter came back to Mayo, before returning to his home at Minto Bridge, to wait for a wire from his brother Robert "Bob" Potter. Bitterness seized Pete Nord and he became obsessed with the events surrounding the Johnson Creek prospect. Nord began to circulate rumours that Potter was a claim jumper, and because prospectors consider claim jumping absolutely despicable, he suffered considerable verbal abuse from Nord's acquaintances.

The retort used by Potter, Nord and others to crush the ore samples from Johnson Creek. *L. R. Bleiler Collection*

The situation became intolerable and Potter took Nord to the R.C.M.P. barracks on April 7, 1946, where he made a statement clarifying his position. Potter told Const. Mast that Nord was publicly accusing him of claim jumping and was calling him a thief and a sneak. He requested that Const. Mast ask Nord to cease these accusations. Nord interjected saying that he had blanketed the claims and Potter was indeed a sneak, claim jumper, and dog poisoner; and then left the office. Potter remained with Const. Mast to explain how the claims on the new discovery had been staked and stated that he had paid Nord for the information which led to the staking.

On April 9, 1946, Carl Otto Nord, Pete's brother, arrived in Mayo from Whitehorse and got a room at the Chateau Mayo Hotel. When Const. Mast saw Pete and Carl in the Chateau Mayo beer parlour the brothers declared that Potter was a sneak and a claim jumper. Mast warned the pair to cease further name-calling and ordered them to stop molesting Potter, whereupon the brothers heartily agreed to do so. Mast further stated that all questions regarding claim jumping would be settled when Mining Recorder Sam Mason-Wood returned from Dawson City.

On the morning of April 11, 1946, Potter was struck by prospector Ellis Johnson who had heard the Nord brothers' accusations of claim jumping. Const. Mast encountered Potter shortly after the attack and saw blood on the man's face but he assured Mast that everything was alright. Potter went to see Ed Kimbel and told him that he would soon have to move out of the hotel because some of the "boys" were abusing him. Kimbel told him he could stay with him until he was ready to return to Minto Bridge; however, on the evening of April 11, Potter remained at the Chateau Mayo. Other guests at the Chateau Mayo that night were Jack Hawthorne, Patrick McLaughlin, and Carl Nord.

Potter retired to his room early that evening. He fell asleep on his left side, facing the wall, with his back toward the door.

Pete Nord went up to his brother's room and invited Carl to come down to the hotel's tavern for a bottle of beer. The pair went downstairs and sat at a table with Mr. and Mrs. Breaden. Pete left the table at some point but Carl Nord did not notice his brother's absence.

Nord left the tavern and went to his cabin where he loaded a British .303 calibre rifle. He walked back to the Chateau carrying the rifle, ascended the back stairs, and moved down the hallway to Room 20 where he kicked in the locked door, stood in the doorway and shot seventy-year-old George Potter in the back. Potter rose and struggled unsuccessfully with Nord in an attempt to take the gun. Hawthorne, next door in Room 19, heard the commotion and opened his room door in time to see Potter stagger into the hallway and collapse. Nord was standing behind Potter at the head of the stairs and Hawthorne moved in to try and take the gun but Nord was still crazed and said "You keep off of me or you'll get it too." At that moment Jim Mervyn, who had heard the noise from his tavern, arrived at the top of the front stairs and ran to assist Potter who had somehow got back to his bed and was calling for Mervyn to help him.

"He walked back to the Chateau carrying the rifle, ascended the back stairs, and moved down the hallway to Room 20 where he kicked in the locked door, stood in the doorway and shot seventy-year-old George Potter in the back."

Nord ran down the upstairs fire escape, crossed the street, and stood beside the Silver Inn Hotel. At that moment, his brother Carl spotted him through the window of the Chateau. Carl walked over to the Silver Inn with the intention of taking the gun from his brother but Pete said, "I am going to blow my top off." This frightened Carl and he ran back to his room at the Chateau and stayed there.

Nord left the Silver Inn and went back to his cabin with the intention of shooting himself. Still confused, he entered Ben Cameron's cabin thinking it was his. When he realized his mistake he mumbled something to Cameron and then went to his own cabin.

When Mervyn saw that Potter had been shot, he raced downstairs and ran to the R.C.M.P. barracks, but no one was there. He returned to the hotel and went into his store (which was housed in the Chateau) where he saw Bill Cathro. Cathro said Constable Mast was visiting Northern Commercial manager John F. Donald at the manager's residence above the N.C. store, and he volunteered to fetch him. He arrived at Donald's at 11:35 p.m., and he, Cathro and C.E. "Bud" Fisher (who was also evidently visiting at Donald's) went to the hotel.

The frightened Hawthorne left the hotel and went to Kimbel's home, arriving "shortly after 11:00 p.m." When he told Kimbel what had happened, Kimbel told him to lie down and then went to the hotel to see if he could be of any help.

Patrick McLaughlin heard "shots" from his room at the Chateau but did not leave his room immediately. He waited for some time and then went out in the corridor and had moved to the head of the stairway when he noticed Potter standing in the doorway of his room. Potter called out to McLaughlin, "Get the police, I am shot." McLaughlin went downstairs where he met Mervyn and LeClair who said the police were being notified and, almost immediately, Const. Mast, Cathro and Fisher arrived and McLaughlin accompanied the three men to Potter's room.

Const. Mast entered the room alone, then came out again and went to the first-aid room. He returned with bandages and Mast and Fisher began to wrap Potter's wounds. Kimbel arrived at that time and said he would fetch Nurse Reynolds (there was no doctor in Mayo in 1946) who he thought was staying at the Van Cleave residence. When he reached Van Cleaves', he learned Nurse Reynolds lived elsewhere. Kimbel found the nurse's residence but Reynolds had already retired for the night and it took her some time to come to the door. Kimbel relayed his message and then returned to the hotel.

At midnight, following regular electrical company practice, the town was plunged into darkness.

Nurse Reynolds arrived at 12:15 a.m. Potter

was in great pain and the nurse administered 1/4 grain of morphia but it did not lessen the pain and so she gave him another 1/4 grain at 1:20 a.m. At that moment there was a noise downstairs. The hotel was in darkness except for the light in Potter's room and a few flashlights. Const. Mast went to the head of the stairs and saw Pete Nord coming up the stairs toward him. Nord said, "Mast, I want you to hang me, I have shot the wolverine."

Const. Mast, accompanied by Fisher and Kimbel took Nord to the R.C.M.P. barracks. On the way there, Nord said that he had shot Potter and that he had also shot himself. He was taken to a cell and an examination revealed that he had shot himself under the left arm. The wound was not serious but Reynolds was called, at approximately 1:30 a.m., to bandage the wound and administer a sedative to calm him.

McLaughlin remained with Potter while Reynolds was at the barracks and he remained in the room after the nurse returned. At approximately 3:30 a.m., Reynolds went back to the barracks to fill the hot water bottles, leaving McLaughlin with Potter. George Ferguson Potter died at 4:00 a.m. on April 12, 1946, just as Nurse Reynolds returned to his room from the barracks.

Dr. Barrie Potter Duncan was summoned from Dawson to perform an autopsy which he carried out on April 12. Inspector H.H. Cronkhite, in charge of the Whitehorse Sub-Division, arrived to assist Const. Mast with the investigation.

The Coroner's Inquest on April 13, and the Preliminary Hearing [Rex vs. Peter Nord] on April 15, 1946, was held in the I.O.D.E. building in Mayo with Sam Mason-Wood administering the proceedings. Jurors selected for the Preliminary Hearing were: J. Bonner Kunze, John F. Donald, H.E. Miller, Wilfred R. Gordon, Rupert S. Steeves, and Clifford E. Fisher.

Witnesses called at both the Coroner's Inquest and the Preliminary Hearing were Dr. Barrie Duncan, Const. Ivor A. Mast, Phoebe E. Reynolds, Patrick McLaughlin, John Hawthorne, James H. Mervyn, Carl Otto Nord, and Edward Kimbel. Additional witnesses called at the Preliminary Hearing were Inspector Cronkhite and Ben Cameron.

Each of the witnesses told what he knew about the events leading up to, and including, the night of April 11, 1946. (Discrepancies appear in the testimony of witnesses regarding such matters as the time the shooting occurred, the number of shots that were fired, and the time at which persons appeared on the scene following the shooting. George Potter's room number, for example, is referred to as both 2 and 20 in the transcripts.) Jurors concluded that Peter Leonard Nord did unlawfully murder George Ferguson Potter on April 12, 1946.

When the case came to trial in Dawson, Nord was formally charged with manslaughter and sentenced to ten years in jail.

The first R.C.M.P. Commanders Office, built by Isaac Mallette in the early 1920's. The building was located on the site of the present [1989] NorthwesTel building. *R.C.M.P. Archives Collection, Ottawa*

THE ANGLICAN CHURCH

The Stewart River-Fraser Falls district was originally assigned to Anglican clergy based in Rampart House. Work in this district really began in 1892 under the Anglican Church Camp Mission. Archdeacon Canham and later the Rev. Canon B. Totty visited the district on their rounds until 1914 when J.H. Brown, a licensed layreader under the Church Camp Mission, held services in the Mayo schoolhouse and at Minto Bridge.

In 1915, Anglican Catechist Julius Kendi arrived at Fraser Falls, from the Head of Porcupine-Peel Rivers, to begin missionary work in the district. He was accompanied by Rev. Frank Buck.

Many of the Mayo Indian people were at a fish camp at the falls at the time and some of them were baptized by the missionaries when they arrived. Kendi went to Mayo and asked the Indian people to select a site where they could establish their own town. A location was selected two miles down the Stewart River from Mayo (the site is now referred to as the "Old Village"). It was on this site that St. Mark's, the first Anglican Church (and rectory) in the district was built. Rev. Kendi, the first resident minister, was ordained deacon at a ceremony at the Bishop Bompas Memorial Church in Moosehide on July 28, 1918.

Residents of Mayo went to Rev. Kendi's church in the village for baptisms and marriages until a church was built in Mayo in 1922.

The original church at the village was destroyed by the 1936 flood and rebuilt a little further from the river bank. It was in use until 1958.

In 1915, the Rev. Frank H. Buck was ordained deacon in Vancouver and licensed as a missionary in the district of the Mayo and Klondike creeks from 1916 to 1924. In 1916, he joined the Yukon Infantry Company, as a private under Captain George Black, and went overseas in 1917. He was promoted to captain and appointed Chaplain of the 46th Battalion. He served with distinction and received the Military Cross. He returned to Mayo in the summer of 1920.

St. Mary's Anglican Church and rectory in 1922.
Mabel McIntyre Collection

In the summer of 1921, lots in Mayo were bought for the future church and rectory by Bishop Stringer, and services were held in the Pioneer Hall and at various mines by clergy from Dawson. The following summer, Buck returned and the church and rectory were built. The buildings were constructed from local lumber milled at Lefebvre and Binet's sawmill. The church was built at a cost of $3,680, and the rectory $2,555. The first service at the new St. Mary's Anglican Church was held on October 15, 1922.

In the late 1920's, a furnace was installed in both the rectory and the church, a garage was built, and the grounds were fenced and landscaped. Gifts presented to the church included a unique font, built by Mrs. Brame of Duncan Creek, from a burl of a pine tree, hollowed out, and placed on a stand. The brass bowl in the font was given by the George Andison family in memory of Mrs. Andison. The brass cross on the Holy Table was given by Mrs. McCullum, wife of Rev. A.C. McCullum, in appreciation for a Diocesan Life Membership in St. Mary's W.A.

In 1922, the Women's Auxiliary (W.A.) was started with Mrs. Binet as president and Mrs.

Buck, vice-president. Their first bazaar netted $400, a princely sum in 1922.

In 1936, during the Rev. W. Valentine's incumbency, Mayo suffered a disastrous flood during spring run-off. The only buildings that remained high and dry were the Roman Catholic Church, the liquor store, St. Mary's Church, and the school and library. A steamboat and other craft were tied up to St. Mary's and it was used to accommodate refugees from the flood.

In 1946, Rev. G.N.W. Wareham arrived in Mayo. He designed and built a community hall, known as the Wareham Hall. Funds were raised by borrowing from local business firms, obtaining a grant and loan from the diocesan building committee, and having labour and materials donated. After the hall was completed, it was used for recreation such as basketball, roller skating, and moving pictures. Proceeds from these activities left the hall debt free.

In 1953, during Archdeacon Kirksey's term, oak pews were obtained from the Presbyterian Church in Dawson, which was no longer in use. That same year, UKHM offered to build a church in Elsa if the Anglican Church would guarantee regular services there. In 1953, Archdeacon Kirksey began services, followed by a student, Fred Lapham, who was employed at the mine. These services were held in the school until the church was built. St. Peter's Church was completed during the Rev. R.N. Alcock's incumbency and consecrated by Bishop Greenwood in 1956.

When it became apparent, in 1959, that the Mayo rectory was beyond repair, UKHM built a rectory in Elsa under the same generous terms that the church had been built. Since that time, the rector has lived in Elsa, coming to Mayo for services and visitations.

All Saints Church in Keno City is also part of the parish of Mayo. Under the leadership of the Rev. R.N. Alcock, ground was secured and a small building was purchased from UKHM and moved onto the lot. It was adapted for church use and furnished through the efforts of the rector and residents of Keno City. All Saints Church was consecrated by Bishop Greenwood in 1956.

When the Indian people moved from the "Old Village" back to Mayo in 1958, the Anglican Church in Mayo became St. Mary's with St. Mark's.

A view of the McLaren house far left and St. Mary's Anglican Church (centre) and Rectory (right) during the Mayo spring flood of 1936. The water rose to the knoll of the Anglican Church during this period. Boats are beached on what was formerly dry land. *W. S. Hare photograph, E. C. Bleiler Collection*

St. Mary's Sunday School, 1949. From left, front row: Leslie Somerton, Francis Morberg, Mary Jean Boyle, Bobby Andison, Baxter Wareham, Michael Kerr, Sybil Bleiler, Lowell Bleiler. Second row: Johnny Jenson, Henry Arthur, Mary Profeit, Betty Gordon, Hope Mast, Marie Fisher, Norman Morberg. Third row: Mary Andison, Patsy Fisher, Margaret Mason-Wood, Larry Profeit. Back row: Anne Kerr, Nancy Whitney, Ila Fisher, Rev. Norman Wareham, Eileen Mason-Wood, Adelia Besner, Ruth Batty. *G. A. McIntyre Collection*

The following ministers served in the district to 1957:

Stewart River-Fraser Falls

Totty, Rev. B.	1893 (based in Rampart House)
Canham, Rev. T.H.	1894–1911 (based in Rampart House)
Tizya, J.	1911–1919 (based in Rampart House)
Buck, Rev. F.H.	1916–1924
Notik, Rev. G.	1930–1933
Martin, Rev. G.	1930–1933
Martin, Rev. J.	1935–1936, 1938

Left to right: Rev. Valentine & wife Mary, Bishop Geddes with Anne Geddes, Mrs McCullum (rear), boy David Geddes, Mrs. Geddes, Nancy and brother John Whitney. *F.&T. Whitney Collection*

The "Old Village"
(All Resident Ministers)

Kendi, Rev. J.	1915, 1919–1928 and 1941–1946
Martin, Rev. J.	1929, 1935
Connor, Miss L.	1950–1952
Hellaby, Deaconess H.	1952–1959

Mayo, Keno, Elsa
(All Resident Ministers)

Buck, Rev. F.H.	1920–1924
Leigh, Rev. G.F.	1925–1928
McCullum, Rev. A.C.	1929–1934
Hughes, Rev. O.W.R.	1935–1936
Valentine, Rev. W.	1937–1938
Boyd, Rev. R.S.	1938–1940
Vance, Rev. J.R.B.	1941
Wareham, Rev. G.W.N.	1946–1951
Kirksey, Rev. C.W.P.	1953
Alcock, Rev. R.N.	1954–1957

The original St. Mary's Church in Mayo has been well-maintained and is still in use today. The church looks much as it did when it was first built.

St. Mark's, although in a state of disrepair, still stands at the abandoned "Old Village."

THE CATHOLIC CHURCH *by Father Henk Huijbers*

CHRIST THE KING MISSION

Father Rivet and Mr. Tremblay of Dawson City demolished the old St. Mary's Church at Dawson City (which was built by Father S.J. Judge in 1898 near the Sister's Hospital) in order to get materials to build a church in Mayo. The materials were taken to Mayo by boat.

The church in Mayo was built by the old-timers and was opened on Sunday, November 3, 1923, by Reverend Bishop Bunoz.

In this extract from *The Cross in the Rockies*, by Lechevallier OMI, the details surrounding this undertaking are sketched out:

Christ the King Mission in Mayo surrounded by flood waters in June 1936. *G. A. McIntyre photograph*

On October 7, 1922, Rev. Father Rivet OMI arrived in Dawson City as assistant to Father Gagne OMI, who had been alone since the departure of Father Plamandon OMI.

On November 29th, he left to visit the Mayo district some 150 miles east. He slept and ate outside, but accepted these sacrifices generously to be able to bring the consolation of the religious rites of the Christmas season to a handful of Catholics. After having been on the trail for some 50 days, Father Rivet returned, after his difficult trip, convinced that it was a little premature to think about building a church at Mayo and to station a permanent priest there. Some 31 persons assisted at Mass, and the cost of materials and transportation to Mayo and the camp was far too high. Therefore, he intended to wait for the Bishop's visit.

But during the months following this, someone solved the question of a church for Mayo. When the district at Mayo developed enormously, without a Catholic Church, it was noticed that the town of Dawson (with a day to day declining population) had the luxury of having two churches. The oldest one was outside the town by the hospital and was only used occasionally. So, Father Gagne thought if the building could be taken down carefully, it would take a load off Dawson's shoulders, and it would give Mayo enough materials to build a large chapel there according to the needs of the population. So said; so done.

Father Henk Huijbers celebrating Mass at a mining camp near Keno, 1948. Middle row: Barry O'Neill, John R. O'Neill, Et'tsì ts'ò. Bottom: Miles Flynn. *Herb Foster photograph, O'Neill Collection*

When the most Reverend Bishop Bunoz OMI arrived in Dawson on July 24, 1923, on a pastoral visit, he was very disturbed to see the building taken down and called it an act of "vandalism." "This church," he said, "was built with the founding of the Town of Dawson, and had seen many sorrows and joys, many anxieties and successes; and is the place where the remains of the good Father and benefactor William Judge, its founder, now rest. This church should have had the dignity to be kept as a historic monument of great value for posterity." The grave of the good missionary Father Judge still remained in the same place, but was now very isolated.

Before leaving Dawson on August 22nd, Bishop Bunoz OMI appointed Father Rivet OMI, as pastor of St. Mary's parish in Dawson downtown, and Father Gagne OMI as chaplain for the Sisters and the hospital.

Following is a list of priests who have served at Christ the King Mission in Mayo to 1958:

From 1920	Father Rivet OMI–travelled to Mayo from Dawson
1930–1934	Father Leray OMI–travelled to Mayo from Dawson
1937–1940	Father A. Monnet OMI–first resident priest in Mayo
1940–1942	Father Leo Bosset OMI
1942–1948	Father J. Plaine OMI
1948–1952	Father Henk Huijbers OMI
1952–1953	Father E. Turenne OMI and Father F. Van Rooy OMI
1953–1954	Father E. Turenne OMI and Father O. Pauwels OMI
1954–1958	Father E. Turenne OMI

MAYO SCHOOL

by Linda E. T. MacDonald, Gordon McIntyre, & Suzanne Laberge

In July of 1913, George Cunningham wrote to the Honourable George Black asking if it would be possible to establish a school in Mayo. Cunningham stated that a school in Mayo would enable him to bring his two children over from Scotland.

Black agreed to the establishment of an Assisted School. An Assisted School meant that there must be a minimum enrollment of seven children, the community had to provide the classroom, furniture, heat and light, and the government would hire and pay the teacher (on a month-to-month basis) and send some basic school equipment and supplies.

Alex Profeit was hired as teacher and the Mayo Landing School opened in October of 1913 with seven pupils registered. These first pupils were the three Kelly children, Phyllis, Gerald and Brian; the three Ross children, Mary, Lena and Nora; and Olive Kinsey. Olive Kinsey did not come in from the creek until late March or early April and by that time the Ross children had gone back to Minto Lake.

To coincide with the placer mining season, the school was only open for six months a year for the first three years (1913–14, 1914–15, and 1915–16).

Profeit taught for the first two years and then Emma Allen was hired in October, 1915, at a monthly salary of $125. The major operating cost of the school, other than the teacher's salary was cordwood to heat the school. Wood was $10 to $11 per cord at the time. In 1918, parents and guardians were asked to pay $2 per family per school month to help cover costs.

ABOVE, Mayo School. *Flora Middlecoff photograph, Middlecoff-Bleiler Collection*

BELOW, Mayo School 1919. Back row, left: Kathleen McLaren, Florence Cunningham, Suzzane Turgeon, Dorothy McLaren (teacher), and Lillian McLaren. Middle row: Josephine Cantin, Nora McLaren, Maggie Cunningham, John Windsor, Hilma Miller, and Henry Miller. Seated: Rose Turgeon, Annette Cantin, and Jock Cunningham. *D. Walloch (McLaren) Collection*

The Mayo Landing School was for non-native and metis children only. Indian people were a federal responsibility and their education was in the hands of the Church until the end of the 1950's, at which time the Federal Government contracted the Yukon Government to educate them. (See *The Na-Cho Nyäk Dun*).

The first school building was located near a knoll at the end of Duncan Avenue on Lot 1, Block 12. In those days, Duncan Avenue ended at Montreal Street. Next to the school building was the office of J.M.E. Henry who was the Agent for the Minto Bridge Mining Recorder. A short distance down the hill from the school was a small cafe, possibly run by Mrs. McCloskey.

During Dorothy McLaren's tenure (from 1918 to 1920), school was held in Alex Nicol's first cabin. Classes were conducted in the living-room of the house. It is probable that this was the same building in which the school opened at the end of Duncan Avenue. Among McLaren's pupils in 1918 were Henry and Hilma Miller,

Suzzane and Rose Turgeon, Florence and Maggie Cunningham, Nora McLaren and John Meyers.

In September of 1919, the school was moved to a cabin belonging to Frank Cantin. In 1920, Cantin began to charge a monthly rental of $25 for the building. That fall, the supply of coal oil destined for Mayo had been left by the boat at McQuesten River. With no coal oil to light the classroom, Mr. Letourneau agreed to furnish electric lights for the school if the government would supply the transmission wire.

The Mayo Landing School Board of Trustees, (Mrs. Alex Nicol, D. A. Matheson, and J. E. Ferrell) asked the Gold Commissioner to secure a government school for Mayo and to establish high school classes in 1920. However, it was not until 1923 that the government agreed to pay for high school instruction. That year, Rev. Frank Buck was contracted to instruct the four students

Mayo School. *William Hare Collection, Yukon Archives*

Mayo school children and their teacher, 1923-24. Teacher Dorothy Hoggan is standing on the left of the school house door and Commissioner George MacKenzie and his wife are standing in front of the door. Some of the children in the group are Annie Churchward, Henry Miller, Yvonne Besner, Emelia Besner, and Hilma Miller. Rose Turgeon is standing third from the left in the middle row. In the front row, the fifth girl from the left is Isabel Profeit and the seventh girl from the left is Ella Profeit. *William Hare Collection, Yukon Archives*

Top, School in Mayo 1917–1918. Back row from left: Ethel and Evelyn Middlecoff, Suzzane Turgeon, Edith Murphy (teacher). Front row: Cunningham girl, Annette and Josephine Cantin, Jock Cunningham, his sister behind him, Rose Turgeon, Elmer Middlecoff Jr., and Ted Sunderland. *Middlecoff-Bleiler Collection*
Middle, Students displaying Valentines, 1938. *McIntyre Collection*
Bottom, Mayo School children, 1941. *G. A. McIntyre Collection*

(Annie Churchward, May Gillespie, Henry Miller and Nora McLaren) who were ready for high school. The students met in a small room to the right of the sanctuary and altar at St. Mary's Anglican Church and the parents of the children each paid $6.25 per month towards tuition. Several high school students attended school in Dawson, but not all parents of high school students could afford room and board for their children in Dawson.

In 1921, the government entered into a lease agreement for a school with Cantin for use of the building on Lot 1, Block 16, at the corner of 4th Avenue and Centre Street. This building was used as the school until 1941.

The Mayo Landing School ceased to be an Assisted School in 1922 when the town was assessed for the first time and land taxes applied.

In May, 1929, the Ex-Service Club started work on a playground for school students.

Gordon McIntyre taught school in Mayo from 1932 until 1941. He has written the following account of his teaching days.

> I arrived in Mayo in August 1932, just before the Discovery Day celebrations on August 17th. School began the following Monday.
>
> The school was located on the site of what is now the North Star Motel (formerly Tim-O-Lou Motel). It was an old log residence that had had all the partitions removed. The walls, which had been broad-axed on the inside, were painted white. There were windows on the south and east sides, green chalk boards on the west and north. There were four drop lights with bare 100 watt bulbs. In the centre was a wood stove made out of an oil drum. There were three doors, one between each pair of windows, and one in the west end of the north wall. The latter door led into a frame lean-to addition to the school. This was used by the Imperial Order of the Daughters of the Empire (I.O.D.E.) as a lending library. There were coat hooks on the east wall.

On the right of the south entrance was a small shelf equipped with a galvanized water can, a dipper, and a hand basin. This provided both drinking and washing water.

On the north wall, in the east corner, was a rough set of book shelves which contained the school supplies and library. The supplies consisted of some white chalk, a few sticks of coloured chalk, six chalk erasers, ink powder, and a small supply of construction paper. The books were used and out-of-date textbooks, Canadian and British histories, geographies, and readers.

The desks were old ones from the Dawson school, about half were double desks. The doubles were for the lower grades and were to the north of the stove, the single desks were for the upper grades and were to the south of the stove. They faced west. In front of the desks was an old-fashioned teacher's desk with hinged lid, duly equipped with a handbell and a strap made from a piece of harness trace.

When school opened in August 1932, there were thirteen children present, but this was soon increased to eighteen by the addition of the three Briba and two

Mayo School children and their teacher Gordon McIntyre, 1933–34. Left to right, back row: Alex Hare, Charles Profeit, Vera Breaden, Doris Fisher, Gordon McIntyre, John Whitney, Joyce Hartnell, Kippy Fisher. Middle row: Doreen Lee, unknown, Bessie Profeit, Henry Breaden, Bill Jeffrey. Front row: Gail Kimbel, Joan Curry, Annie Close, Neil Keobke, Mildred Besner, Yvonne Besner, Marjory Nagano, Edward Besner. *G. A. McIntyre Collection*

Mayo school children and teacher Gordon McIntyre, 1932–33. Front row, from left: Nancy Whitney, Ruth MacDonald, Edmee Briba, Vera Breaden, Charlie Profeit, Earl Briba, Kippy Fisher, Philip Briba, John Whitney (with visor). Back row: Doris Fisher, Maisie Profeit, Vera Maclennan, Alice Maclennan, Audrey Breaden, Gordon McIntyre, Ian Alcroft, Jimmy Profeit, Norman Hartnell, Billy Profeit. *G. A. McIntyre Collection*

Whitney children from Wernecke. These families moved to Mayo because the Treadwell Yukon mine on Keno Hill had shut down and everyone had to leave the company town except the watchman, Jack Faulkner, (Victoria Faulkner's father). Apart from these two families, there were two Fishers, four Profeits, two Maclennans, two Breadens, Ruth MacDonald, Ian Alcroft, and Norman Hartnell.

Shortly after school started, the councillor for the Mayo District, Thomas McKay, asked me if I could teach four teenage girls who were anxious to further their education. Since I had completed a commercial course just the year before, I offered to teach them bookkeeping, shorthand and typing. Mr. McKay supplied one typewriter and I supplied a second one. The four commercial students were: Pearl Kimbel (Kazinsky), Flora Middlecoff (Bleiler), Margaret Mervyn (Wood), and Betty Maclennan (Taylor). This meant teaching Grades I to VIII from 9:00 a.m. to 3:00 p.m. and the commercial class from 3:00 to 5:00 p.m. This was the first vocational course offered by the Yukon school system. Other students who took this course were Vera Maclennan (Dunnett) and Alice Maclennan (Smith).

In 1932 the Alberta curriculum was in use in Yukon, but in 1934 Superintendent of Schools, Harry Menzies was replaced by a Mr. Paterson from British Columbia, and under his direction Yukon changed over to the British Columbia curriculum. The school year, which had been from the first Monday after Discovery Day to the second Friday in June, now changed to the first Monday after Labour Day and ended the last Friday in June. Now the beginners learned to read about Dick and Jane instead of Tom Tinker and Betty Pringle.

There was little provision for recreation during those years. Emil Forrest

TOP, Mayo school children 1940-41. Left to right, back row: Ruth MacDonald, Doreen Lee, unknown, Henry Breaden, Dick Mervyn, Kippy Fisher, Gordon McIntyre, Bessie Profeit, Billy Jeffrey, Edward Besner, Yvonne Besner, Neil Keobke. Middle row, kneeling: Norma Palmer, Joan Curry, Alex Mervyn, Dick Gillespie, Annie Close, Margie Nagano, unknown, Robert Close, George Nagano. Front row, seated: Joanne Keobke, Keith Gillespie, Mildred Besner, Dorothy Profeit, Gail Kimbel, Ila Fisher, Thelma Jeffrey. *G. A. McIntyre Collection*

MIDDLE, Mayo school Christmas concert staged in the Pioneer Hall, 1933. *G. A. McIntyre Collection*

BOTTOM, Mayo school children standing in front of the Mayo school with teacher Gordon McIntyre in September, 1940. *G. A. McIntyre Collection* and friends put some homemade playground equipment in a vacant lot west of the school, but the main playground was the street intersection in front of the school. That is where we played softball, tag, circle dodge ball, and other games needing space. We also played Andy-Eye-Over either over the Catholic Church or the big wood pile in front of Cantin's house north of the school, Run Sheep Run within a block's radius, and cap ball on the sidewalk.

Apart from the usual holidays there were two days which were observed. If the ice went out during school hours, or if the first boat arrived, we would break off classes and go down to the White Pass and watch. In addition, if the temperature was to fall below –50°F the school was closed, and that happened every winter.

With the approach of my first Christmas I found out that the school was expected to put on a Christmas concert on the night of the community Christmas tree. I must admit that my first concert was not all that good, but the rest were great, not on my account but because of the help of two people, Mrs. Alex McCarter and Mrs. Millar. Of course there were others who helped, but these two women in particular literally took over the production. All I had to do was look after the carols. During the first year admission

Mayo Commercial Education class, 1933. From left: Pearl Kimbel, Margaret Mervyn, Betty Maclennan, Flora Middlecoff. *G. A. McIntyre Collection*

Mayo school children and teacher Gordon McIntyre, 1936–37. Left to right, back row: Alex Hare, Charles Profeit, Vera Breaden, Joyce Besner, Gordon McIntyre, Mary Forbes, John Whitney, Nancy Whitney, Joyce Hartnell, Bill Jeffrey. Front row: Edward Besner, Henry Breaden, Betty Hare, Bessie Profeit, Bobby Gallagher, Yvonne Besner, Annie Close, Doreen Lee. *G. A. McIntyre Collection*

was free, but in order to get funds to buy books for the library, I decided to have the concert separate from the Christmas tree and charge an admission to raise money for library books. We raised enough money to buy a set of *Encyclopaedia Britannica* Junior and other books for the library; but the people did not like the idea of having the concert and "tree" separate, so we went back to the original combined "tree" and concert together.

As the years went by, the school population gradually increased, the big impetus being the opening of the Elsa Mine on Galena Hill in 1935 by the Treadwell Yukon Corporation. By 1941 there were over forty children in attendance—still only one teacher for the eight grades. Perhaps it was to escape this that I joined the Army in 1941, returning after five years but never to teach again.

When I look back on the nine years that I taught school in Mayo, my fondest memories are of all those wonderful children, and their parents, who made my job workable. The children were respectful with no discipline problems and the parents were supportive and cooperative. A teacher could ask for no more.

In May of 1941 the Mayo community petitioned for a new school building. Both Ed Kimbel and Frank Cantin submitted bids on the building construction. Cantin, the successful bidder, built the school and rented it to the government. This building remained the school until 1959. It is now known as the Masonic Lodge and still stands in Mayo on Lot 6, Block 30. Mrs. Alice Carthum began classes in the new school on September 29, 1941.

Students and teachers on the steps of Wareham Hall. 1949–50. Front row from left: Frances Morberg, Alan Gatey, Leslie Somerton, Albert Pelland, Mary Jean Boyle. Second row: Marie Fisher, Mike McGeachy, Bobbie Andison, Frank Taylor, Lowell Bleiler. Third row: Johnny Jensen, Fenton Williams, Henry Arthur, Malcolm Smith. Fourth row: Mary Andison, Catherine "Cathie" Germaine, Clara Besner, Betty Gordon. Fifth row: Patsy Fisher, Margaret Mason-Wood, Ila Fisher, Eileen Mason-Wood, Mrs. Paula Burtt (teacher). Sixth row: John L. Burtt (Principal), Art Wajski, Jacquie Foley, John Ewing, Adelia Besner, David Andison, Jeanie Nagano, Jerry Somerton. *G. A. McIntyre Collection*

In August 1941, Mrs. Livingston Wernecke sold her piano to the Mayo school for $50.

During December, 1942, lack of heat in the school due to janitorial problems forced the teacher and children to wear overcoats in the classroom to stay warm. The following January, 16 of the 18 students contracted whooping cough and the school was closed until the epidemic passed. When the school reopened, janitorial problems continued. The light supply was also a problem. The light plant powered the electric lights only until noon so afternoon classes were lit with Coleman lamps borrowed from the Road Department.

In 1943, Cantin built a new outdoor toilet for the girls. It was 6 ft. by 8 ft. and equipped with a small wood heater. The boys continued to use the old, unheated variety.

Mayo school children, 1935. From left: Henry Breaden, Dick Mervyn, Kippy Fisher, Nora Mervyn, Nancy Whitney, Norman Mervyn, Charles Profeit. *G. A. McIntyre Collection*

In 1947, the government set the teacher's salary at $2,700 per annum and bought the school building and land from Cantin for $5,000.

The teaching staff was increased from one position to two in 1948. That summer the old Mining Recorder's office was rented from Mabel McIntyre for $25 a month for use as a high school classroom. Ruth Batty was appointed principal of the Mayo school at a salary of $3,600 per annum, and her sister Margaret was hired to teach the junior room at a salary of $3,000 per year.

In 1948, the Polar Bear Club was established by Ruth Batty and the students. The first executive of the Polar Bear Club was: Jerry Somerton, President; Delia Besner, Vice-President; Ila Fisher, Secretary-Treasurer; and Margaret Batty, Staff Advisor. In 1949, the Polar Bear Club chose school colours of cherry and grey and a gramophone and records were purchased for use at school dances.

In 1949, the first *Aurora* yearbook was published by J. L. Burtt. *The Aurora* has been published, in various forms, almost continuously since that date. Finally, in 1949, the school was equipped with a septic tank, water well, and indoor plumbing.

At the end of the 1948–49 school year Batty resigned her teaching position to open Ruth's Coffee Shop and Mr. and Mrs. J.L. Burtt were hired to fill the two teaching positions commencing the 1949–50 school year.

The 24th of May parade (Queen Victoria's Birthday) was a popular annual event. In 1950 Miss Dougall and some of her students from the Old Village participated in the parade. The I.O.D.E. supplied ice cream and chocolate bars during the celebration.

In 1954, a Panabode school was erected just behind the location of the present Yukon Government Administration Building. It was a T-shaped building with a teacherage in the back and the washrooms and two classrooms across the front. The primary grades remained in the original frame building (later the Masonic Lodge) and for a time attended school in the former Treadwell Yukon office which was later the Baptist Mission building, (the green building located on Lots 1 and 2, Block 2, until 1989).

In 1956–57, a new two-storey school was built on the site (present Yukon Government Administration Building). The frame buildings were moved off the property and the Panabode was moved to the back of the lot and converted into a teacherage with three separate living quarters. The new school had two regular classrooms, an industrial arts shop, home economics room, art room, science lab, library, washrooms and a gymnasium. In 1963, the gymnasium was converted

ABOVE, Norman Hartnell and Jimmy Profeit playing ball in front of the Mayo school, 1935. *G. A. McIntyre Collection*
BELOW, Pupils sitting at their desks in the Mayo school, 1935. *G. A. McIntyre Collection*

into four new classrooms and a new gymnasium was built. Kindergarten was offered for the first time in 1968 with Eileen Shilleto as Kindergarten teacher and Hazel Baker assisting. In 1973–74, additional classroom space was required and portable classrooms were added for primary students.

This school burned to the ground on April 10, 1975. Only the new gymnasium and portable classrooms were saved. The fire caused a loss of only two days of instruction. The students who were attending the portable classrooms did not miss any school days. Temporary quarters for schooling were arranged in the Catholic Church, community hall, curling rink, the basement of the group home and, once again, Frank Cantin's building (present Masonic Lodge).

During the summer of 1975, the current school site was secured and a new school was constructed of Atco prefab trailers. The school has eleven regular classrooms, a fully equipped home economics room, science lab, art room (complete with dark room facilities for photography), industrial arts shop, library, shower rooms, nurse's room, storage and maintenance rooms. The school was officially named the J.V. Clark School in honour of Dr. Clark, Mayo's longest serving resident medical practitioner.

In 1977–78, the gymnasium from the old site was moved up the street and attached to the new J.V. Clark School.

LIST OF MAYO SCHOOL TEACHERS
1913 to 1957

1913–14	Alex Profeit
1914–15	Alex Profeit
1915–16	Emma Allen
1916–17	Florence McLaren
1917–18	Edith Murphy (Married name Mrs. Murphy Reid)
1918–19	Dorothy McLaren
1919–20	Dorothy McLaren
1920–21	Dorothy Hoggan
1921–22	Dorothy Hoggan
1922–23	Dorothy Hoggan
1923–24	Dorothy Hoggan Rev. Frank Buck (High school, part-time)
1924–25	Laura Gillespie Rev. Frank Buck (High school, part-time)
1925–26	Laura Gillespie (first term) Sybil Martin (second term substitute) Lieut. Harold Taber (High school, part-time)
1926–27	Laura Gillespie Rev. G.F. Leigh (High school, part-time)
1927–28	Laura Gillespie Rev. G.F. Leigh (High school, part-time)
1928–29	Vera Gillespie
1929–30	Vera Gillespie
1930–31	Noel Manuel
1931–32	Noel Manuel

1932–41	Gordon McIntyre (Mr. McIntyre also taught part-time high school Commercial courses)
1941–42	Mrs. Con Carthum (née Alice Fisher)
1942–43	Mrs. Con Carthum
1943–44	Mr. H.V. Hildidge
1944–45	Mr. H.V. Hildidge
1945–46	Josine Stone
1946–47	Ruth A. Batty (Married name Mrs. Gordon McIntyre)
1947–48	Ruth A. Batty
1948–49	Ruth A. Batty – High School Margaret Batty – Elementary
1949–50	John L. Burtt – High School Paula Burtt – Elementary
1950–51	John L. Burtt – High School Paula Burtt – Elementary
1951–52	John L. Burtt – High School Paula Burtt – Elementary
1952–53	John L. Burtt – High School Paula Burtt – Primary Doris M. Batty – Intermediate
1953–54	Ruth Kerbs – Primary Mae Reilkoff – Intermediate Don Machan – Grades 9—11
1954–55	Ruth Kerbs – Primary Marilyn White – Intermediate Don Machan – Grades 8—12
1955–56	M. Doerksen –Primary M.J. Nicholson – Intermediate Ben R. Scambler – Grades 7—12

MAYO GENERAL HOSPITAL *by Linda E. T. MacDonald*

With two major mining companies, Keno Hill Limited and Treadwell Yukon Company Limited, operating in the district and the population surge brought about by their activities, the need for a district hospital became evident.

Patients at the old hospital in Mayo with nurse Alice (née Hryciw) Boyes in 1954. Eddie Kimbel has a neck brace on. *J. & A. Boyes Collection*

In July of 1922, it was decided by the Territorial Council that the Samaritan Hospital in Dawson would be dismantled and the salvaged materials be used to build a hospital in Mayo. The Council voted $4,000 toward the hospital construction project.

On July 25, 1922, a meeting was held at the Pioneer Hall in Mayo to appoint a hospital location committee and a committee to administer the finances. Appointed to the hospital location committee was C.A. Boerner, Livingston Wernecke, Oscar Letourneau, Jim Ferrell, and Frank Short. The finance committee appointments were Isaac Taylor, Gene Binet, Dr. Randy Maclennan, William Douglas, and Alex Lesperance.

The hospital location committee arranged to purchase four acres, near the Mayo River, from Joseph Cantin for $150 per acre. A roadway access was also required and Cantin agreed to deed to the hospital a strip of ground 115 feet wide at the lower end of his property line, nearest the Mayo River. At the suggestion of George P. MacKenzie, Gold Commissioner, an additional plot of ground, which lay between the Cantin homestead and the Mayo River, was purchased from B.C. Sprague to enhance the scenic aspect of the hospital location. It was necessary to pay John Krantz for the cabin he had built on Sprague's property.

The materials salvaged from the Samaritan Hospital totalled 60 tons and they were brought to Mayo on the steamer *Nasutlin* and barge. In a spirit of generosity, the White Pass & Yukon Route agent wrote to the Gold Commissioner outlining the following freight rate reduction:

> In view of the fact that a portion of the funds necessary to carry on the hospital work is to be raised by subscription, this company is willing to contribute thereto in the way of reduction of the charges for carrying the freight to the extent of 25%. We would prefer it to be looked upon as a donation from this company of approximately $600, towards the erection of the hospital.

"Hospital occupancy rates were fixed by the Gold Commissioner at $5.00 per day for ward patients and $7.50 per day for private rooms."

The term "subscription" referred to hospital funds received from the mining companies and the general public.

The hospital was constructed under the supervision of James H. McNeill, YTG Superintendent of Works and Buildings, and Gideon Letourneau who was Foreman of Construction. It was to be 24 feet by 64 feet with accommodation for 12 to 15 beds, just as the Samaritan Hospital in Dawson had been. It was necessary to supplement the materials salvaged from the Samaritan with other second hand materials in order to construct a building of that size. Construction workers received $1.75 per hour for foreman of construction, $1.25 per hour for carpenters, and $1.00 per hour (or 63 cents per hour with room and board) for labourers.

A Delco lighting plant was purchased for the hospital and Howard Sutherland, from Dawson City, installed the heating plant. The fireman of the steamer *Keno*, under Chief Engineer John Scotland, cleaned the boiler without charge. A Laundrette Electric washing machine and mangle (pressing machine) was also purchased for the hospital. A sewer passage was dug in the slough of running water at the back of the hospital.

On September 3, 1922, a public meeting was called by George MacKenzie, Gold Commissioner, to appoint a five member board to oversee hospital affairs. The members appointed by Mr. MacKenzie were Sgt. W.J.D. "Jack" Dempster, position of Secretary; Dr. W.W. Chipman, J.A. Fairborn, Oscar Letourneau, and Rodolph Rosmusen. One member of the board was to represent Keno Hill Limited and Treadwell Yukon. The mining companies were responsible for hiring a physician and therefore, their representative on the board was always the doctor in the district at the time. The hospital board was to act in an advisory capacity during construction and assume full management control when the hospital was completed.

The board opened an account at the Canadian Imperial Bank of Commerce in Dawson in the name of the Mayo General Hospital and all donations and collections from patients were to be deposited in that account. By September 18, 1922, $1,512 in subscriptions and $1,500 from Keno Hill Limited had been paid to the account.

Hospital occupancy rates were fixed by the Gold Commissioner at $5 per day for ward patients and $7.50 per day for private rooms.

Dr. Chipman was charged with hiring the matron, one nurse, a cook and an orderly. He selected Matron Martha Morkin and Nurse Elizabeth McDougall. (It appears that Matron Morkin did not stay long as Elizabeth McDougall was promoted to the position of matron early in 1923.)

On September 20, 1922, the Mayo General Hospital was ready for occupancy.

An ice house was built for the hospital in March of 1923. That summer, after complaints about the excess mud being carried into the wards on visitors shoes, a sidewalk was built to the hospital. The Territorial Council voted $7,000 for hospital maintenance for the year ending March 31, 1923.

The hospital board was responsible for patient billing and collection. The board had great difficulty in collecting many of the bills owed by patients and, as a consequence, the bank account would become overdrawn. Mr. Maltby, Territorial Treasurer, would tactfully cover the bank overdraft and issue a mild reprimand to the board.

In 1924, local fund raising initiatives were

begun to raise money to buy a sterilizer. Soon, over $800 had accumulated and the sterilizer was purchased for $400. It was decided to put the remaining money toward a fund to purchase an X-ray machine. An X-ray machine was very expensive and in 1925, a citizens drive was launched to raise money for the project. With the assistance of residents such as Mrs. Wallace Jones, who collected monies from miners on Keno Hill, a substantial sum was raised. X-ray fund contributors were as follows:

Mayo Donations	$ 592.00
Keno City Donations	165.00
Treadwell Yukon Donation	500.00
Treadwell Yukon Employees Donations	610.00
Keno Hill Miners	245.00
Galena Hill Miners	120.00
Cash left over from collection for sterilizer in 1924	435.80
Mayo Order, Daughters of the Empire	143.75
Mayo Baseball Club	22.50
Proceeds from Excursion on Steamer *Thistle* to Fraser Falls	80.00
Proceeds from Moving Pictures obtained from Mrs. George Black	22.75
Proceeds from sale of 8 Delco Batteries	25.00
Dawson Order of Eagles Donation	100.00
Territorial Government Donation	1,000.00
Total	$4,061.80

By September of 1925, another $1,000 was still needed. The Gold Commissioner was petitioned by district residents to guarantee the balance of the X-ray machine purchase price. Finally, in August of 1926, the machine was purchased and installed in the Mayo Hospital.

When the terrible flu epidemic hit Mayo in July of 1925, the hospital's eight public ward beds and four private rooms could not handle the flood of patients and it became necessary for nurses to attend flu patients in their homes.

In 1929, the Mayo General Hospital account was moved from the Bank of Commerce in Dawson to the recently established Bank of Montreal in Mayo. That same year the Department of Indian and Northern Affairs paid $1,750 for a small addition to the hospital. In 1935, the hospital's foundation was replaced.

The Mayo hospital did not go unscathed in the 1936 flood. There was considerable damage to the building's interior and furnishings. The linoleum in the main ward was completely de-

A view of the two storey Mayo Hospital, circa 1922–23, with two nurses and a patient on the hospital porch. Miss Page, the Matron, is in the middle and Tommy Jones is in the wheelchair. *Ralph Rogers Collection*

The Mayo Hospital circa 1935. *G. A. McIntyre Collection*

stroyed, and some of the mattresses and most of the chairs were ruined. The Delco pump and X-ray machine required repairs.

Over the years, Mayo hospital staff treated all manner of illnesses as well as accidental injuries, silicosis cases from the mines, alcohol poisoning, dog bites and wild animal bites, flu and whooping cough epidemics, and numerous maternity cases. They fretted over cleanliness, water pollution, and the dreaded infectious disease tuberculosis.

From 1936 to 1938 there were 37 babies born in the Mayo hospital. One child, born in 1938, could not wait to be born in the hospital. An account of his birth was recorded by Archie Gillespie in *The Mayo Miner* on July 15, 1938.

> At midnight on July 13, 1938, Dr. Nash received a call to hurry to the Indian village where a prospective mother was in need of medical attention. So Dr. and Mrs. Nash left for the village in a boat driven by Simon John. Mrs. Steven Louis (formerly Sally Lucas), the expectant mother, was put into the small power boat and, accompanied by her husband, Dr. and Mrs. Nash and Simon John, and the little group started up river on the two mile trip to Mayo. About three quarters of the way had been made when Mrs. Louis gave birth to a son. Putting in to shore at once, the doctor tended to Mrs. Louis and the baby on the river bank. The party then continued on into Mayo to drop off Dr. and Mrs. Nash. Mr. and Mrs. Louis, their baby son, and Simon John then returned to the village.

With the declining population and depressed economic conditions brought about by World War II and the closing of the Treadwell Yukon mines, G.A. Jeckell, Controller of the Yukon Territory, decided to close the hospital in Mayo. He stated that it would remain closed until conditions in the Mayo district improved to such an extent as to warrant the reopening of the institution. On June 30, 1942, the Mayo hospital was closed. Patients in the hospital at the time were transported to the Dawson hospital and the nursing staff returned to Vancouver.

Enid Boerner née McIntyre, wife of the Northern Commercial agent and a former nurse at the Mayo hospital; and Mrs. Phoebe Reynolds (formerly Miss Burns) also a former nurse at the Mayo hospital, offered to provide assistance in the case of medical emergencies and first aid cases. A small stock of medical supplies from the hospital was delivered to Mrs. Boerner's home.

The entire stock of medical supplies and equipment from the Mayo hospital was sent to the Dawson and Whitehorse hospitals. The Dawson hospital received the bulk of the medical supplies, while the Whitehorse hospital was the beneficiary of the beds, mattresses, bedding, linens, surgical instruments, washing machine and mangle. The latter items were shipped on the steamer *Keno* in September, 1942, and placed in the newly built addition of the Whitehorse hospital. The newest of the two X-ray machines was stored in the Mayo liquor warehouse. The microscope and accessories were sent to M.W. Warwick, Principal of the Whitehorse High School.

"The newest of the two X-ray machines was stored in the Mayo liquor warehouse"

Dr. Allan C. Duncan was practising in Dawson at the time and he requested that the electrical suction pump unit and the two infrared heat lamps be shipped to St. Mary's Hospital in Dawson.

With the purchase of Treadwell Yukon's claims and assets by Keno Hill Mining Company Limited[41] in 1946 the district's population began to increase. This growth necessitated the employment of a doctor and the reopening of the Mayo hospital.

The Territorial Government agreed to cover the $10,000 hospital start-up cost and Keno Hill Mining Company Limited agreed to hire a doctor. Unfortunately, many of the items loaned in 1942 could not be returned and replacement items had to be obtained. Due to general shortages at the time, great difficulty was encountered in replacing the supplies necessary to equip the hospital. The microscope was returned to the hospital in April, 1947.

In March, 1947, Dr. Phillipe Maisonville, from Toronto, was hired by Keno Hill. The doctor's salary was guaranteed by the Territorial Government, Keno Hill Mining, and Wm.V. Smitheringale for Northwest Exploration Ltd. (who was also working in the district at the time). The salary was supplemented by the doctor's private practice. A new hospital board was struck and Dr. Maisonville hired Marion Warren to assume the duties of matron. Members of the hospital board were Sam Wood, Dr. Phillipe Maisonville, Ernest J. Corp, and Gordon McIntyre.

Numerous people worked as orderlies, cooks, and janitors at the Mayo hospital over the years. Some of them were as follows: Mr. and Mrs. Portlock, cook and orderly in December, 1922; Massa Sakata, cook, early 1920's; Norman Niddery, janitor, 1920's; Frank Morishiga, cook, 1924; George Palmer, orderly, 1925; Mr. Walsh (former pantryman on the steamer *Casca*) cook, 1925; Fortuna Dube, orderly, 1920's; Antonette Hobbis and many more.

Unfortunately a complete list of the hospital's nursing staff does not exist. Some of the nurses who worked at the hospital are: Martha E. Morkin, the first matron hired in 1922; nurse Elizabeth McDougall, promoted to matron in 1923 (married Lou Titus); Margaret A. Kenney (married Dr. Randy Maclennan); Mrs. Nora Jones, Mrs. Collins, Sarah Catherine Smith (married Sgt. Dempster in 1926), Veronica Page, matron; Enid McIntyre (married C.A. "Kippy" Boerner), Phoebe Burns (married George Reynolds), Miss McLeod, Lillian Ross, Alma B. Jones, Miss Roll, Winona Spence (married Dr. Duncan), Katherine McDonald, Alberta Neale, Frances Waugh (married Tom Portlock), and her sister Mildred Waugh (married Chuck Harrison), Marion Warren, matron, (she later married Dr. Maisonville).

Persons who served on the Mayo Hospital Board, for various terms, from 1922 to 1947 are: Sgt. W.J.D. "Jack" Dempster, J.A. Fairborn, Oscar Letourneau, Rodolph Rosmusen, Stanley White, D.A. Matheson, J.E. Ferrell, R.L. Gillespie, Frank Carscallen, Harry Colley, L.H. Titus, J.C. Philip, Joseph S. McIntosh, J. A. Longtin, J.D. Gallagher, Gordon McIntyre, Sam Wood, and Ernest J. Corp.

Before a doctor was established in the district, Dr. Randy Maclennan provided assistance in medical emergencies. Although Maclennan was a Doctor of Medicine, a graduate from McGill University, he was reluctant to practise medicine and did so only when absolutely necessary. He operated a small drugstore in Mayo beginning in the 1920's and continued with the drugstore until his death in 1935. Dr. Maclennan came to Mayo around 1904.

For many years, doctors were required to purchase a business license from the Town of Mayo to practise in the district. From those records, a list of doctors who worked in the town was compiled. (The list does not include the names of doctors who may have relieved resident doctors during vacation leave.)

LIST OF MAYO DOCTORS

Dr. Wm. W. Chipman	November, 1922–July, 1926
Dr. C.C. Ross	August, 1925–1927
Dr. D.R. "Randy" Maclennan	1926
Dr. R.K. Behrns	April, 1927–1929
Dr. H.S. Evans	September, 1930–1932
Dr. Gordon Ferguson	1932–1933
Dr. Allan C. Duncan	September, 1933–1935
Dr. E.T.W. Nash	1935–1941
Dr. Cyril M. Robertson	1941
(hospital closed)	
Dr. Phillipe Maisonville	1946–1950
Dr. Rothwell	1951
Dr. John Barker	1952
Dr. J.V. Clark	Elsa 1952; moved to Mayo in 1954 where he still practises

Dr. Allan C. Duncan recalls some of the lighter moments at the Mayo Hospital in the early days:

> When Jack Dempster was squiring Miss Smith, a Mayo hospital nurse, he found the road to the hospital too rough [to promenade] and the new sidewalk [built in 1923] too narrow for he and 'Smithy' to walk side-by-side. Being on the hospital board, he got money passed to widen the same by one six-inch plank for the full distance. This rather thicker board was called 'Dempster's Plank' for years.
>
> Alma B. Jones was the hospital matron—a gal who knew her own mind. She was our ether anaesthetist also. Many of the men patients (there were few women except pregnant ones) got very restless after long periods in hospital. They did not play cards and were not conversationalists by nature.
>
> Jones initiated an amusement for them—no less than a rifle range from an upstairs hospital window to targets placed outside at various distances from the hospital. The miners were all good shots and they enjoyed the contest. This is probably a first in hospital social service.
>
> Jones also allowed the men pet cats in the large ward. They would sleep beside the bed pan on the side tables.
>
> "Shorty" Carter also deserves a write up. In addition to all other duties he was 'enforcer' in the hospital to see that the men did not annoy the nurses.

Just as it is today, dental care was provided by "travelling" dentists, and with the advent of air travel "flying dentists," from Dawson City. Dentists who serviced the district from 1923 to 1935 were Dr. F.L. Mellor, 1923; Dr. I.A.J. Gillis, 1923; Dr. J.R. Franks, 1925 and 1926; Dr. D.S.R. Martin, 1926; Dr. Snider, 1928 and 1929; Drs. Franks & Snider, 1930; Drs. Cozza and Johnson, 1930; Drs. Herkenrath & Johnson, 1931; Dr. J.R. Franks, 1931; Drs. Franks & Snider, 1932–1935.

In 1956, a modern new hospital was built in Mayo complete with operating theatre. The building was opened in March of that year and is still the hospital in Mayo today. The "old" Mayo hospital was torn down by George Besner and the materials used to build the house that still stands on Block 16, Lots 13, 14, and 15.

When hospitals became a federal government responsibility, the hospital board was dissolved. Dr. Clark was the last doctor to be hired under the old system. He chose to remain in Mayo and thereafter UKHM hired doctors specifically for Elsa. These doctors lived in Elsa and maintained hospital privileges in Mayo. In the absence of a doctor in Elsa, the responsibility fell to Dr. Clark who drove to Elsa once a week to hold a clinic.

RECREATION *by Linda E. T. MacDonald & Gordon McIntyre*

An incredible array of activities lined the Mayo recreation roster in the past. There were athletic events, musical and dramatic presentations, organized clubs and celebrations of all kinds. They were at times remarkable, often zany, frequently amusing, somctimes hilarious, always entertaining.

The first athletic event ever recorded in the district occurred at Fort Nelson Post at the mouth of the Stewart River. It was a game of "shin-maiming" hockey played on the slough ice in 1884.

There is no written record of leisure activities in Mayo from the time the town was established in 1903 until 1919. In 1920, the population surged, as a result of the silver discovery on Keno Hill, and recreational activities were recorded and photographed.

Dinner parties, community picnics and dances were popular. Vehicles and airplanes were still a novelty and residents enjoyed rides in the two cars and one truck in town. Barnstormers offered airplane rides and people happily paid $1 a minute for the thrill of a ten minute flight. Ray and Gordon Stewart brought in what could be called the first snow machine. It was a side-car equipped motorcycle mounted on skis.

The Yukon Order of Pioneers established Lodge No. 3 in Mayo in 1921 and built the Yukon Order of Pioneers hall (see *Buildings and Artifacts*). This was the first hall in the community suitable for dancing and local musicians played to capacity crowds. Kippy Boerner began to show 35mm silent films in the new hall.

Community picnics were held, during the summer season, at Alex Nicol's homestead or on the "sunshine sand bar" just upstream from Sam Wood's residence on the Stewart River.

In 1924, a golf course was built on the "sunshine sand bar" by a group of golf enthusiasts. Editor Marie Fotheringham wrote a humorous account of the golf course opening on July 29 in the *Mayo-Keno Bulletin* (August 1, 1924).

Ralph Rogers skiing down the river bank in front of the Chateau Mayo Hotel, 1922. *R. Rogers Collection*

Mayo residents at the "sunshine sand bar" near Alex Nicol's homestead. *Ralph Rogers Collection*

GOLLUFF' The worst has happened' Last Tuesday evening was an epoch making event for Mayo. With little fuss or splutter the Mayo Country Club and Golf Course was opened by four enterprising and adventurous Mayoites, to wit; Sgt. Dempster, Jim Fairborn, Wm. Burls, and Allan Fraser. While only the four principals were present at the take-off; a small gallery soon gathered and followed the pioneers over the course. To Allan Fraser fell the honour of the first stroke; for it was he who disembarked from the first boat this spring with the "Golf-bug" already raging fiercely in his blood; due, no doubt; to his adventure on the famous Lower Lebarge Course this spring.

Allan teed the ball in the most approved fashion; eyed it with the air of a professional, took a healthy swing and landed with a mighty swat. When the sand had been dug out of the eyes of those present, it was seen that the ball had actually moved all of THREE FEET' Allan's remarks on this occasion has been reported to the Bulletin thusly "Well, that's that —***! But knowing Allan as we do, we ha'e oer doots."

Some time later our worthy magistrate was observed arguing the point in the deepening dusk with the difficult approach to hole number two. He was labouring under the tremendous handicap of numbering among his audience, a lady. If you don't think this IS a handicap, ask Jim Fairborn;

Fred Kilbride on his motorcycle at Williams Creek, July 1926. *J. B. Kunze Collection*

Leo Fielder and Pat Pattison on the Mayo golf course, 1936. *May Gillespie Fielder Morrell Collection*

or try a round on the Mayo course.

The last hole was fought out in almost inky blackness, but the faithful four stayed with it, completing the course one way. When asked for scores, the guilty parties were most singularly reticent, probably due to their well known shyness.

Next afternoon, another hole was added, and now Mayo can boast a full 9 hole golf course. Think of it people — Dawson came, bloomed and boomed, then died, and never had the like. One more mark of Mayo's progressiveness.

The "bug" seems to be epidemic in Mayo now and club officials report a brisk business to date with every expectation of a hectic boom in the near future when those who have threatened, receive their outfits from the outside. — FORE

The first airplane hangar built in the Yukon stood a short distance from the golf course and pilots were frequent golfers.

Remarkably, a wooden tennis court was built

in Mayo in 1926 and the club, boasting 23 members to start with, had 60 members in 1938. The club remained active for about 20 years. The court was located on radio station property (near Block 5, Lots 1 and 2). It was rebuilt, because of rotting lumber, in 1935. In the summer of 1936, the new court was destroyed during the flood. After the flood the court was rebuilt on Joe Cantin's homestead about where Mrs. McCoomb's house is located today (Title 192QQ). This was a better court than the old one and it was in use from 1937 until the mid 1940's. It was neglected during the war years and by the summer of 1946, was in poor condition. There was no interest in repairing the court and it was abandoned together with the clubhouse. In 1951, McCoomb's bought the property from Cantin and took over the clubhouse and what was left of the court.

Mayo tennis court, 1933–34. From left: Roy Thomas, Gordon McIntyre, and William Hutchings (Manager, Bank of Montreal). *Charles Taylor Collection, donated by Rick Cocks*

Church clubs were formed following the building of St. Mary's Anglican Church in 1921 and Christ the King Catholic Mission in 1922.

Steamboats that delivered supplies to Levac's sawmill at Fraser Falls, would pick up townspeople in Mayo and take them to the falls for a picnic while the crew offloaded supplies for the sawmill.

Canoeing and horseback riding were commonplace. Dog teams were used mostly for winter hauling and transportation but occasionally, small teams of one or two dogs pulled sleighs of children for fun. Cross-country skis and snowshoes were used for travel as well as recreation. Ralph Rogers, early Treadwell Yukon employee, was known to flash down the river bank in front of Binet's Hotel on his downhill skis.

All manner of card games were popular, espe-

Mayo tennis court and club members. Photograph was taken at 12 o'clock midnight on June 21, 1933. Front row, left to right: Kippy Boerner, Pat Pattison, May Gillespie, Joe, William Hutchings, Catherine McDougall, Florence Wasson, Lillian Ross, Pearl Kimbel, Emelia Besner, Maclennan girls, Mabel McIntyre. Back row: Randy Maclennan, J.J. McCarthy, Dr. Duncan, Wilma, Chappie Chapman, Roy Thomas of the R.C.M.P., Everett Wasson, Mr. & Mrs. Andison. *May Gillespie Fielder Morrell Collection*

This is what the flood did to Mayo's tennis court in 1936. *G. A. McIntyre Collection*

cially bridge and a game called pan (panguingue) played with nine decks. Although pan was an extremely complicated game, some people played it every evening.

Indian people played a card game called "Guhju" (pronounced goojew). Winners of the game painted losers' faces black with charcoal. Traditional stick gambling was frequently played.

Mayo Badminton Group, 1939. L to R: Laura Roche, Doris Fisher, Unknown, Theos Whitney, Margaret Clifton, Mrs. Valentine, Agnes Andison, Flora Bleiler, Daz McCarter, Frances Portlock, Mildred "Millie" Waugh, Mrs. Andison. Boy in front, David Andison. Second row, L to R: Veronica Page, Rose Mason-Wood, Thelma Jeffery, Vera Breaden, Nancy Whitney, Mabel McIntyre, Mrs. Palmer, Fred Marshall. Back row, L to R: Jack Guthrie, Con Lakness, Johnnie Kazinsky, Bill Jeffery, John Whitney, Alf Richardson, Jack Gallagher, George Andison, Chuck Harrison, Rev. Bill Valentine. *G. A. McIntyre Photograph, F.&T. Whitney Collection*

In this game, participants sit in a circle with their hands behind their backs and the object is to guess who has the stick in his hand.

Baseball, softball, and horseshoes were all-time favourites.

By the 1930's, clubs of every description had been organized. There was the Yukon Order of Pioneers, church clubs, a badminton club, dramatic club, I.O.D.E., tennis club, golf club, poker club, pan club, debating club, Ananias club, birthday club, sports clubs, and the Masons.

Recreation in the 1930's is wonderfully documented in Archie Gillespie's newspaper.

The drama club staged four to five plays a year in the 1930's.

Laura Roche directed "Sad About Europe," a one act play, with the leading roles played by Sam Wood and Hilma Miller. Gillespie supplied a sensational headline for a melodrama presented in 1938.

VILLAINS ROAR, VAMPIRES VAMP AND HEROINE EMOTES AS MAYO DRAMATIC CLUB PRESENTS "MELLER-DRAMER"

Mayo Dramatic Club ushered in its new season on Wednesday night when a selected cast presented a thrilling melodrama for the enjoyment and edification of the other members of the club. It was the first regular meeting of the fall and the thirty or more members present augured well for the enthusiasm that is being taken on the club's new program.

The "first night" performance was held in the IODE house which proved a very suitable and comfortable "theatre."

Those taking part in the play presented Wednesday night were: Betty Taylor, Elsie [Elsy] Ray, Mary Broadfoot, Gordon Lee, Gordon McIntyre, and Jack Guthrie. Dr. E.T.W. Nash directed the play and arranged the "opening night" entertainment. Tea and Peak Frean biscuits were served after the performance. (*The Mayo Miner*, 1938)

Skilled organization and large numbers of people were required to stage a major production. An elaborate three act comedy titled "Nothing

ABOVE, Mayo costume party, 1922. Left to right: unknown, unknown, Jim Fairborn, Laura Gillespie, and Ted Sunderland. *Ralph Rogers Collection*

BELOW, A birthday party in Mayo. From left: Katherine Lee, Alan Kimbel, George Nagano, Leila Palmer, Marjory Nagano, Nancy Jeffrey, Dick Gillespie, Patsy Fisher, Keith Gillespie, Ila Fisher, Norma Palmer, and David Andison. *G. & E. Lee Collection*

"On our way to a dance at Keno, February 22, 1924." *D. Walloch (McLaren) Collection*

But the Truth" was presented by the club at the Yukon Order of Pioneer's Hall on December 24, 1937. The play was directed by Dr. E.T.W. Nash and the leading roles were played by W.E. McElhone and Margaret Clifton. Supporting actors were Jim Fairborn, Daz McCarter, Gordon Lee, Isabel Murdock, Mary Broadfoot, Thelma Jeffrey, Gordon McIntyre, and Ed McKenzie. Set construction was carried out by Corp. A. Mason-Rooke and Jack Andison; set decoration by Mrs. George Andison. J.D. Gallagher, Sam Wood, and Jack Bellerby handled the seating arrangements and ticket sales. Refreshments were served by the I.O.D.E., after the play. A dance followed with music supplied by Betty Taylor, Mrs. M. Sullivan, Eddie Kimbel, Albert Pelland, and Wm. Jeffrey Jr.

Group of Mayo residents in 1922. Front row: unknown, Kathleen McLaren, Olive Powers (née Kinsey), Vera Gillespie. Second row: Alice Fisher, Margaret Gillespie (Sheardown), unknown, Mrs. Kinsey, unknown, Eddie McKenzie, Harry Whiteside is seated at the piano. Back row: unknown, Walter Murphy, Ralph Rogers, Laura Gillespie, Yorke Wilson, Irvin Ray. *Ralph Rogers Collection*

In his review of the play, Gillespie wrote, "From the time the curtain rolled up on the opening scene until the final denouement the audience was kept in a perpetual state of mirth, merriment and riotous laughter. At the end of the play the audience echoed and re-echoed its enthusiastic acclaim with round after round of well-deserved applause."

Children's Christmas concerts were held every year in the Pioneer Hall. This was a special event and well attended by the entire community. Santa Claus was always on hand to distribute a present to every child from Mayo and the Old Village. The concerts were arranged by Gordon McIntyre with the assistance of Daz McCarter and Mrs. Millar as well as volun-

Nothing But the Truth, December 24, 1937. From left: Margaret Clifton, Const. McElhone of the R.C.M.P., Isabel Murdock, Gordon McIntyre, Gordon Lee, Dr. E. T. W. Nash. Seated: Daz McCarter and Jim Fairborn. On extreme left: Thelma Jeffrey, Ed McKenzie, Mary Broadfoot. *G. A. McIntyre Collection*

teers from the community. Christmas Concerts were normally held in conjunction with the Christmas Tree. The concert was put on by the Mayo school, the tree and hall were provided by the Yukon Order of Pioneers, and the presents were the responsibility of the I.O.D.E. Long before Christmas, a list of all of the children in the district was compiled. An order was prepared for Spencer's store in Vancouver, with a list of the age groups for both boys and girls and the number of children in each age group. The presents were paid for with money collected from the community. The old bachelors (half the population of Mayo in the 1930's seemed to be old bachelors) never failed to contribute most of the money collected for the childrens' Christmas gifts. The program for the December 24, 1937 Christmas concert has been preserved in the pages of *The Mayo Miner*.

1. Play: Vera Breaden, Doris Fisher, Nancy Whitney, John Whitney, Henry Breaden, Billy Jeffrey, Kippy Fisher.
2. Carol: "All that Wondrous Xmas Night"
3. Littlest of All: Joan Currie
4. Turkey Chorus: Margery Nagano, Norma Palmer, Joan Curric, Dick Gillespie
5. Run and Scrub: Annie Close, Bessie Profeit, Yvonne Besner, Doreen Lee
6. Eleven Surprises: Pupils
7. Carols: "Polish Carol" and "Kolyada"
8. Recitation: "His Xmas Tree" Buckie Keobke
9. Skit — John and Susan: John Whitney and Joyce Besner
10. Piano Solo — "Star of the East": Vera Breaden
11. Skit — Confident Boys: Kippy Fisher, Henry Breaden, Billy Jeffrey
12. Recitation — "Good Night": Nancy Whitney
13. Recitation — "Farewell": Edward Besner
14. Final Chorus: Jingle Bells

On New Year's Eve a gala masquerade ball was held in the Yukon Order of Pioneers Hall. The Grand March, led by Oscar Letourneau each year, took place shortly after eleven, followed by the unmasking at 11:30 p.m. Dancing continued until midnight when the Old Year was roundly cheered

A masked ball in Mayo, 1947. *M. H. Ewing Collection*

Applesauce, starring Dr. E.T. W. Nash, Neil Keobke, Laura Roche, Hector Morrison, Daz McCarter, Mrs. Nash, Gordon Lee, Gordon McIntyre, and Const. McElhone of the R.C.M.P., 1937. *G. A. McIntyre Collection*

out and the New Year gaily ushered in by everyone in the hall. Dancers and spectators joined hands all around for the singing of "Auld Lang Syne." Dancing continued until 3:00 a.m. Costume prizes were awarded in various categories. Prizes awarded on New Year's Eve 1937 are as follows (*The Mayo Miner*, January 7, 1938):

Best Dressed Lady:
Miss Mable Sullivan, dressed to represent a rainbow.

Most Original: 1st Prize – Miss Catherine McDougall as a Sandwich Man
2nd Prize – Charles Taylor as "Robert the Robot"

Best Sustained Character:
Alex Smith as the "Absent-Minded Professor"

Best Dressed Comic Couple:
Mrs. Wm. Jeffrey Jr., and Mrs. Wesley Palmer dressed as "Pop Eye the Sailor" and the child "Sweet Pea"

Children's Prize: Miss Vera Breaden dressed to represent a paint box.

Among those whose costumes and characterizations won much favourable praise were: Mrs. Robinson, dressed as "An Apple Seller"; Mr. and Mrs. James Wood, dressed as a "Spanish Couple"; Miss Laura Roche, dressed to represent "Santa Claus"; Mrs. Alex Smith and her sister Mrs. Sinclair Dunnett Jr., dressed as mechanics; Miss K. MacDonald dressed as a Gypsy; Pete Buckway as "Tarzan of of the Apes"; Frank Buckway dressed up as a lady; and numerous

New Year's Eve, 1937. *G. A. McIntyre Collection*

others, whose names we cannot mention for lack of space, in this write up.

On New Year's Eve, 1938, several well known residents from the Silver King came to the Mayo ball dressed as Snow White and the Seven Dwarfs. That year the Lambeth Walk was danced in Mayo for the very first time by Betty and Charles Taylor. The couple started the Lambeth Walk and soon all the dancers in the hall were following their example. The Silver King Orchestra played that night. The musicians were Mrs. H. Lovdahl, H. Lovdahl, Reg Millar, Don Millar, Phil McKay and Don Kozi. Fred Marshall was the Master of Ceremonies.

Mayo musicians who played for dances included Betty Taylor, Mrs. M. Sullivan, and Malcolm Matheson, piano; Daz McCarter, sax; Mrs. Ted Richards, traps; Glenn Sullivan, banjo; Tommy Portlock, banjo; Alf Richardson, violin; and Albert Pelland, drums. Other musicians were Wlodga Kondroski Kimbel, piano; Sgt. C.B. Tidd, piano, trombone, and saxaphone; Gordon McIntyre, drums and piano; and Ed Kimbel, jazz whistle and banjo. This list is by no means complete, as a considerable number of people could play musical instruments in those days.

The Mayo Miner describes the annual Armistice Dance (November 11) staged by the I.O.D.E. (Imperial Order of the Daughters of the Empire):

> The big hall [Yukon Order of Pioneer's Hall] was ablaze with appropriate decorations, flags, colored bunting and pictures of the King and Queen which lent a patriotic air to the setting. Former familiar and popular patriotic dance scores and songs also helped to highlight the Armistice spirit. Dancing got underway at 9:00 p.m. and lasted until three in the morning. At eleven [p.m.] an intermission was called in the dance program when the usual two minute silence was observed. Refreshments were served at 11:30.

Kippy Boerner continued to operate his weekly commercial theatre in the Pioneer Hall through the 1920's and into the early 1930's. Mayo, like Dawson, was the end of the line for

Nothing but the Truth, put on in Mayo in 1949. Back row, from left: Rev. Norman Wareham, Fred Warren, Jack Andison, Flora Bleiler, H.E. "Butch Miller, Dale Robertson, and Syd Leigh. Front row from left Anne Taylor, Jean Gordon, Vi Lakness, Margot Taylor, and Betty Robertson. *G. J. Gordon Collection*

A Christmas Card, circa 1930, from Joseph Eugene "Gene" Binet with a photograph of his establishment, the Chateau Mayo. *Middlecoff-Bleiler Collection*

ABOVE, Men gathered in George Kazinsky's cabin listening to the radio. "What a treat, it came in night after night over the radio waves via super 10 heterodyne." Left to right: Norval Lochore, George Kazinsky, Johnny Kazinsky (jitney operator), Ted Bleiler, Nick Grerier. *T. R. Bleiler Collection*

BELOW, Boxing match in Mayo. Ed Bleiler in white shorts he made from flour sacks. Referee Eddie Kimble. Ed Lesosicks black shorts, Jack Sullivan standing at ringside. Margaret "Maggie" Mervyn standing by the window. *E. C. Bleiler Collection*

A skating party at Mayo 1922. From left: Frank Short, Ethel Middlecoff, Elsie Schellinger, Warren McFarland, Bunny McFarland, Ruth McFarland, unknown woman, Kathleen McLaren holding onto Joe Schellinger. *A. K. Schellinger Collection*

the films. The films accumulated in one of the Northern Commercial Co. warehouses in Mayo. Sometime in the late 1940's the films, believed to be in metal boxes, were taken to the old dump south of the airport. Charlie Taylor and Gordon McIntyre brought in 16mm silent films which were shown in the Pioneer Hall with the proceeds going to some worthwhile cause.

Amateur radio sets were in evidence in the 1930's. Alec Berry operated his set (VE5-ADW) on Galena Hill every evening. Berry offered a talk show format with readings from *The Mayo Miner*, guest speakers, and a children's hour which included poetry readings and messages to friends and family. Roy Hallin also operated a local set (VE5-AFB). The local stations were able to talk to sets in Dawson City. Jimmy Wood and Malcolm Matheson operated a low power bootleg broadcasting unit, in the early 1930's, broadcasting popular music. Their listeners recall Jimmy's recurrent, "We hope you're listening and we hope you like it." The enterprise was short-lived as someone complained to Ottawa and Wood and Matheson received orders to shut the operation down.

People skated on the Stewart River on a stretch of ice extending from the old Keno Hill Ltd. warehouse to a point opposite the Taylor & Drury store at the foot of Congdon Street. Some years the ice froze thick before the snow fell, providing ideal skating conditions. Skating by moonlight was popular. After the snow fell it was cleared and the ice flooded with the fire hose. *The Mayo Miner* describes the snow removal process.

> The "rink" on the Stewart River was cleared of snow thanks to the efforts of Pete Keobke. With the aid of a horse, kindly lent by Eddie Kimbel, and a plow, Pete cleared off a big space on the river, opposite Moreland Junction, and Fire Chief Joe Longtin, Charley Taylor, Jack Guthrie and E. Lakness flooded it with the fire hose.

Badminton and ping-pong were played in the Pioneer Hall from October to April. The thirty or so members of the club leased the Pioneer Hall for two nights a week, Tuesdays and Fridays. On

Fridays juniors played at 7:30 p.m., and adults at 9:00 p.m. The club wound up its season in April with a social evening of cards, refreshments, speeches and dancing.

Boxing was a popular sport among men. Any man who felt he understood the sport tried his hand at it. Ed Kimbel frequently bested Tom Jones in amateur matches in Keno and Mayo. At these fights, even seats at hotel windows overlooking the "ring" were sold for $4 a head. During the Discovery Day celebrations in Dawson in 1937, Clint Nicholson, the "Silver Cyclone" from Calumet, fought the undefeated champion of the Yukon, "Dawson Flash" Art Fry, to a draw in a six round bout.

ABOVE, Mayo ladies championship softball team, 1938. Back row, from left: Mabel McIntyre, Maggie Wood, Hazel "Toots" Linn, unknown, Florence Palmer. Front row, from left: Rose Mason-Wood, Ella Rear, Emelia Aylwin, Florence Pelland. *G. A. McIntyre Collection*

BELOW, ball game in Mayo, circa 1934. *Charles Taylor Collection, donated by Rick Cocks*

Men's baseball and women's softball teams were organized in both Mayo and Elsa. Frequent tournaments occurred among the various teams during the season. In addition to the teams in Mayo, there were the Kimbel Mill Lumber Kings, Treadwell Yukon All-Stars and the Longshoremen. Even the S.S. *Keno* crew raised a team on occasion.

Emil Forrest coached the women's softball team in 1939. The players that year were Mabel McIntyre, Emelia Aylwin, Daz McCarter (manager), Rose Mason-Wood, Hazel Linn, Ella Rear, Thelma Jeffery, Florence Palmer, Vera Breaden, Florence Pelland, Margaret Wood, and Anne Goulter.

Hiking was a popular pastime with both families and individuals. Groups of people often walked out to the Ringle Ranch (Group 1004, Lot 88) or to Crocus Hill (Group 1004, Lot 87) for a picnic. In April 1938, while a group of picnickers were playing bridge on the slopes of Crocus Hill, a small band of caribou emerged from the timber and began to graze along the ridge just downhill from where the group was camped.

Swimming at Five Mile Lake and Halfway Lake (known as Sullivan Lake) was as popular in the past as it is today. A few hardy swimmers dipped in the chilly waters of the Stewart River at the foot of the White Pass slip. The first display of "aquaplaning" in the Yukon was at Halfway Lake in 1939.

The big lake at the Halfway House [currently the site of Drago and Anna Kokanov's Silver Trail Inn], 20 miles from Mayo on the Silver King road, resembled an aquatic meet last Sunday when some 100 holidayers from Mayo and Galena congregated there to enjoy one of the biggest swimming days in the history of the silver camp.

From early morning, until late at night, the swimming and aquatic fans were in and out of the lake, diving, swimming, splashing, boat riding and aquaplaning.

Don Poole, Manager of Kelowna Acquatic for the past number of years, took a surf

Back row, from left: Yorke Wilson, unknown, Louis Kazinsky, Pat Pattison (Signals), Leo Fielder (R.C.M.P.). Middle row, from left: Ken Schellinger, unknown, Bill Bromley, unknown, Tom Portlock, Johnny Kazinsky. Front row: Darrel Bissel. Circa 1933-34. *Charles Taylor Collection, donated by Rick Cocks*

board out to the lake and gave an exciting exhibition of aquaplaning. Eddie Kimbel provided the motor power with his speed boat and a busy afternoon it proved for Skipper Kimbel. The surf board proved a popular attraction and several Mayo and Galena swimming stars tried valiantly to ride the careening plank as it was towed swiftly through the water. Among those who tried their skill at aquaplaning were "Red" MacParlon, Bill Lamb, Sam Wood, and "Slim" DeMers. Mrs. A. A. Gillespie was the only lady swimmer to try her skill on the surf board. (*The Mayo Miner*, July 21, 1939)

The annual Mayo Discovery Day celebration held on August 17th was the biggest event of the year. In addition to the Mayo crowd, it was attended by residents of the Old Village, Keno, Elsa, Highet Creek, Duncan Creek and Minto Bridge. Events of the day included sprints for children and adults; adult relay races, nail driving, and tug-of-war; children's spoon and egg race, wheelbarrow race, 3-legged race, sack race, and bicycle races. Ball tournaments were played among the various district teams. The tournament winner was awarded the Yukon Order of Pioneers silver cup and in the late 1930's a silver cup was donated by Coast Breweries [donated in 1939 and designated for the men's champion baseball team]. Candy, lemonade, ice cream, hot dogs, sandwiches, tea, and coffee were available

Mayo Discovery Day, 1920. *L. Cheatham (McLaren) Collection*

in booths at the ball field.

In the late 1930's, an Agricultural Fair became part of the day's events. A large tent, erected at the ball park, housed the numerous garden and floral exhibits. The entries were judged and prizes awarded. MP Martha Louise Black encouraged the Agricultural Fair. The day ended with a dance at the Yukon Order of Pioneers Hall. The Elsa Swing Band played for the Discovery Day Dance in 1938. Band members were Mrs. George Kazinsky, piano; Johnny McDonald, sax and clarinet; Ted Bleiler, violin; Dick White, traps; and Bill Thomey, guitar.

In the early 1950's, the dangerous sport of rafting the Mayo River during high water began. Groups of men on a single raft would ride through the river gorge and end up behind the old hospital where a waiting crowd would cheer their safe arrival.

Dale Robertson and Alex Wark were the promoters of the Mayo curling rink. Brodie Hicks, the Mine Manager for United Keno Hill Mines, who was an ardent curler and a good friend to Mayo, donated UKHM's old warehouse to the community to use as a curling rink in the early 1950's. The building was made of corrugated iron and open on the street side. The open side was closed in with iron from Lefebvre's sawmill. All work was volunteered and almost every able-bodied man in Mayo put some time into the project. Probably the man who worked the most was Larry Requa, the Park Warden, who literally put days of work into the rink. The founding members each put in $200 cash to get the club going. The stones came from the Dawson Curling Club. The curling rocks were donated or the Mayo club got a very good deal on them. The first year or two the rink was insulated but there was no heat in the rink. At –50 degrees most women could not throw a rock the length of the rink but it was great fun and it was all done without government money.

TOP, a group swimming in the Stewart River with the town of Mayo in the background. *T. McKay Collection*
MIDDLE, a swimming group circa 1920. *D. Walloch (McLaren) Collection*
BOTTOM, Discovery Day, 1920. *L. Cheatham (McLaren) Collection*

TOP, Mayo residents picnic on August 17, 1922, at Alec Nicol's homestead. *Ralph Rogers Collection*
MIDDLE, Group picture taken in the recreation hall, Wernecke Camp, circa 1922. *F. & T. Whitney Collection*
BOTTOM, A "Hard Times Dance" in the Wernecke Camp Recreation Hall, circa 1926. *Middlecoff-Bleiler Collection*

Picnic at the Middlecoff farm two miles from Mayo. *Middlecoff-Bleiler Collection*

Rev. Wareham initiated a new community hall in the early 1950's. It was known as Wareham Hall and was located on low ground near the Anglican Church. Wareham Hall was poorly constructed and did not last. The maple floor was salvaged and is in the current community hall.

Gardening

Gardens flourished in and around Mayo during the long hours of summer light and the harvest was of great importance. Grant Huffman sold potatoes and root crops grown on his farm near Mayo Canyon to the Keno Hill mine. Cora Rich sold fresh vegetables and berries to the steamboats that docked in Mayo. The Saskatoon bushes she planted still grow on her old lot in Mayo.

In the 1920's, Theos Whitney brought strawberry plants to Mayo. These strawberries continue to grow in many gardens in Mayo today, having adapted themselves well to northern conditions. They are extremely hardy and easily grown, with fruit approximately three quarters of an inch in size that is sweet, tender and densely formed.

Archie Close was a major producer of hay,

Mrs. George (Christena Matheson) Andison with her garden produce in Mayo. *G. J. Gordon Collection*

clover, oats, sweetclover, alfalfa, potatoes and vegetables. He supplied the Keno Hill mine with produce. Close experimented with pumpkins, corn and apples. In 1937, Hughie Monohan grew a two pound potato in his garden just past the Anglican rectory on the hill. Gene Binet displayed it in his store window for a time. Mrs. George Andison grew tomatoes, pumpkins, and vegetable marrow, among other things, in raised beds in her back yard. Rhubarb plants and larkspur flowers continue to grow on old cabin sites throughout the district. Rhubarb planted at Lansing in the early days has gone "wild" and now grows to heights of five or six feet.

Garden produce and wild berries were canned in jars for winter use. Root crops were stored in bins in dirt cellars under cabins or cellars dug into hillsides. A large harvest of peas was easily shelled by running them through the wringer of a motorized wringer washing machine and the shelled peas were caught in the tub portion of the machine.

The numbers of gardeners and the variety of produce they grew was evident at the Discovery Day Agricultural Fair. This event was held in conjunction with the annual Discovery Day celebrations beginning in the late 1930's. Following is

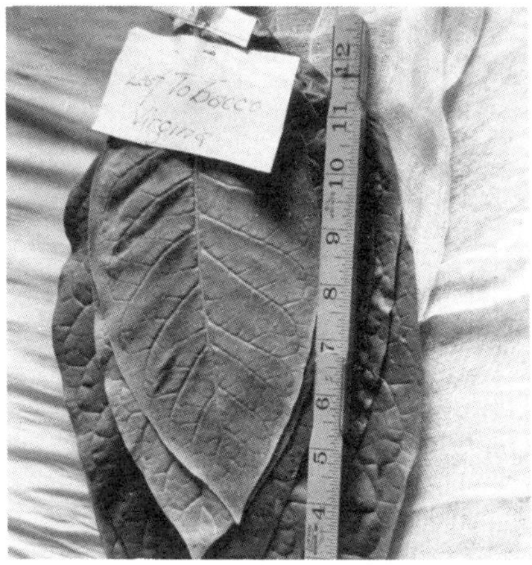

Tobacco leaves for an agricultural display. *G. A McIntyre photograph*

a list of prize winners for the 1939 show:

Lettuce	— Mrs. Lee
	— Mrs Cora Rich
	— Mrs. William Jeffery Jr.
Cabbage	— Joe Lelievre
	— Mrs. Cora Rich
	— Johnny Vanicot
Cauliflower	— Mrs. Wm. York (Elsa)
Swiss Chard	— Mrs. Jeffrey Jr.
Beets	— S. M. Wood
	— Mrs. Jeffrey Jr.
	— Johnny Vanicot

Vegetables grown in Mayo. *G. A. McIntyre photograph*

Carrots	— Sam Wood
	— Alex Nicol
	— Johnny Vanicot
Potatoes	— Sam Wood
	— Johnny Vanicot
	— Joe Lelievre
Rhubarb	— Hugh Monohan
Peas	— Hugh Monohan
	— Archie Close
	— Mrs. Steeves
Cucumbers	— Mrs. Boerner
	— Mrs. Sullivan
	— Mrs. Jeffrey Jr.
Tomatoes	— Mrs. Boerner
	— Mrs. Sullivan
	— Mrs. Jeffrey Jr.
Turnips	— Johnny Vanicot
	— Sam Wood
	— Wm. Jeffrey Sr.
Celery	— Frank Whitney
Strawberries	— Frank Whitney
Chinese Cabbage	— Jim Fulton
Scotch Kale	— Mrs. Wm. Jeffrey Jr.
Onions	— Sam Wood
Vegetable Marrow	— Mrs. M. Sullivan
Squash	— Mrs. Yorke Wilson
Sweet Potatoes	— Mrs. M. Sullivan
Watermelon	— Mrs. M. Sullivan
Peppers	— Mrs. Boerner
Beans	— Archie Close
Horse Beans	— Mr. Brame
Tobacco	— Mr. Brame
Baya Beans	— Mrs. M. Sullivan
Avocado	— Mrs. J. Bellerby
Raspberries	— Sam Wood
Barley	— Archie Close
Wheat	— J.W. Ellis
	— Joe Lelievre
Oats	— J. Cantin
	— Archie Close
Clover	— Archie Close
	— Elmer Middlecoff
Broom Grass	— Joe Lelievre

THE MAYO FIRE DEPARTMENT *by Linda MacDonald & Gordon McIntyre*

At a meeting of the Mayo Board of Trade on August 10, 1922, the matter of a fire hall for Mayo was discussed and a committee was struck to select a site for the fire hall and stationary engine. David Ballentine was recommended for the position of fire chief, a position he held until 1926.

Committee members James Cody, Frank Short, and Gene Binet, together with territorial employee J.H. McNeill, selected a site on the waterfront midway between Congdon Street and Centre Street, near Binet's pump house. The location permitted a gentle, easy lift and allowed the suction end of the hose to be in the river.

Mayo waterfront, circa 1925. The man is standing beside the water scow which contained a large pump and a bed. *A. K. Schellinger Collection*

The fire hall consisted of an anchored barge with a large gasoline driven pump. This was lifted out in the fall when ice formed in the river. Years later, a well was dug near the fire hall on Front Street, where a large electric pump was utilized to draw the water.

The fire hall was constructed with a tower design to allow a loft for the hose reels and drying process, with easy grade out on to the street. The building was erected with the intention of pulling it away from high water in the spring. It was large enough to allow Ballentine to use a portion of it for a work shop. The Territorial Government voted $5,000 toward the Mayo Fire Department project.

Alex Maclennan became Fire Chief in 1926

Binet's Chateau Mayo Hotel and Store burning to the ground on July 13, 1930. *Middlecoff-Bleiler Collection*

and held the position until his death in 1936. His daughter, Betty Taylor, recalls that he was hired at $175 a month, and reduced to $75 a month during the Depression.

During the 1930's, Maclennan's crew consisted of Charlie Taylor, George Aylwin and Gordon McIntyre. After Maclennan's death, Joe Longtin was appointed Fire Chief. He was succeeded in the mid-1940's by Jack Andison and later Art Shaw and J.J. Van Bibber.

Following the 1936 flood, the Mayo fire hall assumed quite a tilt; however, it was restored and was finally demolished in 1959.

While records of the Mayo Volunteer Fire Department are unavailable for corroboration, oldtimers say that Mayo was not subject to the kind of conflagrations experienced by other frontier towns. The Mayo Volunteer Fire Department has had to contend with major fires at Binet's Hotel on July 13, 1930, at the Bank of Montreal in the 1940's and at the Radio Station, also in the 1940's.

BUILDINGS & ARTIFACTS

Today there are very few reminders of the past in Mayo. Most historic buildings have burned down, artifacts have disappeared and others have yet to be unearthed.

The following is information on various buildings and artifacts that has been recorded.

BANK OF MONTREAL BUILDING

This structure was a small log building on First Ave. between the N.C. store and Taylor & Drury's. The Bank of Montreal was the first to establish banking services in Mayo on May 2, 1928, and remained until May of 1942. The building was owned by Eddie McKenzie, Dick Major and Malcolm McCown. Betty Taylor recalls that in 1926 it was an ice cream parlour run by McKenzie and Major.

CHURCHWARD'S SHEET METAL WORKS

Block 12, Lots 23 & 26
(Bob Adair's Garage – 1989)

The lumber from which the place was built came from the old Broadway Hotel in South Dawson. Dad bought the hotel in 1921, had it torn down, and shipped the lumber to Mayo. The lumber arrived on the first boat to Mayo in 1922.

Dad had left the family in Tacoma in August 1921, and had gone to Dawson to his old tin shop and there he put in all winter making stoves—airtight heaters, Yukon camp stoves, stove pipe, etc. The items were to be taken to the Mayo mining district where a lot of activity was going on.

To others it's a garage, to me it's the old tin shop and the apartment upstairs was where we lived. My brother Phil helped fill the walls with sawdust which was considered good insulation in those days (I think I helped a little too).

Phil made the sign atop the building which you could see from downtown. The sign was made of metal letters about two feet high, painted black, and ran the length of the building. Years later, a few years before the place was sold to Bert Klippert, my husband George took the sign down, as it had become so rickety we were afraid it might fall and hurt someone. *–Annie (Churchward) Besner*

GEORGE BESNER'S HOUSE

Block 12, Lots 5, 6 & 7
(Bob Adair's House – 1989)

This house was built by George Besner for himself and his family. He built it on his two corner lots adjoining the residence he was living in at the time, which was referred to as the Henderson house. He owned the Henderson house and the lot it was on. The Henderson house had once belonged to Bob Henderson of Klondike fame. The new house was a two-storey frame building. Besner completed the framework and the roof in October, 1937, and finished the house completely in the spring of 1938. He then tore down the old Henderson cabin because the logs were being destroyed by wood-eating insects. Verna and Bert Klippert bought the house when the Besners moved to British Columbia in 1965. Klippert sold the house to Buddy Rich who sold it to the current owner, Robert "Bob" Adair.

WERNECKE'S MAYO LAKE COTTAGE

The Werneckes had a cottage at Mayo Lake where they had a summer fishing camp. Later someone bought it, hauled it into Mayo, and set it up next to the Pioneer Hall. It was painted green. *–Annie Besner*

THE ANGLICAN CHURCH RECTORY

Block 15, Lots 8 & 9
(The Ewing House)

This house was formerly the Anglican Church Rectory and was located next to the Anglican Church. The rectory was built in the early 1920's. In 1976, Harry Ewing purchased it and moved it to its present location. *– Dorrie Ewing*

RICHARD O'LOANE'S LOG HOUSE

Block 1, Lots 23, 24 & 25
(Harry Ewing's Log House)

This house was built by Grant Huffman in 1921 for Richard O'Loane. O'Loane came to Mayo to manage the newly built Taylor & Drury store in Mayo. He came from Fort Selkirk, Yukon, where he had managed the Taylor & Drury store there.

In 1923, O'Loane was hired to replace Jim Cody as manager of the new log store built by Northern Commercial, and as living quarters were supplied above the Northern Commercial store, the O'Loanes sold their house to the Treadwell Yukon Company, Ltd.

Treadwell Yukon bought the house to use as a residence for company doctors. Harry Ewing then purchased the house and rented it out for a couple of years. He and his family took up residence in it in 1945 and remained until July of 1976. *– Mabel McIntyre*

FIRST MINING RECORDER'S OFFICE

Block 6, Lot 30
(Mabel McIntyre's Cabin)

This cabin was the first official Mining Recorder's office in Mayo and was built in 1921 by Samuel Blackmore for Dick Gillespie who was the Mining Recorder at the time. In 1933, Blackmore built a frame building (currently Joan & Dick Ewing's shed) in front of this cabin and the Mining Recorder's office moved into it.

Blackmore then took up residence in the cabin. Mabel McIntyre inherited the cabin and property upon Blackmore's death in 1946. She lived in the cabin from 1964 to 1981. The frame building was closed during World War II and the Mining Recorder's office was moved to the Liquor Store on Front Street. Sam Wood was the Mining Recorder at the time. *– Mabel McIntyre*

YUKON ORDER OF PIONEERS HALL

Block 3, Lots 4 & 5

The Yukon Order of Pioneers Lodge No. 3

was formed in Mayo in 1921.[42] Lodge Presidents from 1921 to 1932 were: 1921 Jim Ferrell, 1922 Oscar Letourneau, 1923 J.V. Smith, 1924 N.A. Lefebvre, 1925 J.A. Anderson, 1926 W.J.D. Dempster, 1927 G.S. Churchward, 1928 C.D. Matheny, 1929 D.A. Matheson, 1930 J.E. Ferrell, 1931 W.J.D. Dempster, 1932 Elmer Middlecoff (list supplied by Laurent Cyr). Laurent Cyr has told us that the sign from atop the hall and a Y.O.O.P. banner, found at Highet Creek, are on display at the Elk's Hall in Whitehorse. Harry Ewing and Mabel McIntyre remembered that, at one time, mirrors in molded frames, ringed the inside of the hall. When the hall burned down in 1984, some of the old mining equipment stored there survived.

This hall was built in 1921 by Grant Huffman and his crew. It was the last original Yukon Order of Pioneers hall left in the Yukon until it burned down in 1984. The sign from the front of the hall was taken away for restoration by Laurent Cyr, Yukon Order of Pioneers Whitehorse, many years ago and it is now preserved in the Y.O.O.P. Corner of the Elks Hall in Whitehorse. – *Mabel McIntyre*

The building was erected by the Yukon Order of Pioneers in 1921. For many years this hall was not only used as a meeting place for the Pioneers Lodge No. 3, but also as a community hall. It was used for dances, meetings of all kinds, a court room, picture shows, etc. The picture shows were of 35mm silent films and the show was operated by Kippy Boerner, Manager of the Northern Commercial Co. The films had to be brought in on the river boats as there were no planes, and in order to have shows during the winter months, a large shipment of films had to be shipped before the last boat came up the Stewart. The films were stored in the hall until needed. The instructions from the companies who supplied the films stated that these winter films were to be destroyed after they had been shown. I used to assist by taking tickets at the door. With the advent of sound films, the shows were discontinued until someone else brought in sound films and projectors.

During meetings of the Historical Sites and Monuments Board, the Yukon Order of Pioneers Hall in Mayo was mentioned as being a historic building worthy of being preserved, but due to the small budget it was not possible to do anything with it. The board tried several times to get a grant to preserve this historical building but were unable to do so.

We did eventually get a few dollars to purchase roofing paper to repair the roof which had to be put on by volunteer labour. We had the Territorial Government engineers make a survey of the repairs needed. This survey showed that the

Yukon Order of Pioneers Hall, Mayo. *L. Cheatham (McLaren) Collection*

repairs needed would amount to far more than the board's budget could finance. The engineers pointed out that the hall would have to be repaired so that the building could handle a capacity crowd. It would need a completely new foundation as the building was of log construction with the wall logs being mortised into large upright logs. This type of construction allows water to enter the mortised joints and causes rot. – *Charles "Chappy" Chapman*

WAREHAM HALL

Near Block 13, Lot 31

The Wareham Hall was built in the late 1940's. It was located opposite our old house [currently Bob Adair's house]. Rev. Wareham initiated the plan for the hall and insisted that it had to have windows. After the hall was built, wooden shutters had to be made for the windows so that badminton could be played in the building. – *Annie Besner*

Construction of the Northern Commercial Store in Mayo, pre-1930. *C. A. Boerner Collection*

Wareham Hall was probably started in 1949. It was planned by Rev. Wareham and built under his direction by Dave Ballentine. It was of frame construction, but because it was built without a properly trussed roof there was too much thrust against the side walls. The walls began to bulge the first year after construction and had to be shored up to prevent the building from collapsing. Eventually it was condemned and demolished. – *Gordon McIntyre*

WAREHAM HALL PIANO

This piano was a baby grand. It came to Mayo just prior to World War II. Herb Lovdahl, a shift boss for Treadwell Yukon Mines, had made a trip to his native land (Norway, I believe) and on his return journey had met a lovely German lady. They were married and he brought her back to Mayo. Since Mrs. Lovdahl was a talented pianist, a baby grand piano was purchased in Vancouver and shipped to Mayo. As married accommodation at the mine was not provided in those days, Lovdahl bought the C.D. Matheny house in Mayo (Tony Blais's residence in 1988) and made it into a comfortable home. When the mine closed down in 1941, the Lovdahls left Mayo. Subsequently, Fred Taylor, who was mining on Dublin Gulch, bought the Matheny house with contents from Lovdahl. In 1949 or 1950, Rev. Wareham, the Anglican Minister, built the ill-fated church hall. Since the hall was used for dances a piano was required and Taylor made the piano available. It is my impression that at the time it was a donation.

The Wareham Hall was a disaster waiting to happen. It was a miracle that it didn't collapse during one of the wingdings held there during those boom days of the 1950's. It eventually had to be torn down and at the time the piano was taken out of the hall and leaned up against the back or side of the church, where it remained open to the elements.

Sometime in the early 1960's, Harry Gordon-Cooper, Clerk of the Territorial Court (and author of *Yukon Tales*), was in Mayo on court business, and while taking a walk, discovered the beautiful old piano, battered by the weather, and made some kind of a deal with the then minister. He had the piano shipped to Whitehorse, restored and tuned, and eventually sold it to Sally Halliday, a CBC reporter living near Whitehorse. Sally has since moved to Alberta. – *Gordon McIntyre*

[Sally Halliday sold the piano to Archie Lang, owner of the Watson Lake Hotel, before she left the Yukon. Lang placed the piano with John Ross, piano restorer. The piano has not been restored as yet (1988) and at some time, has been through a fire.]

TAYLOR & DRURY BUILDINGS AND RECORDS

Block 4, Lots 10, 11 & 12

The Taylor & Drury store and warehouses stand in Mayo today looking much as they did when they were owned by Taylor & Drury. They are currently owned by Rick Cocks who operates Mayo Craft Supplies. In 1985 Cocks donated the Mayo and Keno store records to the Mayo Historical Society. The records had been stored in the Taylor & Drury buildings since operations ceased in Mayo.

Yukon Archives agreed to care for the records until such time as an appropriate facility was available in Mayo to house the records. A total of 89 boxes of records were packed and shipped to the Archives. Many of the boxes containing the store records were in original wooden store crates. In addition to the store records, Cocks also gave the society a few small artifacts: 1 Ernst Leitz, Wetzlar Yellow Camera Filter #2 in box; 1 door key; 2 cardboard powdered milk barrels; 1 small display case in poor condition. The cardboard powdered milk barrels and the display case are at the McBride Museum in Whitehorse.

Cocks also donated the 22 photograph negatives he found laying about the buildings, to the Society. Some of the prints made from these negatives appear in this book. Charlie Taylor was the photographer.

ALEX NICOL'S WAREHOUSE

Block 9, Lots 7, 8 & 9

(The Legion Building)

This two-storey log structure was built by Alex Nicol in 1936 or 1937. He planned to rent the building to the Bank of Montreal but Gene Binet built a frame structure next to Nicol's building and the bank opted to lease from Binet. The building remained unfinished until approximately 1950 when the Yukon Territorial Government rented it to store liquor. It was completed at that time.

GENE BINET'S BANK BUILDING

Block 9, Lot 10

(The Royal Bank Building)

Gene Binet had this building constructed in 1937 to rent to the Bank of Montreal. Excavation work on the cellar and foundation of the building commenced during the first week in August, 1937. It was located on Front Street. The building is of frame construction and was built by James Mervyn. The building adjoined the general store of Binet Bros., and was (and is) situated on the high knoll of ground midway between the store and Alex Nicol's two-storey building (the Legion Building) next to the liquor store. Fittings for the inside of the structure and finishing materials were brought in from the "outside." Lumber for the exterior and roof of the building came from the Lefebvre and Kimbel sawmill located in Mayo. Archie Close scraped out the cellar and leveled the ground surface with his team of horses. Mervyn was assisted in the construction by Albert Pelland and Kaiser Mervyn. On Tuesday morning, October 12, 1937, manager A. T. Hall, and accountant C. W. Harrison opened the bank for business. (The bank, up to this time, was located in the McCown building.) The business office occupied the front part of the main floor. The counter was constructed in a continuous L-shape, with the teller's desk to the left as one entered the front door. The manager's desk was to the right of the teller. A ledger desk was also installed. The old-style tellers cage used in the McCown building was done away with. The counter was faced with insulite hardboard and the walls were lined with a neutral-shaded canec board. Heavy, inlaid linoleum was used as a floor covering in the customer lobby. The window casings, wall strips, doors and mouldings were of the finest outside fir. The main door was of cedar. Artificial light was provided by two large hanging lamps of modern design. A metal mail box was attached to the outside wall near the main entrance. With the exception of the material used for doors, window casings and finishings, native lumber was used in the construction. The entire building is 28 x 32, two storeys high with basement running the whole length of the main floor. A central heating unit in the basement was to insure uniform temperature both upstairs and downstairs. In addition to the main office and vault room, there were spacious living quarters both on the main floor and upstairs. The upstairs quarters were provided in the event that the bank should send a married manager. On the day the bank opened in the new building, customers were presented with compli-

Gene Binet's home, yard, and pens, winter of 1926. *Isabelle Poirier Collection*

mentary gifts: cigars for men and chocolates for lady customers. – *The Mayo Miner*, Oct. 15, 1937

THE GENE BINET HOUSE

Block 8, Lot 12

This house was built for Gene Binet circa 1920–25 by a carpenter named John McDonald. Binet had it built for his young bride, Jewel, whom he brought to Mayo from the U.S. It was considered one of the finer homes in Mayo at the time. It was said that the wallpaper was imported from France.

In 1937, Mr. and Mrs. Neil Keobke lived in the Binet house. That year, Hugo Seaholm, carpenter, and G.S. Aylwin sided the building with lumber to make it warmer during winter months. Robertsons and later Cambiottis lived in this house as well.

DICK MAJOR'S HOUSE

In November of 1937, Mr. and Mrs. Alex McCarter moved from the Lovdahl house (Block 6, Lot 10), where they had been living since the previous spring, into another residence located on 4th Avenue. The residence they moved to was the former home of Mr. and Mrs. Dick Major [it later burned down]. At the time the McCarters moved into it, it was owned by Bud Fisher who purchased the property early in 1937. In August, 1938, the McCarters built their own home on Congdon Street on property which they purchased from Joe Longtin (Block 6, Lots Part 16, 17 & 18). The carpenter in charge of building the house was W.D. Malcolm, assisted by J.J. Robinson. Malcolm and Robinson flew in from Dawson to construct the house.

Mr. and Mrs. Bon Kunze lived in the house later, followed by the Gordon McIntyres, who lived in the house for many years.

Joan Shaxon owned and lived in the house during her teaching years in Mayo. She sold the house to Cal Lindstrom in the 1980's.

DR. NASH'S FIVE MILE LAKE COTTAGE

The cottage was built by Ed Kimbel in Mayo and moved by bulldozer out to the property the Nashes had at Five Mile Lake in July 1938. The cottage was referred to as a bungalow. Mayo Utilities installed a telephone in the cottage for the doctor, as the Nashes lived there during the summer months. Dr. Nash commuted between his lake residence and Mayo by bicycle.

WES PALMER'S HOUSE

Wes Palmer's home was located on 4th Avenue. In September, 1938, his brother Elmer Palmer built a two-room addition onto the house.

This house was later occupied by the Yorke

Wilson family followed by Rupe and Harriet Steeves and then Winnie and John Shandro.

THE "OLD" AIRPORT TERMINAL BUILDING

Lot 37, Group 1004

To accommodate passengers during the cold winter months, Northern Airways Ltd. erected a waiting room at the airport in October, 1938. Elmer Palmer constructed the building with his brother George assisting. This building continued as the Mayo airport passenger terminal until 1984.

JOE MCCAFFERY'S CABIN

This log building on Centre Street, just past Third Avenue, was a landmark in Mayo for many years. It was the scene of many gay and rollicking times in the era of the camp when it was the "gathering place" for the clan from the hill. McCaffery dismantled it in 1939 and built a smaller cabin on the south corner of the lot.

THE DEMPSTER HOUSE

The Dempster residence was located next to Lot 10 Block 6. The house was later owned by the I.O.D.E. In September, 1938, the Mayo Public Library, which was operated by the Mayo Chapter of the I.O.D.E., was moved from its former location in the school building to the Dempster residence. In May of 1938 a new foundation was put under the house so that it would be ready for the library. The large timbers used for the foundation were donated by Elmer Middlecoff. The residence had a large verandah in front.

This building was also used as a temporary classroom for grades four to six in 1953–54.

The Dempster residence was eventually sold. It was moved and the structure altered and can be seen today as the building known locally as the "games room," located on Block 4, Lots 25 and 26. It bears no resemblance to the original Dempster home.

TREADWELL YUKON'S WAREHOUSE-GARAGE

Block 3, Lots 1, 2 & 3

(Genier's Warehouse-Garage)

This large tin-covered storehouse was built and owned by Treadwell Yukon. It is currently owned by Joanne Genier.

Mayo Airport, circa 1940. *G. A. McIntyre photograph*

TREADWELL YUKON OFFICES
Block 2, Lots 1 & 2

This is where Treadwell Yukon's Mayo office was located. J.J. McCarthy later lived in this house followed by Dale and Betty Robertson circa 1947, and then the Jack Foleys lived there. It became a temporary school for grades four, five and six in 1953–54. It was bought by the Baptist Church. In 1989, this building was demolished and the lot sold.

JEAN BOYLE'S CABIN
Block 17, Lots 3, 4 & 5

Frank Cantin constructed this house from fire-killed logs, in the 1950's, and rented it to various people. Jean Boyle lived in it for many years. It was later bought by Bert Klippert who sold it to Dan Sabo.

"SPOT CASH" BREADEN'S CABIN

This cabin was built by Spot Cash and he lived in it for many years. It stands on Block 2, Lot 11.

JIM MERVYN'S HOUSE
Block 9, Lots 23, 24 & 25
(The Wallingham House)

Jim Mervyn built this house in the 1930's so that his children could stay in Mayo to attend school. In 1936 Mervyn rented it to the R.C.M.P. to use as a barracks. Mervyn sold the house to Martha Noel Mast who continued to rent it to the R.C.M.P. until the early 1950's.

Maggie and Dick Wallingham lived in the house for many years and raised their family in it.

EDDIE KIMBEL'S CABIN
Block 23, Lot 39
(Irene Melancon's House)

This house was built by Eddie Kimbel. Dunc & Ruth McGeachy lived in it for many years. It is now occupied by Irene Melancon.

SAM BLACKMORE'S FRAME HOUSE
(Joan and Dick Ewing's Shed)

This building was built by Sam Blackmore before 1930 and was located next to Mabel McIntyre's cabin (to the south). The building was used as the Mining Recorder's office until 1942 and was later used as a high school classroom before becoming an annex to Ruth's Novelty Shop. It now sits at the back of Dick Ewing's lot, near the alley.

THE C.D. MATHENY HOUSE
Block 6, Lot 10
(The Tony Blais House)

This house was originally built by Thomas Collins in the 1920's. Matheny lived in the house until he died. In 1938, Herb Lovdahl bought the house and renovated it. At one time it was the only house in Mayo with hardwood floors.

ABOVE, I.O.D.E. Hall in Mayo. *A. K. Schellinger Collection*
BELOW, A view of the front of Treadwell Yukon Co.'s dispatch office and warehouse without a foundation. The foundation was washed away when the Stewart River flooded Mayo in June 1936. *W. S. Hare photograph, E. C. Bleiler Collection*

Treadwell Yukon offices in Mayo, 1926. *Isabelle Poirier Collection*

Fred Taylor bought the house and contents in the 1940's. Dorothy Allan lived in the house in the 1950's. More recently Bonnie and Jack Smith, followed by Tony Blais, lived in the house.

THE GIDEON LETOURNEAU HOUSE
Block 17, Lots 11, 12, 13
(The Michael McGinnis House)

Gideon Letourneau built this two-storey house for himself and his wife in the 1920's. Mrs. Anthony Kondroski lived in the house in the 1930's. The house stands next to the Christ the King Mission. The house was later owned by Evelyn and Gordon Lee, George and Marge Smith, Chuck and Florence Beaumont and Ed and Jean Kunze. Ed Kunze then rented the house to C.N.T. as a residence and switchboard location It is currently owned by Michael McGinnis (1990).

THE CLIFTON HOUSE
Block 1, Lots 1 and 2

This house was built for Joe and Margaret Clifton in the early 1920's by the Kimbel brothers. After Clifton moved to B.C., the house was occupied by Gordon and Evelyn Lee and family, Geoff and Mary Bidlake and family, Ed and Flora Bleiler and family and Cliff and Maureen Greig and family. The Greig family had the longest residency, living in the house from 1955 to 1976.

JOE HOYNE'S CABIN
Block 15, Lots 1 & 2
(Cooper's Cabin)

This cabin was built by Martin J. "Joe" Hoyne in the early 1920's. Hoyne was a prospector who lived in the cabin for around 20 years. The cabin was built of local logs and is approximately 16x20 feet. The roof was made of flattened gas cans.

There is an adjoining shed of similar construction beside the main cabin. In the 1940's, the cabin was bought by Phoebe and George Reynolds. Mrs. Reynolds served "high tea" in her little cabin using a silver tea service encrusted with ivory.

Dr. Hilda Hellaby (Anglican Deaconess) lived in the cabin for about four years while she was teaching at the Old Village.

In more recent times, Sylvia Hackbarth, a German lady lived in the cabin. She is remembered for her kindness to children. She decorated the trees in her yard with balloons so the children would enjoy it when they played there. She died alone, without family, and her things were there just as she had left them, when Ken Cooper bought the cabin in 1972.

In the mid 1970's the cabin was renovated.

The canvas partitions were removed, the flooring was replaced, water and sewer lines were installed, it was reinsulated, electricity and electric baseboard heating were installed, and a porch and loft were added. The cabin was jacked up in the summer of 1975 and the deteriorating bottom row of logs was removed. These were replaced with two rows of new logs. The roof remains with the white gas cans as cover. – *Shannon Cooper*

LANSING CREEK TRADING TOKENS

Jim Mervyn used trading tokens rather than currency at his Lansing Creek post as the traders of old had done. (This was a common practise in the days of the Hudson's Bay Company, the Northwest Co., Lamson Hubbard, and Taylor, Drury & Pedlar.) There were three different denominations of Mervyn tokens:

$ 1.00 token of aluminium	(36mm in size)
$10.00 token of brass Octagon	(30mm in size)
$20.00 token of brass Octagon	(36mm in size)

In 1938 when Mervyn closed Lansing post, he advertised his tokens for redemption. When they were redeemed, he threw them into the Stewart River in front of the Chateau Mayo. The few tokens which were not turned in are now highly prized by collectors.

THE OLD CEMETERY

Lot 78, Parcel A

This was the first cemetery in Mayo. It is thought to have been used from the time the town was founded until the early 1920's.

When the site was first selected, the water table was low. Later, the water table rose and the site was abandoned for a new cemetery near the airport. The names of only 4 of the approximately 15 people buried in the old cemetery can be identified.

Mrs. Joseph "Joe" Cantin was buried there in 1917. Mr. and Mrs. Joe Beaupre are also buried there. Mr. Beaupre was the last person to be buried in the cemetery.

There is one perfectly preserved marble headstone on the site. This belongs to Garnet Banister, the man who was killed in a saw accident, while working for Thomas Aitken at the Silver King, in 1916. Banister was born in Buenos Aires, Argentina.

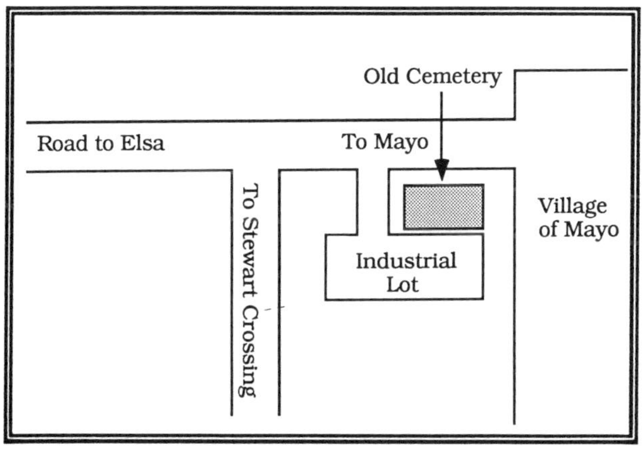

TOP AND MIDDLE, James Mervyn's brass token for $20.00. *Eileen Shilleto Collection*
BOTTOM, Map of the Old Cemetery.

JIM FERRELL'S PHOTOGRAPHS

When Jim Ferrell died in July 1941, money was scarce and his wife, when she left, was not able to take along all of the possessions the couple had accumulated over a lifetime. She asked several people if they would like to have any of the items, but no one was interested. A couple of crates of Ferrell's photographs were among these items. Unwilling to take them to the garbage dump, she hired Jack Smith to dig a deep hole between Ferrell's house and store. She put the photographs in the hole, as well as some dishes, a few suitcases which had been left at the store for safekeeping by drifters (who never returned to retrieve them), the hide of the grizzly that mauled Jim Christie at Lansing and one other grizzly hide, and any other household items that could not be packed. Smith filled in the hole and Mrs. Ferrell left for Oregon, where it is believed her brother lived. It is assumed that the hole was approximately 6 feet deep because it was as deep as Smith was tall and it is estimated that he was between 5 feet 8 inches and 5 feet 9 inches. Years later, (1950's) approximately 4 feet of fill was hauled onto these lots and it is probable that the artifacts are now as far as 10 feet down. Very likely, Ferrell's photographs were glass plates, since pieces of glass plate photographs were later found in the buildings. The Ferrell's house and store were located on Lots 11, 12, 13, 14 and 15 in Block 7. Ferrell purchased them in June 1923, and they remained in his name until 1949.

Whether or not anything remains of the artifacts after so many years, remains a question. Every effort has been made to pinpoint the location but, so far, without success. These are privately owned, developed lots and the location must first be determined within a few feet before digging. An aerial photograph of the Ferrell lots (circa 1937), shows the buildings, and a map indicates where the artifacts are buried in relation to the buildings. Mayo's streets and landmarks themselves have changed considerably since the time the artifacts were buried.

ABOVE, The Chateau Mayo Hotel in the 1930's. *G. A. McIntyre Collection*
BELOW, Mayo from the air, 1937. *F. & T. Whitney Collection*

CHATEAU MAYO HOTEL

Block 9, Lots 12 to 15

The first hotel on this site was owned by J.E. Binet who built a 30x60 foot, two-storey log structure, with dining room, in 1903. It was known as Binet Bros.' hotel and it was the first commercial establishment in Mayo, and, other than a few cabins, it was the first structure erected on the townsite. He later enlarged the hotel to 60x60 feet. The second floor had 24 rooms, with a total of 44 beds. In 1921, he built a new 30x60 foot store building on to the hotel. With the addition, the building had a total frontage of 90 feet and a depth of 60 feet. By the early 1920's he had invested $50,000 in buildings and equipment for the complex. The entire place was lighted by an electric light plant located in the large basement and supplied 60 lights. Barns, outhouses, and some adjoining buildings were also lighted by the plant. Under the hotel and store complex were two large cellars, one 30x26 feet, the other somewhat larger, and both 10 feet deep. The cellars were used for winter storage of vegetables and perishables. The complex was supplied with water pumped from a well on the edge of the Mayo River, and filtered through a heavy bank of gravel. Large tanks in the garret supplied water for fire and general purposes. Sewage was drained through a heavy pipe into the Stewart River. Binet also had a waterfront warehouse, 30x60 feet, another warehouse for oil and several stores and buildings in town which he rented to other business concerns. – *Dawson Daily News – Yukon and Gold Fields Edition, 1922.*

The sign on the front of the hotel and store read "Hotel — Binet Bros. Merchants — Store" but Binet referred to the hotel as the Chateau Mayo on his business stationery.

On July 13, 1930, the log hotel burned to the ground and Binet immediately erected a new frame construction hotel. In 1938, he sold the hotel to Jim Mervyn, former owner of Lansing Post. Mervyn leased the hotel to Fred Matthews and Dick Mercure from 1947 to 1950 and then sold the hotel to Ed Barker, Elsy M. Rowan and Robert J. Rowan of Tourist Services. Dunc and Ruth McGeachy and their sons, Mike and Brian, moved to Mayo from Whitehorse to manage the hotel for Tourist Services.

Joseph A. "Alex" Wark and Dale Robertson bought the Chateau Mayo from Tourist Services in 1953. Roxy Carriere managed the hotel for them. When Robertson became Territorial Secretary and moved to Whitehorse in 1954, Wark bought him out.

Dick Wallingham joined the Chateau staff in 1954 and worked as bartender until 1956 when he and Carriere became partners and leased the Chateau from Wark. When Carriere retired in 1962 and moved to Carmacks to open a store, Wallingham and his wife, Maggie, leased the Chateau Mayo and then bought it in 1967. They owned the hotel until it burned down on March 23, 1987.

Overview drawing of the cabins past the Anglican Church. *Drawing by Linda MacDonald*

CABINS UPSTREAM FROM ANGLICAN CHURCH

Information supplied by Gordon McIntyre (Sketch is not to scale but it is representative.) From left to right on sketch:

1. Anglican Church and Rectory (within fenced enclosure).
2. Jack Royston 1920's and 1930's.
3. Hugh Monohan 1920's, 1930's and 1940's. "Hotstove" Douglas late 1940's and the 1950's
4. Fred Swanson 1920's and 1930's. Mabel Huffman rented the cabin in the 1930's and Ed Barker (purchased cabin from Swanson estate) 1940's, 1950's and 1960's.
5. Isaac Mallette owned the cabin in the 1920's, 1930's and 1940's. It was rented to various tenants including Bill Taylor and Charlie Taylor.
6. Ben Cameron 1920's, 1930's, 1940's and 1950's.
7. Possibly Bob Levac's cabin. Jim Anderson's residence in the 1930's.
8. (not shown on drawing) Snell's cabin and the Cox, Snell & Cox sawmill.

This cabin was situated on the river bank past the Anglican Church and belonged to Hugh Monohan from the 1920's to the 1940's. "Hotstove" Douglas owned the cabin from the late 1940's through the 1950's. *G. A. McIntyre Collection*

Fred Swanson's cabin which was later rented by Mabel Huffman and then bought by Ed Barker. *G. A. McIntyre Collection*

VETERANS

World War I

Shortly after war was declared in 1914, Joe Boyle raised approximately 55 recruits in Dawson City. This group was known as The Yukon Machine Gun Detachment. In October of 1914, the men left Dawson for training in Vancouver and became part of the Second Canadian Mounted Rifles Regiment (in the First Canadian Mounted Rifles Brigade). The section sailed for England aboard the S.S. *Grampian* on July 1, 1915. Upon their arrival in England they became the #4 "Boyle" Battery of the Eaton Motor Machine Gun Brigade. In 1916 the Battery was recruited as a motor machine gun company with the Fourth Canadian Division and sent to France. In 1918 it became part of the Second Motor Machine Gun Brigade, along with the Eaton Battery, the 17th Machine Gun Company (the former Yukon Infantry Co.) and the 19th

Machine Gun Company. Boyle's Contingent were at Somme, Passendale, and the retreat at Amiens. (Rodney, William *Joe Boyle, King of the Klondike.* Yukon Archives Search File – *Biography of Joe Boyle.*)

In 1916, George Black raised 275 men in Dawson City under the name The Yukon Infantry Company. The men left Dawson on October 16, 1916, for training in Vancouver and Victoria. On January 24, 1917, the Yukon Infantry Company, part of the Canadian Expeditionary Force, embarked from Halifax on the S.S. "*Canada*" for England. In England, the company was renamed the Yukon Motor Machine Gun Battery and it became part of the Fifth Canadian Division known as The 17th Canadian Machine Gun Company. In January, 1918, it was sent to France as part of the First Motor Machine Gun Company. In June of 1918, the company was transferred into the Second Canadian Motor Machine Gun Brigade (the Yukon and Eaton Batteries forming its nucleus) along with the 19th Canadian Machine Gun Company. (Martha Louise Black's Scrapbook – *World War I*, Yukon Archives Manuscript Collection.)

Dave Fotheringham. *McLaren-Swensen Collection*

The Second Canadian Motor Machine Gun Brigade was formed in June, 1918, from the Yukon Motor Machine Gun Battery (Boyle's recruits), the Eaton Battery, the 17th Machine Gun Company (Black's recruits) and the 19th Machine Gun Company. The King's and Regimental Colours were presented to the 2nd Canadian Motor Machine Gun Brigade on April 3, 1919, at Seaford, England. The headquarters of the Brigade was in Vancouver, B.C. When the Brigade arrived home in May, 1919, its colours were given to Vancouver City Hall to be placed temporarily in the mayor's office. The regimental colour, maroon, carries upon it the badge of the brigade and on the scroll beneath, the motto of the brigade "Straight and True" with the battle honours of the unit "Arras, Cambrai, Amiens." (*The Daily Province*, Vancouver, May 20, 1919.)

Following is a partial list of World War I veterans who were either in the Mayo district when war was declared or came to the district at the end of the war. (Unfortunately, this list does not include men from the Mayo district who may have joined in locations other than Yukon and came to Mayo after the war.)

BREADEN, James
BUCK, Rev. Frank Hepworth
DAGLISH, Hamilton
DOUGLAS, Walter Scott
DOUGLAS, William Lawrence
DRAPER, Peter
ELLIOTT, William James
FOTHERINGHAM, David H.
GLEESON, Daniel
GREAVES, Robert Bruce
GREENAWAY, Thomas
HENDERSON, Harvey

JOHNSON, Frederick John
JOHNSTONE, James Kennedy
LeBOEUF, Albert
MEREDITH, Edward
MILTON, John Joseph
O'BRIEN, James Jonathan
O'BRIEN, William John
NOYD, Amos
RAE, Samuel
REID, George Thomas
SMITH, Clarendon Clark
STEWART, Alexander
SUTTLES, John Jackson
TOWNSEND, Norton Turner
VUCER, John
WILLIAMSON, William Hodge

A few of the men involved in the Mayo district who were known to have joined in places other than Yukon are Ed BURNELL, Arthur CARTER (the first man from Yukon to join the British forces when World War I broke out), Dr. W.W. CHIPMAN, Howard "Harry" COLLEY (who served with Dutch Intelligence under Viscount Alexander during World War I), and Alexander FRASER, who joined the 54th Battalion from Kelowna.

Don Baker, July 1942, CN Station, Edmonton, Alberta.

Louis Bouvette struck silver on Keno Hill in July, 1919, and this discovery brought numbers of returning veterans to the Mayo district.

World War II

Following is a list of men from the Mayo District who enlisted during World War II. This list does not include former residents of the Mayo area who were in the forces, nor does it include veterans who took up residence in Mayo after World War II.

BESNER, Gus
BOYLE, William A.
BUYCK, Wes
CLOSE, Robert
EVOY, Alex
FISHER, C.E. "Kippy"
HARTNELL, Norman
HAYTER, Orville
KALDAL, George
KAZINSKY, Louis
LAMB, William
MCINTYRE, Gordon
MCPARLON, Lynn
NICHOLS, Jack
NICHOLSON, Clint
NIXON, Charles
PALMER, Elmer J.
PALMER, George
PALMER, Jimmy
PELLAND, Albert
PROFEIT, Charley
PROFEIT, Jimmy
PROFEIT, William
RICHARDS, Ted
SEEL, Ernie
SNIDER, I. E.
WASHBROOK, W. B.
WESTMAN, Robert
WIGHTMAN, Norman
VAN BIBBER, Archie

(List compiled by Gordon McIntyre and Wes Buyck.)

The Royal Canadian Legion

Information Compiled by Don Baker

The Keno Hill Branch of the Royal Canadian Legion was founded by a group of service veterans who were employed by United Keno Hill Mines Ltd. The first meeting with recorded minutes was held in the Elsa cook house on February 14, 1954. An explanation of what had been done to establish a Charter Branch of the Canadian Legion of the British Empire Service League, was offered at this meeting. Motions were carried "to form a Legion Branch at Elsa" and "to have the Branch known as the Keno Hill Branch." A temporary slate of officers was elected (MacDonald - Acting President, Roberts – Acting Vice-President, Thompson – Acting Treasurer, and Chisholm – Acting Sergeant at Arms). Charter Members appointed at this first meeting were C.H. Carveth, R.J. Bold, J. Myers, A. MacKillop, L. Curry, B. Mills, and L. Lance.

At a meeting in the Elsa cook house on May 10, 1954, the Branch Charter Certificate Keno Hill (YT #3) Branch, Canadian Legion of the British Empire Service League, was displayed for the very first time.

On July 8, 1954, the first executive was elected: A. Roberts – President, J. Powell – Vice-President, W. Thorpe – Secretary-Treasurer, C. Carveth – Sergeant at Arms. This meeting was held at the cook house in Calumet.

During a meeting at Calumet on August 12, 1954, a motion was carried "to declare an Honorary President and an Honorary Vice-President." Clare White was unanimously elected Honorary President and Alec Berry and Al Pike were nominated Honorary Vice–Presidents.

Through the years, the Legion has played an active role in community events, sports, and the annual winter carnival in Mayo. The first Mayo Winter Carnival Queen sponsored by the Legion was Mary-Ann Curry. The first children's curling team sponsored by the Legion entered a bonspiel in Dawson City on March 7th and 8th, 1959. Transportation for the curlers was provided by "Frenchy" Raymond and the Legion supplied the gas.

Every business in Mayo, as well as United Keno Hill Ltd., were always very supportive of Legion activities. Hutton's Service is remembered for transporting hockey and baseball teams, the Klippert Bros. for supplying wood when it was needed to keep the skating shack warm, and Bill MacIntosh and K.J. Proctor for supporting youth sports programs.

An amusing story was recorded in the early Legion minutes. It concerned a member who asked the Legion to help him get his job back after he was fired by United Keno Hill for drinking and missing shifts. His defence was that he had not been fired by the Army while stationed in Gibraltar, when he stole a fishing boat, took it to Spain, and was AWOL for some time. There is no record as to the outcome of his plea.

LEGISLATORS

The following information is part of a paper prepared and written for the Yukon Archives by Cheryl Conant in 1981, titled *History of Administration in the Yukon Territory*.

Thomas "Tommy" McKay. *L. Cheatham (McLaren) Collection*

The Yukon Territory Act was passed in Parliament on June 13, 1898 and the Yukon District of the North West Territories, proclaimed the year before, became a separate territory. The Act provided for a chief executive called Commissioner, responsible to the Governor in Council or the Minister of the Interior. The Governor in Council could appoint up to six persons as a Council to assist the Commissioner. In 1899, the Yukon Act was amended to allow for the election of two members to Council. Advances were made in representative government in 1902. The Yukon Act was amended again to allow for five elected, in addition to six appointed members of Council, and to allow for the election of a Member of Parliament. On December 2, 1902, James H. Ross was elected as Yukon's first Member of Parliament.

In 1908 an amendment to the Yukon Act made Council fully elective. Its ten members were elected for three-year terms. The first representative Council met in 1909.

The end of the Klondike Gold Rush and the introduction of large scale mining resulted in a drop in population from 27,000 in 1901 to 10,000 in 1908. The population continued to decline. The census of 1911 showed 8,512 residents and by 1921 it had dropped to 4,157.

In 1919 the Yukon Act was amended to reduce the Council to three elected members. The Commissioner's office was abolished and its duties transferred to the Gold Commissioner. George Black, Commissioner from 1912–1916 was the last person to hold the office until 1948 [when it was resumed].

The amendments of 1908 continued to be in effect in so far as the Council could still present bills to the Gold Commissioner for assent. The Gold Commissioner initiated all bills regarding revenue and retained power to summon juries, control liquor traffic, game laws, establish and maintain schools, and administer other local matters. Administration dwindled with the population; in 1929 revenue totalled $220,000 consisting of a $118,000 federal grant. By 1941 the grant was less than $50,000. During the Depression, liquor profits became the largest source of territorial income.

The office of the Gold Commissioner was held by G.P. MacKenzie from 1918–1925, Percy Reid from 1925–1928, and George MacLean from 1928–1932. In

Keno area in winter. From left: Warren McFarland, Eddie Kimbel, Ernest J. Corp, and Gordon Lee.
G. & E. Lee Collection

Councillors and M.L.A.'S

1903—1905	Rev. J. Pringle (Klondike)	1920—1922	Paul Shakespeare Hogan (Klondike)
	Maxime Landreville (Klondike)	1922—1925	James Emmett Ferrell (Klondike), Mayo resident
1905—1907	George Black (Klondike)		
1907—1909	George Black (Klondike)	1925—1928	James Emmett Ferrell (Klondike), Mayo resident
1909—1912	Maxime Landreville (Klondike)	1928—1931	Charles Bossuyt (Mayo)
	Angus Wm. McLeod (Klondike)	1931—1932	Frank Carscallen (Mayo)
		1932—1934	Thomas McKay (Mayo) appointed
1912—1915	Eugene A. Hogan (Klondike)		
	Archie N. Martin (Klondike)	1934—1937	Ernest Joseph Corp (Mayo)
1915—1917	Archie N. Martin (Klondike)	1937—1940	Ernest Joseph Corp (Mayo)
	John F. McCrimmon, Speaker (Klondike)	1940—1944	Richard Gordon Lee (Mayo)
		1944—1946	Ernest Joseph Corp (Mayo)
1917—1920	William Campbell Lowden (Klondike)	1946—1949	Ernest Joseph Corp (Mayo)
		1949—1952	Ernest Joseph Corp (Mayo)
	James Singleton Wilson (Klondike)	1952—1955	Alec F. Berry (Mayo)

(*Information provided by Piers McDonald,* M.L.A.)

1932 the Gold Commissioner's office was abolished and government became the responsibility of the Controller. G.A. Jeckell held this position until 1947, then John E. Gibben to 1948.

During the 1940's Council had only indirect control over administration which was generally handled by federal departments.

An amendment passed in 1948 provided for the appointment of a Commissioner and Administrator. The indemnity for councillors was increased and the territorial civil service enlarged. During the late 1940's the population in the territory began to grow, due evidently to mining activity and augmented military establishments. To reflect population growth and shift to the southern part of the territory, in 1951 the Council was increased to five members to be elected for three year terms. In 1953 the capital city was moved from Dawson to Whitehorse.

In 1960 Council was increased to seven members, and the Commissioner was allowed to sit in Council sessions again.

The Mayo district was represented by the member for the Klondike until 1928.

TOP, A view of the sternwheelers at dock in Mayo. Left to right: the *Nasutlin* (only partially in view), the *Canadian* and the *Alaska*. May, 1923. *Wm. Hare photograph, Ralph Rogers Collection*
MIDDLE, The *Canadian* docked near Treadwell Yukon's ore pile, July 1926. *Isabelle Poirier Collection*
BOTTOM, The *Canadian* docked at Mayo unloading lumber and materials from Dawson, circa 1923. *Ralph Rogers Collection*

Chapter 7

Transportation, Communication & Electric Power

STERNWHEELERS

STEAMBOATING BEGAN ON THE STEWART in the late 1800's when bar gold was found along the river. McQuesten and Mayo of the Alaska Commercial Company operated the little steamer *Yukon* on the river, bringing supplies and transporting miners to and from the district.

In the spring of 1883, the *Yukon* was sunk by an ice jam and McQuesten and Mayo purchased the *New Racket* from the Scheiffelin party who were leaving the country. The *New Racket* was operated by both McQuesten and Mayo along the Yukon and Stewart Rivers.

After the discovery of gold in the Klondike, there were more than 60 boats and barges on the Yukon River. With the completion of its railroad to Whitehorse, The White Pass purchased the *Canadian*, the *Columbia*, the *Sybil*, and the *Yukoner* from the Canadian Development Company, and in 1901 created The British Yukon Navigation Company to operate boats between Whitehorse and Dawson.

Steamer *Canadian* and two others at Mayo docks. *F. & T. Whitney Collection*

With the discovery of gold on Duncan Creek, boats began to service the Stewart River. The riverboat *Lightning* brought miners and supplies up the Stewart River circa 1901–1909, as did the Barrington brothers' Side Stream Navigation Company boats. Their boats, the *Prospector* and the *Vidette*, brought miners and supplies to Gordon Landing, initially the disembarking point for the overland trek to Duncan Creek, and then to Mayo Landing when it was established in 1903. The *Vidette*, originally named *May West*, was 134 tons and was built in 1896. When the *May West* was bought by the R.N.W.M.P. (from the original owner), it was renamed *Vidette*, and later sold to the Barrington brothers. It transported the first shipment of silver-lead ore mined from the Silver King by Alverson and Huffman. The *Prospector* (165 tons) was built in Whitehorse in 1901. The machinery was eventually taken out of the *Prospector* and installed in the *Nasutlin*. The Side Stream Navigation Company supplemented the White Pass's British Yukon Navigation Company from 1909 to 1919. Other side stream boats such as the *Tanana* also operated on the Stewart.

The development of the Silver King and Louis Bouvette's major discovery of silver ore at Keno Hill in 1919, greatly added to steamboat revenue. At low water, ore was taken from Mayo to the Yukon River on barges pushed by the shallow draft sternwheelers *Vidette* and *Pauline*, and left at Stewart City to be pushed up river by the larger boats. Sometimes a sternwheeler would push two barges loaded with ore up river to Whitehorse. The ore was then shipped by rail to Skagway and waiting cargo ships. When the water was high, larger boats such as the *Canadian*, the *Aksala*, the *Nasutlin* and others were able to navigate the Stewart. This method of transporting ore continued until 1950. In 1922 the sternwheelers hauled some 12,000 tons of sacked, high-grade ore from the Yukon Gold Company and the Treadwell Yukon operations.

The Mayo dock after a steamboat has unloaded its cargo, circa 1920. *G. A. McIntyre Collection*

Wood was the principal source of fuel used by steamboats. Each of the main river steamers burned 8,000 cords of wood at an approximate cost of $55,000 per season. A boat could burn two

A view of the sternwheeler *Canadian*, three barges, and silver-ore sacks on the Mayo river front. The sternwheeler and barges are carrying 600 tons of silver ore. The corrugated metal building is the White Pass freight shed and the building with the false front is the Northern Commercial Co. store. May 19, 1923. *W. S. Hare photograph, Ralph Rogers Collection*

cords an hour going upstream. Wood camps were established approximately 30 miles apart along rivers with boat traffic. Hundreds of men were hired through the years to cut wood and stack it along the river. Wood was usually stacked and dried for one year before it was used.

In 1919, Taylor & Drury bought the *Thistle* from White Pass. The *Thistle* was built in Puyallup, Washington. It brought supplies from Whitehorse to Taylor & Drury's Mayo–Keno operations three or four times a year. The *Thistle* was approximately 225 tons and pushed a 60 foot barge, permitting loads of 80 to 85 tons.

Unloading the S.S. *Keno* at the Mayo dock. *Middlecoff-Bleiler Collection*

In 1922, The Alaska Railroad and The White Pass joined forces to haul silver-lead ore from the Mayo district. Soon their cooperation spread to include a joint freight rate agreement, and later tourist schedules were worked out in conjunction with the Alaska and Pacific Steamship companies and the principal Richardson Highway carrier.

In 1929, the *Thistle* sank in Lake Laberge with a full load of cargo destined for Mayo. There was no insurance in those days and Taylor and Drury had to absorb the loss not only of the boat but its cargo as well. Afterwards, the company negotiated an acceptable freight rate with The White Pass and purchased the *Yukon Rose* to supplement their freight hauling requirements. The *Yukon Rose* was a side stream motor-powered boat, built for Taylor

S.S. *Keno* hooking up to a fully loaded barge at Mayo. *Ken Schellinger photograph, A. K. Schellinger Collection*

S.S. *Keno* pushing a barge on the Stewart River in front of Mayo. *G. A. McIntyre Collection*

and Drury in Vancouver in 1930. They operated the *Yukon Rose* on their Whitehorse to Mayo run until approximately 1943 when the boat was bought by a man named Jack McDonald who worked for the mine in Keno.

Men who captained boats on the Stewart River in the 1930's include Campbell, McMasters, Charlie Gardiner, "Kid" Marion, and Bill Bromley. Captain Hugh Morrison was the skipper of the *Keno* throughout the 1930's.

The *Keno* was built at Whitehorse and completed in August 1922 as a side stream freight hauler. In 1937 she was made ten feet longer to accommodate more freight. She worked mainly the Stewart River, hauling silver-lead concentrates from Mayo to Stewart City, a distance of 180 miles. The *Keno's* draft was less than two feet, making it easy to navigate the Stewart at low water. The boat worked the river from 1922 to 1951. When there was a large ore load to transport, the boat pushed a barge, both loaded with ore. The *Keno* could haul 120 tons, and the barge

Dave Fotheringham's boat *Klondyke* on the Stewart River in 1923. *Ione Christensen Collection*

S.S. *Klondike* going by the Old Village. *Dave Hager Collection*

an additional 225 tons (each sack of silver-lead ore weighed 125 pounds). The *Keno* had a crew of approximately 21 men, plus captain and pilot.

Boats operating on the Stewart in the 1940's included the *Keno*, *Aksala*, *Nasutlin*, Taylor & Drury's *Yukon Rose* and the tunnel boats *Hazel B* and *Neecheah*; among others. The *Neecheah*, a 65 foot boat with a 60-horsepower gasoline engine, began operation on the Stewart about 1922. She carried a horse who was trained to step off the gang plank, drag out cables for lining the boat up over five swift places on the Stewart River, and then step back on the boat, thus earning the nickname the "Hayburner."

S.S. *Aksala* and *Keno* with barges docked in Mayo. *Dave Hager Collection*

When trucking ore proved more economical than barging it, the steamboat era ended. The *Aksala* was docked for good in 1953, and the *Casca, Klondike, Whitehorse* and *Keno* in 1955.

In the 1950's, Captain Hughie (Angus) Morrison brought the *Klondike* up with a barge. It was one of the last runs of the large boats. The last sternwheeler to travel a river in the Yukon was the *Keno* when it steamed to Dawson on August 23, 1960, under the command of Captain Blacky, a Mackenzie River captain, and Frank Slim, a famous native Yukon river pilot. In Dawson City, the *Keno* became a museum and National Historic Site under the care of Parks Canada.

The sinking of the S.S. *Nasutlin* in shallow water in the Stewart River near Mayo, May 1920. *McLaren-Walloch Collection*

TRAILS AND ROADS

In 1902 The White Pass & Yukon Route was contracted by the Yukon Territorial Government to construct a 330 mile overland trail from Whitehorse to Dawson City. It was built by a small crew with horses, axes, cross-cut saws, ploughs, and horse-drawn wheel scrapers.

The trail led from Whitehorse, crossed the Yukon River at Mackays, continued to the Pelly River (crossing about three miles above the mouth), then north to the Stewart River (crossing at Valley Creek), up Valley Creek and over Wounded Moose summit, continuing to the Indian River and thence via Eldorado and Bonanza Creeks into Dawson City.

The first overland stage left Whitehorse for Dawson City on November 2, 1902, with passen-

Dawson Trail, April 1911. George (front left) and Hugh Coffey in coon coats. George was Hydraulic Superintendent and later Dawson's resident manager of the Yukon Gold Co. *A. K. Schellinger Collection*

gers G. A. Jeckell, later to become Controller of the Yukon Territory, and H. Wheeler, the Superintendent of Stage Service.

When Mayo became a busy mining camp, the trail north of the Pelly River was changed to go via Scroggie Creek and Blackhills. The road branched off at Minto, crossed the Pelly below Mica Creek and joined the Stewart River at Crooked Creek. A branch ran from that point to Mayo, while the Dawson City road followed the Stewart River for some distance, then went up the Slough Creek Valley and down Flat Creek into the Klondike Valley.

During freeze up and breakup, wheeled stages were used. The stages were built by The White Pass & Yukon Route when it became apparent that their imported models could not stand up to the gruelling wear and tear of the trail. The company had its own carriage, harness and blacksmith's shop, as well as a horse hospital in Whitehorse.

Stage drivers were known as "skinners" and wore coonskin coats with a long red sash tied at the waist, and soft buckskin gloves with silk or wool liners.[43]

In 1904 and 1905 the first winter stage route was constructed between Dawson City and Duncan Creek. When this trail was built a Liberal government was in power and the route was dubbed "The Liberal Trail." From Dawson, it travelled the Dominion Creek road for some distance, then down the Tintina Trench to Barlow (with a branch to Clear Creek), then to McQuesten, Moose Creek, and Mayo. Later, when a Conservative government was elected, the route was changed and the new road became "The Conservative Trail." The trail remained the same from Dawson to Barlow, then moved to Clear Creek,

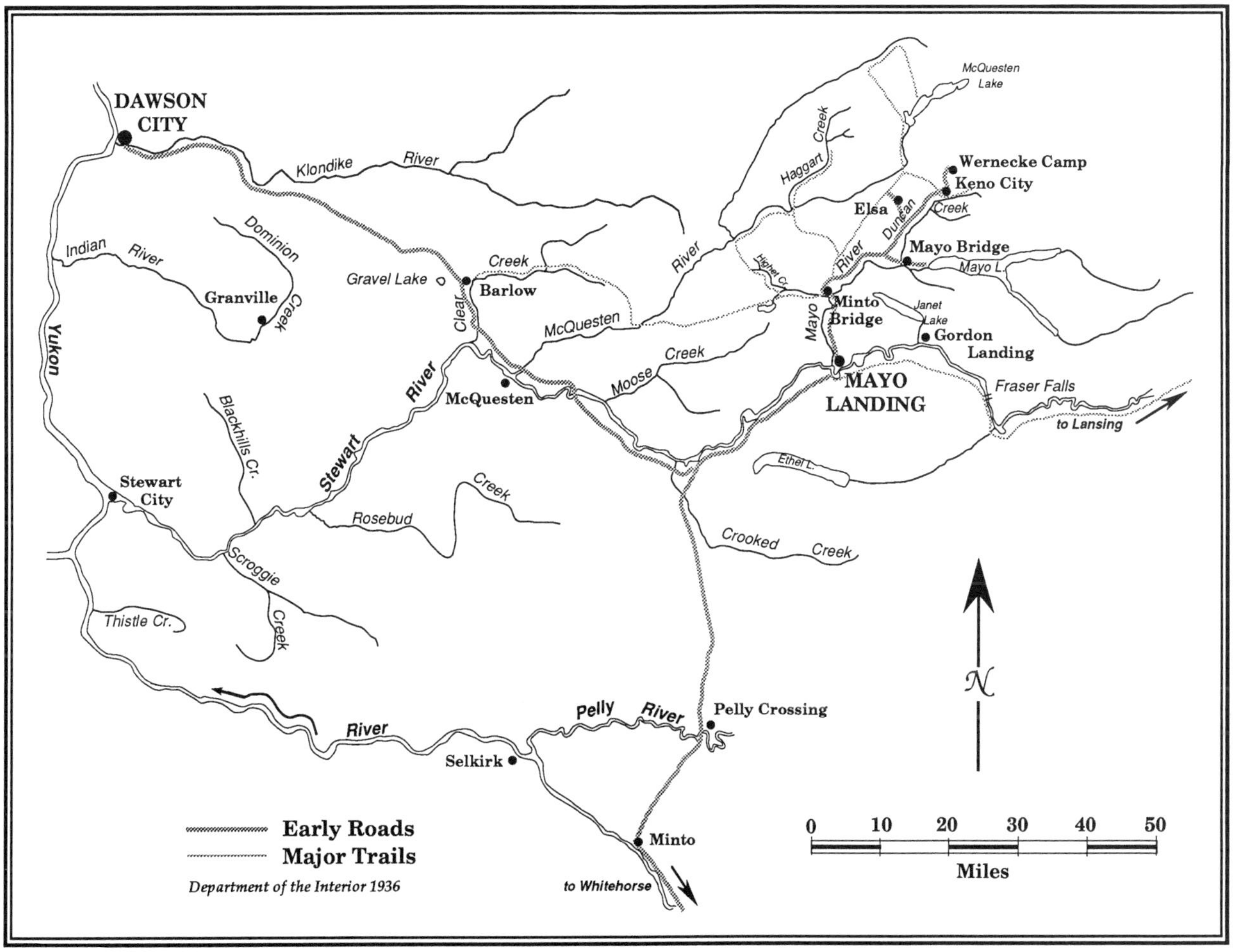

Transportation routes in the Mayo area *Map by E. L. Bleiler*

Teams doubled up (two six-horse teams) for the pull up Glacier Hill, 1922. *Ralph Rogers Collection*

down Vancouver Creek, crossed the McQuesten River and up Bear Creek to Minto Bridge. The merits of the two trails remain debatable. Some said the Conservative Trail was drier and faster while others insisted the Liberal Trail was shorter.

The government also constructed a good wagon road from the mouth of the Mayo River to Duncan Creek, a distance of 24 miles. Roadhouses were established at several points along the route. Travellers to Duncan Creek usually took the steamboat from Dawson to Mayo and then travelled the wagon road to Duncan Creek. In winter, dog teams travelled via Dominion Creek to Clear Creek and then up the Stewart River on the ice.

By 1915, government roads lead from Minto Bridge up all of the main creeks and a branch had also been extended to the silver-lead discovery at

Travellers standing around the horse-drawn sleigh outside the post office and the Hotel Royal Alexandra. *T. McKay Collection*

Galena Creek.

With Louis Bouvette's discovery on Keno Hill and the involvement of Guggenheim's Yukon Gold Co., a road to Keno City was built. It was constructed at a cost of $18,000 using 60 men with picks and shovels. Horses were used to haul logs for corduroy sections (soft areas filled with logs laid in a crosswise manner). With the completion of this road, the hauling of Keno Hill ore began.

Ore could only be hauled in winter. This was accomplished using "drag-rack and rawhides." A drag-rack was a large bob-sleigh with a six foot wide, heavy bunk. Two eighteen-foot, eight inch logs were eye-bolted at one end to each outer end of the bunk. The other ends dragged in the snow at the rear and were held together by a piece of eight foot timber, bolted across the drag ends. A platform was made with cross sills and planks on which four tons of ore could be loaded. Behind the drag, three horse hides or raw cowhides were fastened, each carrying a ton of ore, with a thick pole laced across one end. After the ore was loaded on the hides, the sides and rear were laced through with rope drawn up over the load. The load was brought down the mountain by a four horse team, rough-locked with chain links on steep slopes and dragging rocks as brakes. This method of transporting the ore from the top of Keno Hill, four miles down the mountain to Keno City, was referred to as "rawhiding."

In Keno City the ore was reloaded on to Minnesota logging sleighs, 18 tons to the load. The loaded sleighs were hauled by three teams, hitched tandem, with one driver. There were four relays on the 40 mile trip from Keno City to Mayo. Each outfit made ten miles a day and returned empty to the starting point. There were eight six-horse teams used along the route, with eleven spare horses.

Eighteen men were employed to handle them. After a time, two four-horse teams were used between each of the relay points (a total of 90 head) and cordwood and mine timber were carried on the back haul. Initially, the cost was $50 a ton to haul ore to Mayo, which dropped to $30 a ton as the road improved and the operation became more efficient. Ore hauling with horses lasted approximately three years. After 1923, Cat trains

ABOVE, Minto Bridge, 1910. *Middlecoff Collection*
BELOW, Cat train arriving from Whitehorse, on the Stewart River in front of Mayo, April 1922. A steamboat barge is frozen in the river ice in the background. *Besner Collection*

were implemented.

In 1924, The White Pass & Yukon Route ran their last Royal Mail stage, unprofitably. Horse-drawn stages were replaced by contractors using trucks or "Cat" trains. Cat trains could haul sleighs and cabooses and travel nonstop. Winter passenger service became less frequent, slower and more uncomfortable. Many roadhouses, no longer maintained by White Pass, closed down. Among firms who held contracts to carry mail along the Dawson–Mayo trail were Coates & Kastner, Greenfield & Pickering, and Klondike Airways, owned by T.C. Richards and W. Phelps.

Mayo's White Pass management, 1922–23. Left to right: Jim Fairborn, Ralph Rogers, Allan Fraser, and Jack Bellerby. *Ralph Rogers Collection*

When motorized transport took over the ore haul, Greenfield & Pickering utilized their horses and obtained the Royal Mail contract for Whitehorse, Dawson and Mayo from Coates & Kastner. Greenfield operated the Whitehorse end and Angus McIntyre ran the Dawson end. Greenfield and Pickering held the contract from 1925–1929 and then Klondike Airways successfully bid on the contract. Richards employed Caterpillar tractors on the run. "Slim" Keobke developed a half track truck which was used with some success. Richards lost the contract to the air carriers.

In 1923 Emil Forrest and Cam Smith brought the first Cat train of supplies from Whitehorse to Mayo. Each tractor could haul four sleighs with no relays and no stops. Cat operators became known as "catskinners." Joe McLellan, Scotty Mowatt, Jack Graham, Roy McDermid, Charlie Rankin, and Ed Barker, all former horse teamsters, became Cat operators.

The cold, miserable trip by Cat train in the

"The Galena stopping place," circa 1921. Roadhouse at Silver King. *D. Walloch (McLaren) Collection*

"Cat" train of 1926–27. Silver-lead ore to be unloaded at Mayo for summer transfer to barges for shipment to Kellog, Idaho smelter. *Ed Kunze Collection*

1930's took seven to eight days from Whitehorse to Mayo, and was almost unbearable at –60°F and colder. The Cat trains hauled mainly freight at a cost of 35 cents a pound. In later years, passengers usually went by plane in winter and by steamboat in summer. In the 1930's, scheduled road service ceased.

In 1949 a new, 125-mile road link from Whitehorse to Carmacks was completed. In October, 1950, the highway from Elsa to Mayo and

The road-building crew using a tractor pulling a scraper. This could possibly be the building of the Duncan Creek Road. *A. Berger Collection*

Early model vehicle and driver outside the Hotel Royal Alexandra in Mayo during the winter months. *A. Berger Collection*

Whitehorse was also finished. Ferries were established on the Stewart, Pelly, and Yukon Rivers in summer and ice bridges were built in the winter.

With the new road, the first trucking of sacked ore and concentrates began. To counteract the high freight rates, United Keno Hill organized their own transport and supply division in Whitehorse, and by late 1950 all concentrates were being trucked in the company's own trucks. They were the sole distributor of International trucks and tractors in Yukon. Each truck carried 20 to 30 tons of ore concentrate.

Bridges over the Pelly and Stewart Rivers were officially opened on October 9, 1960.

Early model mud covered vehicle with chains on rear wheels shows that travelling the roads was not easy. *A. Berger Collection*

In the winter of 1959–60, Leo Proctor of Whitehorse built the Wind River Trail (also referred to as the Amerada Road). This was a winter road built for truck haulage of oil well drilling equipment and supplies. It began a few miles beyond Mayo and ran north to the Peel Plateau at a point on the Bell River just a few miles short of the Arctic coast. Proctor was instrumental in the conception and construction of this trail, built by the Arctic Oilfields Transport Ltd., a subsidiary of The White Pass & Yukon Route. Construction was started on October 12, 1959, and completed on January 12, 1960, covering a distance of 385 miles. Over 3,000 tons of material moved over the road after it was completed. It was the most northerly truck road in North America at the time.

Ferry carrying United Keno Hill's concentrate and truck across the Stewart River, late 1950's *A. Aho Collection*

THE MAYO VALLEY RAILWAY, LTD.

On March 29, 1921, an Act to incorporate The Mayo Valley Railway, Limited (Bill C-20) was passed by the House of Commons.

Shareholders in the company – formed to gain access to massive deposits of lead ore at MacKay Hill, up the North McQuesten River – were Alexander J. Gillis of Dawson City, Robert

S.S. *Keno* pushing barge on Stewart River loaded with Treadwell-Yukon Company Ltd. equipment. Taken by Ken Schellinger. *A. K. Schellinger Collection*

B. Young, William Alfred H. MacBrien, and Norman J. Robinson of Toronto, Ontario and John McDougall of Ottawa, Ontario. The capital stock of the company was $750,000 and its head office was to be in Dawson City.

The company was authorized to lay out, construct, equip and operate a railway line from a point at or near the junction of the Mayo and Stewart Rivers, in a northerly direction, following the valley of the Mayo River to its junction with Duncan Creek, thence by the Duncan Creek Valley to its junction with Lightning Creek, thence via Lightning Creek and Christal Valleys to the McQuesten River, thence by the most feasible route to connect with and include the valleys of the McQuesten, Ladue, and Beaver Rivers; or, alternatively, from a point at or near Gordon Landing on the Stewart River via Janet Lake to the junction of the Mayo River and Duncan Creek. The company was also authorized to lay out, construct, equip and operate branch lines not exceeding in any one case ten miles in length from the main railway line.

The company was granted the power to transmit telegraph and telephone messages for the public, and collect tolls.

The company could also construct, acquire, charter and navigate steam and other vessels and ferries, for the conveyance of passengers, goods and merchandise and could construct, acquire, lease and dispose of terminal stations, wharfs, docks, elevators, warehouses, offices, and other structures to be used to facilitate the carrying on of business in connection with railroad activities and could carry on the business of warehousemen and wharfingers, and could charge wharfage and other dues for the use of any such property.

The securities issued by the company were not to exceed thirty thousand dollars per mile of the railway, and were to be issued only in proportion to the length of railway constructed or under contract to be constructed.

The railway, in fact, was never constructed. There is no further record of the Mayo Valley Railway, Limited, beyond the company's Act of Incorporation. It would appear that the law requiring the filing of annual returns either was not in effect in 1921 or was not enforced, in respect to companies incorporated by private Act.

Former Yukon Member of Parliament, Erik Nielsen, maintains that Bill C–20, an Act to Incorporate the Mayo Valley Railway, Limited, still exists exactly as passed by the House of Commons in 1921, and that there have been no amendments to the Bill nor has it been repealed.

AVIATION *by Linda E. T. MacDonald*

Yukon Airways & Exploration Company, based in Whitehorse, was the first commercial airplane company to operate in the Yukon. It was a private company founded in 1927 by Jim Finnegan, Clyde Wann (both from Alaska) and pilot D.A. "Andy" Cruickshank who resigned as corporal in the N.W.M.P. (on April 4, 1927) to form the company with his partners. Cruickshank had flown with the Queens Own Oxfordshire Hussars Royal Flying Corps during World War I.

The first airplane owned by the company was a Ryan B-1 Brougham. It was a sister ship to Charles Lindbergh's famous *Spirit of St. Louis*. Lindbergh was making plans to fly the Atlantic with a Bellanca aircraft at the same time that Yukon Airways' monoplane was under construction at the Ryan Aircraft Company's factory in San Diego, California. Lindbergh approached Ryan Aircraft following a disagreement with the Bellanca's manufacturers. As he was pressed for time, Ryan asked Yukon Airways if they would allow Lindbergh to take over their plane and they agreed. When Yukon Airways' plane was built, many of Lindbergh's innovations were incorporated: the wingspan was increased, the cabin enlarged to carry four passengers and the landing gear was reinforced.

Yukon Airways named their plane *Queen of the Yukon*.

> The 'Queen of the Yukon' attracted attention wherever it went; everyone wanted a ride on "Lindbergh's" plane. Cruickshank put the interest to good use, barnstorming during the summer [of 1927] in California, Nevada, Texas and Washington to raise the customs duty required to bring the plane into Canada. In September, after introducing more than 1,000 passengers to the thrills of flying, he landed in a cow pasture in Vancouver.[44]

Esme Trevor-Bulkley was in the Vancouver crowd that turned out to see the sister ship of the *Spirit of St. Louis*. She met pilot "Andy" Cruickshank and they were married on October 17, 1927. The Cruickshanks were to spend their honeymoon on the adventure north.[45]

The plane was dismantled in Vancouver and shipped on the Canadian Pacific Steamer *Princess Alice* to Skagway where she was reassembled. On October 25, 1927, Cruickshank and his new bride flew to Whitehorse accompanied by Clyde Wann and one other passenger (probably mechanic J.E. Smith).

On November 24, 1927, the first airborne mail arrived in Mayo from Whitehorse. Cruickshank offered Mayo and Keno residents ten minute exhibition flights for $10.

Cruickshank left the company in 1927[46] and T.G. "Tom" Stephens took over as pilot in April, 1928. The *Queen of the Yukon* was demolished in an accident at the Whitehorse airport on May 5, 1928 when Stephens bounced in a gusty crosswind while landing and crashed into Billy Puckett's truck. Mayo residents Elizabeth McDougall Titus and Thomas McKay were passengers on the *Queen* when it crashed and Mrs. Titus was seriously injured.

Following the loss, Yukon Airways & Exploration bought a Whirlwind-powered Alexander Eaglerock biplane (G-CAUZ) in Colorado Springs, Colorado, and named her *Northern Light*. Rocky Mountain Airways chief pilot John Melvyn Patterson was hired to fly the *Northern Light* to Whitehorse and then to stay on as Yukon Airways pilot. Patterson and the *Northern Light* arrived in Whitehorse on December 8, 1928.

Sam Peter and Dave Hager, hire Yukon Airways plane *Northern Light* for beaver hunting at Lake Creek, circa 1928. *Dave Hager Collection*

In 1928, Treadwell Yukon Co. Ltd., bought a Fairchild FC2-W2 (G-CARM) christened the *Flame* while Greenfield and Pickering formed Klondike Airways and bought a de Havilland D.H. 60X Moth (G-CAUM) which they christened the *Rilla Martha*. Klondike Airways did not operate the *Rilla Martha* (or any other aircraft); instead, Treadwell Yukon took over the operation of the Moth. The Fairchild arrived first and Wernecke hired Stephens as pilot and had pilot Everett Wasson act as the Fairchild's mechanic until the Moth arrived. Stephens piloted the Moth when she arrived, Everett Wasson became the Fairchild's pilot and Emil Forrest was hired as aircraft mechanic. The planes were based in Mayo and Treadwell Yukon built a hangar for them on the east bank of the Stewart River, south of town.

Treadwell Yukon's *Flame* was renamed the *Claire* by Wernecke for his daughter who became the first woman pilot in the Yukon. (Claire Wernecke soloed the Moth *Rilla Martha* in July, 1931.) The *Claire* was considered the first "corporate" airplane in the Yukon.

In a letter to R.L. Gillespie, Mayo Mining Recorder, Wernecke wrote, "The Fairchild and Moth made 263 flights in 1929 flying 358 hours covering 31,240 miles of territory, carried 131 passengers, 11,351 pounds of express and 1,877 pounds of mail." Wernecke also conducted a Yukon-wide air reconnaissance prospecting program from 1928 to 1930 for Treadwell Yukon.

On August 17, 1929, Yukon Airways & Exploration Company took delivery of a new plane. It was a Ryan B-5 Brougham, an improved version of their first airplane, and they named her *Queen of the Yukon II*. She was piloted by 22-year-old John Melvyn Patterson who went to pick it up in St. Louis. Patterson and the *Queen of the Yukon II* (CF-AHD) arrived in Whitehorse on September 26, 1929. This new arrival bought the number of aircraft in the Yukon to four; Yukon Airways *Northern Light* and *Queen of the Yukon II*, Treadwell Yukon's *Claire* and Greenfield & Pickering's *Rilla Martha*.

Queen of the Yukon, Keno, November 1927. *Ione Christensen Collection*

On November 2, 1929, the first aviation fatality in the Yukon occurred. Patterson was killed when the *Queen of the Yukon II* suffered an engine failure while taking off from the Stewart River, in front of Mayo, crashing nose-first into the river ice. George Besner was the first person to reach the crash site and he recalls the overwhelming smell of gasoline fumes when he opened the door of the plane. As the pilot had no family in the Yukon, an emergency meeting of the Mayo Yukon Order of Pioneers was called. John Patterson was made an honorary member and was subsequently interned in the Pioneer's plot at the Mayo Cemetery. (See *John Melvyn Patterson* on the *Pioneer List.*) A large metal-tipped wooden propeller stands as a monument over his grave.[47]

Queen of the Yukon and Archie Close's dog team on the Stewart River ice in front of the White Pass & Yukon Route warehouse in Mayo, circa 1928. Andy Cruickshank is on the far left and William Hargraves is on the extreme right. *Ralph Rogers Collection*

Yukon Airways continued to operate their remaining plane until pilot Perry Nelson and passenger Clyde Wann crashed the *Northern Light* on the shore of Coal Mine Lake near Carmacks, during a routine flight from Mayo to Whitehorse. The aircraft was damaged beyond repair, but the engine was salvaged and the plane was abandoned on the north shore of the lake. [Mrs. Helen Brooks (née Goulter) who grew up in Carmacks, remembers seeing the abandoned wreck lying on the beach of Coal Mine Lake in the 1930's and 1940's.] The loss of the *Northern Light* brought an end to Yukon Airways & Exploration Company.

In 1931, Treadwell Yukon acquired a radio equipped Bellanca Skyrocket monoplane, formerly owned by screen actor Wallace Beery, for Wernecke's use and he made frequent flights with his pilot Chuck Gropstis throughout Yukon and Alaska. Later, when Wernecke was based in San Francisco, he flew back and forth from there to the Mayo district in the Bellanca.

In 1934, Skagway-based White Pass Airways introduced scheduled air service to the North, linking Skagway, Whitehorse, Mayo, and Dawson. The company hired Vern Bookwalter, an experienced pilot, and their first plane touched down at the Whitehorse Airport in June, 1934. It was a Loening Keystone Commuter, nicknamed *The Duck*. (It was this aircraft that was beached in front of Binet's hotel during the 1936 flood in Mayo.) White Pass purchased a Ford Tri-Motor (NC-5092) in Seattle, to be used for the scheduled air service. It carried 10 passengers and a 2-man crew. Pasco Boyle was Bookwalter's co-pilot and mechanic.

White Pass Airways established a base of operations in Whitehorse in July, 1935. For the rest of the 1930's and into the early 1940's, it continued to provide Mayo with scheduled air service using Ford Tri-Motors, Fairchild 82's, a Bellanca Aircruiser, a Curtiss Kingbird; and a 15-passenger Curtiss Condor, the largest airliner in Canada at the time. James Fairborn was the White Pass Airways agent in Mayo.

Pilots during the 1930's and early 1940's included Everett Wasson, Lionel Vines, Jesse Rice, Norm Stone, Vern Bookwalter, Bill Holland, and Les Cook. Everett Wasson flew the Fairchild FC2-W2 (G-CARM) for Treadwell Yukon from 1928–1935 and then became White Pass Airways general manager and chief pilot from 1935 to 1940. Wasson was involved in the very first B.C. aerial search and rescue while he was a pilot for Treadwell Yukon. Following a long and arduous search in winter conditions, Wasson and Mayo prospector Joe Walsh found Paddy Burke's

Crash of the *Queen of the Yukon*. *M. H. Ewing Collection*

downed plane on the Liard River in 1930 and rescued the survivors. (See *Everett Wasson in the Pioneer List.*)

In addition to scheduled mail, express, and passenger service, pilots flew freight to posts such as Lansing, and ferried prospectors and dignitaries to various points in the district. Martha Black, M.P., was a frequent passenger when vacationing in the wilderness areas of this district and one of her favourite locations was the magnificent Bonnet Plume. White Pass Airways received fierce competition for these non-scheduled air services throughout the Yukon from Northern Airways.

George Simmons formed Northern Airways Ltd. in Carcross in 1935 with the purchase of Treadwell Yukon's Fairchild and assigned Wes Palmer as his Mayo agent. Wasson was briefly involved in the founding of Northern Airways with Simmons. Simmons advertised his company in *The Mayo Miner* and encouraged potential customers to "Fly with the Golden Fleet." In addition to charter flights, Northern Airways Ltd. offered weekly plane service between the communities of Mayo, Dawson, Whitehorse, Carcross, Atlin, and Juneau. The fare from Mayo to Dawson was $25 and from Mayo to Juneau, $60.

Grant McConachie's Edmonton-based United Air Transport acquired the mail contract between Edmonton and Whitehorse and the service was inaugurated on July 5, 1937. The company also began weekly service from Edmonton to Mayo and Dawson City. On August 27, 1937, McConachie and his co-pilot Len Waagen landed on the Stewart River in a 550 H.P., Wasp-powered, Noorduyn Norseman, cabin plane. McConachie stated that his firm would be flying through Mayo northbound on Tuesdays and returning to Mayo from Dawson on Thursdays en route to and from Edmonton. Wilfred Gordon, then of Dawson City, was the very first through passenger from Dawson City to Edmonton. The distinction of being the first Mayo passenger flying directly to Edmonton via United Air Transport went to George Kaldal, Treadwell Yukon employee. The one way fare from Mayo to Edmonton was $160.

In 1938, United Air Transport successfully outbid Northern Airways on the Vancouver to Yukon air mail contract. Following this, Northern Airways was to advertise, "Maybe we didn't get the Vancouver to Yukon air mail contract but

Crash of the *Queen of the Yukon II* on the ice of the Stewart River in front of Mayo, November 2, 1929, which resulted in the first fatality in Yukon aviation history. Undertaker Alex Lesperance is standing, hand on the plane. *Sam Wood Collection, Yukon Archives*

we're still flying passengers speedily—comfortably—reliably." White Pass Airways maintained the Yukon mail contract for Whitehorse, Dawson, and Mayo, connecting with United Air Transport in Whitehorse.

Following a fire in a maintenance shack at their Carcross base and a forced landing of their Waco on the Taku River in 1939, Northern Airways decided to withdraw from central Yukon and operate only in the southern Yukon and Atlin, B.C. district. Gerald Hansen, the firm's accountant of only a few months, was sent to close the Dawson and Mayo agencies during the first week in February, 1939.

Treadwell Yukon's Bellanca crashed off the coast of British Columbia on October 21, 1941, and both the pilot, Chuck Gropstis, and Treadwell's General Manager and Consulting Engineer Livingston Wernecke, were killed. *(See Livingston Wernecke in the Pioneer List.)*

In 1942, White Pass Airways and several other companies were amalgamated by McConachie's Yukon Southern Air Transport and within that same year became Canadian Pacific Airlines. It continued to serve Mayo with scheduled air service using Barkley-Grows, Lockheed Lodestars, and finally, DC-3's.

Harvey Johnson was one of the C.P. Air pilots who flew into Mayo in the 1940's. He brought the first DC-3 into Mayo. This event was watched by all who could get to the airport to see the "huge" plane land. Johnson later became a Chief Pilot for C.P. Air overseas flights.

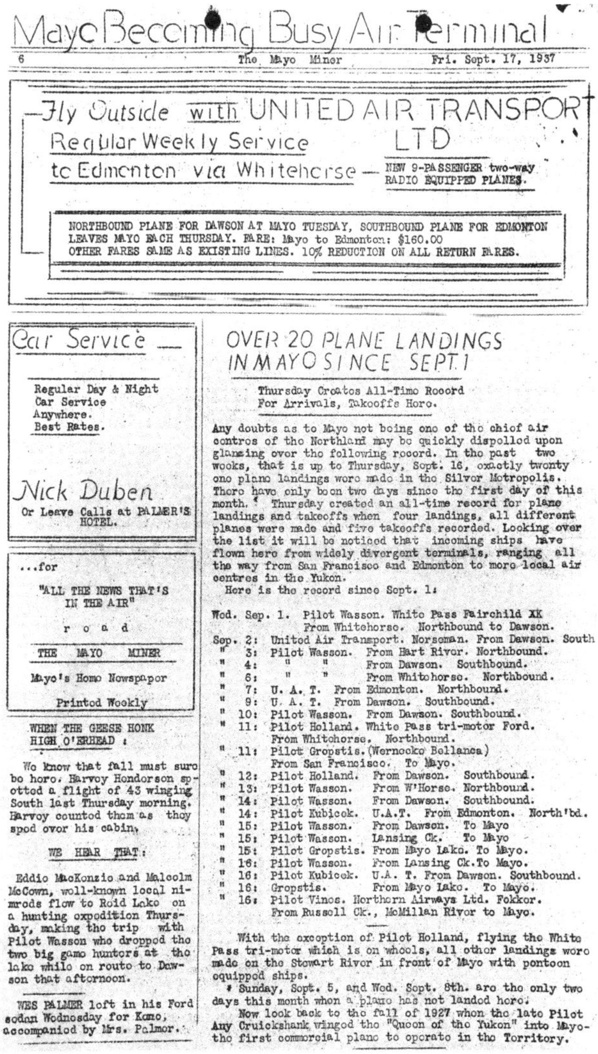

Mayo Becoming Busy Air Terminal

6 The Mayo Miner Fri. Sept. 17, 1937

Fly Outside with UNITED AIR TRANSPORT LTD

Regular Weekly Service to Edmonton via Whitehorse — NEW 9-PASSENGER two-way RADIO EQUIPPED PLANES.

NORTHBOUND PLANE FOR DAWSON AT MAYO TUESDAY, SOUTHBOUND PLANE FOR EDMONTON LEAVES MAYO EACH THURSDAY. FARE: Mayo to Edmonton: $160.00
OTHER FARES SAME AS EXISTING LINES. 10% REDUCTION ON ALL RETURN FARES.

Car Service —

Regular Day & Night Car Service Anywhere. Best Rates.

Nick Duben

Or Leave Calls at PALMER'S HOTEL.

...for

"ALL THE NEWS THAT'S IN THE AIR"

read

THE MAYO MINER

Mayo's Home Newspaper

Printed Weekly

WHEN THE GEESE HONK HIGH O'ERHEAD :

We know that fall must sure be here. Harvey Henderson spotted a flight of 43 winging South last Thursday morning. Harvey counted them as they sped over his cabin.

WE HEAR THAT:

Eddie MacKenzie and Malcolm McCown, well-known local nimrods flew to Reid Lake on a hunting expedition Thursday, making the trip with Pilot Wasson who dropped the two big game hunters at the lake while on route to Dawson that afternoon.

WES PALMER left in his Ford sedan Wednesday for Keno, accompanied by Mrs. Palmer.

OVER 20 PLANE LANDINGS IN MAYO SINCE SEPT. 1

Thursday Creates All-Time Record For Arrivals, Takeoffs Here.

Any doubts as to Mayo not being one of the chief air centres of the Northland may be quickly dispelled upon glancing over the following record. In the past two weeks, that is up to Thursday, Sept. 16, exactly twenty one plane landings were made in the Silver Metropolis. There have only been two days since the first day of this month.* Thursday created an all-time record for plane landings and takeoffs when four landings, all different planes were made and five takeoffs recorded. Looking over the list it will be noticed that incoming ships have flown here from widely divergent terminals, ranging all the way from San Francisco and Edmonton to more local air centres in the Yukon.

Here is the record since Sept. 1:

Wed. Sep. 1. Pilot Wasson. White Pass Fairchild XK From Whitehorse. Northbound to Dawson.
Sep. 2: United Air Transport. Norseman. From Dawson. South
" 3: Pilot Wasson. From Hart River. Northbound.
" 4: " " From Dawson. Southbound.
" 6: " " From Whitehorse. Northbound.
" 7: U. A. T. From Edmonton. Northbound.
" 9: U. A. T. From Dawson. Southbound.
" 10: Pilot Wasson. From Dawson. Southbound.
" 11: Pilot Holland. White Pass tri-motor Ford. From Whitehorse. Northbound.
" 11: Pilot Gropstis. (Wernecke Bellanca) From San Francisco. To Mayo.
" 12: Pilot Holland. From Dawson. Southbound.
" 13: Pilot Wasson. From W'Horse. Northbound.
" 14: Pilot Wasson. From Dawson. Southbound.
" 14: Pilot Kubicek. U.A.T. From Edmonton. North'bd.
" 15: Pilot Wasson. From Dawson. To Mayo
" 15: Pilot Wasson. Lansing Ck. To Mayo
" 15: Pilot Gropstis. From Mayo Lake. To Mayo.
" 16: Pilot Wasson. From Lansing Ck. To Mayo.
" 16: Pilot Kubicek. U.A. T. From Dawson. Southbound.
" 16: Gropstis. From Mayo Lake. To Mayo.
" 16: Pilot Vines. Northern Airways Ltd. Fokker. From Russell Ck., McMillan River to Mayo.

With the exception of Pilot Holland, flying the White Pass tri-motor which is on wheels, all other landings were made on the Stewart River in front of Mayo with pontoon equipped ships.

* Sunday, Sept. 5, and Wed. Sept. 8th. are the only two days this month when a plane has not landed here.

Now look back to the fall of 1927 when the late Pilot Andy Cruickshank winged the "Queen of the Yukon" into Mayo—the first commercial plane to operate in the Territory.

"Mayo Becoming Busy Air Terminal," Archie Gillespie, *The Mayo Miner*, September 17, 1937. *Mayo Historical Society*

A crowd gathers around an early model plane in Mayo. *A. Berger Collection*

White Pass & Yukon Route's first airplane, a Loening Keystone Commuter nicknamed *The Duck*, in front of the Chateau Mayo hotel during the flood of 1936. This plane was based in Skagway. *Besner Collection*

In the early 1960's Canadian Pacific Airlines sold the Yukon domestic routes to Whitehorse-based Great Northern Airways. They continued to serve Mayo with DC-3's until the service was taken over by Northward Airlines in 1970.

Charter air service out of Mayo included Callison's Flying Service, operated by Pat Callison in Dawson, (Beaver, Cessna 180) from 1948-1955. Connelly-Dawson Airways, based in Dawson City and owned by Ron Connelly, bought out Pat Callison and operated charter air service from 1956 to 1962. Ron Connelly is currently the president of Trans North Air, which operates helicopter charter services out of both Mayo and Dawson.

CANADIAN PACIFIC AIR LINES, LTD.

Tax - $ 35⁰⁰

Full Single Fare $

PASSENGER'S COUPON

NOT GOOD FOR PASSAGE

From MAYO

To WHITEHORSE

Via

☐ ONE WAY

Date

Flight No. a.m. p.m.

AIR PASSAGE CONTRACT

In consideration of the issuance of this ticket, the purchaser agrees on the following conditions:

LIMITS: Tickets shall be valid for a period of four months after date of issue and only between points named on the Flight Coupons and for flights on which space has been reserved. Flight Coupons must be used in order and are not valid if detached.

STOP-OVERS: Stop-over will be allowed subject to regulations for reservations or release of space.

REVOCABLE: The Company reserves the right to revoke this ticket upon refunding the purchase price.

NOT TRANSFERABLE: This ticket is not transferable.

REFUNDS: The Company will refund the amount paid for this ticket if the flight is cancelled. Refunds will not be granted for other reasons, unless the ticket is surrendered at least three hours before departure of plane.

ASSUMPTION OF RISK: I AGREE THAT THE COMPANY SHALL NOT BE LIABLE FOR ANY INJURY OR DAMAGE UNLESS OCCASIONED BY ITS OWN NEGLECT OF DUTY IN THE OPERATION OR CONTROL OF THE AIRPLANE.

(CONTINUED ON REVERSE SIDE)

Form CPA-1 240654

Passenger's coupon from Mayo to Whitehorse, circa 1950. *J. B. Kunze Collection*

MAYO UTILITIES CO. LTD.

The Mayo Utilities Co., Limited, was incorporated on May 7, 1920, and had a franchise to operate a telephone company in this district. D.A. "Dan" Matheson was the president and manager of the company, and his wife, Adolphine, was the secretary-treasurer. Mayo Utilities telephone line ran to Minto Bridge, Field Creek, Huffman's Roadhouse, Keno City, and beyond to the headquarters of the Yukon Gold Co., and Treadwell Yukon's Wernecke Camp. The company also offered a messenger service. Messages received by the Royal Canadian Corp. of Signals wireless operator in Mayo, to be passed on by telephone or hand delivery, were handled by Mayo Utilities messenger service.

J.E. Binet also had a franchise to operate a telephone company between Mayo and Keno Hill but he did not pursue it.

The total number of telephones installed in

Mayo Utilities office building following the flood in 1936. *Hare Collection, Yukon Archives*

the 1920's by the Mayo Utilities Co., Limited, is uncertain. Telephones were, of course, installed at the Yukon Gold Company and Treadwell Yukon offices. Other establishments mentioned in the records include Burns Butcher Shop in Mayo, Binet's Hotel, and Tom Greenaway's store and post office in Keno City. On August 9, 1922, a telephone was installed in the Mining Recorder's office in Mayo and in the Sub-Mining Recorder's office in Keno City. Telephone rental appears to have been $25 per month.

The Mathesons operated the company until 1938, when their son, Malcolm, a graduate of the Milwaukee Institute of Technology in Electrical Engineering, became manager. After returning to the Yukon, Malcolm worked for the Dawson Electric Light and Power Co., Ltd. (a subsidiary of the Yukon Consolidated Gold Company).

Under Malcolm's management, the company expanded its service to include Elsa. It is believed Malcolm moved to Vancouver in the early 1940's. During World War II, officials asked Malcolm to return and start a telephone service in downtown Whitehorse, which he did in 1942. In 1944 Mayo Utilities negotiated the purchase of the water and sewer system in Whitehorse from the U.S. Government.

The company's telephone system was sold in 1950 to Stan "Telephone" Barker, who named his company The Yukon Telephone System. Mayo Utilities Co., Limited remains a registered Yukon company (1989).

Barker used short wave radio transmitters to connect long distance calls between Mayo and Whitehorse. There were still problems with this system, as daylight atmospheric interference meant calls had to be placed during hours of darkness, making communication during the summer months difficult. Barker sold the Yukon Telephone System to Canadian National Telecommunications (now NorthwesTel) on November 1, 1958. C.N.T. subsequently installed telephone lines, linking Mayo with the "outside" world.

THE MAIL SERVICE & RADIO COMMUNICATIONS *by Anne Meaney-Leckie & Lynette Bleiler*

The discovery of gold and silver in the Mayo District brought many people to the area in the late 1800's. By 1905 settlements had been established at Gordon Landing, Duncan Creek, Mayo Landing and Minto Bridge. The population increased as new gold and silver strikes were made. The development of resources in this district however, were impeded by a lack of communication with the outside world. There were no telegraph lines or radio stations in the area. Information and mail were exchanged via the infrequent steamboat service during the summer months and over an ever changing land route during the winter months. The North West Mounted Police with their dog teams provided the latter service.

The Mail Service

Prior to 1897, the mail service in the Yukon was organized by the trading companies at their own expense. Woodall[48] indicates that the companies probably recovered their costs by charging the miners for private letters. The mail was taken up the Yukon River over the Chilkoot Pass to Dyea, Alaska and then to Juneau where it was picked up by the U. S. postal services. (See map on page 18.)

The earliest effort to formalize the distribution of mail fell to the North West Mounted Police. In September of 1897 Inspector D'Arcy Strickland was appointed postmaster and opened a post office at Tagish Lake. In May of the following year, another post office was established at Lake Bennett. Mail arrived there from the Dyea Post Office on the Lynn Canal via the Chilkoot Pass. From Lake Bennett, the winter mail was sent by N.W.M.P. dog teams to Tagish, and from Tagish it was distributed to points along the Yukon river, ultimately terminating at Dawson City. There is no indication that mail was formally distributed during the summer until steamboat navigation came to the area in 1899.[49]

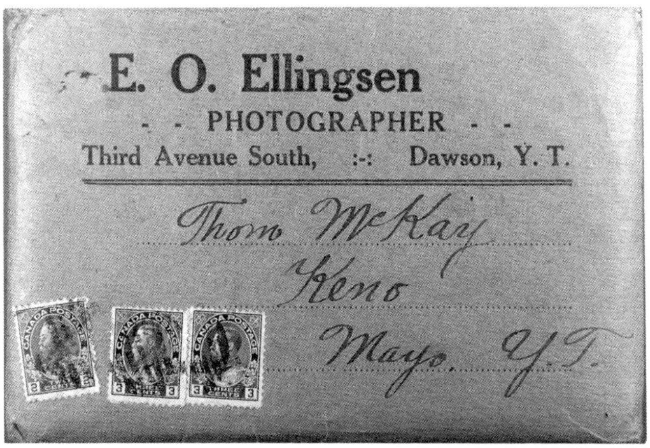

Photograph envelope from E. O. Ellingsen to Thomas McKay. *McLaren-Swensen Collection*

Efforts to distribute mail in the Mayo District were first made in December of 1899 by Corporal Holmes[50] of the N.W.M.P. Corporal Holmes was stationed at Stewart City and was the volunteer acting postmaster of the Stewart River Post Office for several years.[51] His volunteer status did not allow him to handle registered mail, which had to be sent through the postmasters at either Tagish or Lake Bennett. Therefore, Holmes requested that a full-service post office be established at Stewart City.

The need for a post office at Stewart City was verified by Dawson City's Postmaster, I.J. Hartman, in a communication to the then Commissioner, William Ogilvie. He commented that there was more mail sent and received from the Stewart River Post Office than any other office between Dawson and Skagway. He also added that a Mining Recorder was soon going to be stationed at Stewart City, thus increasing the activity in the settlement.

Ogilvie promptly wrote the Postmaster General in Ottawa with the request that a post office be established at the mouth of the Stewart River. Ogilvie also proposed that the post office be named "Stewart City" because there was a "Stewart" on the "Official List." "Stewart Mouth," he maintained, would be too difficult to pronounce.

In late January, 1900, the Deputy Postmaster General responded to Ogilvie's letter with a request for an estimate of the potential revenue to be generated by the proposed post office. Ogilvie replied in March that it would be impossible to estimate the potential revenue because of the transient nature of the population in the area. For example, during the winter of 1898–99 there were between 2,000 and 3,000 inhabitants, but in the winter of 1899–90 the population was only about 300 adults.

In terms of costs to run the operation he suggested that someone be appointed to "act" as postmaster, which would be an unpaid position. Ogilvie pointed out that there was a report of a rich strike on the Stewart River which would bring more people into the area. Several months later, payable gold was discovered on Clear Creek and another stampede began.

By December, 1900, in spite of increased activity in the area, a post office had still not been established at Stewart City. The Commissioner continued to urge the matter with Ottawa, but did not think the Clear Creek discovery increased the urgency. The justification for this position was that mail for Clear Creek could be sent from Dawson via road more quickly than it could be sent from Stewart City.

Additional pressure for the installation of a post office at Stewart City was brought to bear from a number of different sources. For example, the Mining Recorder's office at Stewart City was experiencing difficulties in transacting business because of the lack of an official post office.

In January, 1901, the Superintendent of the N.W.M.P. wrote the Commissioner that the police should not be responsible for mail service. He pointed out that members of his force were constantly changing and mistakes could easily be made. He asked the Commissioner to appoint Bowen Smith, owner of the Stewart River Grocery, as acting postmaster. The Commissioner advised that the police should continue to handle the mail.

W. Meed, Manager of the Stewart River Steamer Company, noted to James Hamilton Ross, M.P., Ottawa, that in addition to the services provided by the N.W.M.P., his company had been voluntarily collecting and distributing mail in the absence of an official post office. Meed said that postmasters were needed at the mouths of Clear Creek, the McQuesten River and at the foot of Fraser Falls. He offered the services of his company to carry mail between those places and Dawson City at a mutually agreed price.

In December 1902, N.W.M.P. Major Z.T. Wood wrote Member of Parliament Ross with a petition signed by 87 residents of Duncan Creek asking for a mail route between Stewart City and Discovery on Duncan Creek. There were some 500 people in the district whose only means of getting mail was by the occasional police patrol. Duncan Creek residents Gordon and Hoy, offered a bi-monthly mail service during the winter months between Dawson City and Duncan Creek and the Upper Stewart District for $500 per month.

As a partial solution to the problem, the

Mail arriving at Mayo, date unknown. *Mayo Historical Society*

N.W.M.P. agreed that in the future, mail for the Duncan District would be sent via Dominion Creek to McQuesten and on to Duncan Creek by patrols. Despite these concessions, complaints about unsatisfactory and irregular mail service in the region in general continued to be made. For example, by January 1903, mail was taking over a month to get from Dawson City to Mayo Bridge (near Duncan Creek), via Stewart City.

The police remained dissatisfied in their role as postmasters and letter carriers, so Ottawa appointed the Acting Mining Recorder, Wilfred Adam to act as postmaster. However, as Adam was also the Mining Inspector, and frequently out of the office, he was advised to decline the position. Ottawa suggested the telegraph operator for the job, but the Commissioner pointed out that this was an impossibility as the telegraph office was on one side of the Yukon River, and Stewart City, the Police Station and all the inhabitants were on the other side.

In an effort to solve the conflict as to who would serve as postmaster, the Deputy Postmaster General advised Commissioner Congdon in February, 1904, that local Postmasters should be appointed and the police should help only with the mail delivery. However, Major Wood stated that he had no intention of instructing the police to discontinue their postal duties at Stewart City. In February 1905 he ordered that the N.W.M.P. at Stewart City act as postmaster because no one else would accept the position for the remuneration allowed. The police continued to find this unsatisfactory because members of the detachment were often away for several days at a time. An informal arrangement was made with the new owners of the General Store, the Smythes.

In August of 1905, M.L. Daley offered to set up a mail service from Dawson City to Mayo. He proposed leaving Dawson City every two weeks and Mayo the alternate week for seven months on the following schedule (see map on page 185):

Leave Dawson City say Monday Morning:

Destination	Time	Cost (Man & Horse)
Gold Bottom	Noon	$2.00
Dominion	Overnight	$6.00
Hunter's Cabin	Noon	$2.00
Gravel Lake	Overnight	$6.00
Clear Creek	Noon	$2.00
McQuesten	Overnight	$6.00
Moose Creek	Noon	$2.00
Crooked Creek	Overnight	$6.00
Square Cabin	Noon	$2.00
Mayo	Overnight, Saturday and Sunday off	$18.00
		$52.00
Return Trip		$52.00
Total Round Trip		$104.00
Total Expenses for two round trips	per month	$208.00
Wages, Wear and Tear		$200.00
Total per month		$408.00
Seven winter months and five summer	months @ $400.00 per month	$4,800.00

The Mayo Post Office with sacks of mail stacked in front. This building housed the Post Office during the 1920's and 1930's. *G. A. McIntyre Collection*

There is no evidence that Daley's offer was accepted; therefore the residents of the Stewart River District continued to be dissatisfied with the mail service that was provided. In December of 1905, 31 miners and residents signed a petition requesting a weekly mail service from Stewart City to Stewart Crossing,[52] to replace an existing fortnightly service. In a letter to the Commissioner, they stated that this service could be provided by the N.W.M.P. for very little extra cost. The request was supported by Corporal A.G. Haddock of the Stewart City Detachment. In February, 1906, the Stewart Crossing and Stewart City detachments of the recently renamed R.N.W.M.P. were outfitted with an extra horse to provide a more frequent mail service during the winter months. However, by the winter of 1907, the force was reduced to such an extent that it had to give up the mail service. After a hiatus in mail service lasting several months, Waldron Smythe took a contract to carry mail from Stewart City to the creeks.

Mail service continued to improve in the Mayo District with the establishment of a post office at Minto Bridge in November 1910, with George Cunningham as the first postmaster. The mail route between Stewart City and Thistle Creek brought many outlying areas, such as Black Hills and Scroggie Creek into more frequent communication. In March 1914, Post Office Inspector E.A. Fletcher of Victoria suggested to Commissioner George Black that the post office at Minto Bridge be relocated to Mayo Landing, as the Mining Recorder, who was also the postmaster, was now stationed there. The post office at Mayo Landing opened in June, 1914, with Phillip Kelly as postmaster.

Having agreed to the change, Black suggested that instead of closing down the post office at Minto Bridge, one of the residents should be named postmaster. He also pointed out that there was a need for a post office at Haggart Creek as important developments in placer and quartz mining were taking place in that area.

In a subsequent communication, Fletcher noted that a petition had been forwarded to him from the postmaster at Dawson City asking that a mail courier be employed during the summer season to carry mail from Mayo Landing to Haggart Creek via Highet Creek. This service comprised eight trips during the open season (May 24–September 30) for a total cost of $500, or $62 per trip. The winter population on Haggart Creek in 1914 was about 50 miners, three with wives and families. In lieu of establishing a post office there, Black arranged for the Minto Bridge winter mail contractor, Allan McIntosh, to bring mail to

Evelyn (left) and Helen Buyck in front of the Mayo Post Office. *Buyck Collection*

Haggart Creek for an additional $75 a trip. The mail was then distributed by Mr. Abbott, one of the miners on Haggart Creek.

In December, 1916, the transportation company of Greenfield and Pickering bid $265 per trip from Dawson City to Mayo Landing and was awarded the Royal Mail contract. They were replaced in September, 1917, when Dan Moriarty of Dawson City acquired the contract at a price of $200 per trip and unlimited weight.

The first airborne mail arrived in Mayo from Whitehorse on November 24, 1927 via the Yukon Airways & Exploration Company with mail for Mayo Landing, Keno, and Wernecke Camp. James K. Johnstone was postmaster until he was replaced in 1932 by D.R. (Randy) Maclennan.

During Maclennan's time, the post office was a log cabin located on the corner of Fourth Ave. and Centre St. due south of the old school (Block 6, lots 29 and 30). He bought the Fournier family home just across the street (Block 7, lots 16 and 17) after his marriage in the mid 1930's, as he required larger quarters for both the post office and his own accommodation. After Maclennan's death, Hilma Miller was appointed postmaster. She remained on the same premises until 1942 when she left Mayo to marry Roy Hallin.

Mabel McIntyre was appointed postmaster for Mayo Landing in 1942 and maintained the position for 30 years. It was during her term that the post office was moved to the Chateau Mayo. On August 2, 1958, the name Mayo Landing Post Office was officially changed to Mayo Post Office. In 1983, the Mayo Post Office was moved to the new Government Building on Center Street.

Mayo has had a continuous postal service since 1914 because of the importance of the town as a supply and service centre for the area. However, the silver boom of the early 1920's brought large numbers of people into the Keno, and later Elsa, areas requiring additional postal services. The types of postal services offered were related to the mining activities taking place. On July 1, 1921, Robert Greaves opened a post office at Keno City which closed in 1950 but was reopened in 1952 with renewed mining activity in the area. For example, in 1952, mineral production topped $11 million, with silver, lead and zinc providing most of the returns.[53] The post office was closed again on June 25, 1968, with a decrease in activity.

In addition to the post office at Keno, E.H. Barker opened a post office at Treadwell Yukon's Wernecke Camp which operated from November 1, 1926, until August 22, 1933. The existence of this post office roughly corresponded with the height of activity at the Wernecke Camp. Likewise, the first post office was opened in Elsa on May 26, 1949 by UKHM Mine Manager H. Brodie Hicks and closed in June, 1989. This post office operated differently from most as United Keno Hill Mines received funds from the federal government for its operation. However, the employee was hired and paid by UKHM and was not a federal government employee.

Stewart River telegraph office. Dawson DeWett is standing in front of the Dominion Telegraph Office which he operated at Stewart River, April 1920. Claude *Tidd Collection, Yukon Archives*

Radio Communication

Although mail was the major form of communication during the early days of mining activity on the creeks, and it was also possible to telegraph Dawson from Stewart City, exploitation of the major silver deposits in the Keno area after 1919 necessitated a better system of communication. The contracted mail system was not adequate for the activities being conducted by the large, hardrock mining corporations. Major efforts were made on their behalf to install a radio or telegraph system. These efforts, however, were stymied by lack of government assistance and it was not until October, 1923, that the first wireless message was sent from Dawson City to Mayo Landing.

From a series of letters and telegraphs exchanged between the Gold Commissioner, G.P. MacKenzie at Dawson City and J.B. Hunter, Deputy Minister of Public Works in Ottawa,[54] it is possible to piece together the series of events that led to the installation of wireless service at Mayo Landing.

As early as 1920, MacKenzie, on behalf of the territorial government and local businesses, requested that Ottawa consider the possibility of constructing a telegraph or telephone line to the Mayo district.

The issue was considered by the Public Works Department and in June, 1921, Hunter informed MacKenzie that the installation of a wireless station to communicate with Fort Egbert, 180 miles from Mayo, would cost $42,500 with an estimated $10,000 per annum required for maintenance. A land line from Granville, south of Dawson City, to Mayo would cost about $36,000, with maintenance about $6,000 per annum. Hunter maintained that either proposition was too expensive to be considered and also noted that Parliament had not appropriated funding for such a project. In a later communication, Hunter suggested that some of the wealthy corporations in the Mayo area might be interested in supplying their own means of communication.

MacKenzie continued his push for a wireless

The Signals Corps of Engineers building, built in Mayo in 1925 on Block 4 of the Government Reserve on Second Avenue. *G. J. Gordon Collection*

station through the summer of 1921. His efforts in themselves brought home the need for a system at Mayo Landing in a very interesting manner. MacKenzie noted that telegrams he was receiving from Hunter often left Ottawa for Chicago, then Seattle, and ultimately came by way of Eagle, Alaska. He pointed out that not only was this an unbusinesslike way to conduct transactions, but it also jeopardized the security of government communications.

MacKenzie was joined in his efforts to make the government install a wireless in the Mayo district by Livingston Wernecke, the manager of the F. W. Bradley mining interests (Treadwell Yukon Company, Ltd.) on Keno Hill. Wernecke threatened to install his own wireless if the government continued to procrastinate. He noted that such an installation would decrease revenue earnings for any future government lines.

In 1922 the federal government had several opportunities to become involved in joint ventures. In January, the Mayo Utilities Company, which operated a local telephone system within a 50 mile radius of Mayo, asked the government to grant them a subsidy of $5,000 a year for four years. In return, the company would install a link of their telephone to government telegraph lines 100 miles away. If the line was laid while the snow was still on the ground, it would be in operation by July of the same year. The government waited until well into the spring before giving a negative reply to the request.

“ MacKenzie wrote a strongly worded letter to Hunter asking for a definitive answer as to the government's future intentions for the area. ”

At the end of April, 1922, MacKenzie wrote a strongly worded letter to Hunter asking for a definitive answer as to the government's future intentions for the area. He noted that, in addition to mining interests, the J. A. MacDonald Company was proposing to open a branch store at Mayo Landing and would install radiophones at the Mayo Landing and Dawson City stores if no other form of communication was forthcoming.

Finally, in June of the same year, Wernecke offered to construct a land line from Fort Selkirk to Mayo if the government would agree to take it over in a year or two. Hunter replied that the government could not accept the proposal and added that they did not plan to bring communication into the area in the near future or aid the territorial government to do so.

The future of communication in the Mayo District was now firmly in Wernecke's hands. On July 14, 1922, Marconi Wireless offered Wernecke a radiotelephone system for $10,000, but by July 26th, MacKenzie had found a better deal for him. A K&W wireless plant belonging to the Fortymile Power and Dredging Company was on the dock in Dawson City. But, as the company was out of business and the freight and duty charges were not paid, Wernecke could make a deal for it. MacKenzie, on behalf of the territorial government, and Dawson City merchants were willing to cover the cost of installation at their end if Wernecke would set it up at the Mayo end.

However, due to customs regulations, the government could not release the K&W plant to Wernecke. By September, 1922, it became obvious that another winter would come and go without the Mayo District having reliable communication with the outside. It would have to rely on mail. MacKenzie attempted to have the Dawson City mail routed through Crooked Creek, giving Mayo weekly service, but failed in his endeavour.

Support for the installation of a wireless at Mayo Landing came from an unexpected source. In April, 1923, the Director of the Northwest Territories Office of the Department of the Interior in Ottawa, O.S. Finnie, contacted G.J. Desbartes, Deputy Minister of National Defence, asking if it would be possible to install a wireless station at Mayo Landing by summer's end. He stated that he had received a number of requests over the years for such an installation and, having visited the region, he felt the high level of activity could be the source of considerable business. Finnie stated that the station would not have to be high powered because it only needed to be able to communicate with Dawson City, 100 miles away, which could dispatch the messages via its link with

Hazelton, B.C.

On June 9, 1923, a sum of $19,000 was secured from the Department of the Interior to install a wireless station at Mayo. It would be part of a relay system which would include stations at Dawson, Fort Simpson and Fort McMurray.

However, as the Mayo Landing line had been added to the system at the last minute, the Department of Defence planned to hold off on the construction of that portion of the line until the following summer. A series of communications between George Black, Member of Parliament for the Yukon, Finnie and W.W. Cory, Deputy Minister of the Department of the Interior, established the need for a wireless link to be established before the close of navigation in the fall. The reasons cited were the large population in the Mayo area, about 700–800, the important silver mining industry that was being handicapped, and the fact the area would be inaccessible after freeze up.

MacKenzie was made the federal government agent to oversee the site selection and construction of the stations. Major (later Lt-Col) W.A. Steel, a Signal Corps Engineer and six men were dispatched from Ottawa to establish the stations and remain for the winter to maintain and operate them. A standard wireless set was installed in late August in Dawson, while a temporary 120-watt transmitter was installed in a rented building at Mayo Landing. The building was located on Second Avenue on Block 4 immediately adjoining the Government Reserve.

The first wireless communication was sent by MacKenzie in Dawson to J.E. Ferrell, the Mayo representative on the Yukon Council, on October 20th, 1923. A rate of $1.50 per ten-word message with 10 cents for each additional word, was established. George Black went on record as objecting that the prices were twice as high as they should be. On September 23, 1924, approval was received from Ottawa to construct a new building to house a permanent standard radio set.

The building was constructed on Block 4 of the Government Reserve and was the same type of building used at all stations with the equipment occupying the first floor, and personnel accommodation on the second floor. Its source of power and light was a small, gasoline engine powered, electrical unit. The building was completed in the spring of 1925.[55]

The Signal Corps continued to operate the wireless in Mayo until February 10, 1960, when the service was taken over by the Department of Transport and confined to a weather station.

HYDROELECTRIC POWER

The concept of hydroelectric power for the Mayo district was first introduced by a couple of promoters in 1900. Frederick W. Dillabough and D.H. MacKinnon of Dawson City made an application to the Department of the Interior for water power rights on Fraser Falls on the Stewart River. Their plan was to place turbine wheels on the left bank of the Stewart above and at Fraser Falls in order to generate electricity. The scope of their project was outlined in a letter to F.K. Gosselin, Crown Timber Agent in Dawson.

> A joint-stock company will be formed to place a plant in operation at the falls and the question of the development of power and the supplying of power for the purposes of thawing, pumping, heating, transportation and general mercantile purposes, will, the promoters think, be solved. The intention of the promoters is, to install a plant somewhat similar to the turbine power plant at Niagara Falls, and to transmit power to the various mining districts which are at present being developed in the Stewart River country, and if necessary even, to transmit the same as far as the Klondike.

Dillabough and MacKinnon's application was rejected by the Commissioner because it was felt that there was not enough activity in the district to warrant such a plant and therefore the power would not be utilized but would only be held for speculative purposes.

Fraser Falls was studied again in 1919 as a possible source of hydroelectric power by Keno Hill Limited.

A hydroelectric plant was finally built in the district in 1951 but it was placed on the Mayo River rather than at Fraser Falls.

N.C.P.C. in the Yukon

by Joe Long and E. W. (Ted) Humphrys

The Commission's activities in Yukon originated with the Mayo River Hydro plant commissioned in November, 1952. This development was a first cousin of the Commission's initial project, the Snare Rapids development in the N.W.T. The project was sparked by Ventures Ltd., a mining development company that had also become interested in the United Keno Hill mine at Elsa, about 28 miles north of Mayo. This mine had operated for several years using old style, somewhat inadequate diesel power. In keeping with the objective of improving and expanding the United Keno Hill Mine operation, Ventures Ltd. initiated a search for an appropriate source of hydro power within economic proximity of the mine site. An experienced engineer, W.G. Stewart, was commissioned in the fall of 1948 by Ventures Ltd. to carry out a hydro reconnaissance survey of the Mayo area. Stewart quickly identified attractive hydro potential on the Mayo River a few miles north and upstream of Mayo Landing.

Encouraged by the Federal government's participation in the Snare River project, and by the fact that the Northwest Territories Power Commission had been established to "facilitate the construction and operation of power plants for mining and other interests to encourage the development of mining properties," the Ventures-UKHM officials suggested that the Federal government develop a source of hydro power in the Yukon, primarily to supply their mine site at Elsa. The government acquiesced and directed that this initial Yukon power project be undertaken by the Northwest Territories Power Commission. To that end the NWT Power Commission Act was promptly amended in March, 1949 to extend the Commission's "domain" to include the Yukon Territory, but the corporate name was not changed until the Act was further amended in 1956.

Montreal Engineering Company carried out a full scale feasibility study and engineering investigation of the hydro potential of the Mayo River, with a capacity objective of 2500–3000 kW. Topographic surveys, water measurements and geological studies were carried out in the summer of 1949. The study selected the site of the existing hydro plant, some five miles upstream from Mayo on the Mayo River, as the optimum location for a 2500 kW development. Mayo Lake would be developed as a storage reservoir by constructing a dam at the lake's outlet. The main dam upstream of a bend in the river would raise the river level, making it necessary to raise the highway bridge between Mayo and Elsa.

ABOVE, Mayo Lake storage dam construction, 1951. BELOW, Main dam construction, 1952. *J. Long–N.C.P.C. Collection*

The prime contractor for the project was Northern Construction and J.W. Stewart Ltd. of Vancouver, B.C., with Andy Seraphim as superintendent. Northern Canada Power Commission, then known as Northwest Territories Power Commission, was represented by Joe Long.

Construction commenced in March 1951, with the building of access roads and camp facilities for the construction personnel. Rock excavation for the tunnel, powerhouse, spillway and intake was completed during the first year while part of the Mayo Lake storage dam and survey for the transmission line to Elsa was also completed in 1951. Construction activities were closed down for the winter.

In March, 1952, construction resumed with the building of form work and pouring of concrete for all structures. Installation of equipment began in the summer when the powerhouse building, transmission line, substations, and the Mayo Lake storage dam were completed. The plant, with one turbine and generator, was commissioned in December, 1952.

In 1957, the powerhouse was enlarged to accommodate a second turbine generator unit, bringing the capacity of the plant to 5 MW. This extension was built to reduce coal consumption at the United Keno Hill Mines.

Long was assigned to the Commission head office staff following completion of the Mayo project, and held various senior positions, becoming

ABOVE, Powerhouse and surge tank, December 1952
BELOW, Hydro construction staff, 1952. From left, front row: Joe Long, Les Calhoon, Phil Veniot, Jack Boomhower, Dave Dogoid, Dave McKirdy. Back row: Frank McGowan, Bill Hardcastle, Con Mulherin, "Tom" Tomkinson. *J. Long–N.C.P.C. Collection*

Transmission line construction, 1952. *J. Long–N.C.P.C. Collection*

Generator installation, 1952. *J. Long–N.C.P.C. Collection*

general manager in 1978. He retired in June, 1982 and died in September, 1983 after an eight month battle with cancer.

The late Jesse Barwise was transferred from Snare Rapids to become the first superintendent of the Mayo development. Lorne Vance, now on Commission head office staff, was a Mayo operator-lineman in the early years of the Mayo plant.

The original project included a short transmission line into Mayo Landing to provide hydro power on a wholesale basis to the Mayo Light and Power Company. The Commission purchased this distribution company in 1956. During the 1950's, the main transmission line was extended from Elsa to the Keno City area to supply a short-lived milling operation, along with wholesale supply to a private company that had developed a small distribution system serving the Keno City community. In 1957, a second unit was added to the Mayo plant, increasing the installed capacity to 5150 kW.

Mayo Hydro and Power Dam. *G. A. McIntyre photograph*

Chapter 8

Other Settlements

FROM THE LATE 1800'S TO THE 1920'S, numerous settlements sprang up in the Mayo area. Many flourished, though only briefly, and almost all were mining-related. Settlements were established at Dublin Gulch, Highet Creek, Minto Lake, Haggart Creek, Scroggie Creek, and Thistle Creek with the discovery of placer gold.

By 1902, Duncan Creek was staked from its headwaters to the Mayo River, and a settlement developed at the mouth of Duncan Creek which was referred to as Mayo Bridge. It is believed that there was a N.W.M.P. post at Mayo Bridge although the only record of it is an indication on maps at the Mayo Mining Recorder's Office (Plan #7025, Lot of A, Group 1054).

Bob Levac operated a general store and trading post at Fraser Falls from 1915 to 1931. A trading post existed at Russell Post in the early 1900's (circa 1900–1910). A small settlement was listed at Steamboat Bar in the 1920's.

After the Hector Calumet mine was found in 1934, the town of Calumet was established near that site. During the 1970's, the abandoned town of Calumet was burned down for fear the old buildings might spark a forest fire.

Barlow

From *Yukon: Places & Names* by R. Coutts

Barlow was located at the junction of Clear Creek and Barlow Creek. It was a small settlement started in 1903 when placer mining began on Clear Creek and its tributaries. A steamboat landing, roadhouse, and trading post to supply the miners were built. Barlow was a stopping place on the Dawson-Mayo-Whitehorse winter road. It was abandoned in the 1920's.

Beaver City

From *Yukon: Places & Names* by R. Coutts

Beaver City was located at the junction of Carpenter and Settlemier Creeks. It was established in 1923 when silver-lead veins were found in the vicinity. It was a distribution point to supply the prospectors. The ore found was not rich enough to mine in this isolated area and the place was abandoned by 1925. It received its name because many of the prospectors came to the area by way of the Beaver River.

Gordon Landing

Gordon Landing was established by George

Gordon in 1902 following the Duncan Creek stampede. By 1903, the settlement was abandoned.

Gordon Landing was located on the Stewart River 17 miles above Mayo and was the disembarking point for miners on their way to Duncan Creek. From Gordon Landing they hiked two miles to Janet Lake and another twelve miles north to the mouth of Duncan Creek where the Mayo Bridge settlement was located. There was a N.W.M.P., post at Gordon Landing and Mining Recorder John Rowley had a recording office there in 1902. It was soon discovered that the river was too shallow to make Gordon Landing suitable for steamboat dockage. This initiated the survey and settlement of Mayo townsite early in 1903. By the time the road from Mayo Landing to Duncan Creek was completed in 1904, Gordon Landing was completely abandoned.

Lansing

Lansing was located at the confluence of the Stewart and Lansing Rivers. Lansing was named for Lansing River which received its name from Samuel Lansing, who prospected the Lansing River in 1883. Lansing Post was established in 1902 by Frank Braine and Percival Nash who brought along a group of Hare Indians from Fort Good Hope to settle at Lansing. The Lansing country was rich in furs. Braine and Nash built a trading post and supplied it from Dawson City. In 1908 Jim and Helen Ferrell took over Lansing Post, operated it until 1915, and then sold it to Jim Mervyn. Mervyn closed the post in 1937 and moved to Mayo. Lansing was completely abandoned by 1940. Today, some evidence of Lansing Post can still be seen at the old site. The river has washed away some of the river bank near the cemetery and at some point, graves may begin to wash into the river. (See: *The Na-Cho Nyäk Dun, Buildings and Artifacts, J.E. Ferrell, and J.H. Mervyn.*)

McQuesten

McQuesten was established on the south side of the Stewart, opposite the mouth of the McQuesten River, after bar gold was discovered on the Stewart River. By the early 1900's (circa 1901), a N.W.M.P. post had been established there as well as a trading post operated by Crowell and Higgins. A number of cabins had been erected by miners and approximately forty Stewart River Indian people lived at McQuesten. A winter trail ran past McQuesten and after 1902 the post was serviced by steamboats on their way to Mayo. After Mayo was established, McQuesten was gradually abandoned, except for a roadhouse which was operated in connection with the winter stage road. Roger Connor/s ran the roadhouse at McQuesten in 1916. (See *N.W.M.P.–R.N.W.M.P.–R.C.M.P.*)

Minto Bridge

Minto Bridge settlement was named for Minto Creek and it was located ten miles north of Mayo at the junction of Minto Creek and Mayo River. (Minto Creek was named for the Earl of Minto, Governor General of Canada from 1898–1904).

The settlement gained prominence as a pivotal distribution centre when the road was built from Mayo to Minto Bridge and the 175 mile overland trail from Dawson City to Minto Bridge was completed in 1905. Government roads led from Minto Bridge to all of the main gold producing creeks and later to the silver discovery at Galena Creek. Mayo and Minto Bridge were the distribution points for the Mayo area during the summer and winter months, respectively. In summer, steamboats brought passengers and freight to Mayo Landing and they were then transported over the wagon road to Minto Bridge. In winter, passengers and supplies travelled over the trail from Dawson City and dispersed at Minto Bridge.

In 1904 the Mining Recorder's Office was moved from Duncan Creek to the more central Minto Bridge settlement. Thomas H. Hinton was the Mining Recorder. The first post office in the district was established at Minto Bridge in 1910 with George Cunningham as Postmaster. Amy Patterson operated a roadhouse at the settlement in the early 1900's (circa 1901–1919) as did partners Frank Chasni and Gus Schogren, whose roadhouse also operated in the early 1900's (circa 1901–1919) and in the 1920's. Mr. and Mrs. Rupert Steeves operated a roadhouse at Minto Bridge for Treadwell Yukon in the late 1920's through to the early 1930's. It was the last stopover for cat skinners before Mayo Landing.

Persons listed as having a Minto Bridge address in the mid-1910's and the early 1920's are as follows: (This list is by no means complete. These names are the only ones that have survived in historic records.)

John "Jack" Adair
William Bramley
William Burrell
Frank Chasni
Arthur Davidson
William Hayden
Hans Johnson
Otto Kastner
Joe Laska
Elmer Makela
Martin Malesich
John Mawhinney
Katherin McRae
Gustav "Gus" Schogren
James Scougale
John Turner
Joe Walsh

When ore was hauled by horse and team from Keno Hill, Minto Bridge was one of the stopping places on the relay to Mayo. As modes of transportation changed and improvements were made to the road systems, activity at Minto Bridge settlement dwindled.

In 1952 the Mayo River Hydro project permanently flooded the old settlement. The bridge across the Mayo River there is still referred to as Minto Bridge and carries a sign to that effect. The Mease family operates a farm near the old settlement location.

Old Stewart Crossing

In the early 1900's (circa 1903–1906), Stewart Crossing was a tiny settlement at the point where the Whitehorse-Dawson winter trail crossed the Stewart River. It was located between McDiarmid's wood camp and Rosebud Creek. It was described in Polk's 1923–24 directory as "a station on the Whitehorse Dawson winter trail, 75 miles southeast of Dawson City and 19 miles southeast of Black Hills the nearest post office." By 1918 it was referred to as "the Old Stewart Crossing."

Stewart Crossing

by Wendy Andre

Stewart Crossing was originally known as "Frenchman's Bar." It was named for a bar in the

Minto Bridge in summer. *Middlecoff Collection*

Stewart River where bar gold was mined in the early days. Until the highway was built, Frenchman's Bar was serviced by steamboat during the summer months and by stage in winter. The settlement was populated by Indian people, miners, and trappers. Beaver were plentiful in the area.

In 1950, the highway from Whitehorse to Elsa was completed and a ferry was floated on the Stewart River at Stewart Crossing. There were three shifts of ferry operators, mostly Van Bibbers and Jack McDiarmid. Proctor operated a sawmill where the government garage is located today. The first highway lodge was a tent with a dirt floor. It was run by Red Nickles and served coffee and meals.

When the Proctor sawmill closed down, Jack and Mary McDiarmid bought the buildings and moved them down by the ferry landing. Their first residence was comprised of two cabooses placed together. There were so many truckers, highway crew, and ferry crew stopping at McDiarmid's for coffee and snacks that the couple moved into one of the larger buildings and converted their caboose residence into "Mary's Lunch Bar."

Construction of the Stewart River bridge started in 1959 and continued to the fall of 1960. The population swelled with construction workers and Stewart Crossing was bustling. As civilization crept into the settlement, it became necessary to comply with building codes and water and sewer regulations. In 1957, McDiarmid built a new lodge near the ferry docks. When the bridge was completed, the Stewart Crossing Lodge was moved to its present location. Gas pumps and tire repair were added to the business (these are the frame buildings that still stand behind the lodge).

United Keno Hill Mines maintained the highway from 1953 to 1957 and then the government took it over. The government compound at Stewart Crossing was built in 1958. Yukon Electric installed a power plant in 1964.

Bobbie and Cliff Desjarlais bought the Stewart Crossing Lodge in May of 1966. The cafe was enlarged and kitchen facilities were improved. The large metal garage was built in 1975.

ENDNOTES

[1]Memoir 284, Yukon Territory, Selected Field Reports of the Geological Survey of Canada 1898–1933: compiled and annotated by H.S. Bostock. Excerpts from pages 26, 128, 129, 130, and 161 reprinted with permission of the Minister of Supply and Services Canada, 1989.

[2]Julie Cruikshank, *Through the Eyes of Strangers: A Preliminary Survey of Land Use History in the Yukon During the Late Nineteenth Century*. Report to the Yukon Territorial Government and Yukon Archives, 1974. Appendix I, p. 8.

[3]Catharine McClellan, *Part of the Land, Part of the Water* (Vancouver: Douglas & McIntyre Ltd., 1987) p. 156. Reprinted by permission of Douglas & McIntyre.

[4]Subsequent chiefs were Simon Mervyn, Richard Moses (acting chief for three months), Peter Johnson Lucas, Robert Hagar (1976–1986, 1987–present), and Albert Peter (1986–1987).

[5]R. C. Coutts, *Yukon: Places & Names* (Sidney, B.C.: Gray's Publishing Limited, 1980), p. 174.

[6]R. C. Coutts, p. 102.

[7]R. C. Coutts, p. 121.

[8]Richard Poplin, Pete Wibourg, Frank Morphat, Joe Macheal, Thomas Boswell, John Fraser, J. Chapman, Charles Powell, Joe Ladue, Howard Franklin, Thomas Williams, Mike Hess, Ike Powers, Steven Custer, and the two brothers Al and Hugh Day.

[9]R. C. Coutts, p. 252.

[10]Allen A.Wright, *Prelude to Bonanza: The Discovery and Exploration of the Yukon* (Whitehorse: Arctic Star Printing, 1980), p. 250.

[11]R. C. Coutts, p. 251.

[12]*Memoir 284, Yukon Territory*. Selected field Reports of the Geological Survey of Canada, 1898 to 1933, compiled and annotated by H.S. Bostock, p. 141.

[13]*Memoir 284, Yukon Territory*, p. 143.

[14]Yukon Archives Aaro Aho Collection, 82/161. *Keno Hill: An Era of Individualism in Yukon's Great Silver District*," ca. 1972.

[15]*Memoir 284, Yukon Territory*, p. 398.

[16]*Memoir 284, Yukon Territory*, p. 398.

[17]H. S. Bostock, *Pack Horse Tracks - Recollections of a Geologist's Life in British Columbia and the Yukon 1924-1954* (Ottawa: G.S.C., 1979), p. 164.

[18]R. C. Coutts, p. 187.

[19]R.L. Debicki, *Yukon Mineral Industry: 1941 to 1959* (Whitehorse, 1982).

[20]*Memoir 284, Yukon Territory*, p. 127.

[21]*Memoir 284, Yukon Territory*, p. 388.

[22]Yukon Archives Aaro Aho Collection, 82/161. *Keno Hill: An Era of Individualism in Yukon's Great Silver District*, ca. 1972.

[23]*Memoir 284, Yukon Territory*, p. 133.

[24]*Memoir 284, Yukon Territory*, p. 389.

[25]Yukon Archives Aaro Aho Collection, 82/161. *Keno Hill: An Era of Individualism in Yukon's Great Silver District*," ca. 1972.

[26]Livingston Wernecke. Letter to Mayo Mining Recorder, R.L. Gillespie, 2-9-28.

[27]Yukon Archives Mining Recorders Records (Placer) Series 10 1-7. Vol. 267.

[28]Yukon Archives Aaro Aho Collection, 82/161. *Keno Hill: An Era of Individualism in Yukon's Great Silver District*," ca. 1972.

[29]Mayo Mining Recorder's reports of 1934 and 1935.

[30]*Memoir 284, Yukon Territory*, p. 138.

[31]Rudolph Gulch and Rudolph Pup are both named after Rodolph Rosmusen.

[32]Yukon Archives Aaro Aho Collection, 82/161. *Keno Hill: An Era of Individualism in Yukon's Great Silver District*," ca. 1972.

[33]Yukon Archives Mining Recorders Records (Placer) Series 10 1-7. Vol. 267.

[34]*Memoir 284, Yukon Territory*, p. 397.

[35]*Memoir 284, Yukon Territory*, p. 397.

[36]Letter from White Pass & Yukon Route to G. P. MacKenzie, Gold Commissioner.

[37]Diane Johnston, *Carrying the Yukon: Silver-Lead Mining in the Mayo District 1919–1942*. Unpuplished paper, Yukon Archives 81/100 F-166, 1981.

[38]Diane Johnston, *Carrying the Yukon: Silver-Lead Mining in the Mayo District 1919-1942*.

[39]Thayer Lindsley was a mining engineer and promoter. He was born of American parents at Yokohama, Japan on August 17, 1882. He returned to the U.S. at 15, graduated from Harvard in engineering and worked for the New York City Subway. After serving in World War I, he worked up a mine in Oregon, selling it for $30,000 in 1923. With this stake he set up Ventures Ltd. in 1928. Ventures prospected, developed and financed mines, and by the mid-1950's owned mines or shares of mines on every continent except Asia. Lindsley's best-known venture was Falconbridge Nickel Mines near Sudbury, Ontario, which he and a partner, Joseph Errington, incorporated in 1928. He and his brother, Halstead, also helped establish Sherritt Gordon Mines in 1927. Thayer Lindsley died at New York City on May 29, 1976. Robert Bothwell, *The Canadian Encyclopedia*, 1st edition, 1985.

[40]Cockfield, Geological Survey of Canada, Memoir 248. Compiled and annotated by H. S. Bostock.

[41]Renamed United Keno Hill Mines Limited in 1947.

[42]Information taken from Alex Nicol's YOOP Mayo Lodge membership certificate, 1927.

[43]W. R. Hamilton, *The Yukon Story*, 2nd ed. (Vancouver: Mitchell Press Ltd., 1967), p. 100.

[44]R.C.M.P. Quarterly, Fall 1985.

[45]*The Islander*, (Saltspring Island, B.C. newspaper). Article by June Lunney [Cruickshank's daughter]

[46]Andy Cruickshank was killed on June 29, 1929, in a crash northwest of Great Slave Lake, N.W.T., while flying a Canadian Airways Fokker Super Universal. R.B. Cameron, Yukon Aviation Historian.

[47]This propeller has always been referred to in the Mayo district as the propeller from Patterson's plane *Queen of the Yukon II*; however, R.B. Cameron, Yukon Aviation Historican, has stated that although the propeller is genuine, it is not the propeller from the *Queen of the Yukon II* but from a much older aircraft most likely of World War I vintage. The propeller from the *Queen of the Yukon II* was made of metal rather than wood and as the plane crashed nose first into the ice of the Stewart River, it is doubtful the propeller could have survived. Jack Andison who was in Mayo at the time of the crash, recalls that the propeller from the *Queen of the Yukon II* was removed from the wrecked plane and taken to Wernecke Camp where it was repaired and brought back to Mayo to be placed on Mr. Patterson's grave. Alex McCarter who was also in Mayo when the crash happened, believes that the propeller is not from the *Queen of the Yukon II* but from a Moth owned by Treadwell Yukon which was a much older aircraft than the *Queen of the Yukon II*. (The Moth is believed to have been wrecked when the engine was over revved while the plane was on the ground, causing the Moth to flip onto its nose.) George Besner recalls that Treadwell's Moth had a wooden propeller and when the pilot wanted to start the engine he would shout "contact" and George would crank the propeller by hand. R.B. Cameron has said that although it is possible that the propeller on Patterson's grave is from a Moth, the propeller appears to be too large for that type of aircraft.

[48]Robert Woodall, *The Postal History of Yukon Territory, Canada*. (Lawrence, Mass:: Quarterman Publications

Inc., 1976).

[49]W. R. Hamilton, p. 97.

[50]The data on mail services to the region was obtained primarily from letters and telegrams available in Yukon Archives, Whitehorse. *Yukon Lady* by H .D. MacLean was also consulted.

[51]The Stewart River Post Office was located at Stewart City. In many of the documents consulted Stewart City and Stewart River were used interchangeably. We have used Stewart City consistently in this document.

[52]There were various locations for "Stewart Crossing." One was at Rosebud, the other at the McQuesten Post, and the last before the road was built was 30 miles below the present day Stewart Crossing.

[53]K. Coates and W. Morrison, *Land of the Midnight Sun: A History of the Yukon* (Edmonton: Hurtig Publishers, 1988), p. 269.

[54]Copies of these communications were made available to the Mayo Historical Society by Yukon Archives in Whitehorse.

[55]Yukon Archives 971.21ICA *The Yukon Territory* 1926, Department of the Interior, Ottawa, 1926.

PART TWO PIONEER ACCOUNTS

KENNEY & DAISY ADAIR

by Robert Adair, November 1989

Ken Adair was born in Alberta on April 8, 1914. He came to the Yukon from Barrhead Alberta, in the middle 1940's and worked with Jean Jocquot at Burwash Landing and drove trucks on the Alaska Highway. He then went to Dawson City where he worked as a woodcutter for the Yukon Consolidated Gold Company building a cabin at Flat Creek hill during the winter of 1948–49. Ken also worked for Clear Creek Dredging with Bill Scott until the dredge was mothballed in 1950.

Daisy was born in Oxbow, Saskatchewan. Her older siblings were all born in Wales and Daisy was the first Powell child to be born in Canada.

Daisy and her three sons, Robert, Jim and Tom, arrived in Whitehorse in September 1949 and Daisy left the boys in the convent at Whitehorse while she looked for work. She got a job nursing at the old hospital in Dawson City and was on duty when the old hospital caught fire and burned down that winter. Daisy helped move patients to Whitehorse and to the new temporary hospital in Dawson. While Daisy was nursing at Dawson's temporary hospital, she met Kenney Adair who was brought to the hospital with a broken leg.

She was reunited with her boys in June 1950 and in November 1950 Daisy married Kenney Adair. Daisy and Ken came to Mayo to be married and the ceremony was performed by Gordon McIntyre at the Chateau Mayo Hotel where the couple spent their honeymoon.

Ken moved the family from Dawson City to Mayo at Easter time in 1951. They moved into a home on the mill road which was formerly Dick Kimbel's residence. Ken went to work for the sawmill running rafts of logs down the Stewart River from the falls to the sawmill (which was located on what is now the Mehaffey home) and he also worked the edger in the mill.

Daisy was a relief nurse at the old Mayo Hospital from 1951 until the family left in 1956 and also acted as Dr. Clark's operating room nurse.

In June 1952 Fred Adair was born in the old Mayo hospital. Daisy was Dr. J.V. Clark's first maternity case in the Mayo district.

In 1953 the family moved to Keno City where Ken worked for Cole Bros. until he went to work for MacIssac Construction in 1955.

Daisy's oldest son, Robert "Bob" Adair, drove a truck for Keno City's Cole Bros., hauling gravel on the Mayo-Dawson Highway in the summer of 1954 and during the winter of 1954–55 he drove a taxi for Peter Duke in Keno. In the summer of 1955, Bob again drove a gravel truck for Cole Bros., this time hauling ore from Mackeno Mines (Galkeno), Belle Keno and the Shamrock mines. In the winter of 1955–56 Bob drove a taxi for Vic Cherny's Silver City Taxi and in the summer of 1956 drove Cole Bros. truck again.

The Adair family moved to Dawson Creek, B.C., in August of 1956.

Robert Adair returned to the Yukon in the summer of 1961 and worked for Roy McDiarmid until October and then worked in Mayo for the Yukon government until November when he landed a job with United Keno Hill Mines. He remained with United Keno until Christmas 1963 and then moved back to B.C. Bob returned to work for United Keno in 1966 (and remained with the company until July 15, 1971).

After Kenney Adair died of cancer on February 10, 1967, Daisy and her youngest son, Fred, moved back to Mayo. Daisy went to work at the new hospital in Mayo under Nurse-In-Charge Alice Smashnuk. Daisy bought the former Besner residence in 1969.

In 1970 Daisy accepted the position of Nurse-In-Charge at Macaulay Lodge in Whitehorse and worked there until her retirement in approximately 1975. Daisy was killed in a motor vehicle accident on the Alaska Highway in 1983.

Robert Adair operates Mayo Taxi in Mayo (1989) and continues to live in the home Daisy bought in 1969.

DOROTHY ALLAN (née Dignam)

Interview and transcription by Linda E. T. MacDonald, 1983

I was born in Toronto on September 10, 1920, and attended school in Oshawa at the Sisters of St. John and in Toronto at St. Clements. After my stenographer's course I worked for the Imperial Bank (now the Royal Bank), IBM, Workmen's Compensation, and GE.

I always admired my Grandmother Dignam, and felt she provided inspiration to the women in my family. She was an artist of considerable renown who painted Dutch and Canadian scenery. Her works were shown in the Arthur Lizmer Gallery in Toronto and she made well over fifty trips abroad to paint there. I believe she also took the first group of Toronto debutantes abroad. Her contemporaries often said she should have been one of the "Seven."

World War II had a tremendous impact on my family. One of my brothers was at Dieppe and the other was in the Italian campaign. My sister was interpreter for the British Embassy in Santiago, Chile, and later Earl Mountbatten's interpreter. I joined the RCAF (WD) and went to England as a Clerk Stenographer. While in England, I met and married my husband Fredrick Robert Allan. Fred was a paratrooper. He was born and raised in Nova Scotia.

After the war my husband went to work for N.C.P.C., and in 1954 we were posted from Snare River, N.W.T., to Mayo, Yukon. My daughter Linda was only eight or ten days old at the time. We flew from Yellowknife to Edmonton and then from Edmonton to Mayo. The plane followed the Alaska Highway from Edmonton to the Yukon, and the flight was quite rough. I remember looking at my daughter and seeing that she looked slightly blue, and I said to the stewardess, "Could we have some oxygen here." Then I looked at my son who was sitting behind me and he looked positively ill.

That same plane flew into Mayo for years. The plane had DIG on the wing. I will always remember that, as my nickname when I was a kid was "Diggy"; which was a play on my maiden name Dignam.

" Sometimes we didn't have Christmas until the end of January because we couldn't get anything in. "

We lived in the N.C.P.C. plant compound, about five miles from Mayo, as did all of the N.C.P.C. personnel and their families. When I used to drive from the plant compound to town, the road was so narrow that if I met any of the big Elsa ore trucks, I would just pull off the road and stop. The sacks of ore off the trucks would be piled on the riverbank, and once the river was open to navigation, the steam boats would come in and the ore would be loaded and the booze unloaded. That was always a big day. Archie Close used to do all the hauling when the boat came in.

The only stores that were in Mayo when we lived there were the Northern Commercial and Taylor and Drury. There were also no bridges on the route to Mayo in the early 50's. Sometimes we didn't have Christmas until the end of January because we couldn't get anything in.

I remember Mr. Barker's telephone system. You could only make calls until 8 or 9 in the evening and if you called after that it cost $1.

I first met Dr. Clark during the big flood in Mayo, although I didn't know who he was at the time. We were all out sandbagging the area and he was the fellow next to me who was shovelling while I held the bag. The bags we used were ore bags from Elsa. The bar remained open during all of this activity. There was water in it up to our knees but we would go in and get a drink and then go out and bag some more. We all had to have typhoid shots as the water was so polluted.

Some people were moved out near the airport and the hospital staff went out and checked on them every day. The first time I ever saw Dr. Clark professionally, I went to the old hospital for an appointment. There was a dog team tied to a tree beside the hospital and there was this guy standing there whom I'd met while sandbagging during the flood. He was wearing a big pair of R.C.M.P. suspenders and a bright red shirt.

We moved out of the Yukon for one year in 1959 and went to Ontario. The following year we came back and moved into the old rectory behind the Anglican Church. It had a big pot-bellied stove in it for heat, which was a good thing as the ceilings in the rectory were so high it was very hard to heat.

Later we moved into the house that Jack Smith recently owned [it is now owned by Tony Blais, 1989]. It used to be the only house in Mayo with hardwood floors. I also had a little cabin down by the river. Mag Wallingham was going to help me fix it up. The inside appeared to be done in cardboard covered with cheese-cloth or something like that. While looking around the cabin for turpentine we found a bottle of cherry brandy which the previous owner had made. Needless to say we didn't get anything done for the rest of the afternoon.

My mother came up for a visit one summer and I took her on a tour around town. She pointed to a building and said who owns that cute little cottage. I said, "Mother that's the airport!"

We used to have some great baseball players in Mayo: Dunc McGeachy, Woody, Dick Wallingham, and others. We used to go and sit on the grass and watch them play.

"It's funny, in all the years I was working, I used to save like crazy to take trips 'outside.' Now that I am retired and could move 'outside' any time I wished, I find I prefer to live in the Yukon."

Sometimes in the spring people would ride rafts through the river gorge and end up over behind the old hospital. My husband thought they were all crazy. One year a couple of the nurses' husbands came through the gorge and one of them fell out of the raft but managed to get back in. When they arrived we were all waiting on the riverbank and gave them a big cheer.

I started work as secretary at the Mayo Hospital the day my daughter Linda started school. Linda and Elaine Wallingham started school at the same time and Mag Wallingham took them on the first day. I worked at the hospital for seventeen years and made $2,500 per year. When I first came to work at the hospital, all of the X-rays were in a big box with no year, no name, or any identification on them at all. Dr. Clark, however, could recognize each one of them on sight.

One time I took a trip to Alaska on the Sunset Trail. I had a Mini Minor (1957 model); my first car. I had my daughter Linda and Elaine Wallingham with me; they were about twelve years old at the time. We ended up running out of money and had to have some wired to us. I remember taking my car in to get a tire changed and the attendant said, "How do you expect me to change a tire on a wheelbarrow like that?"

The Legion used to be very active in Mayo. We had around 100 members, many of them from Elsa. At one time I tried to have a war memorial erected in front of the hospital. It was to have been a cement cairn with three rifles (bayonets fixed) leaning together with a helmet on top of them. I did get permission from Ottawa to have it erected but in the end it didn't materialize.

My husband died in 1977; he was working on the Dempster Highway at the time.

My daughter Linda Mae Allan is a Dental Therapist now and is working out of Whitehorse. My son Peter Robert Allan is in the Navy and based in Nova Scotia. I have one grandchild.

In 1978, after suffering a stroke, I moved to Whitehorse; and, with the exception of a few trips "outside," I have lived there ever since.

It's funny, in all the years I was working, I used to save like crazy to take trips "outside." Now that I am retired and could move "outside" any time I wished, I find I prefer to live in the Yukon.

I've always loved the Yukon summers.

[Dorothy Allan passed away in Whitehorse on September 14, 1987.]

JACK & AGNES ANDISON

by Kevin Shackell, Yukon News, June 23, 1982

It all started in 1929 when Jack, on the urging of his father who was already working in Mayo, set out for the tiny Yukon town called Mayo. It was the beginning of the Depression and work was scarce so Jack decided to leave Vancouver for the wilds of the North.

Nothing he was told down south prepared him for the adventure he was to encounter on his way up. It was a cold, snowy October day when Jack rattled over the White Pass and Yukon railway in a drafty passenger car. Already disgruntled by the weather, Jack soon found himself conversing with an overly talkative woman, who identified herself as Mrs. Wernecke, wife of the Treadwell Yukon mine manager. When Jack said he was destined for Mayo and was hoping to catch the last boat out of Whitehorse, Mrs. Wernecke told him it had already departed.

The well-meaning woman told him not to worry and said he could travel with her. When the train pulled into Carcross, Jack's luggage was off-loaded. Wondering what other form of transportation she had arranged, Jack was told he'd be flying to Mayo in a private plane, an old canvas-covered Fairchild. "And here's me with a city suit on . . . damn near frozen," he said of his woefully inadequate attire.

Jack, the mine boss's wife and three other passengers boarded the rickety float plane which attempted to take off from Nares Lake. It failed to clear the water on the first run and had to taxi back to base to off-load excess baggage. After some of the gear was removed, the pilot made the passengers lean forward to bring the weight off the tail, and the group was air bound for Mayo. "About two and a half hours later we landed on the Stewart River and boy, was I relieved."

Jack had left his sweetheart behind, to seek his adventure in the North and he had found it even before arriving. He joined his father who was running the cold storage warehouse for Burns Foods in Mayo. Jack had asked for Agnes' hand in marriage before he left but her father wouldn't agree to the idea. "He wouldn't let me come north in the fall but if I was of the same mind in spring I could come," she explained. "There's no way he'd let me leave Vancouver for the ice and snow."

Correspondence throughout the winter had assured both that nothing had changed and in the spring Agnes was northward bound to meet her husband-to-be. On her way north Agnes had a few adventures of her own. The S.S. *Keno*, the river boat she was travelling on, was way-laid at Stewart City for three days while the vessel's boilers were cleaned out. Although she didn't think much of Stewart City, Agnes whiled away the time playing cards with other passengers and people of the community.

Meanwhile, Jack, who learned of the extended stop-over, made plans to meet his fiancée as the steamer resumed its journey upriver. He and partner, Jack Sharke, paddled down the Stewart in an Eskimo kayak his friend owned.

Unbeknownst to Agnes, Jack had wired the skipper he would be awaiting the boat upriver and would signal his presence by waving a white flag. The steamer hove into sight in the twilight hour of 2 a.m.

Jack stepped on board and the two were reunited after a year apart. The steamer docked at 9:00 a.m. and the couple were married later that morning by the Anglican minister.

The first year in Mayo the young couple put down roots which were to last much longer than either expected. "When I came in I intended to stay two years and the first time I left the country was 27 years later," Jack recalled.

Settling into married life went smoothly for the Andisons. When Agnes arrived, Jack went to work at Wernecke Camp. "Mayo was a friendly little town, you knew everybody. There were dances in the winter and it wouldn't matter if it was 60 below," Agnes fondly recalled.

But in the early 1930's, the Treadwell mine closed down for a year and Jack found himself out of a job. Agnes moved back to Vancouver for a year to stay with her sister while Jack and his partner moved on to Dawson in search of work.

Jack hiked overland to Dawson and found employment with the Yukon Consolidated Gold Corporation on Dominion Creek. He also held down a job as a night cook at the Arcade Cafe and tended bar at the Rochester Hotel on his days off.

After a year, the economic climate improved and the mine in Mayo reopened.

Agnes gave birth to their first son in 1932. "The day my son was born the whole long-shore crew went on a bender," Jack grinned.

The couple has three sons, a daughter, eight grandchildren and a great-grandchild.

In 1952 the Andisons left Mayo for Vancouver. Agnes stayed for one and a half years and Jack lasted less than four months. "He came out in October and went back in January, he just couldn't stand the big city," Agnes explained.

The Mayo-Keno district has figured heavily in the Andison's lives, although its not the same area they once knew, they contend. Two of their sons work for the United Keno Hill mine in Elsa.

At 98 cents an hour, wages were far from good, but then people were much more self sufficient, explained Jack. "Meat was expensive so in the fall we'd go out to get moose, caribou and grouse. We grew almost all our vegetables," Jack said. Neighbours all pitched in to help one another out. "If somebody got hurt or sick they didn't want for anything, the whole community dived in," Jack recalled. In those days there was no U.I.C.

Social gatherings, informal cups of tea and full-fledged parties were always taking place. "You socialized a lot more in those days," said Agnes. For example, the year the Christmas turkeys didn't arrive on time, a neighbour donated a large rabbit for the Christmas feast. Agnes said those are the kinds of memories that will always be cherished.

Agnes, 74 and Jack, 76, are now living their retirement years to the fullest. With children and grandchildren scattered hither and yon they don't lack for places to visit. Although they've been in retirement since leaving Mayo eleven years ago, they don't look on it as the end, but rather as a beginning.

GEORGE & EMELIA AYLWIN

by Kevin Shackell, Yukon News, October 13, 1982

Emelia [Besner] Aylwin, whose parents came to Yukon at the turn of the century, was raised in Mayo. She was working as a switchboard operator when she met her husband-to-be.

George Aylwin, who worked at the Bank of Montreal in Trail, B.C., was transferred to the Mayo branch in the spring of 1930. "I came up with the Bank of Montreal for two years and that was 52 years ago," he chuckled. "When I joined the bank the first thing they asked me was whether or not I was interested in working at a northern branch." Having accrued some seniority after ten years of service with the company, Mr. Aylwin never expected to be sent North. It was the beginning of the Depression, however, and he wasn't about to argue.

Their relationship was kindled on the old, wooden tennis court in Mayo during the summer of 1930. In the days before they were married Mr. Aylwin lived in quarters which were attached to the bank. In addition to his $900 a month salary, he received a northern living allowance. "And I got my own meals when I didn't bum off Emelia's mother," he laughed. The couple soon found that their interest in each other extended beyond the tennis court. After a two-year courtship, they were married October 15, 1932.

The newlyweds couldn't afford a honeymoon, so they went grouse hunting instead, Mr. Aylwin joked. Shortly after they were married Mr. Aylwin left the bank for the Northern Commercial store.

"The bank and I had quit company so I had to find another job. I went over to the Northern Commercial Company," he explained matter-of-factly. The NC store, which supplied all the mining camps, had the lion's share of the retail business in Mayo. It sold everything from hardware to dried goods and fresh produce in the summertime. Mr. Aylwin remained with the company store for ten years.

Their two children, Kenneth and Fred, were both born in Mayo.

Mayo in the 30's was a small, friendly town where everyone knew everyone else. The hospital, stores and hotel were the only establishments that had the luxury of telephones. "You didn't have to phone to say you were coming over, you just went over," Mrs. Aylwin said.

Mr. Aylwin used to join his buddies on their annual moose hunting trip and often hunted grouse and rabbit to supplement the family's larder. Mrs. Aylwin maintained a vegetable garden in the summer.

In 1942, the price of silver and lead dipped way down, forcing [Treadwell Yukon] to close down. That's when the couple decided to move their family to Whitehorse.

Mr. Aylwin was offered a job with the Taylor and Drury store which was booming as a result of the Alaska Highway construction. Even though she had lived in the Yukon all her life, it was the first time Mrs. Aylwin had ever been to Whitehorse. The sleepy little river town called Whitehorse was transformed almost overnight into a city of 20,000 soldiers and construction workers. "It was kind of exciting to see all the activity after Mayo," Mrs. Aylwin commented.

The family lived above Taylor and Drury's store and Mr. Aylwin recalled the noise the U.S. army trucks made when they were started up at night in 40 below weather to keep the engines from freezing.

After a year of residing above the shop, the company built employee housing at Third and Jarvis Streets. The couple then bought their own place on Steele Street, which they sold in 1967 to the Roman Catholic Bishop, after they decided to leave the Yukon for Victoria.

"The sleepy little river town called Whitehorse was transformed almost overnight into a city of 20,000 soldiers and construction workers."

But their experiment with the "outside" didn't last long. "I looked for a job for 364 days and couldn't find one so we came back," laughed Mr. Aylwin.

For George, 78, and Emelia, 69, their brief experiment with living outside the Yukon has taught them that, indeed, the grass is not always greener on the other side. "We've decided this is where we are going to stay for the rest of time."

DON & HAZEL BAKER

by Don Baker, 1984

In July of 1960 we motored to Mayo from Wembly, Alberta, to visit Hazel's brother Wilfred Gordon and his wife Jean.

We liked the country and the Gordon's wanted us to move north and help "Wilf" with his firewood and timber business.

We went back to Wembly, packed up, and were back in Mayo in time for our four girls to start school in September.

Things went along smoothly until early one morning in 1963 when I lost my truck in a garage fire. Wood sales were falling off due to oil fired furnaces taking over, so I went to work as a plumber's helper at Elsa for United Keno Hill Mines. Later on, Keno Hill needed a maintenance man to look after the boilers and boiler room equipment. I was offered the job and accepted it.

When my old car died a natural death, I rode to work with Don Curry or Renny Burian. We never missed a shift though the temperature was down to –80°F on several mornings one winter.

In 1965 the custodian job in the Mayo School came open so I applied for it and was accepted.

During the time I worked at the school I also took part in community events and affairs such as a member of the Community Club taking part in erecting the community hall. I also served a term as fire chief and fire-fighter, helped out with the Winter Carnival, served a term as a member of the L.I.D., and was a member of the Royal Canadian Legion.

That's enough about what I did. Hazel was here too. Besides keeping a house and our four girls in order, she found time to bake for bake sales, make donuts for Legion social evenings, help out with the first kindergarten class, serve on the school advisory board and look after three lovely little foster children.

I retired in April 1982, and Hazel and I have been visiting mostly with our daughters Theresa and Joan and their families in Whitehorse, and Mary, Frances and Nancy and their families in Camrose, Alberta.

DAVID WESLEY BALLENTINE & FAMILY

(Lived in Mayo from 1923 to 1928)

by Ione Christensen, May 1988

Family members who lived in Mayo:

David Wesley Ballentine		1876–1956
Elizabeth Jane Symington Moreland Ballentine	– Wife	1876–1947
Sara "Ellie" Wright (Ballentine)	– Oldest Daughter	1901–1971
Margretta Gaundroue (Ballentine)	– Third Daughter	1905–1986
Jim Ballentine	–Only Son	1909
Edith Nielson (Ballentine)	– Fourth Daughter and Youngest Child	1916–1976

Elizabeth Jane Symington Moreland Ballentine, Mayo, 1926. *Ione Christensen Collection*

The second daughter, Martha (Ballentine) Cameron (1904–1990) wife of Gordon I. Cameron, stayed in Dawson. Martha Cameron is the mother of Ione (Cameron) Christensen, and grandmother of Paul and Philip Christensen.

David Ballentine was born in St. John, New Brunswick, in 1876. There were six boys and two girls in the family. His father, Wesley, came to Canada from Ireland.

In January of 1898, Wesley Ballentine and his four sons, John, Allen, David and Jim, left St. John for the Klondike Gold Rush.

Of the five men, only David remained in the Yukon. He had met his Irish bride in Boston in 1897. They were engaged and David had planned to go back to Ireland where they would be married in her family home. The Yukon changed all that. David wrote Elizabeth Moreland, saying that if she still wished to marry him, she was to come back to Boston where he would meet her and they would then come north. She did and in the early summer of 1900 they were married. In a short three months Elizabeth had travelled from Belfast, to Boston, to Dawson, to a small cabin on Hunker Creek. She never left the Yukon again until 1945 when she went to Vancouver for a short holiday. All of their years were spent in Dawson City but for one period of six years when they lived in Mayo.

In 1919 Ellie (Sara) Ballentine left her home and family in Dawson to work in Mayo for a family by the name of Portlock. That first year she met and married George Wright. George was a miner and they lived in a log cabin back of where the Mayo Pioneer Hall was later built. Her younger brother, Jim, came to Mayo for his summer holidays in 1920, returning to Dawson City for school in late August.

The Pioneer Hall was being built at that time and Jim was able to get summer work helping with the construction. The first Lodge meeting was held in August, 1920. Jim remembers the August 17th Discovery Day celebrations that year. All the events were held on Front Street by Binet's hotel. There were foot races and a big parade but it was the horse race he remembers best. It started at the north end of the street and ended at the Lefebvre sawmill. Most of the riders had been celebrating since early morning and as a result the majority of horses arrived at the finish line without riders, the riders having been lost along the way.

In the meantime, David Ballentine was on the Chisana Gold Rush up the White River, but luck was not with him and when he returned to Dawson he was looking for work. The White Pass needed a builder to construct a new warehouse and wharf in Mayo. Dave was a carpenter by trade so he was hired and travelled to Mayo in the summer of 1923. He worked on a number of buildings around town that summer.

There was a need for a fire chief and Ballentine was appointed. He built a fire barge with living quarters for the chief. The barge was docked at the north end of the Taylor and Drury warehouse. There was a Fairbanks–Morse gas engine water pump. This provided protection for all of Front Street. They also had a hand pulled hose cart and a horse-drawn wagon. The horse was owned by Joe Longtin who delivered water to Mayo homes. It does not appear there was ever a conflict, the horse was never needed on both jobs at once.

Mrs. Ballentine moved to Mayo in 1924 with her youngest daughter Edith and son Jim. They came on the Steamer *Klondyke*, which was built and owned by Dave Fotheringham. After two weeks aboard the steamer, they arrived in Mayo. The trip would usually have taken only five days; however, Fotheringham did not have the money to pay for wood which was already cut, so frequent stops were needed to cut wood for the water tube boiler. Young Jim worked as fireman feeding the boiler with three foot

From left: Elizabeth Ballentine; Joe Bronett, R.C.M.P.; and Margretta Gaundroue, Mayo, 1926. *Ione Christensen Collection*

logs. It was a very small boat with few crew members so they were required to tie up each night.

This boat made excursions from Mayo to Fraser Falls on the upper Stewart and Jim continued his duties as fireman all summer.

Jim was going to school in Mayo. In Grade 6, his teacher, Dorothy Hoggan, told Mrs. Ballentine Jim would be much better off gainfully employed, and not sitting in school disrupting the class. Jim was delighted—Mrs. Ballentine distraught. The last straw was when Jim and Gus Besner absconded with the teachers leather strap and cut it into little pieces. Obviously they felt it was getting too much use!

There was lots of work and Jim went to work for Treadwell Yukon at Wernecke Camp. First he worked in the machine shop as helper and later in the mine as nipper, then in the mine stopes, and lastly as a hoistman at the Lucky Queen.

Margretta Ballentine went to Mayo from Dawson City in 1928. She worked as a waitress and helped Mrs. Fotheringham in the publishing of the *Mayo–Keno Bulletin*. There was a young mill operator working for Treadwell Yukon at Keno Hill by the name of Elmer Gaundroue. Margretta had known him when he had first come to the Yukon in 1925. Elmer had worked in Dawson and then moved up to Mayo. He and Margretta met again in Mayo and the romance of the Stewart Valley claimed yet another Ballentine daughter. Elmer and Margretta were married in Mayo and moved up to Elsa. They lived there until the late 1930's. Both of their boys, Bob and Jim, were born in Mayo.

Cutting wood was always a way to make a dollar. In the summer of 1926, George Wright, Ellie and Jim, with a bartender from Mayo by the name of Frank, went down the Stewart River to McDiarmid's Point. There they cut wood for the Taylor and Drury Company steamboat *Thistle*. The 16 foot logs were cut in the bush then hauled 10 miles to the river. Here they were cut once again into 4 foot lengths. All of this was done by hand for the princely sum of $3.50 a cord. When the contract was finished they moved down river to Rim Rock. At this camp they cut 150 cords of 16 foot logs which they made into a raft and floated to Dawson City. There it sold for $9 a cord on the beach.

By now Dave Ballentine was on the overland mail run from Whitehorse to Mayo. This lasted for one winter. He went to work for Treadwell Yukon at Wernecke as a miner and then on to driving one of the big new ten-ton Holt tractors hauling ore from the mine into Mayo. Here it was stacked on the riverbank awaiting the boats in the spring.

Burns Meat Market in Whitehorse would drive cattle from Yukon Crossing to Mayo overland. On one such drive, when the cattle came to the Stewart River, across from Mayo, try as they might, the drivers could not get the lead steers into the river. Ballentine saw the problem on the far bank of the river, so he got a small boat and rowed across. (Dave was noted for his massive strength. He had carried the heaviest load ever recorded over the Chilkoot Pass and there was no such word as "can't" in his vocabulary. If you wanted a job done you would "Ballentine" it!) Tying one end of the rope to the horns of the lead steer, the other to the boat, he rowed the animal into the river and across, disregarding all of its efforts to resist. The other animals followed without protest.

In 1928 the company stopped using cats for the ore haul and switched to trucks. Dave Ballentine didn't like the trucks and he missed Dawson City and the gold creeks of that area. So, in the summer of 1928, Dave, Elizabeth, and young Edith left Mayo. Ellie and George Wright had already moved on and were living in Fairbanks, Alaska. Jim was working and Margretta was well settled in Elsa. For the two senior Ballentines Dawson remained their home. Elizabeth passed away at Dawson City in 1947 and Dave in 1956.

ALEC F. BERRY

from Yukon Chamber of Mines Newsletter of January, 1983

Alec Berry arrived in Nelson, B.C., from England, in 1912, at age 16. He worked at the Granite Poorman mine in 1913, and being too young for underground work, began as "cookie" at $40 per month. He later worked in the ore sorting operation, acted as powder boy, and learned the operation of the "twenty-stamp" mill. Later on, Alec worked at the Motherlode mine in Greenwood, B.C.

During World War I he served with the Infantry Signal Corps and with the Royal Engineers. He was wounded in the battle of Hill 60 and discharged for recovery under medical care back at Nelson, where he worked with the Provincial Assayers without wages for awhile. He applied to the Greenwood Smelter and worked on the boilers at Copper Mountain Mine. He was soon promoted to the Engineering Department, and later went to Anyox where he worked in the flotation department.

Alec first came to the Yukon in 1924, to demonstrate the new "wet drilling machine" to Livingston Wernecke's staff at Keno. He returned to the Yukon in 1925 and took a lay on the Settlemier property on Galena Hill, where he mined 600 tons of good ore; but the depressed silver prices left him with an operating loss of $10,000. He then went to work for wages and eventually paid all of his accounts.

Alec travelled and worked with Wernecke, learning many skills from inspection of prospects to surveying. After the death of Mr. Wernecke and the later closing of the mines, he remained to oversee local properties and explain the workings to examining engineers from many companies. In 1945 he worked for F. M. Connell and the Conwest organization in charge of the 140 ton mill and assay office at Keno Hill. He was present at the purchase of the Elsa Mine property by his employers, and received authority from F. M. Connell to use the company's funds as required to engage in property inspections when he heard of interesting prospects. This eventually led to the purchase of the Clinton Creek asbestos property from Fred Caley and partners. Later, while examining copper deposits on the Turnagain River in northern B.C., he heard of an asbestos prospect in the Cassiar district and learned the details of an examining option about to expire. He obtained company permission to pursue the prospect, which he successfully acquired—the prospect that is now the Cassiar Asbestos Mine.

Alec was the member for Mayo on the Yukon Council from 1952 to 1955 and was instrumental in drafting mine safety legislation. He was an honorary life member of the Yukon Chamber of Mines and of the Prospectors and Developers Association. He always felt that the history of silver mining in Yukon deserves more historical attention and credit for its part in the development of the Yukon.

Mr. Berry died on December 10, 1982.

ED BLEILER - MEMORIES OF MAYO AND VICINITY 1933–1990

by Edwin Charles Bleiler

Ed and Flora Bleiler with their children Lowell and Sybil. *Bleiler Collection*

It was late May 1933, the infamous Hungry Thirties, I had made the decision, that at age 22, it was time to seek my fortune elsewhere. I had come to a dead end. I'd finished high school and completed two years in pharmacy with no money to go any further. It seemed to me the terrible plight of so many would last forever. Therefore on May 23, 1933 I left Edmonton, Alberta with the grand sum of $90, a pack-sack on my back with a few extra clothes and some food.

There were no coast to coast highways, airlines or credit cards, so to stretch my $90 as far as possible I chose to travel in the same manner as thousands of other so called "hoboes"—by slow freight as far as Prince Rupert, B.C. Passage was free but depending on the weather there could be much discomfort, and a good hot meal was just something you dreamed about.

After about ten days of new adventures, some quite amusing and others not so funny, and good companionship with others in the same "boat" or should I say same box car as myself, I arrived in Prince Rupert. It was my first experience of almost perpetual rain. My slow freight was not timed to make connections with the boat up the coast, so after a delay of more than a week the beautiful S.S. "*Princess Norah*" arrived. It seemed like a floating palace after the box cars of the C.N.R.

I had a lovely trip to Skagway even though I travelled second class, sharing one large cabin below deck with several other men. However, meal time found us second classers eating in the same dining room as the first class passengers They were marvellous meals with the evening dinners made more enjoyable by the added luxury of beautiful dinner music played by a live orchestra. I arrived in Skagway and then took the slow but very scenic train ride to Whitehorse, which at that time was just a small village having quite a long sleep after the hectic days of '98.

There was yet another delay of more than a week in Whitehorse where I met many young Yukoners. I also became acquainted with some of the old-timers who participated in some of the copper discoveries of the area. I rented an equipped log cabin while in Whitehorse, bought my first groceries at the Taylor and Drury Store and my first moose steaks from Jim Boss, the tall, handsome Indian Chief.

The trip down the Yukon on the S.S. *Casca* was a delight, the long days and the perfect weather added to the enjoyment. It was all a different world from where I had come. There seemed no end to the adventures in store for me, new sights, new sounds, so many new and very colourful characters. The prospectors, trappers and steamboat men were like the heroes in the western movies I had seen before I left Edmonton.

At Stewart I switched from the S.S. *Casca* to the S.S. *Keno* which also pushed a barge in front similar to the *Casca*'s. The trip up the Stewart was again very interesting as we sighted lots of game, mostly moose and bears. The boat would stop where necessary to unload freight at points along the river and also to take on wood for fuel for the steam boilers that powered the boats.

On arrival at Mayo, about a month from the time I departed Edmonton, I was greeted by the sight of every man, woman and child in Mayo on the dock awaiting the arrival of the steamboat. The whistle on the boat was always sounded within sight of the town so that everyone knew that it would soon be docking. I was the newcomer on that trip and my fellow passengers were two young fellows coming home after being at school all year. One was Kenneth Schellinger from Palo Alto, California, and the other was Tommy Portlock from British Columbia. Ken is the son of A.K. Schellinger who represented the Guggenheim mining interests in Dawson and Keno Hill, the same A.K. in whose honour a cairn was erected on Keno Hill. Tommy was the son of early day miners on Haggart Creek. I had had the pleasure of their company from Whitehorse to Mayo, however we didn't know that our paths were to cross several times over the succeeding years.

The present generation and future ones might wonder what Mayo was like in 1933, so I will give my impressions as I remember them.

In 1933 Mayo was predominantly a village of log cabins, with an occasional house or business constructed of sawn lumber, these were built after the installation of the two sawmills. One sawmill was right in town close to the riverbank and the other was upstream a mile or two. The only hotel, the Chateau Mayo, was of lumber and was owned by "Gene" Binet, a French Canadian. There were a number of families in the town, however quite a large percentage of the log cabins were occupied by bachelors, many being French Canadian. The town was neat and clean and just about every cabin or house had a well-tended garden. There were three general stores: Taylor & Drury, Northern Commercial, and Binets. There were the Anglican and Catholic Churches, one restaurant and a beer parlour both of which were in Binet's hotel building. The post office, liquor store, undertaker, shoemaker etc. were all in log buildings. The one room schoolhouse with grades one to eight was, I believe, a log building sitting on land where the motel is now. Just about all the stores and businesses were on Front Street facing the river because the river and the steamboats were the main life line to the outside world.

Very few homes had electricity and the outdoor biffy was in evidence everywhere, although each home usually tried to hide it as much as was possible. Some homes had inside wells with a hand pump. Those that didn't, either bought their water by the bucket from Joe Longtin who made deliveries with his horse, Eli, and a wagon equipped with a not very large water tank, or they went down to the river with a bucket and got their water there. Since everyone, without exception, burned wood for heat and cooking the wood pile in every yard was an important feature.

Horse power was still very much in use in 1933. There were very few privately owned automobiles, I can only recall about six. There were about four privately owned small trucks, but there wasn't much business for them because the silver mine had closed down in the Depression when silver reached a low of 25 cents per ounce. Dog-power was still quite common when I arrived, every trapper and some placer miners had their dog team, strictly for utilitarian reasons.

My brother, Ted, had preceded me to the Yukon several years earlier, 1929 I believe. He taught school in Keno and later at Wernecke Camp but in 1932 he and a partner got involved in placer mining on Dublin Gulch. Through his letters I got the itch to follow him and get involved. Ted knew that I was coming, in fact my $90 was just about gone by the time I reached Whitehorse and I had to wire him for assistance in order to reach Mayo.

Ted arranged to have an old ninety-eighter named Jack Smith meet me at the boat. Jack gave me a meal and found me a ride that same day as far as Minto Bridge where there was a roadhouse. At Minto Bridge there were six or seven log cabins up and down the Mayo River, all occupied by bachelors.

When I arrived at Minto Bridge I was met by another old timer, Bobby Burns, an Australian who would guide me on the final leg of my journey to Dublin Gulch. This we accomplished on foot with a break about halfway at the "log jam" on the McQuesten River where Bobby had his hunting and trapping headquarters. There was no road as such, just a narrow trail which branched off the road to Keno about three miles north of Minto Bridge. The trail was good in some areas while other quite long stretches were very wet and swampy and of course, there were clouds of extremely aggressive, hungry mosquitoes.

" It was a land of plenty but the cold and hardships made it difficult to survive "

That summer began my education, learning the rudiments of placer mining which on Dublin Gulch was a never ending battle with boulders. There was always a shortage of water as well. Everything was pretty well hand labour, pick and shovel, crow bar and sledge-hammer. Yes, sometimes we would beat a boulder down to size by breaking it up with a sledge-hammer. Unfortunately that summer was an extremely dry one—there was not much gold in the poke come fall, so wages were out of the question for another year.

The winter of 1933–34 was a very, very severe one so I learned about cold and the never ending battle to acquire an adequate supply of wood to keep from freezing. We felled and cut it all up by hand and hauled it in with dog teams. The dogs required food too, so one of us was either cooking rice or corn meal enriched with grease or lard. We hunted moose or caribou, they were fairly plentiful, and set snares for the numerous rabbits. It was a land of plenty but the cold and hardships made it difficult to survive. The dogs, the snowshoes and the rifle, not to mention the axe and saw, were the tools of survival.

Our meat supply was about one hundred percent wild except for the occasional luxury of ham or bacon. Canned meat was rarely used. In late January of that first winter we started freighting a year's supply of food for five people to Dublin Gulch with two dog teams. Ted had one team and I had the other. We relayed the supplies, taking small loads on the uphill stretches and larger loads on the downhill inclines. We used four log cabins which were still habitable on the Haggart trail at that time. It was early April before the several tons of freight were delivered to Dublin Gulch.

The second year of mining was on a bench deposit on Haggart Creek. Bobby Burns and I teamed up and again we were quite content to end the season with enough for another year's outfit. I tried my hand at trapping that second winter on the Little North Fork of the McQuesten River. Considering how inexperienced I was I did not do too badly. However in spite of the wonderful adventures, I decided that in the future trapping would be a last resort. I was far to compassionate towards the animals that I had to trap.

In 1935 silver prices began to improve so Treadwell Yukon, the company that owned Wernecke Camp and other well regarded properties decided to develop the Elsa. A new camp and mill was built there and I decided it might be more rewarding to work for wages than to continue placer mining. In the 1930's the going wage of $5 a day plus room and board was really good money. In the meantime I had met Flora Middlecoff, my future wife, whose family had been in mining near Dawson City since 1900 and then on Highet Creek since 1903. Flora and I became engaged and I got a job in the mill at the Elsa Camp. Everyone worked eight hours per day, seven days a week at the mine. I would put in my shift and labour several hours after each work day building our future home near the Elsa Camp.

While I was building our home, I lived in the company bunkhouse, which was one big room upstairs and the same downstairs, housing about 40 men on each floor. There was no plumbing, we washed in one common washroom and dipped water from 45-gallon drums. Again the outdoor biffy, a six- or eight-holer was near the bunk house as well as others strategically located in different areas of the camp. Wood was the sole source of fuel for cooking and heating. Nevertheless, it was a happy time for most of the men, [with] security, good food, and a roof over our heads. We made our own recreation and there was little discontent or complaint.

With the help of friends our two room matrimonial shack at Elsa was completed by the fall of 1936. Flora and I were married in Mayo and moved right into our new home. We had several happy years there but in 1938 we were drawn back into placer mining. Flora's brother, Elmer Jr., was tragically killed on Highet Creek so Mr. Middlecoff, who was approaching 75 years of age, needed help to continue his operations. We joined him for four seasons and had some success, but with the outbreak of the Second World War, Middlecoff eventually had to cease operation as all the young men either went to war or to Whitehorse where there was much prosperity and opportunity.

For close to forty years Elmer Middlecoff Sr. had been one of the main employers in the area having had anywhere from five to 35 men on the payroll.

After the war the Mayo–Keno area was suddenly a hot spot again. Silver was increasing in price and a new company, United Keno Hill Mines (UKHM) bought all of the Treadwell holdings plus many other properties, all of which were idle during the war. UKHM was very well financed and soon the camps were humming, bigger and better than ever. A highway was built to connect Mayo—Elsa—Keno with the Alaska Highway in 1952, and the hydro was built to supply the Mayo, Elsa, and Keno areas with electricity.

In 1949 I became involved in the trucking business, timber and lumber being in great demand. Soon I had four trucks and the only wood

Ed and Flora Bleiler on their wedding day, 1936.
G. A. McIntyre photograph

saw in town. At the peak of operations I had 14 men on the payroll. By 1957 most of the small silver mining companies which had established themselves in the boom years of 1950 to 1955 had disappeared reducing the business for truckers, so Flora and I decided to return to Highet Creek. Our children, Lowell and Sybil, were old enough to help at the mine. After a few years of disappointment we began to show a good profit, especially after acquiring more suitable equipment. We stuck to hydraulic mining for quite a few years and were able to work two ten hour shifts with Lowell on one shift and myself on the other.

Towards the end of the 1960's the mining was good to us. Lowell left for Australia, married and settled down, Sybil did the same on the outskirts of London, England. Flora and I were finally free to travel and enjoy the rewards of hard work and persistence. We had some marvellous winters trailering in Mexico but in 1968 Flora died very suddenly and so I was alone on Highet Creek.

In 1973 the price of gold started to take off and my body was sending me messages that I could not go on by myself forever. Lowell and his wife, Lyn, came on vacation in 1974 to size things up, after which on their return to Australia they had several months to make a decision whether to be placer miners or not. No doubt the decision came easily for Lowell as he had the experience, and a father and grandfather with placer mining in their blood.

Looking back in retrospect to that day in Edmonton in 1933 when my family and friends thought it was a crazy idea for me to think I could go to the far off Yukon and make my fortune. In a way it was crazy when I recall how inexperienced I was and how much I had to learn, but I've never been sorry I did it. Life has been interesting and full of adventures in this glorious Yukon. No doubt Elmer Middlecoff senior's family must have wondered about him leaving the sure thing of farming that wonderful Missouri land, and heading for the far, far North.

In a way it was the beginning of the whole thing away back in 1903 when he planted his first claim posts on Highet Creek. So I am forever grateful to Elmer Middlecoff, Carmack and all those other wonderful characters who broke the trail for us and proved to us who came later, that it was possible to make things happen the way we wanted.

Editor's Note: Ed Bleiler is in his 79th year (1990) and still mining on a small scale on Highet Creek accompanied by his wife Martha. He no longer winters in the Yukon as they also have a home in Mazatlan, Mexico, Martha's homeland.

THEODORE R. "TED" BLEILER

by Theodore R. Bleiler

KENO CITY 1929–32

As a farm grown youth and admirer of authors Oliver Curwood and Jack London, I would often fantasize that I would go to God's Country and discover another bonanza. That I would build a castle in some remote valley in the mountains, hidden away from the world and accessible only by aeroplane. There I would author books about an idyllic existence in the wilds; compose vibrant music for violin or piano all under the adoration of a damsel whom I'd rescued from a fate worse than death. Little did I believe that I would actually go north of 60 and actually hunt for ellusive treasure. It was never in my plans to do such things; so engrossed was I with the idea that I would settle down as a farmer or rancher in sunny Alberta along with a country bred lass.

That I did go to the Yukon was a case of serendipity. At the end of the spring term at the school where I was the teacher about 20 miles north of Vermilion, Alberta, I was returning to the family farm and home, on the outskirts of Wetaskiwin when it happened. One of my former classmates, Hector Whitson, of the Mannville School, chanced to be on board the same train. The meeting was a pleasant surprise and it was a delight to sit together and reminisce about our childhood days and bring ourselves up-to-date. Upon the exhaustion of interesting items of the past and present, we got around to discussing plans for the future.

"I think that it would be great if we could get a job teaching in the Yukon," remarked Hector, and he went on to relate how a friend of a friend succeeded in getting employment as a teacher in Dawson City, earning double the going rate for the same work in Alberta, and who after a couple of years came out with a large enough stake to purchase a farm. The outcome of those remarks led me to send an application to the chairman of the Dawson City School Board and the ultimate assignment to teach school at Keno City.

When I first laid eyes on Keno City in the fall of 1929 my mind conjured up "Sleepy Hollow" I'd read about at some time during my childhood.

Keno City is surrounded by high hills on three sides; Keno Hill on the north, Sourdough with Lightning Creek at its foot on the west, and Galena Hill southwards. The terrain drops away in a southeasterly direction along the valley of Christal Lake and Creek which empties into the South McQuesten River about ten miles from the settlement.

I think that according to the memory I have of the village, if I said there were a total of thirty buildings which served as dwellings besides the government and business establishments would be much closer to the truth than the 10,000 souls reported in a recent historical account.

The business establishments were as follows: the Northern Commercial Co. Ltd. (N.C.) managed at that time by Dick O'Loane; Taylor and Drury with William Jeffrey in charge, a small bakeshop with Mrs. Erickson; Jimmy Sugiyama's Hotel and Eatery and Jackson's Saloon with a hall on the top floor; operated under the guise of a social club. There were four government controlled entities: the Post Office with Tiny Greaves (she wasn't so petite) officiating, the R.C.M.P. under Corporal Coleman, the liquor store with Vic Grant as its manager, and finally the government assay office with Billy Sime as assayer and oft times as Government Agent acting for the Gold Commissioner, George Jeckell at Dawson City.

I met Mr. Jeckell when I went to Dawson at the outset to sign a contract and to receive the only briefing I was to get during the time I worked in Keno. I also took in the Discovery Day celebrations on August 17th; an annual event which is still observed.

On my way to Keno via the Stewart River and Mayo Landing I spent a week at Stewart City and did some long-shoring as the regular crew had missed the boat back from the festivities of the 17th. It was here that I made my first acquaintance with high-grade and silver-lead concentrates from the mines north of Mayo. All of this helped to kill the time I had to wait for the S.S. *Keno* to return from the Landing where it picked up a barge load of the ore to be trans-shipped up river to Whitehorse where it proceeded by rail to Skagway, Alaska and thence to the smelter by ocean freighter. When the S.S. *Keno* dropped off a barge load of ore, exchanging it for a barge loaded with freight for Mayo, Keno and the mines I continued on to my destination.

The following excerpts from the Journal which I kept will portray first hand impressions made during my first three years in the mining area of the Mayo region:

The silver mine at Wernecke Camp (Treadwell Yukon Company), was running at full capacity in 1927 and 1928. Keno City was born to reap a fair share of the returns being a town situated just under four miles from the mine site. It was also a base for a few trappers, hunters, prospectors and a trading place for Chief Louie's Indian Band. There were a few cabins up-stream on Lightning Creek which housed a number of women who had "no visible means of support." They were a great attraction for the men working in the mines. Occasionally the R.C.M.P. raided their cabins searching for signs of bootlegging, only a few charges were laid.

Besides the operation at Wernecke Camp there were a few small outifts high-grading adjoining properties some of which had their headquarters in Keno. During shift changes and weekends Johnnie Kazinsky's jitney (taxi) did a land-office business bringing men down and returning them back to the camp. Those on night shift would come down during the day, and those on day shift would come down at night making the town a beehive of activity.

Even though there was a sort of company store at Wernecke Camp the half dozen families who lived in their own houses near the mill and the mine did quite a lot of shopping at Keno City, adding much to its commerce. Once a week movies were shown in Jackson's Hall, which was an added attraction. On change days the place was more than alive with activity, spirits flowed and overflowed, hilarious gossip, rowdiness and the occasional brawl were in evidence. Card tables as full as the players; money flowed like water. Visitors from Mayo regularly added to the numbers of those who came to Keno for exciting recreation.

The bread-winners of the few families who collectively had enough children to warrant a school eked out a living by doing all sorts of odd jobs, cutting wood, trapping, hunting, and doing part-time work with the small operators who were high-grading rich pockets of ore which they were lucky enough to find on their own claims.

Things were going strong in 1929. I was able to supplement my teaching salary unloading rawhides and stacking the ore sacks, each one weighing over one hundred weight. Rawhiding was a method used by the high-graders to move their ore down the mountainside to a spot where it could be picked up by truck and hauled to Mayo. Up to a dozen of the small jute sacks of ore would be laced into a hide, hair side in, and a single horse could manage to drag three or four such hides, one behind the other, over the not so steep stretches. On steep stretches a horse has been known to sit on the leading hide, with front feet acting as brakes, and with the miner riding the hide at the end of the string; altogether they made quite a sight. In this manner tons of ore were moved during the winter months when there was plenty of snow to develop a suitable slide.

My life as a teacher at Keno was a happy one because of the brightness and enthusiasm of the children. It was especially delightful to see them like sponges, soaking up any knowledge I was able to impart. These children were so exposed to adults, a great source of their amusement was listening to the comments and conversations relating to the events of the day. The school, being the chief diversion from the daily routine of home living, was the place where the action took place; thus the attendance

August 1929, Ted Bleiler on the far left stacking ore at Stewart City from barges brought down by the *Keno* from Mayo Landing. It was then loaded onto the *Casca* and barged to Whitehorse. The ore could be handled up to five times from Wernecke Camp to Whitehorse. *T. R. Bleiler Collection*

was excellent and only serious illness could prevent the children from making an appearance. I became their teacher on the second year that the school was in operation and the names of the children were as follows: Jack, Laura, Dolly, Frank and Mable Sullivan; Barbara, Mildred and Dick Hargraves; Albert and Harvey Pelland and Theodore, Mrs. Erickson's son. Miss Alice Fisher was the first teacher to conduct classes in Keno City. At the end of my second year some of the families with their children left for the greener pastures of Mayo, Dawson City and Whitehorse. I transferred the school equipment to a roomy log building at the Wernecke Camp and completed a third year conducting classes where they had enough school aged children to warrant government approval and support. A few of the former students who were in Keno struggled up the hill in all kinds of weather to attend classes; a true indication of their dedication. I served as teacher at this point for one more year after which I was followed by another who filled the position for a term after which with the closure of the mill at Wernecke Camp, families dispersed and there was no further need for a school.

Canada was by this time occupied by World War II which disrupted the distribution of manpower. Shortly thereafter Keno City's population dwindled to next to nothing, and in the decades which followed managed just to survive by the grace of flash exploration on various properties located in the hills surrounding Keno City.

Often the names of some of the characters who peopled these historical sites come to mind: Tiny Greaves, "Hot Stove" Douglas, "Hard Rock" MacDonald, Brefalt and the Big Swede who were numbered amongst the high-graders. There were partnerships such as Corp and Ryan, Colley and Boon, Carpenter and McLean. There was "Spotcash" Breaden, intrepid hauler of freight from the Landing, about forty miles from Keno, through all kinds of weather and conditions.

There was the memorable Massa Sakata who in his later years acted as host to the occasional visitors who paused in Keno on excursions to the summit of Keno Hill where a cairn honouring A.K. Schellinger, "Shelly," is now located. There are few survivors who can bear witness to the fact that with its population of about one hundred souls, Keno City had a most colourful career.

PLACER MINING ON DUBLIN GULCH, 1932–1937

After only a few weeks in Keno City my interest in mining and prospecting was aroused, due to the fact that all of the talk of my associates was about mines or the prospect of mines; mostly the silver and lead properties in the area. I spent many hours with Billy Sime, the assayer, so had a sort of inside track on prospects, and I actually staked claims on Galena Hill.

One of the employees in the mill at Wernecke Camp, Norval Lochore, and I became holiday partners and made an extensive hike together into the Bonnet Plume country in 1931 on the strength of a Lost Mine story told by one of the Dawson City characters. It was also a good excuse not to attend Summer School as most certificated teachers are expected to do through July and August.

There was little mention of placer mining. Occasionally one heard about unemployed roving types rocking the bars of the Stewart River during the low water season and making as much as five dollars per day. There was mention of an elderly chap who was on Thunder Gulch gophering around the huge boulders on that creek not very far from Keno, and the Ross Family that trapped during the winter, and washed out enough gold on the benches of Davidson Creek, (a tributary of the Mayo River), to pay for the next year's grub stake during the summer.

With the arrival of July 1932, I was again champing at the bit for adventure. Once more Norval took his holiday and we made a trip to do assessment work for Bobbie Fisher, a noted hunter and prospector, and Archie Martin, a sort of promoter, on some tungsten claims in the shadow of the Potato Hills. After this had been done we slipped down into the nearby Olive Pup (a small tributary of Dublin Gulch), where the crown granted claim "The Olive" with its exploratory tunnel was located, and where rich ore samples had been discovered on the surface at the turn of the century.

Several oxidized surface stringers of the arsenic group containing considerable free gold were discovered along Dublin Gulch by prospectors who were attracted by the boom in the Klondike, and came into the country via the Mackenzie River and Stewart Pass Route, and finding signs of precious minerals in the mountains north of Mayo some of them paused and became part of that scene. The only extensive work on the Olive Claim was a tunnel where it was proven that as one got into the vein the gold was no longer free and required uneconomical cyanide processing.

Both Bob Fisher and Archie Martin suggested that if we had the time and grub to continue down Dublin Gulch to the Cantin diggings we could do some panning for gold.

With the aid of dammed water, which came out of the melting snow on the higher slopes, a makeshift reservoir and a manually operated gate, we moved the desired yardage for the assessment work in a much shorter time than we had expected so we were able to take that advice. We followed the stream down to a cabin which was in fair condition with stove, dry bunks, utensils and even desiccated foods stored in metal drums tied to the rafters to protect them from squirrels. So the two of us and the three pack dogs settled down late in the afternoon, about the second week in July. We had quite an amount of food which we supplemented with rabbits and grouse, both of which were plentiful, making it possible for us to stop long enough to have a good look at the creek with its diggings. We also wondered about some haunches of bear that were hanging in nearby trees in advanced stages of putrefaction.

The next day we learnt the answer. In another cabin about a quarter mile further downstream, Jim Gibson, a real sourdough, had set up housekeeping in what he termed as a "drying out" period. He had shot the bears and regaled us with more bear stories plus stories of his experiences mining and prospecting in Alaska and the Yukon.

While we were thus regaled, Bobbie Fisher arrived on the scene carrying a moose liver in his improvised pack sack. Bobbie had worked with the Cantin brothers the last summer that they mined on Dublin Gulch. They mined for the scheelite which they shipped at a profit. It was worth $75 a unit (20 lbs.) during the final years of World War I. Bobbie remembered that he panned good colours where they left off on the right limit, so late that afternoon we took pans and went over to explore the area.

> **"Suttles was upset enough to put his 'curse' on the property, so that the Cantins wouldn't gain anything from it."**

There was an exposed eight foot face opposite the Jack Suttles' cabin. Jack was one of the first to stake Dublin Gulch in 1898 and had gophered out quite a pit next to it. Fisher knew him quite well and said that he had seen some of Jack's cleanups, which consisted of coarse gold compared to the gold that the Cantins found with the scheelite.

As the story goes, Suttles had unintentionally neglected to renew his claim and the Cantins being aware that it had lapsed on the books at the Mining Recorder's Office "jumped" it. Suttles was upset enough to put his "curse" on the property, so that the Cantins wouldn't gain anything from it. Strangely enough the Cantins also managed to allow their lease to lapse and the ground was open at the time that Norval and I were looking at it that summer of 1932. To make a long story short, after some panning we got "gold fever," so having quite a lot of grub and enough tools, such as picks and shovels which were in the cache next to Walby's cabin, we started sluicing under the eagle eyes of Gibson and Fisher, and as fate would ordain we struck a rich spot below the blue clay and took out over $1,400 in coarse gold in about three weeks. The largest nuggets were about half an ounce.

The die was cast, we had proven that there was gold in the gravel below the clay hard-pan which occurred at depths between seven to ten feet. Since the placer-scheelite appeared to be in the newer deposits of gravel, the Cantins not concerned about gold did not go any deeper. With the idea that we had a bonanza waiting in the depths, we staked a two mile prospecting lease, beginning about a quarter of a mile from the mouth of Dublin Gulch, much to the chagrin of the Cantins—so it seemed that Suttles curse was at work.

Even though the Cantins let it be known that they intended to continue mining if their business in town slacked off and the price of scheelite improved, Frank Cantin sloughed the idea with, "We didn't really care, the price of scheelite is too low to make a profit, and there is not enough gold to pay the price of moving all those huge boulders. The mining tools and other equipment in our cache out there is worth about a thousand dollars, you can have it for one hundred and fifty."

I gave him a cheque on the spot. Their equipment included a dozen 10 inch pipe in 12 foot lengths, heavy canvas hoses of various lengths and sizes, nozzles, picks, shovels, a wheelbarrow, blacksmith tools and coat, and quite a number of telescoping 12 foot sluice boxes.

When Norval and I were weighing our first clean-up, on regular gold scales which were included with the stuff in the Walby cabin, we had quite a number of spectators. Ed Barker, Kit Watters, Jack Hawthorne, Louis Kazinsky as well as Jim Gibson and Archie Martin who had arrived just the day before. There were fellows just fishing for grayling or hunting grouse which were plentiful while others were hunting a small band of a large species of woodland caribou that summered above the mouth of Haggart Creek on Lynx Fork. They were surprised to see that kind of gold on Dublin Gulch and as a result claims were staked on Haggart and a new life for the area began. Jim Gibson staked the stretch below us to the mouth after he made the decision to try his luck in his 70th year, his last effort in the mining game.

During the following winter a change of role from teacher to prospector took place. We freighted tons of freight by dog team over the divide between Keno City and Lynx Fork. We cut logs for a new cabin, plus more for fire wood and cleared the brush from the place where we planned to mine the next spring and summer, which rolled around in no time at all. I also bought up cabins and equipment belonging to Walby, Stewart and the partners, Jahnke and Gill including a cabin at Minto Bridge which served as a way station on our trips between Mayo and the creek. I could have opened up an antique shop, as I ended up with many items in triplicate including gold scales.

When Ed Barker and company settled on Haggart Creek they used the Portlock cabin which they acquired and which was also in good repair.

Three of us, Fisher, Lochore and myself worked Dublin Gulch together until 1933, the year that brother Ed arrived on the scene. Sad to say, the fabulous pay streak just petered out, much to our disappointment. Since the right limit was permanently frozen we followed the thawed centre and left bank (looking upstream) where during subsequent seasons the gravels proved to be pretty lean. It wasn't in the cards that I would make my fortune on Dublin Gulch. Bobbie Fisher gave up after the second summer, and Norval Lochore took up fruit farming after his parents lost their lives in an automobile accident in 1936. Since I was the only member of the trio who put personal capital into the venture (about two thousand dollars) I ended up with sole ownership. I mined one season by myself and the following year (1937) I mined in a sort of partnership with Fred Taylor. That was also the year I married Almuth who agreed to cook for us during the mining season.

During an average season the creek virtually dried up after a period without rain. The actual number of hours that one could sluice or even shovel was usually very disappointing. We never did reach a solid well defined bedrock, but worked at times to a depth of eighteen to twenty feet using near primitive methods. It seemed back breaking work for few rewards. Almuth and I decided that life would be much easier working for wages at the Elsa Mine.

Subsequently I turned the ground over to Fred Taylor who with the use of earth moving equipment was able to make a success of mining on Dublin Gulch.

I had begun my venture a few years too soon, before such equipment came into common use to revolutionize the placer mining industry throughout the Yukon.

Fred Taylor stripped the insulating cover from the right limit enabling him to harvest the more profitable part of the creek. My one satisfaction was in the fact that my efforts had proven the extent of the deposit of gold bearing gravel on Dublin, and like the character in Robert Service's poem,

"... Yet it isn't the gold that I'm wanting

So much as just finding the gold. . ."

At this point I might add that Fred Taylor was able to profit from the sale of scheelite which had soared to $45 a unit (20 lbs.) with the beginning of World War II. Tons of scheelite had been mined in my time but it was just left in piles on the mined sites. It was difficult to separate the very fine flour gold from the scheelite which has a similar specific gravity.

Only the smelter was able to do this. As it turned out there was sufficient gold content in abandoned scheelite bearing sand to pay the shipping charges.

When the mines began shutting down I worked the 1939 season for Fred Taylor. It was the only summer that he hired a crew and we were still hand mining. It was a fairly good year for water, so he was just able to pay off his crew's wages. Almuth helped with the cooking, for which she was paid, so actually financially, we came off better than the boss.

It was only after Fred Taylor returned from his stint overseas that he began making extensive use of heavy earth moving equipment and did quite a lot of experimenting, especially in methods of mechanically separating the flour gold from the scheelite.

Out of all my efforts I can say that adding the experiences of placer mining to that of farm hand and school teacher prepared me to face life as I took on more adventures in diverse activity.

To conclude the account of my ten year sojourn I must add; there are the memories of the many "first times" which will come to mind from time to time, until the day I draw my last breath.

I remember the first time I travelled through the Rockies to the western sea and took the ocean voyage along the inside passage on the *Princess* ship to Skagway where I travelled the breathtaking bends of the narrow gauge railway to Whitehorse. I recall the first time I cruised down the Yukon on the sternwheeler *Casca* and up the Stewart River on the *Keno*. I'll never forget the first time I picked five gallons of blueberries on the south slope of Keno Hill in two hours; and shot a mess of ptarmigan with a 22 rifle and later on in the season ate the luscious steaks of my first caribou.

I recall the awe when I first witnessed the migration of the great herds and broke malamute pups to fly on the command "mush," or extracted gold from a shovel full of gravel by the swirl and wash of the pan. I well recall the excitement of the first cleanup and the thrill of the sight of nuggets; the fear of disaster when accidently wetting moccasined feet at 40° F below zero in an overflow (underneath snow, at the edge of a mountain stream) miles from the shelter of a cabin.

I remember the first time I experienced the "silence," enveloping existence of a remote trapping cabin at the head of the McQuesten and I recall the breath-taking view of the Ladue and McQuesten valleys the first time I reached the top of Keno Hill and the awe inspiring Northern Lights as seen on a late summer's night from the Dublin Dome of the Potato Hills. These are just a few of the first times culminating with a saddened heart on the final departure for the inviting meadows of other pastures.

C.A. "KIPPY" BOERNER

C.A. "Kippy" Boerner[1]

I was born in Stockton, California, on June 21, 1875, and in 1883 our family moved to Seattle for the "big boom" there. My father went into business and inside of a month had a big shipment of goods coming in on the San Francisco boat. To take up this shipment, he drew all of his money out of the bank in Seattle, and that was the last seen of him alive. His body was found in Belltown, after he had been shot and robbed, leaving my mother, brother and myself broke until rescued by an uncle from Stockton.

We returned to Stockton, and I started my boating career running between Stockton and San Francisco as wheelsman, learning to be a pilot, at a salary of $25 per month. Working hours were from four in the afternoon until seven the next morning.

Word arrived in 1897 that gold had been struck in the Klondike, and the whole San Francisco area could talk of nothing else. We were then working on the Steamer *Mary Garrett*, and when the captain

1. Capt. C. A. "Kippy" Boerner. NC Flag, June 1959. (NC Flag was the Northern Commercial Co.'s quarterly magazine)

said he had signed up with the Alaska Exploration Co., for two years and that he could take his own crew with him, the two mates, steward, deck hands, and yours truly were all raring to go north.

We left San Francisco June 13, 1898, on the steamer *Leelanaw*, a former collier which had a house built on the upper deck to hold about 250 passenger. We reached Unalaska on June 21.

There, instead of four steamers and six barges ready to go to St. Michael, we found only the Steamer *Linda* and one barge waiting to take the big load of passengers and freight from the *Leelanaw* to Dawson. About the middle of July, the Steamer *Herman* and the barge *Seal* were ready, so hitching these two to the steamer schooner *Brunswick* we started for St. Michael. It was a long tow as we were about 800 feet apart with the barge on the tail end. Three of us were on the barge, Capt. J. Laggas, one deck hand and myself. It was so rough that we had to put the coffee pot in the oven to get a cup of coffee. Captain Laggas and I stood six-hour watches steering while the deck hand did a little cooking. Mostly we ate from tin cans.

After loading at St. Michael, we started for Dawson and arrived on August 14th. We left again on the 16th, and on the way down river, I was transferred to the Steamer *Arnold* to return to Dawson and then down river to winter quarters 24 miles above Eagle City.

There were 22 of us aboard, and we had fresh meat only three times all winter. But we got rid of six barrels of salt horse that had wintered for several years at Pt. Barrow. Living aboard a sternwheel steamer is a little colder than a refrigerator, and we burned about 250 cords of wood to keep warm. The gold strike was made at Cape Nome while we were in winter quarters, and as soon as the ice was out of the river we returned to Dawson to take a full load of passengers to St. Michael. That summer, 1899, I was pilot on the *Arnold*, and in 1900, I was pilot on the Steamer *Linda*.

In 1901, the Alaska Commercial Co., the Alaska Exploration Co., the Empire Transportation Co., and the Seattle Yukon Transportation Co., merged, with the boats under the Northern Navigation Co flag and the stores under the Northern Commercial (N.C.) Co. I was in Dawson when Captain Hanson, the superintendent, announced the merger and stated that the company was long on pilots and short on pursers. That summer I was purser on the Steamer *Rock Island*.

We took a full load of passengers and freight as soon as navigation opened. When we arrived at the mouth of the Yukon, we found the ice still firm in Norton Sound, and before we got out 13 days later, there were 13 boats all loaded with passengers for Nome. The passengers slept all day and gambled all night. When Captain Hibberd arrived on the C.H. *Hamilton*, he had the Coast Guard boat *Nunivak* lay alongside the biggest gambling layout, the *Leon*, and all the gold shipments of the other boats were put aboard the *Rock Island*. All together, there was $2,100,000.

We went out in the ice pack and through the canal, the first boat into St. Michael, bringing the first word of the consolidation of companies. On the beach when we arrived were Volney Richmond, then port steward of the A.C. Co., Jack Hillard, bookkeeper of the A.E. Co., and Captain McNoble, my old captain from Stockton. That summer, Mr. Richmond was sent as agent to the *Koyukuk*. Hillard, after a year with the company, entered the Customs Service and for years was Collector of Customs at Eagle.

In 1902 I was pilot on the *Rock Island* and in 1903, on the Tanana River. First I was on the *Seattle No. 3*, first boat to go to Fairbanks for N.C., then on the *Rock Island* on a trip to Fairbanks, then as captain on the Steamer *Isabelle*, a boat that was taken over from Capt. E.T. Barnette. I ran the *Isabelle* most of the summer, hauling lumber that was used to build the present N.C. grocery and the dry goods departments with a few warehouses thrown in.

In 1903 at Fairbanks, Frank Manley (for whom Manley Hot Springs was subsequently named) took a lease and option to buy, on two side claims on Cleary Creek from Captain Barnette, for $10,000 each. The next year, after Manley had made his "cleanup" he came to Fairbanks loaded down with gold dust to take up his option on the two claims. As there was no bank in Fairbanks, he took the dust to Barnette who asked for gold coin. In those days, all contracts called for "payable in gold coin of the U.S."

The nearest bank that could furnish that much gold coin was at Nome, Manley went to Howard Turner, N.C. manager at Fairbanks, to find out how he could get to Nome and back inside a month, al-

A view of the N.C. Co. with clothing hanging outside in an attempt to dry it after the flood of June 1936 that devastated Mayo. *W. S. Hare Photograph, E. C. Bleiler Collection*

lowing a few days leeway on his option. At the time, I was captain of the Steamer *Koyukuk* and had just arrived in town with a load of freight. To unload and wash boilers would have taken too much time, but the Steamer *Isabelle* was at the dock without a crew. I took my crew over to the *Isabelle* and had a couple of loads of wood brought to the dock so that we would not have to stop before we got to Tanana, where the Steamer *Sarah* was due on her way to St. Michael.

Turner had already wired Tanana to hold the *Sarah* for our arrival, and also wired St. Michael to have the tug *Meteor* waiting to take Manley to Nome and back, ready for the first boat upstream. We never rang a bell to slow down all the way to Tanana, and I think it was the fastest trip ever made from Fairbanks to Tanana. When Manley returned, we were at Chena with *Koyukuk* to bring him back to Fairbanks with just over two days to spare.

From 1906–08, I was captain of the Steamer *Lavelle Young*. Once, in 1907 the Steamer *Seattle No. 3*, stopped us on our way down river from Dawson to say that their mail room had been broken into and a quantity of gold dust taken out of the mail sacks. Missing was $41,000 and their deck boy. I was advised that I had been made a deputy marshal and was to make an arrest if possible. I found the deck boy standing on the bank at Victor's wood yard below Fort Yukon. He claimed that he had fallen overboard, but later at his trial admitted that he and a partner had taken the gold dust and thrown it overboard with a life preserver tied to the sack. He then went over the side and swam to shore with nothing on but a pair of overalls and a shirt.

In 1909 I captained the Steamer *Delta*, and in 1910, took the Steamer *Lavelle Young* from St. Michael to the Kuskokwim River and built a store and warehouse at Boerner on the Tacotna River, two miles below Nixon Fork. We built a store and warehouse at Tacotna in 1911.

I returned to San Francisco in 1912, but came back to the Yukon River in 1913. All the boats were sold to the White Pass & Yukon Route in 1914, and I worked with them until 1917 when I left the river for good. I returned to N.C. that fall as agent at Hot Springs, then went as agent to Iditarod until 1920 when I again went to San Francisco. In 1921 I left for Circle as agent, and a year later was appointed assistant manager to Jim Cody at Mayo where we built a store and warehouses. When Cody was transferred to Dawson, I remained as assistant to R.M. O'Loane. [Boerner met his wife Enid McIntyre in Mayo, where she was a nurse at the Mayo Hospital.]

The summer of 1932, my wife and I travelled around while I took inventory at Whitehorse, Dawson, Eagle, Circle and Fort Yukon, as Mr. Whitney did the auditing. Also that summer, O'Loane was transferred to Hot Springs, and I was made agent at Mayo. I stayed at Mayo until January, 1943, when I was appointed agent at Whitehorse. There were no agent's quarters in Whitehorse and no accommodations for help in the town while the war was on. The Home Office authorized the purchase of a lot on the corner of Steele Street, adjoining the store where we built a two-storey addition to house the agent, and the Caterpillar manager. It also contained four rooms for women and two for men employees.

I had ordered lumber in Mayo for this addition as there was none in Whitehorse. Wanting it on the first boat out of Mayo, my wife and I had flown to Mayo to be there when the *Keno* arrived. On the day before we were to leave for Whitehorse again, Owen Williams, assistant manager, wired us that the Whitehorse store had burned that morning. I then set out to get all the lumber I could in Mayo.

When we arrived back in Whitehorse that evening, the burned store was still smoking. That was June 14. By December 12, the living quarters were finished and the store was opened for business to a crowd that was two abreast for the whole block. Goods were so scarce that the people had stood in line for two and a half hours in 40-below weather.

I continued as manager at Whitehorse until February 1, 1949, when I retired after 50 years and seven months with N.C.

ALEX & BLANCHE BORIS

by Blanche Boris & Doris Forsythe, 1989

Alex Boris was working on the White Pass railway which ran between Whitehorse, Yukon and Skagway, Alaska when one of the Mervyn's from Mayo who was working with him told him about a deceased estate that the Public Administrator had up for sale in Mayo. So sight unseen Alex bid $500 and got it. Not long after that Alex Boris and Blanche were married on June 4, 1949 in Whitehorse, Canada and were on their way to Mayo Landing, Yukon, and their new home. It turned out great, but the house was in need of cleaning and painting. Blanche recalls with horror, the sight of dozens of old socks hanging from the rafters. Alex and Blanche spent the next four days cleaning the house while they camped beside it in a tent provided by Alex Nicol. Their house had belonged to a shoemaker (Olaf Anam), and since Alex was skilled in this and other leather work he fit in perfectly. In fact he made and repaired various harness equipment for big game hunter, Louis Brown for his horses and pack horses for his hunting trips.

Since Mayo is about 150 miles below the Arctic Circle it was sort of primitive when Blanche and Alex first went there—no roads into Mayo yet. People had to come by boat from Whitehorse down the Yukon River then up the Stewart River into Mayo or by plane into the little airport in good weather. Eventually roads and bridges were constructed, but not then.

Alex continued working for White Pass as night watchman on the barges, however in later years he was hired by the Northern Commercial Co, a general merchandising store. These were the good years for them, although they had few luxuries. They were happy in this small town of around 350; everyone knew everyone else, and helped each other when needed.

Blanche and Alex decided to build a new home after a few years. Alex went to a U.S. Army base located at Johnson's Crossing, Yukon to buy materials. They purchased a building lot in Mayo and had the materials delivered there. In his spare time Alex worked on their home with Blanche as his helper. With brace and bit Blanche bored new holes for the bolts; she also learned to hang sheetrock and to insulate the walls with wood shavings. It took an entire summer to get the house finished enough to move into.

Their first home sold fast so they had to move quickly. They then found the water on their new lot was not good enough for their use and so decided to move the house to the other end of town where the water was fine. They bought two lots with good water. Alex dug a hole for a basement, then hired a truck to pull the house across town, over the basement it went. Alex levelled it and all was well.

During the summer months the temperature might reach 100°F and it was light all day. However, the winters were severe. After the temperature reached –30°F the winds died down. It was a dry cold and sometimes moisture in the air would freeze and fall as ice crystals. However when the temperature fell as low as –70°F the ice crystals would become so dense that you would be unable to see a block away.

Cars would freeze up while still running. Blanche and Alex soon learned how to dress for this kind of weather and always kept their heads and faces covered and didn't dare wear glasses out as they would freeze to their face. To keep their feet warm they wore heavy wool socks and mukluks of tanned moosehide that came to mid-calf.

Wood had to be cut for winter, Blanche helped with the cutting and stacking. Each year they would calculate how much wood they would need and then hire a man to cut the logs and load them into their pickup and Blanche would drive the load home. One time, one of the wheels came off the truck and went rolling down the road in front of her! She stopped the truck and was lucky that the minister, Reg Alcock, was out visiting at Horst's and helped her.

They kept their home warm with a wood burning furnace made from a fifty gallon steel drum with metal legs bolted on to it and door cut into it. During the extreme cold Blanche always worried about house fires, she would splash their heating pipes with water to cool them. Their friends accused her of burning more water than wood. Blanche even trained their cocker spaniel, Sandy, to bark when the pipes got hot and began to smell. A neighbour died in a house fire. When a fire broke out everyone dropped whatever they were doing and rushed to help. People had to watch out for one another and did so.

"One time, one of the wheels came off the truck and went rolling down the road in front of her!"

Blanche would pound the stove pipes to get the creosote loosened so it wouldn't stop up the pipe and burn. She beat them so much they were full of dents. People teased her about this too.

Then with spring's arrival was the threat of flood. Blanche and Alex experienced two floods while living in Mayo. When the temperature rose the snow would melt rapidly, and when this happened the entire town, including children would fill sand bags to build a dyke to hold back the rushing river water.

One winter too much water was released from the Mayo River Dam. This dam was five miles out of Mayo. The water glaciered up as it was released, and as more water was released, it flowed over the ice and on towards town. Government crews were sent to help control the flooding. However with only one restaurant in town, feeding the men became a problem. Town's people were asked to help with the cooking and Blanche would bake eight to ten loaves of bread every day for the men. This particular flood never reached their home but many had water in their houses.

Another time, another flood caused many, including Blanche and Alex to leave their home. They moved into tents on higher ground. What an experience! Three families lived, slept and cooked together in one big tent. Again they had no water damage.

During the summer months Blanche had a big garden, plants grew fast in the long daylight hours of summer. She planted radishes, onions, beets, turnips, lettuce, potatoes, peas, cabbage, cauliflower and carrots. One year some of her potatoes weighed 2 lbs apiece. Another year her potatoes turned green as soon as they were dug and exposed to the air. This reaction was attributed to nuclear fallout. The Russians had been testing in the northern area and prevailing winds blew the fallout over Mayo. Most people paid little attention to the fallout. However, the town doctor kept testing the soil and expressed concern over what was happening. Many years down the road several of the residents developed cancer, including both Blanche and Alex.

Blanche also grew lots of sweetpeas and took bouquets to the Mayo Hospital. She sold her excess vegetables at the local store. Once she received ten cents a pound for her cabbages. Blanche still laughs about the year her parsnips froze in the ground. She liked to leave them in the ground as long as she could as the cold added flavour, but she waited too long: and had to use a pick to break the ground to get them out.

Food was canned, frozen or stored for the long winter months. Blanche, with friends and neighbours, went on many a berry hunt, letting the bears leave first! They purposely made lots of noise to scare the bears away so they could pick the ripe berries. Throughout the summer months they picked wild strawberries, blueberries, raspberries, red and black currants, high and low bush cranberries, as well as swamp cranberries. All these in season made good jams and jellies and canned fruit for pies and other desserts.

Living in Mayo provided Blanche with her share of interesting experiences and she recalls a moose hunting trip with friends Clara Van Bibber and Mary McDiarmid one fall. Blanche and Clara drove out of Mayo one Saturday afternoon with the laughter of the men folk echoing in their ears. They picked up Mary McDiarmid at Stewart Crossing and after driving about 100 miles set up camp in the back of the pickup. It wasn't long after setting up camp that Clara spotted a moose and calling to the others to join her, she and Blanche shot the moose. However darkness was closing in so it wasn't until the next day that they knew how successful they had been. It took them most of the following day to dress the moose and the third day to haul all the meat onto the back of the truck. No mean feat for three women as the meat was half a mile from their camp and it was also quite heavy to load onto the truck.

They returned to Mayo in triumph about 10 p.m. that evening with the moose's head sticking out the back of the truck between the canvas flaps. The news of their success reached the Whitehorse radio station and newspaper. So the whole Yukon heard about, "The ladies who went out and beat the men to the first moose of the season," said Blanche.

Blanche canned and salted the moose meat, and also the beef and pork they brought from Whitehorse.

The summer of 1951 was a particularly sad one for Blanche and Alex as their friend and minister, Norman Wareham, was drowned while out in a canoe on Mayo Lake. The waters of the lakes in this area are so very cold that a person cannot survive longer than a few minutes. Norman's wife, Dianna, turned to Blanche for help at this time as they were good friends.

Life in Mayo was not all work and survival centred. Blanche and Alex enjoyed socializing with friends. Annie and George Besner became their card playing partners and special friends. It was customary for friends to visit often in the evenings and almost every evening Blanche and Alex had company. It was also customary to serve some type of sweet, so the women baked lots of cakes, pies and cookies. In the winter months Blanche kept a metal can in the garage full of frozen cakes and pies.

Blanche was also involved in church activities. She was President of the Women's Auxiliary for one year and was treasurer of the Anglican Church Board for two years. She was also in the Drama Club and was an honorary member of the Imperial Order of the Daughters of the Empire. She and Alex attended dances and card parties and many church activities and were both on a curling team.

In summer there were picnics with friends and families. Boating, fishing and of course walking were part of the many outdoor activities.

Another exciting event was the arrival of the steamboat. When the boat blew its whistle everybody went running to the dock.

Blanche especially remembers the year of 1953 when Queen Elizabeth II was crowned. Mayo celebrated the coronation with a big dance. All the women in town had new dresses and Blanche designed and made her own. She did lots of sewing for the townspeople.

On July 19, 1959, when Queen Elizabeth and the Duke of Edinburgh were to fly from Whitehorse to

Dawson and Mayo, the Queen stayed in Whitehorse because of her pregnancy. However Prince Philip made a brief stop at the Mayo Airport having piloted the plane, a four-engine de Haviland Heron, for the entire flight. On this occasion the Royal Canadian Mounted Police and the whole town were at the airport to greet the Royal visitor. Prince Philip chatted and shook hands with almost all the residents of the region before flying back to Whitehorse.

Blanche attended painting classes taught by the town doctor, Dr. J.V. Clark. She hooked rugs in her spare time, and also learned to ride a bicycle.

In the spring people made bets as to when the ice would break-up in the river. This was called an "ice-pool" and the person who guessed the time and date closest to the break-up would win a cash prize.

Blanche's family was always in her thoughts. She had one wall in her home full of "pictures from home" and every time she got a new one from her sister or brothers and their families or her father she would put it on the wall.

One summer her sister Doris with her two daughters, Nancy and Barbara came for a visit. Nan turned six that summer so Blanche gave her a party and invited the town's two sets of twins who were near the same age. The Wareham twins, Mary and Margaret and the Shaw twins, Karen and Sharon with Nancy's little sister Barb who was nearly two. What a nice party!

Doris and the girls stayed long enough to see the Northern Lights. What unbelievable colours! All shapes, like waterfalls, huge wavering fans in the sky and rolling colours of the rainbow, all quite unforgettable. It was a very special summer for Blanche's sister and her little girls.

Blanche believes she and Alex left Mayo in 1965. They then moved to Haney, B.C. where they bought a lovely home on the Lougheed Highway.

Alex died in November 1971. Blanche stayed in Haney for another two years before selling her property and moving back to California to be near her family. Blanche passed away February 18, 1990.

The women shot the first moose of the season! Mary McDiarmid, left, and Blanche Boris. *B. Boris photo*

HUGH S. BOSTOCK

by Hugh S. Bostock, October 28, 1983

In 1929 I was appointed to the staff of the Geological Survey of Canada and directed to do the geology of the southern Okanagan. This suited me well as I came from Monte Creek, B.C. (then a stop on the C.P.R. 18 miles east of Kamloops) and my wife came from Nelson, B.C. Then abruptly for me, the Director of the Geological Survey of Canada appointed me to take charge of the exploration and mapping of the Yukon and I continued in that position until I retired. In addition to this I was directed to write the annual report of prospecting and mining developments in the Yukon and northern B.C. In this, as the B.C. Minister of Mines (in Victoria) sent his mining engineers from time to time to Atlin and Cassiar, the Yukon became my territory.

In 1931, the Geological Survey of Canada had made an uncontoured four-mile map around Whitehorse and I was told to assist Dr. E. J. Lees in completing the topography and geology of the Laberge map area and do the Carmacks area. After this we did the Mayo area and then the McQuesten and so slowly set out to map the whole territory on 4 miles to 1 inch. After the Ogilvie map area around the Klondike, where at that time eleven dredges were working, I went by steamboat (1932) up the Yukon River to Stewart Island and then up the Stewart to Mayo where the Treadwell Yukon Company Ltd. was mining from the Wernecke mine camp on Keno Hill and exploring the Elsa on Galena Hill. The Depression was developing then. Metal prices were falling and generally prospecting and base mining were closing down but the steamboats kept running. There were no roads anywhere except wagon roads Whitehorse to Robinson, Carcross, Kluane, and Carmacks in late summer. A few around the Klondike and Wernecke's tractor and sleigh road to Keno City and thence to the Wernecke Camp town site.

Highet Creek, 1941. From left: Survey Assistant Thompson, who later became Dr. Thompson at U.B.C.; Dr. H. S. Bostock; and Rodolph Rosmusen. *A. K. Schellinger Collection*

There was also the wagon road from Minto Bridge up Highet Creek (Highet Creek was named for prospector W. E. Hiatt by someone who didn't know how to spell his name).

In 1932 I drove up Highet Creek with Mr. Gillespie, the Mining Recorder, with a horse and buggy and stayed two nights as I remember at the Middlecoff's and enjoyed my visit with them. Mr. Middlecoff took me up to see Rodolph Rosmusen and showed me over the whole creek up to the summit. The mining that was going on is described in my reports of the Geological Survey of Canada Summary Reports A 1932 until about 1942.

I did the geology of the Mayo map area, 136 to 138 longitude by 64 to 65 latitude, and also Dublin Gulch. This also took in the Beaver River above its junction with the Stewart. My book "*The Pack Horse Tracks*" gives some account of the Mayo District 1932–1954 and what was going on around Mayo in those years.

I walked around the Yukon more than any other man ever has or ever likely will as I tried to see everything that was being done as well as what had been done.

LOUIS BOUVETTE

by Mary Jean Morrison, 1985

Louis Bouvette (often spelled Beauvette but the correct spelling is Bouvette) was born on a homestead in Poplar Point, Manitoba to a French Canadian-Scottish family. Louis and his sister Helene moved west with the family to the Prince Albert, Saskatchewan, area. During the second phase of the Riel Rebellion they remembered being taken to the Fort for protection.

Louis was raised with horses and they remained his love throughout his life. As a young man Louis worked as a wrangler moving further west across the prairies. In 1898 Louis joined the Glen Campbell expedition as a wrangler and they set out for the Klondike through the Peace River country and up the Mackenzie River. The expedition nearly starved to death and they had to turn back to Edmonton. In 1901 Louis went up the west coast and over the trail to Dawson City. However there was little opportunity for the latercomers in Dawson so Louis joined the stampede to Duncan Creek in 1901.

Louis Bouvette with the McLaren girls on either side.
D. Walloch (McLaren) Collection

Louis, a fiercely independent man, prospected and supported himself by hunting meat for the Duncan Creek placer mining camps. Louis always had a horse and at times worked as a teamster hauling goods over the Duncan Creek Road.

In 1919 Louis found silver on his Roulette claim on Sheep Mountain, later renamed Keno Hill. Many of the prospectors, including Louis, had been assisted in their search for minerals through the informal guidance and instruction in geology by Alfred Kirk Schellinger, "Shelly," a Stanford University mining engineer and geologist who had been sent by the Guggenheims to Dawson and later to the Keno-Mayo area.

When Louis' brother-in-law died in Vancouver in 1923 he brought his niece, Jean, up to the Yukon for a visit. Jean and her brother Charles Nixon later rejoined Louis in Keno in the early 1930's. Louis' cousin, Tom McKay was also in the

Keno area from the early part of the century.

Louis prospected all his life in the Mayo Mining District, however much of his prospecting was done during the 1930's with his nephew Charlie Nixon and Donald Morrison, his niece Jean's husband.

Louis moved to the settlement at Minto Bridge in the early 1940's. He still rode his horse, Star, but his health was failing by 1947. Louis Bouvette was lost on the trail near Minto Bridge, it was presumed he had a stroke and had fallen into the Mayo River, however his body was never found.

HENRY BREADEN

by Kevin Shackell, Yukon News, December 13, 1982

[Breaden was] born in Whitehorse in 1927, [and his] family moved to Mayo in 1929. Although his father decided to work in a transportation career on dry land, his grandfather was a river boat skipper who had come to the Klondike at the turn of the century.

His father had moved to Mayo in 1929 to work for Greenfield and Pickering, the company handling the mail delivery contract for Whitehorse, Mayo and Dawson.

Nestled on the banks of the Stewart River, Mayo, like all other major Yukon communities, revolved around a ribbon of water. River boats were the only reliable link with the outside world in the days before road and air travel became established. "The steamboat, you might say, was the gateway to the North," commented Breaden. "It was the communication, transportation and supply route, the whole works."

It was hard not to be fascinated by the comings and goings of river traffic. Whenever the boat was in, townspeople flocked to the jetty to greet the passengers who had arrived and to catch a glimpse of the mountain of supplies off-loaded on the jetty. "You got a first boat, it was a pretty big thing. Of course by May the eggs were getting pretty high. You know when I tasted my first fresh egg I thought there was something wrong with it," he chuckled.

Grade Eight was as far as a person could go in their schooling unless they wanted to take high school in Dawson, Whitehorse or outside. Breaden wound up his formal education after Grade Eight and it wasn't until years later that he decided to expand his education by taking correspondence courses in the engineering field.

So in 1942, at the age of 15, Breaden started longshoring for the British Yukon Navigation Company, White Pass' river division. As a longshoreman he loaded 60-kilogram sacks of ore onto the S.S. *Keno*'s cargo deck. Even in the 40's it was a manual operation. Two men would load four sacks onto a hand cart and then wheel the load down the gangplank into the river boat. If the crew was feeling particularly strong, up to 16 sacks would be piled on the cart in a single load. "It was okay provided you could get a decent balance and keep it moving," recalls Breaden. "I was amazed when I think back, safety wasn't a thing we thought of, it was just a hell of a lot of fun." On occasion, stevedores would fall off the gangplank in the Stewart River, but Breaden said he never recalled anyone getting injured during the loading and offloading procedure.

Later in the season, an opening came up on the S.S. *Keno* and Breaden joined the boats' crew as a deckhand. The *Keno* had a crew of 26, while the bigger boats like the *Casca* and *Whitehorse* had crews of 35. Deckhands were responsible for much of the manual and sometimes menial day-to-day work. Breaden earned $60 per month and an additional $1 an hour for the time he spent stevedoring in port. The hands spent two hours on wood watch, and 10 off, although they were on call 24 hours a day.

The S.S. *Keno* used to burn 5/8 of a cord per hour bucking the upstream current and only half a cord going downstream. Frequent stops were made at wood camps along the way to replenish dwindling fuel supplies. Deckhands could load about 10 cords of wood in an hour, recalls Breaden. Deck-boys and firemen slept aft of the engine room in the compartment between the engine and the paddle wheel. The skipper, pilot and mate had their quarters on the Texas deck below the wheelhouse, while the engineers lived on the saloon deck.

The purser and saloon crew's quarters were located on the second deck.

The incessant throb of machinery was a comforting feeling, recalls Breaden. It was hard to adjust to the silence of sleeping ashore after the noise of the engine room.

Gradually Breaden worked his way up from deck-hand to mate. He also received his master's ticket, but resigned from the British Yukon Navigation Company before he could put the ticket to use.

Skippers carefully gauged the distance between wood camps because having too much fuel aboard would affect the vessel's trim, as well as take up valuable space.

Piloting the boats through the Yukon River's many channels was no easy feat, especially when pushing a barge ahead. The barges were fastened to the boats by way of a block and tackle rig. When rounding tight bends, skippers used to "jackknife" the barges to get them around. Jackknifing was the process of alternately slackening and tightening the tackles allowing the barge to pivot. But a large barge would sometimes break away and be swept down a shallow side channel. Retrieving the stranded vessel was a time-consuming and back-breaking operation. The river boat would have to anchor and using her steam winch the barge would slowly be pulled back into the main river.

One of the big challenges to a river boat skipper was crossing Lake Laberge, well noted for its sudden and violent storms. On occasion the boats would wallow through five and six foot waves as they made their way up the lake.

Due to the pounding wave action, skippers were continually checking the calking to make sure the vessel wasn't springing any leaks. "Lake Laberge was kind of a funny feeling when the boat quartered across," says Breaden. "They handled alright, but they weren't made for it."

In 1945 Breaden joined the *Neecheah*, a prop-driven vessel plying the Stewart River. Shortly after Breaden went aboard, the *Neecheah's* skipper took Breaden under his wing and as his protege, taught him more about the river boat business. This training helped pave the way for Breaden to get his mate's ticket and after writing an examination in Vancouver, he took over as mate of the *Nasutlin* in 1948.

"...skippers were continually checking the calking to make sure the vessel wasn't springing any leaks."

Five Finger Rapids near Carmacks also proved to be a test of the skipper's skill, recalls Breaden. Depending on how the boat was loaded and strength of the current, boats could steam right through without having to use the winch. If the current proved too much, the boats could always "pick up a line" and winch themselves through the rapids.

Like many of the B.Y.N.C. crew, Breaden spent his winter months in Vancouver, coming back in March for refit prior to spring breakup. During one of these winter layovers, he met his wife-to-be, Alice, and in 1948 they were married.

In 1949 she came up to Mayo to join her husband. By the late 40's Breaden saw the handwriting on the wall and knew the days of the river boat were numbered.

The Alaska Highway and the Mayo and Dawson Roads were making river transportation obsolete. The river boat era was coming to an end so 1950 was Breaden's last season.

"When I left White Pass, I was starting to look ahead at that point and I could see that river transportation was coming to an end," he explains. "Of course the world doesn't stand still and I'm very happy to have seen that era."

It wasn't quite the end of Breaden's river boat career, however. In 1960 Parks Canada decided to refit the S.S. *Keno* after she had been sitting neglected in the Whitehorse shipyard for eight years. The old gal was cleaned up and put back into service for a final run down to Dawson City where she lies today as a National Historic Site.

Appointed mate of the vessel for her final voyage downstream, Breaden oversaw the S.S. *Keno's* refit. The funnel had to be hinged in order to allow her to pass under the Carmacks bridge. In the ten years since river boats had ceased making regular runs, many of the channels had shifted or silted up and at Slack Water Crossing near Minto the *Keno* ran aground. The vessel was extricated from the bar and continued her voyage downstream without mishap.

Constantly shifting and changing, the Yukon River is quick to reclaim her own, but for at least one man the memories of an era long since vanished live on.

RENNY & MARY BURIAN

by Ted Thaler, Yukon News, January 24, 1983
With story additions by Mary & Renny Burian

Among his friends, Renny Burian is known as the last of the woodcutters. For more than 15 years, Renny cut wood in one form or another. There was wood for the steamboats, the home fires, and even the lumber to build the homes. For whatever the purpose, Burian cut wood.

Born in Germany in 1908, Burian emigrated to Alberta with his parents, at the age of six. In 1928 they moved to B.C. Like many young men, Burian was caught in the clutches of the Great Depression in the early 1930's. With no prospects of a job in B.C. he headed north to Yukon in 1936 where an younger brother was already living.

Renny and Mary Burian on their wedding day in Mayo, 1943. *R. Burian Collection*

When he first arrived, Burian held a number of jobs. His first winter in Yukon he worked for T.C. Richards driving a truck. Driving truck between Minto and Pelly, Burian hauled a cargo varying from oil barrels to sides of beef. With no bridges crossing the Pelly River in those days, the job was made more challenging. With the risk that the weighty load would break through the ice, the cargo had to be unloaded and hauled across the river by sleigh.

It was as a truck driver that first winter, that Burian was introduced to the true Yukon cold. With no truck heaters and only a canvas cover over the cab, there was little to protect the driver from the chilly elements. "It sure got cold at 75 F below," recalls Burian. So cold that he remembers days when he would have to throw his sandwich lunches away because they would be frozen solid before he got the chance to eat them.

It was after his winter driving for T.C. Richards that Burian became involved in the wood cutting business. Living in Whitehorse, he cut wood for both George Ryder and Antoine Cyr. The wood he cut was used to fuel both sternwheelers plying the Yukon River and the heating stoves of local residents.

While Burian was cutting wood in Whitehorse, his brother Alfred was operating the wood camps supplying sternwheelers along the upper part of the Stewart River between Stewart Crossing and Mayo. However, in December 1939, Alfred was killed in an unfortunate accident when the tractor he was using to haul a load plummeted into the icy waters of the river.

After the accident, Renny and his brother Rudolph began to operate the camps. With Rudolph covering the lower section of the river, Renny took over the upper part of the river.

"I had four or five camps towards the end there," says Burian. But base camp—where Renny himself lived—was 26 miles below Mayo. From his base camp he would travel the river supervising work in the other camps. Burian tried to man each of the two camps with two sturdy men. If the men stayed the winter, Burian supplied them with a cabin, staked them their grub supply, and paid $6 for every cord of wood cut. "You could figure that one man would cut about 150 to 200 cords of wood in the winter," says Burian.

Cutting the wood was only part of the task for Burian. Once the wood was cut, it was his job to haul the wood to the riverbank. "I had a little Cletrac tractor that did all the work in summer and winter," says Burian explaining how he hauled more than a thousand cords of wood each year. On the banks the four-foot lengths of wood were stacked six feet high in seemingly endless rows.

Despite what seemed like an unlimited supply of wood each spring, it did not take long for the sternwheelers straining against the river current to use copious amounts. "Oh they (the boats) used to use a lot of wood on that run," recalls Burian. "You could figure 200 cords for the (sternwheeler) *Keno* in a round trip from Stewart to Mayo and back." Loaded down with supplies and mining equipment, the would also be pushing a barge laden down with a cargo. Plying the 185-mile stretch of river from Stewart to Mayo, travelling against the current, would take up to four days.

Although it was primarily the S.S. *Keno* that plied the Stewart River, Burian recalls one summer when all the sternwheelers were making the run to Mayo. Low water levels the previous summer had meant that sternwheelers could not travel the river all the way to Mayo. As a result, bags of extracted silver-lead ore piled up on the docks in Mayo. The following summer, water resumed its former high levels and a barrage of sternwheelers were brought up the river to haul out the stockpiled ore. "The *Klondike*, the *Keno*, the *Nasutlin*, the whole works came up and I'll tell you the woodpile disappeared," recalls Burian.

But when the supply of wood got tight, the owners of the sternwheelers, White Pass, would always step in and help. "If I needed men, White Pass would send men from Whitehorse that you could depend on."

It was several years after taking over woodcutting operations on the Stewart River that Renny married Mary.

Mary came up to Yukon as a young girl of twelve, along with her brother. Arriving in 1931, she lived with her sister and brother-in-law who ran the GN Cafe in Mayo. Even as a young girl, Mary helped in the restaurant. In fact, it was while she was waitressing in the restaurant that she first came to know Renny—a customer of the restaurant.

In 1943 Renny and Mary married, with Mary leaving Mayo to live at the woodcutting base camp 26 miles down river.

To travel between camps Renny owned a gas-powered motor-boat or he capitalized on an arrangement with White Pass allowing him to ride free on the sternwheelers. As he rode the old steamboats he gradually became familiar with their operation and the crew. So familiar, that the crew would often let him pilot the sternwheeler up the river. Even Mary tried her hand at the wheel of the sternwheeler during her travels. One day while riding in the pilot house the captain asked her if she had driven a river boat before. Mary told the captain that she had only driven a small gas-powered motorboat. Nevertheless, the captain gave Mary the wheel and sauntered off for a coffee. When Renny met the captain in the coffee shop and asked him who was piloting the boat, it was much to his surprise when the captain nonchalantly replied: "Your wife."

Mary and Renny's son, Harvey, had an eventful arrival into the world. Mary was in labour and had to get to the hospital in Dawson as the hospital in Mayo was closed. Harry Ewing, the radio operator, was able to intercept a freight plane flying from Whitehorse to Dawson with a load of Christmas turkeys and mail. The freight was unloaded in Mayo and Mary, accompanied by Olive Kinsey, and Mrs. Jack Andison, flew to Dawson where Harvey was born on December 20, 1944. The Burian's were delighted but the town of Dawson had to wait until after Christmas for their turkeys to come from Mayo.

Despite their contentment, the life of being a river boat woodcutter did not last for the Burians. By the early 1950's, a roadway was pushed through from Stewart to Mayo, eliminating the need for river boats.

From river boat woodcutter, Burian took the only logical step. As the demand for river boat wood dropped off, he converted his operation into a sawmill, supplying lumber to surrounding areas. Using a machine shop that he operated and an assortment of scraps that he assembled, Renny built the sawmill from scratch in one summer. But the Burians operated the sawmill for only a couple of years before selling out.

After selling the business, says Renny, he made the decision to move out of Yukon for good and move south to the warmer climes of Vancouver. But once a Yukoner, always a Yukoner. "It wasn't six months before we were back," says Renny, adding the brief sojourn quenched for good all desire to live "outside."

Upon returning to Yukon, the Burians took up residence in Mayo. During the next few years they held a number of jobs before Renny finally took a position in the machine shop of United Keno Hill Mines. The couple remained in Mayo until their retirement when they finally sold their house and took up residence in Whitehorse.

In 1977, the Burians were voted Mr. and Mrs. Yukon for the annual Rendezvous Festival—an honour bestowed upon sourdough Yukoners. Each summer the couple still load up their camper and head north along the Klondike Highway to visit their old friends and stomping grounds in Elsa and Keno. For the Burians "home" has become the land of friends and fireweed—Yukon.

Steamboat with refrigeration barge at a woodcamp, possibly on the Stewart River. *A. K. Schellinger Collection*

MATT BUTIJER (BUTYER)

Written by Jean Gordon for the Whitehorse Star, circa 1954

It is not often that a man's faith in himself and his ground pay off after many years of hard work and discouraging set backs. Such a man is Matt Butyer. He made his home for many years at Keno City but many who have lived there in the past will not remember him for he seldom was seen around the bright lights. He was too intent on proving to the world that there was more than one mine on Keno Hill. His days were spent putting in tunnels to verify his faith in his ground.

Matt Butyer first saw the light of day in Topolo, Dalmatia, Yugoslavia in 1886. At the age of eighteen he made his way to the new world and landed in the great city of New York. His first job there was as a busboy in the Waldorf-Astoria Hotel. Here the great and near great of the world were entertained during their stay in that city. One such entertainment of that era remains vivid in the memory of Matt Butyer. It was when Crown Prince Olav and Princess Martha of Norway visited the United States. Sumptuous preparations were made and the busboys were kept busy during that time. It was while working at the Waldorf-Astoria that Butyer first made the acquaintance of Miles Flynn, a man who was later to play an important part in the mining industry of the Mayo area.

Leaving New York city, Butyer made his way west to Cincinnati, Ohio, where he worked at various jobs. From there he gradually made his way further west finding himself at last in 1910 in Spokane, Washington. From Spokane Butyer went to Rossland, B.C. During this period he followed mining as a means of earning a living. From Rossland he came north to Fairbanks, Alaska. In the winter of 1913 he and his partner, Emil Anderson, decided to try their luck in Dawson and so preparations were made and the two men started a 500 mile journey pulling their belongings on a sleigh. Imagine if you can the fortitude required to pull everything a man owned over such a distance in the dead of winter in the north. The thermometer would vary from day to day, up and down. One time it would make for easier pulling on that sleigh and the next it would seem that the runners were fast to the snow. Up hill and down dale with no place to stop and get in out of the cold. At night a camp of sorts would be made, fire wood had to be gathered for the night. While one man slept the other kept the fire going till his turn at the sleeping bag came. Morning would come all too soon and the two men would have to be on the trail again.

"While one man slept the other kept the fire going till his turn at the sleeping bag came."

During the summer of 1913, Butyer worked for the big company operating the placer dredging in the Dawson area. That fall he left the Yukon and went to the south eastern Alaska area and worked for a time at the Chicagoff mine before coming into Whitehorse. Here he and his partner, Emil Anderson, felt that the Mayo area looked good from reports so they prepared to make another trip such as they had made from Fairbanks to Dawson the preceding winter. Loading their sleigh the two men pulled all their belongings over the trail to Mayo arriving in the spring of 1914. The trip was not unusual for these two since they had travelled the same distance the year before, under the same hazards and survived. Their journey brought them by way of Scroggie Creek and then a long pull up the Stewart River to the mouth of the Mayo River where a few settlers had established the town of Mayo Landing. Here Binet operated a trading post and hotel and Ferrells a small store. Butyer and Anderson did not tarry long in Mayo but made their way to Mayo Lake where for the next four years Butyer explored placer ground on the creeks flowing into the lake. Ladue Creek, Gull Creek, and Davidson Creek were all prospected by Butyer but the pay streak only gave up a grubstake.

Of his years on Mayo Lake, Butyer looks back to a happy time. He did not make much money but he had a small garden and there were always moose to provide meat for the pot. Life at the lake in the winter was one of constant surprises. One day the lake would be calm and a skim of ice would form and it would appear that the icy hand of winter had stilled the sometimes turbulent lake for the season when up would come a storm and break up all the ice, blowing it from one end of the lake to the other. When winter would finally triumph this broken ice piled hither and thither on the shores would be extremely hard to travel over. As the ice froze deeper in the lake a thaw would expand it and pressure ridges appeared. When the thermometer registered 40 and 50 below the shrinking ice would crack with the sound of a cannon. In the dead of night these weird sounds would almost make one believe that someone was having a little war of their own around the lake.

At this time there was considerable activity on other creeks near Mayo. Haggart Creek had quite a few prospectors making better than the going wage. Duncan Creek was also being mined for its gold. Among those who explored Mayo Lake, at the time Butyer was prospecting there, were George Crissfield and Arthur Hester. Joe Walsh had some ground on Duncan Creek. In 1919, Louis Bouvette made the first silver strike on Keno Hill and Butyer left his placer workings to stake on the hill. He located the *Stone*, *Scott*, *Viola*, and *Ray* claims and began the long and arduous task of proving their worth. He built himself a cabin and in the ensuing years put in four tunnels. One of these tunnels is believed to be the longest ever put in single-handed in frozen ground by any one man. The first 100 feet from the portal had to be cribbed with timber and the whole length of 700 feet averaged about 22 feet a month! Mining in this area is fraught with many disappointments for many feet of overburden must first be taken away before hard rock is reached. Then a prospector is considered lucky if his find is on ore-bearing rock. Butyer was among those few who were able to strike a vein. The entire length of his long tunnel was in a pay vein.

The years following his strike were ones of hard but satisfying work for Butyer. He shipped ore to the outside and spent his time working on his claims. Keno City was born and grew to be a rousing mining camp with three hotels, three stores and a large population. But it was a town that seldom saw Matt Butyer for none of it's bright lights held any attraction for him. He was content to work his claims and care for his garden. A road was built to Mayo but it meant only a means of shipping his ore and getting supplies in for his work.

Mining companies bought ground on Keno Hill and set up large mining operations but Butyer was happy to take out and ship his own ore even if it was a long, slow arduous task. In the hungry thirties the mining company closed down their operations but Butyer was not affected by the fact that a depression raged in the rest of the country. His property gave him a good living even during those lean years. When base metals reached a new level and the mining industry in the Mayo area opened again, Butyer optioned his ground to the same Miles Flynn whom he had met so long ago at the Waldorf-Astoria in New York. Flynn was a mining engineer for one of the companies which took up ground on Keno Hill. This company optioned Butyer's ground at a figure estimated to be the highest ever paid for a single claim in the Yukon.

Other men who had the same faith in themselves and their ground on Keno Hill who have been able to capitalize on it were Emil Anderson, who had been Butyer's old partner in the early days, as well as Tony Hollenbeck, Oscar Miller, Pete Runer, Tom McKay, Jim Clark and Joe Sutherland [Sunderland].

Butyer still holds ground on Keno Hill and now, after having been a patient in the Mayo General Hospital for nearly eight months and not expected to live, he has made a remarkable recovery and is back in Keno on his property. He has implicit faith in the whole area surrounding Keno Hill and feels that to look forward to at least 100 years of mining on Keno Hill as well as Galena and Sourdough Hills, Mount Albert, Mount Haldane, as well as hills across the McQuesten River; is not looking too far ahead. He feels sure that bigger and better mines will be opened in the not too distant future.

HELEN BUYCK (née Hager)

Interview and transcription by Linda E. T. MacDonald, 1983

A short time ago, I watched a television program about the starvation of Eskimo people in the 1930's. The Eskimos told how Indian Affairs withheld their money while some of their people were starving to death. The government agents cheated them on fur sales. Indian Affairs held back money the Eskimos were entitled to for many years and when they finally told the Eskimo people there was money available to them, it was two years before the people accepted it because they had never received anything for free before. I felt angry watching that program. Although our people in Mayo were not starving, we were told of money available to us in exactly the same way. The R.C.M.P. came to our village and told everyone that groceries were available to us at the Northern Commercial store; but everyone was afraid so no one went to the Northern Commercial store. The R.C.M.P. kept coming back and telling our people that they could take anything they wanted from the store. Everyone kept putting it off because nothing like that had ever been done for them before and they did not understand it. They had always worked hard for their food and clothing. Although still afraid, my mother and Alice Hager finally went to the Northern Commercial store and got groceries. After that they went to the store regularly and the storekeeper would always tell them they still had money left. As I watched that television program I wondered if we had been entitled to money for years and it had been withheld from us just as it was withheld from the Eskimos. I wondered if they had done many of the same things to us that they had done to the Eskimo people. They told us our children had to go to school just like they told them; with the threat that if they did not attend school the government would come after us. So, our parents had to stop whatever they were doing and immediately build a school. Even then, the government sent some of the children to residential schools in other parts of the Yukon. They were pretty lonely kids.

Sometimes I think some of the government people are still trying to do the same things to us today.

We used to have our own village. It was located downstream and across the river from Mayo. To this day I still dream about our village and the house we had there. Sometimes I sit and think about how wonderful it was. Everyone worked hard and made a good living. We were happy, the kids were happy, and everyone got along. If a family was in need we shared what we had with them. There were 12 or 13 families living in the village.

"We were happy, the kids were happy, and everyone got along."

In July and August the people in the village were busy making preparations for winter. King salmon was caught, dried, and smoked. The women picked berries. In September four or five families would go hunting and bring back meat for everyone. Tons of meat were smoked and the moose skins were tanned. All of the food was placed in caches. I guess we felt the same way about our caches that you feel about being able to select food from a store today. People walked miles up the river to collect food and going home everyone packed, even the children and dogs. After delivering the food to the village, they would go back up the river to collect more. Sometimes people would pole a boat up the river. If the beach was good, two people would get out of the boat and pull it while one person remained in the boat to guide it.

If anyone became short of meat in the winter time, my father would go to Ethel Lake with a dog team and bring back loads of fish and sometimes meat. The meat kept well at that time of the year because it could be frozen. We always knew when my father was coming back because the dogs would start barking and then we would all run out to meet him. We knew he was coming home with food, he always came back with something.

In the summer time the steamboats would come by our village about two or three o'clock in the morning. The whistle would blow around the bend and everyone would go out and watch it go by. Some of the crew would throw apples and oranges to the shore. Sometime the fruit would land in the river and some of the older children would jump in a boat and try to catch it. People were happy then. I am sure that if you made noise in the middle of the night today, people would be furious.

The children of the village worked hard. They cut wood, carried water and food, helped the family anyway they could, and always treated their parents with absolute respect.

Sometimes I sit with the old people and we talk about the village. It makes us happy just talking about the way it was then. Some of the people that lived in the village have died now. The few that are left are Peter and Jimmy Lucas's mother, Mary Moses, Jimmy Germaine, my mother Mary Hager, Lucy and Sam Peter, Johnson Lucas, Charlie Jimmy, and Big Dave Hager. (Did you know that Lucy Peter was the first child born in Mayo. She can not remember what part of Mayo her mother's tent was in. At that time Indian women delivered their own babies.) That is all that is left of that generation. My generation were once the children that lived in the village. When I look at our old people I think of how hard they worked to provide for their families. I doubt very much if anyone today would duplicate their accomplishments without a great deal of physical and psychological conditioning. The land was their teacher and the knowledge they have of it is far greater than most people can appreciate.

I guess that is why I get angry when I hear people say Indians are dumb and lazy. In their own way, Indians are clever and hard working people. That is also why I get so angry when people say Indians are drunks. Some people say mean things like that.

When the law was changed to allow Indian people to drink liquor our Elders fought against the change. They did not want liquor to be made available to their people in that way because they felt it would be harmful. [At one time, there was a provision in the Indian Act which made it an offence for Indian people to be in possession of alcohol off of a "reserve." Some Provinces were applying pressure to the Federal Government for not enforcing this law and prosecuting Indian people with alcohol in their possession off the reserve. Following the election that year, the Government of the Right Honourable John Diefenbaker amended the Indian Act and the question of whether or not the possession of alcohol by an Indian person was an offence was left for Provincial discretion. The situation in the Yukon at the time prohibited Indian people from purchasing alcoholic beverages in a cocktail lounge. — Information supplied by Erik Nielsen.] Perhaps the odd person did drink before the law was changed, but they did so in secret and never at the village because the chief was very strict. If they were caught drinking in town they were jailed for a month. In the beginning people were too shy to go to the bar. Eventually, they began to go to the bar and then went back, and back again. After that many people were never the same again. I guess the law had to be changed sooner or later. Perhaps if they had received some education about alcohol while they were growing up, they could have dealt with it properly. Every day when I am out riding my bicycle I see my people drunk. I want to see alcohol education taught in the schools. If our children learn about alcohol when they are young they will respect it as adults.

There was a time when Indian children were not allowed to go to the white children's school in Mayo. Generations younger than myself were eventually allowed to go to the school. For a time Mary Moses taught school in our village. She quit teaching when she got married and then Charlie Jimmy taught us for a while. I guess they would have been teaching about a grade one level. It was not until I was 14 years old that we finally got a real teacher. Her name was Sadie Dougall. She worked very hard with me and wanted to take me back to Vancouver with her so that I could continue my schooling but Mom and Dad didn't want to let me go. That's the way Indian people are, they never want to part with any member of their family. I don't know why because it is not as though children can live with their parents for ever and ever.

I felt a little angry when my father died because he was going to live with me forever and now where is he. But that's the way Indian people are, always close together. Even if one does go away, close contact is always maintained. I think these ties are more easily broken by children today. You cannot hang on to your children they must be allowed to enjoy their own lives.

After Sadie Dougall left, Lillian Connor came to teach. By that time I was 15 years old and my brothers and I had had very little schooling at all. For that reason it has been hard for us in many ways. Following Lillian Connor's term Miss Hilda Hellaby came. Miss Hellaby worked very hard with Indian people.

When our children were finally allowed to attend classes at the Mayo school the white children made nasty remarks to them about Indians. Children can be so mean. When my own children went to school I told them to be proud when the other children called them Indian. I am Indian, I am proud of being Indian, and I want my children to be proud of it as well. I want to make sure my children understand our Indian way of life and so preserve it from one generation to the next. I am now sorry I never spoke to my children in their own language. Today when I talk to my mother in Indian my children ask why I have never taught them to speak Indian. I know many people do not like Indian language classes being taught in the schools today but it's there now and it's going to stay, so they might as well get used to it. The school committee asked me what I thought of French and I said French is no good for Yukon.

If you are going to live here all your life then it is better to learn the Indian language.

Each time Mom and Dad went to Mayo they would take one of the children with them. I was about 10 or 11 when it was my turn to go. I was terrified of white people. Mom and Dad used to visit some of the old-timers they knew when they went to town. I would stand behind Mom's chair and say "Oh Mom, let's go home." I am not sure why I was so afraid. Most of the children were afraid of white people because we grew up in the village and therefore rarely ever saw them. Perhaps white children were afraid of Indians at that time.

From left: Helen and Don Buyck and Harry, 1950's. *Buyck Collection*

Have you ever heard of "bushmen?" Indians used to talk about bushmen quite a lot. Bushmen were afraid to come out in the open. Indians saw things; bushmen were real to us. There used to be a fellow here a long time ago who worked for forestry and I asked him if he believed in bushmen. He said no but they could possibly be people who had escaped from prison or war and were just camping around here.

When we were children the adults used to tell us that if we didn't get home before it got dark the ghosts or the bears would get us. So we grew up fearing these things. I have made sure not to frighten my children in any way because I didn't want them to grow up with the same fear that I did.

Mom and Dad were always very strict with us and they would not let me go to town whenever I felt like it. When I was old enough I went anyway. I was 14 years old when I went to work for myself. I cleaned homes in Mayo for 50¢ an hour or $2 a day. At that time things were cheap to buy so I couldn't really complain. I had a dream that someday I would own a house like the ones in Mayo. White people had nicer houses than we had in the village. We always thought white people were really rich; they must be rich because they owned cars, homes, and many other things. What we did not know was that they borrowed money from banks, made down payments and bought on credit. We thought they paid cash for everything. If only someone had explained it to us we could have done the same thing. No one ever explained enough things to us. Indian people used to make a lot of money selling furs and a long time ago they were allowed to sell moose meat. Selling meat was very lucrative. No one ever told them they could borrow money from the bank or buy things on credit. Now we know about these things and we intend to make the system work for us just as it does for other people.

One day the white people came and told us that the doctor said our water was no good and that we would have to move away from the village. I think that is the worst thing they ever did. I suppose some of our people would have moved to Mayo anyway because they had taken jobs in town and had many white friends even though the older people did not approve. Very slowly, families began to move. First one family, then a second, then two more. Mom and Lucy did not want to move and they were the last to move from the village.

They never told us what we could do about the water, we were given no alternatives. A well could have been drilled; something could have been done because there is plenty of good water there. Maybe at that time government people did not want Indians to get too wise. Today Indians are waking up, we know what's good for us and we can live like anybody else if we are given the knowledge to do it.

It is important that I tell you how we became Christians. My brother, Peter, went through some bad times many years ago and was blinded in a fight. He went to Edmonton for treatment and we did not hear from him for three years. Although we did not know it, Peter had met an Indian evangelist there and had become a Christian. One day while looking out from a forestry tower on top of a mountain, my father saw a vision. A great white road suddenly opened up and two men were walking up it. One was Peter and the other (unknown to my father at the time) was Harry the Indian evangelist. Hearing a noise behind him, my father turned and saw two mountain sheep lying there. They did not run away when he moved, they just looked at him. Then he heard a voice say, "Your son is coming back to Mayo tomorrow." He knew it was a message from God. The sheep lay there for two days and did not run away. He could not harm the sheep even though my mother loved to eat sheep meat. When Toby Anderson came to pick my father up the sheep took off like a shot. When he came back to Mayo and told Mom, she said, "How could you know that, we have not heard from our son in three years." The next day Peter and Harry came to Mayo. They told people how good it was to be a Christian and that is how some of our people were able to quit drinking. I used to think a Christian life would be boring. Then one day I just decided to become a Christian and now I live one day at a time and enjoy myself.

"Today I am 53 or 54 years old and I still dream about our village. I guess it is because I think about it so much."

Today I am 53 or 54 years old and I still dream about our village. I guess it is because I think about it so much. Sometimes I wonder why our lives are so messed up today when we were so happy once. Now I sit on committees—the housing committee, the school committee, and attend L.I.D. meetings. Indians are always fighting and struggling to let people know we are human beings. Sometimes I feel like giving up but I know I never will. My oldest son was just 11 years old when my husband died and I wondered what in the world I was going to do with five children to raise by myself. But I didn't give up and now all of my children are grown and leaving home.

Sometimes I wonder if the government does not want us to learn on our own. Indians have always been treated as second class citizens by government as well as some non-Indian people. Perhaps they are afraid that if Indians raise too high they will not remain second class citizens. Knowledge has always been withheld from us. Today the government says they know we have every right to live the way we want but still they are withholding things from us so that we will remain under their thumb. Are they afraid that if Indians are allowed to go free that we will take all of the jobs or that some other disaster will occur. We are human beings, we cannot do those things to other human beings. Sometimes I wish I was well educated enough to go to Ottawa and tell those government people exactly what I think. Somehow the people of this country must be made to realize that Indian people are their equals.

There is hope. Our children are going to school and they have opportunities we never had. I believe that is why my brother Robert is fighting for what is right for his people today.

Because I married a white man, I am better able to understand the white lifestyle. White people can raise chickens and grow gardens. Indians hunt and gather their food. Not many Indians know how to grow a garden. We have learned a lot from white people and I believe white people have learned a lot from Indian people.

Although I enjoyed living in the village and many times I wish we had never been separated from that way of life; I like Mayo and I think it is a peaceful town. I don't understand people who come here and complain about it.

GORDON I. CAMERON C.M.
(Born January 22, 1900)
by Ione Christensen, May 1988

In 1925 Keno City was a booming mining town with all the inherent problems associated with isolated towns of that day. The R.C.M.P., were endeavouring to do their duty by keeping law and order. Evidence, however, is difficult to get when everyone knows who you are. A request for assistance went to Dawson City Headquarters and from there to Vancouver, B.C. This resulted in a young constable by the name of Gordon Cameron being given a plain clothes assignment to investigate alleged illegal sales of liquor by persons in the towns of Mayo and Keno City. On May 27, 1925, he left Vancouver on the S.S. *Princess Mary* for the Yukon Territory.

Constable Cameron arrived in Whitehorse by train on June 1, 1925 and immediately boarded the steamer *Casca.* He arrived in Stewart City on June 3, and transferred to the S.S. *Canadian.* Cameron spent each day at the rail of the steamers watching the river and the vast land roll by. Little did he know then that this would be home, that he would know these rivers and its people as well as he knew the streets in his home of Three Rivers, Quebec. But that day in June, 1925, as the boat pulled into Mayo, it was one big exciting adventure.

The boat was met by "Spot Cash" (Jim) Breaden. He was to take the new workers up to the mine and his old Ford, which was loaded with live chickens, didn't leave much room for passengers. The ride was a memorable one, one mud hole after another and it was the passengers who did the pushing.

Constable Gordon I. Cameron, Keno Detachment, 1927. *Ione Christensen Collection*

Arriving in Keno City, Cameron registered at the hotel of Mr. Yamasaki and wife Tome. Prostitution and bootlegging were the heinous crimes he had been sent to uncover and gather evidence for prosecutions. Either he was very good or there was little evidence for he only worked for Treadwell Yukon Corporation from June 7 to June 29. On that date he broke cover and proceeded with charges. Complaints were alleged against a number of hotels and people which resulted in their court appearance on July 8 at Keno City. There were a number of appeals, with appeal court hearings set at both Mayo and Keno City, during August of 1925. The Honourable Justice C.D. MacCaulay presided, with Mr. C.E. MacLeod for the Crown and Mr. George Black for the defence in all cases.

There is a note on Cameron in the August 8, 1925, edition of the *Mayo-Keno Bulletin.* It is written by the editor and it is obvious Constable Cameron was not one of Mrs. Fotheringham's favourite persons:

'THIS COULD HAPPEN ONLY IN YUKON'

After the Keno Court was over, the editor of the *Bulletin,* who had been attending the proceedings, hastened to book a passage in the only car going down to Mayo that

> day. It was a four passenger car and the three other passengers were H.G. Blankman, Chas. McLeod and Mrs. McLeod. When that "bright young sprig of the law," Chas. McLeod, learned the identity of the fourth passenger he rushed to the barracks and requested Constable Cameron, the informer from Vancouver, to book his passage, and used the "prestige of their high calling" to induce the jitney people to turn the editor down and give the "Mountie" her seat in the car, although this husky young fellow [Cameron] could have stood the jolting in the big truck much better than an old lady of nearly three score years. However, we should worry, considering driver Ed Kimbel and Bobbie Fisher more decent and better company than the "swell-head nit-wits" who formed what they called the "COURT PARTY." Heaven save the mark!
>
> Just the same, if those sprigs of law and disorder are going to pre-empt all the public conveyances, they had better leave their families at home, and that's that.

Staff Sgt. W.J.D. Dempster was in Mayo at the time and Cpl. Thurgood and Constable Schouard were in Keno City. Cameron was transferred to the Dawson City detachment on August 4, 1925, via the S.S. *Keno*. He was back in Keno City for a short stay in September of 1927. By now he was engaged to Martha Ballentine of Dawson City. Her family were living in Mayo at that time and her father, Dave, was hauling ore from the mines to Mayo.

It was from Mayo that Gordon Cameron was assigned to the first air search to be undertaken in the Yukon. On November 25, 1927, the aircraft *Queen of the Yukon* was reported overdue on a flight from Whitehorse to Mayo. Staff Sgt. W.J.D. Dempster ordered Cameron to leave Mayo with a team of four dogs and search for the downed plane. He left Mayo at 8:00 a.m. on November 26, crossing the newly frozen Stewart River and followed the unbroken winter trail. The going was very difficult but early in the afternoon he met the pilot Andy Cruikshank and passenger A. McGinness walking out. The plane had been forced down the previous evening with a frozen oil line which had broken. Both men were in good condition but very happy to have a warm place for the night. After a good rest they returned to Mayo on November 27th.

On November 29, 1927, Cameron was ordered to return to Dawson City. The weather was very cold with the temperature in the low 40's F. He and Constable Jones were to take the police team of horses as far as Clear Creek where they would be met by the Dawson team.

The first day took them to Stewart Crossing where –60°F kept them cabin bound for the next two days. With a moderation to –40°F they travelled on and met the Dawson patrol. Here "Cam" had a pleasant surprise. Martha was with them. Again the cold weather closed in and it was December 9th before they arrived in Dawson City.

When Gordon was later stationed in Fort Selkirk, Mayo was the closest hospital and anyone needing medical attention would go from Selkirk to Minto, catch the overland stage and then two days into Mayo. From 1935 to 1949 both his winter and summer patrols took him to Mayo.

In 1949 the Camerons moved from Fort Selkirk to Whitehorse where Cam traded in the river boat and dog teams for the automobile and from then on all of his Mayo visits were by car.

Martha Cameron passed away on May 29, 1990.

ALICE CANTIN

Written by G. Jean Gordon for the Whitehorse Star, circa 1954

In every country, city and town there are names of those who make the news, but it is not always these people who are most deserving of the credit for making a town or locality what it has been or what it is. Behind the scenes are those who are never heard of who have carried the load and not been recognized. Here it is hoped you will meet some of those behind the scenes as well as those who made the news. Primarily they are men and women who have pioneered the Yukon and who have resided in the Mayo district for many years.

William and Alice Portlock posed outside their cabin on Haggart Creek in 1916, surrounded by their abundant garden *F. Portlock Collection*

Near the east end of Fourth Street in Mayo is a neat white house with a small attractive flower garden in front which in summer abounds with a beautiful assortment of pansies. At the rear is a tidy vegetable garden with each product in its own plot, and not a weed in sight. In the house live Mr. and Mrs. Frank Cantin.

Mrs. Cantin is the former Mrs. William Portlock. Her first husband came to Yukon with his boyhood friend, William Abbott, in 1898 and after a short time in the Dawson area came to the Mayo district where the two friends began mining on Haggart Creek. During a visit to his home in Preston, Lancashire, England, Mr. Portlock met Miss Alice Stevenson who journeyed from there to share with him the lot of a Yukon miner. Mrs. Cantin arrived in Dawson in late June of 1914 where Mr. Portlock was on hand to greet her. They were married at St. Paul's Church by Rev. Davies. Quoting from the Dawson News of Friday, July 3rd, 1914, the paper states, "The bride was charmingly gowned in grey silk, and was attended by Mrs. George Cunningham of Mayo, who wore a beautiful white silk dress. Mr. Abbott, the groom's partner, supported the groom."

Following the service a reception was held and the couple left on the Steamer *Vidette* for Mayo on July 4th. Arriving in Mayo, which at that time was only a settlement of a dozen or more houses, they stayed at the hotel operated by George Cunningham. The following day they began their long journey to Haggart Creek where Mr. Portlock's mining property was situated. Sam Blackmore operated the only means of getting to the [log] "jam" at the McQuesten River. This was a horse and light buggy. Beyond that point it was a case of walking. All this was strange and new as well as fascinating to a girl newly arrived from England.

The first winter on the creek was one of strange new ways and an entirely new life. Mrs. Cantin looks back now and says it seems as though it were only yesterday that she went for a walk up the creek to visit one of their neighbours. The frost on the trees, the snow on the ground, the crisp cool air made her feel that Paradise could not possibly be any better.

At the time of her arrival on Haggart Creek Portlock and Abbott employed 12 men in their mining operations. Among these men were such old-timers as John V. Smith, Charles Tresidder, Joe Morgan, "Shorty" Carter, and a man known as Champagne Johnson. Other miners on the creek at that time were Frank, Joe and Louis Cantin, John Stewart, Maynard, Gill and Jahnke, and Clarence Kinsey.

Robert Fisher was working his "Olive" claim on a nearby hill.

In all the years she has resided in the Mayo district, Mrs. Cantin has only seen one moose and that was during her first summer on the creek. While helping Mrs. Abbott who cooked for the crew, Mrs. Cantin was looking out over the countryside when she noticed a large animal coming down off the hill across the creek. When she drew Mrs. Abbott's attention to it, Mrs. Abbott cried, "It's a moose, a moose!" and dashed off to where the crew were at work to tell the men. Champagne Johnson took his rifle and was soon hot on the heels of the moose. He returned late in the evening but had not been lucky enough to bring down the animal.

In the fall of 1915 Mr. and Mrs. Portlock started for Mayo on foot. A party also going to town passed them and took word to Sam Blackmore who met [them] with a horse and buggy to bring them the rest of the way in to town. Alice took a steamer from Mayo to Dawson where there was a doctor and hospital. In November Thomas Portlock was born and spent his first three months in Dawson. Mr. Portlock was desirous of his wife's well-being and urged her to stay in Dawson for the winter rather than join him on the creek with a small baby but Haggart Creek was home and the mother and son came by overland stage in March from Dawson to Mayo in weather which was sixty below and colder. Mr. Portlock met them in Mayo. The only means of transportation to the creek was by dog team and after a trip by stage to Minto Bridge, Jack Maynard and his team carried the precious cargo over the winter trail with the proud father walking behind. The forty miles to Haggart Creek were covered in one day with only a stop at the McQuesten for lunch. The party arrived at Haggart Creek about eight that night.

Seven happy years followed the arrival of this little family on the creek. Young Tommy grew from a baby to a sturdy boy. The days were filled with work and pleasure. One picnic on a summer Sunday is still vivid in the memory of Mrs. Cantin. Everybody took the day off, crew and all, to go for a picnic fishing trip to a nearby stream which was always well stocked with grayling. There seemed no need to carry other eatables except bread and a frying pan in which to cook all the fish the party would catch. Arriving in full force, prepared with fishing lines and mouths watering, all had visions of the good feed ahead. It would seem that the fish had also taken the day off, for not a fish did anyone catch! These wily grayling were very plentiful, darting from bank to bank, but bite they would not.

In [approximately 1922] Mr. Portlock was in ill health and made a trip to Dawson to visit the doctor there. On his return in the fall the family moved from Haggart Creek to Mayo. Mr. Portlock's health failed steadily until his death on November 17, 1924. Mrs. Cantin was his faithful nurse during those trying years. The *Mayo-Keno Bulletin* published by Mrs. Fotheringham bearing the dateline November 18, 1924, carries this tribute, "Mr. Portlock endured much suffering throughout his long illness, but was always bright and cheerful, and most appreciative of the kindness and care lavished upon him by his devoted wife." Pall-bearers for Mr. Portlock's funeral on November 21, 1924, were John V. Smith, Frank Cantin, Robert Fisher, George Miller, Frank Whitney, Richard O'Loane, John Parker, and Sam Blackmore.

Following her husband's passing, Alice took in sewing and young Tommy attended school in Mayo. When it came time for him to enter high school he and his mother went to Vancouver. In that city on June 27, 1931, Mrs. Alice Portlock was united in marriage to Frank Cantin of Mayo at St. George's Church by Rev. D. Owen. An excerpt from the *Dawson News* of July 14, 1931, reads as follows: "The bride was attended by Mrs. Axel Erickson, and Miss Marion Vigner was bridesmaid. Axel Erickson supported the groom. The bride was becomingly gowned in a blue ensemble, with hat to match. She carried a bouquet of roses and lilies-of-the-valley." Following the ceremony a reception was held at the home of Mr. and Mrs. G. Vigner and the bride and groom left that evening for a wedding trip to Montreal, returning to Mayo in late August.

Commenting on the friendly Yukon spirit, Mrs. Cantin recalls an incident which is indicative of the people of the Yukon. Before her departure from England, Mr. Portlock asked her to contact the sisters of a Dan Sanderson whom he had heard came from their district in England. She looked up these people and was given some family silver to deliver to Mr. Sanderson. At Dawson the couple were unable to contact him so left the heirlooms at the local post office. It was not until her trip to Dawson the following year that Mrs. Cantin was able to meet Mr. Sanderson.

The years have been kind to Mrs. Cantin who will be 74 years old on January 25th [1954]. In spite of a slight limp which she has had since childhood, she still maintains her household better than many younger women and goes to church as regularly as the weather permits. She has always taken an active part in the work connected with her church and is still on call to help anyone who has a sewing problem.

Two sisters still live in England and wish for Mrs. Cantin to come "over" if only for a visit but as she says, "Mayo is home after all these years and I guess I'll always feel that way." Her son, Tommy, and his wife, Frances, and grandson, Franklin, [moved to] Skagway, Alaska [in 1950].

CHARLES H. "CHAPPY" CHAPMAN

by "Chappy" Chapman, February 1983

I initially arrived in Mayo in the summer of 1928 on the paddle wheel steamer *Keno*. Previously, I had completed five years of service in the R.N.W.M. Police and was stationed in Whitehorse and Ross River. After my five year term with the R.N.W.M.P. expired I took my discharge. I then worked for a prospecting outfit, did some diamond drilling at the Carlisle Mines near the Copper King and at the Pueblo and War Eagle Mines [Whitehorse], as well as placer exploration work on O'Brien Bar on the Teslin River.

My first job in Mayo was with Bob Palmer driving a two ton truck to haul supplies from the steamboat dock to his store and hotel in Keno. After about three or four trips over very bad roads, I was offered a job by C.A. "Kippy" Boerner with the Northern Commercial Co. Ltd., in Mayo.

I accepted the job and worked for the Northern Commercial store for about twenty years. "Kippy" Boerner had been with the N.C. Co., for many years, first as a steamboat captain or pilot with the company (which was known then as the Alaska Commercial Co.). One of the last paddle wheelers "Kippy" operated was the *Norcum*, which is now rotting on the bank of the Yukon River at Hootalinqua. At that time, the N.C. Co., had some 34 branches in Alaska and the Yukon. The N.C. Co., was started before the United States bought Alaska from the Russians.

The work at the store in Mayo consisted of looking after the stock that came in on the boats and in putting up orders to go out to the mines and miners. At that time the N.C. Co., had a contract to supply all of the groceries to the mining company on a cost plus 15% basis. I also had to haul all the merchandise from the dock into the store and get it stowed in the basement for the winter.

Mayo was a very pleasant place to live. I had a small apartment over the store. There was a board tennis court, just back of the store, which was a very popular activity and was used from early in May until late in the fall. I also located a small lake five miles from Mayo on the Elsa-Keno road, and with the help of Alex Maclennan I cut about half a mile of road into the lake and it was used for many years as a swimming hole.

After about two years in Mayo, I was transferred to Keno to operate the store there as the manager Dick O'Loane was retiring. The work in Keno was quite different. There was no electric plant in Keno at that time and every day I had to clean and fill about eight Coleman gasoline lanterns to light the store. I was required to order all the supplies for the store from the Mayo store. Looking after a grocery store in those days was very different from today. There was no refrigeration. Butter came in barrels of brine. Hams and bacon had to be hung out in the back and the wrappings removed to prevent moulding. Gasoline and coal oil came in wooden cases in two five gallon cans. The cans were put to many uses after they had been emptied, as water buckets, slop pails, etc. There was no water or sewer system; water was delivered every day or so in buckets. There was little entertainment in Keno and I was pleased when the announcement came that the Keno store was to be closed and I was to return to Mayo.

After my return to Mayo, I bought a new Ford four door sedan Model A for $600. Roads by that time had been greatly improved and the N.C. Co., had taken on the Ford agency.

I continued to work in the Mayo store until 1933 when the Depression was beginning to affect the silver mines and I was put on half-time. While working half-time I used to go to work at twelve noon so that "Kippy" Boerner could go to lunch. One day I was surprised to meet Mr. Volney Richmond, president and general manager of the Company, who used to make regular visits to most of the branches every year. His first words to me were "Where have you been?"

I explained that I was on half-time, and he informed both "Kippy" and I that no one worked half-time for the N.C. Co. Later that day he asked me if I had ever kept books and said they were looking for someone to take over the books in Dawson. As I had completed a correspondence course with the LaSalle Extension University in higher accountancy, I was transferred to Dawson. So, I sold my car and in a few days took the steamer to Dawson.

In Dawson I did the bookkeeping and eventually was appointed manager. While in Dawson I met and married Mattie Barton. We were very active in community affairs.

In the summer of 1951, my wife and our two children moved back to Mayo. We travelled by White Pass's Fairchild plane. We took our clothing and a small refrigerator. In Mayo we bought a house from Mrs. Kinsey and for the first time we had a wonderful garden with many flowers and lots of vegetables. The house was a two-storey building and was situated on the road to the old hospital.

"...and he informed both 'Kippy' and I that no one worked half-time for the N.C. Co."

Mayo was quite a busy place at that time. The road to Whitehorse had been completed and the ore from the mines was being trucked to Whitehorse instead of being piled on the riverbank in sacks. The construction of the dam on the Mayo River was well under way. I began to sell insurance and also acted as Notary Public. The roads to Keno, Elsa and Calumet had been improved and the trip to Keno or Elsa only took approximately two hours.

The old tennis court which had been located on the Army Signals property was torn up and the Anglican Church built a new one close to the church property. The church had also built a large parish hall [Wareham Hall] and the Kunze brothers were operating a 16mm picture show. The hall had been badly built and the walls were starting to lean out and so they had been braced with timbers for safety purposes. The acoustics in the hall were very bad and as there were severe echoes from the walls it was difficult to understand the dialogue.

Soon after we arrived in Mayo, Ed and Bon Kunze, closed down their picture show and my wife, Mattie, and I took it over. We rearranged the speakers, hung tarps on the walls and made arrangements to get better films. The hall was in severe financial difficulties when we took it over. Soon we started to get better attendance at the shows and were able to pay off the debts against the hall.

The new tennis court never did work as it was constructed with lumber that was too light.

After being in Mayo a little more than a year, I took over Bud Fisher's garage and operated both the garage and my insurance business from the garage office. I hired Jack Besplug as mechanic and Jim Gibb as his assistant. We also bought a two-ton tanker from Ryder's fuel service in Whitehorse and started delivering stove oil. Fuel oil was just starting to replace wood as fuel.

Our son Bill completed a welding course with the B.C. Institute of Technology and we accepted a small welding contract on the bridge at Stewart Crossing. Business was very good for some time due to the building of the dam and the completion of the road from Whitehorse; and later the completion of the road to Dawson from Stewart Crossing. We also had a contract with United Keno Hill Mines to service their trucks which carried ore to Whitehorse.

With the improvement of the road to Whitehorse the ore trucks seldom stopped in Mayo. The White Pass also took over the delivery of fuel oil to all of the large customers. The Government of the Yukon Territory set up their own insurance plan and so we lost that business also.

By this time, our daughter Betty was teaching school in Whitehorse. Our son Bill, and his wife Colleen, had moved to Vegreville, [Alberta] where he had taken an upgrading course and apprenticed in a T.V. repair shop. After completing training as a T.V. repair man, Bill went to work for Alberta Government Telephones on the Channel 13 Access. He has now worked for A.G.T. for over seventeen years. Betty is now a librarian at a high school in Spruce Grove, Alberta, and her husband Tony Baumber is principal of a Spruce Grove high school.

After leaving Mayo I worked as an accountant in the office of the Transport Division of Cassiar Asbestos; United Keno Hill Mines; and later on I was appointed manager for Northern Metallic Sales Ltd., in Watson Lake. We lived in Watson Lake for seventeen years.

We now live in Whitehorse where I started my life in the Yukon 62 years ago. In a short article like this it would be impossible to mention the many friends we have made during the years we have lived in the Yukon, but we always enjoy the many visitors we meet and those who call on us from time to time.

PHILIP E. CHURCHWARD

by Philip Churchward, April 1986

I remember arriving at Mayo one June evening in 1922. As always, a goodly number of people were assembled at the wharf. The arrival of a steamboat was always a notable event; it brought new people, or old friends returning—after all the steamboat was the important connection with the outside. Without the steamboat Mayo would never have existed.

I had just graduated from high school; having attended grade school in Dawson, where I was born. My father had a sheet metal shop there. In 1918, with Dawson's population declining my father gave up the shop there, temporarily storing all of his tools and moved to Tacoma, Washington, where, he thought, I could attend what he expected to be a better high school (a rather doubtful assumption).

From 1918 to 1922 he worked for wages—something he never liked to do. So when I graduated from high school in 1922 he decided to start over again in the newer mining district of Mayo which had optimistically great expectations at that time. When we arrived he was already there, with second hand lumber from an old Dawson City hotel ready to build his new shop. In the meantime we would live in a tent. The new shop was built—a two-storey building with living quarters on the second floor. Not luxurious, but after all this was temporary—Mayo was sure to grow and we would prosper.

The shop was located on the edge of town—about a block away from everything else at that time. I remember building a crude sidewalk of poles and slabs to allow people to walk that far in relative comfort. Still farther away a young Englishman [Samuel Tunnicliffe] was building a two-storey structure which he hoped to run as a hotel. A large nicely painted sign from the old hotel in Dawson was found in the scrap lumber we used to build the sheet metal shop. We were glad to donate the Broadway Hotel sign, and the name of the new hotel was changed accordingly.

Both the shop and the Englishman's hotel were completed and opened in late summer. The shop was moderately successful; there was always a demand for stoves, stove pipe and other sheet metal work. The hotel? I believe it had one paying guest for a night or two; that's all I can remember.

I stayed in Mayo a little over two years. During this time I helped build the shop, helped work in the shop a little, worked briefly at Wernecke mining camp on Keno Hill, and made some small change developing and printing Kodak films and selling a few photographs.

It was an experience I am glad to have had; two years of doing without; two years of high expectations. There was no electricity, except for small generators, at the hotel for instance. No water piped in. The water system was primitive but complex. There was the river—even in winter accessible through a hole in the ice. The water was carried—10 gallons at a time from the riverbank to the shop. Then in summer there was rain—which we collected in a tank as it drained off the roof, this was good soft water—much prized for washing. In winter melted snow gave us good soft water. For drinking water we purchased well water which was delivered by a good natured French Canadian whose name I forget—he was our most important public utility.

There was Binet's Hotel, two general stores (Taylor & Drury and the N.C. Co.), a sheet metal works, a shoe repairman, and a blacksmith. While I was there the Church of England built a church. During this time a hospital was also built. For amusement there were weekly dances. Sometimes outdoor skating in the wintertime. As far as I know nobody had skis. And of course the steamboats would arrive in the summer.

In the summertime roads were so bad that travel by the river was by far the best. Roads were actually better in winter. Thus ore from Keno Hill was brought down to Mayo in the wintertime and stacked up on the riverbank for shipping by boat the following summer.

A change in transportation took place in this period. Prior to 1922, freight from the mines was by horse and sleigh—a four horse team could haul about 15-20 tons in winter. The Treadwell Yukon Co. brought in 10-ton Holt track machines which hauled four times as much at lower cost. By 1923 horse freighting from the mines was finished. The arrival of two of these 10-ton Holt track vehicles was a milestone in overland winter transportation and created a local sensation. Of course everything was changed after the construction of comparatively good roads.

Aviation was beginning at this time in Alaska but was nonexistent in Mayo until about 1927–1928.

By 1924 my parents were able to send me outside to attend college, and I left Mayo, not to see it again until 40 years later in 1964.

SHANON (MCDONALD) COOPER

by Shanon Cooper

I arrived in Mayo from Vancouver on September 28, 1958 with my mother Marguerita. My father, Bill, had already been working as a miner at Elsa for a year.

We lived in Flat Creek and my first recollection of the place was that hardly anybody spoke English, but everybody was very kind and everybody liked kids.

After three years of school in Elsa, my parents transferred me to Christ the King Convent in Whitehorse. I grew up moving to a number of schools in and out of the Yukon (returning home at Christmas and in the summer), because Elsa was a small community without a high school. In 1968 I graduated from F.H. Collins in Whitehorse.

After earning a Bachelor's Degree at the University of British Columbia, I received my first teaching job in Mayo in 1972. The salary was $12,000 per year. My accommodation was an 8x10 foot storage room in the old teacherage. The only furniture in the room was an old bed which the principal had placed there prior to my arrival. I don't recall the room having any heat, it was the coldest room in the place. When I got into the bed the mattress sank down until I touched the floor.

I have many fond memories of those first weeks in Mayo. My Grade 2 and 3 class numbered 28. I remember I had to borrow $40 to buy food until I got my first pay cheque. Most of all, I remember the great feeling of joy at being back at home.

INSPECTOR WILLIAM JOHN DUNCAN "JACK" DEMPSTER

Scarlet & Gold, 1964 Edition

Inspector Dempster was born in Wales on October 21, 1876. He engaged in the N.W.M.P. as a Constable, Regimental Number 3193, on September 9, 1897. He was promoted to the rank of Corporal on January 1, 1911; to Sergeant on May 1, 1911, and to Staff Sergeant on July 1, 1918. He was appointed an Inspector on March 1, 1931. Inspector Dempster retired on September 1, 1943.

During his service with the Force, Inspector Dempster was stationed at: Regina, 1897; Lethbridge, 1897–1898; Dalton Trail, 1898; Bennett, 1898–1899; Halfway, 1899–1905; Bonanza and Caribou, 1905; McQuesten, 1905–1906; Stewart River, 1906; Forty Mile, 1906–1912; Dawson, 1913; White River, 1913; Rampart House, 1914–1917; Dawson, 1917–1921; Mayo, 1922–1931; Dawson, 1931–1934.

He spent the winters of 1907–08, and 1917–18, on the Dawson–Fort McPherson Patrol.

In the summer of 1921, Sergeant Dempster visited Mayo by canoe to ascertain the extent of increased activity caused by the Keno Hill developments. Between Christmas and New Year he moved, by dog team, from Dawson to Mayo and sent a detachment of five men to Keno. He spent nine years in Mayo, his lengthiest single posting. He assisted in organizing and administering the building of the Mayo Hospital and was Secretary of the Mayo Hospital Board from September 1922–June 1925. He held the appointment of J.P. and Coroner in Mayo from December 1921 to 1931.

Sgt. (later Inspector) W. J. D. "Jack" Dempster, R.C.M.P., Mayo, 1921. *Ralph Rogers Collection*

On September 8, 1926, Inspector Dempster married Sarah Catherine Smith of Sydney, Nova Scotia (she was Matron at the Mayo Hospital from 1924–1926). The Dempsters had two children. Inspector Dempster died in Vancouver on October 25, 1946.

A new highway from Dawson to Aklavik was officially named the Dempster Highway on September 3, 1963, in honour of Inspector Dempster.

The saga of the McPherson–Dawson Patrol of 1911 is one of the most memorable and dramatic events in Canadian history. A patrol led by Inspector Fitzgerald left Fort McPherson in late December 1910 with mail and documents for Dawson. It became evident to Superintendent Snyder, the officer in charge at Dawson, that the Fitzgerald Patrol was overdue, and he ordered Corporal Dempster to choose his own companions and leave to search the probable trail for the lost patrol.

Leaving Dawson on February 38, 1911, the Dempster search party proceeded eastward under severe and hazardous conditions, and on March 21 found the bodies of two of the members of the Fitzgerald Patrol. The following day the bodies of the remaining members of the patrol were found.

Corporal Dempster's diary of the search party which he led and the tragic happenings to the gallant patrol have left a milestone in northern police history that has never been equalled.

DORIS M. EWING

by Doris M. "Dorrie" Ewing

I am the seventh daughter of William and Erie Batty, one of eleven children, born at Assiniboia, Saskatchewan and was raised on a mixed farm. I attended senior high school in Assiniboia and Moose Jaw, and took my teachers' training in Moose Jaw.

I taught in country schools for several years and in 1949 ventured into the Yukon, following my sisters, Ruth and Margaret. In the Yukon I taught school in Destruction Bay, Haines Junction, and Mayo.

In Mayo I also helped sister Ruth in her cafe and dress shop, cooked at a mining camp on Haggart Creek and worked several years in the Mining Recorder's Office. In the past I liked to curl, play badminton, and take part in women's organizations such as I.O.D.E. and the Anglican W.A.

I married George Smashnuk in 1953 and we later divorced.

In 1961 I married M.H. "Harry" Ewing and we had 24 good years together. I often went to the sawmill sites with him and also to Minto Creek and Field Creek where Harry and his son John did some placer mining. I enjoyed the excursions we made to Harry's different mineral claims to bring back ore for assaying. I very much enjoyed my life with Harry and his family.

[*"Dorrie" Ewing passed away November 10, 1988.*]

MATTHEW HENRY EWING

by Doris M. "Dorrie" Ewing

Matthew Henry Ewing was born at Bedford, Quebec, on May 25, 1903; son of Samuel and Esther Ewing, one of ten children, and raised on a farm. Matthew was known as "Dooley" in all his growing years and it was not until he arrived in Dawson in the spring of 1926 that he acquired the name "Harry." During Harry's school years in Bedford; he worked on the farm, hunted muskrat and skunks for spending money, and worked in the lime quarries during the summer holidays. After graduating from high school he joined the Canadian Army Permanent Force, Royal Canadian Corp. of Signals and took his training at Camp Borden, Ontario, starting in October of 1924.

Harry very much enjoyed his first posting—four years in Dawson City, where he married Elizabeth Belle Grant in September of 1928. Elizabeth passed away in February, 1959. She was a niece of Robert Henderson, co-discoverer of gold on Bonanza Creek. Elizabeth and Harry had an active life together; playing cards, dancing and curling as Dawson was a social town.

Their first daughter, Esther Mary (Mrs. C. E. Fisher Jr.) was born in Dawson in 1929 and now lives

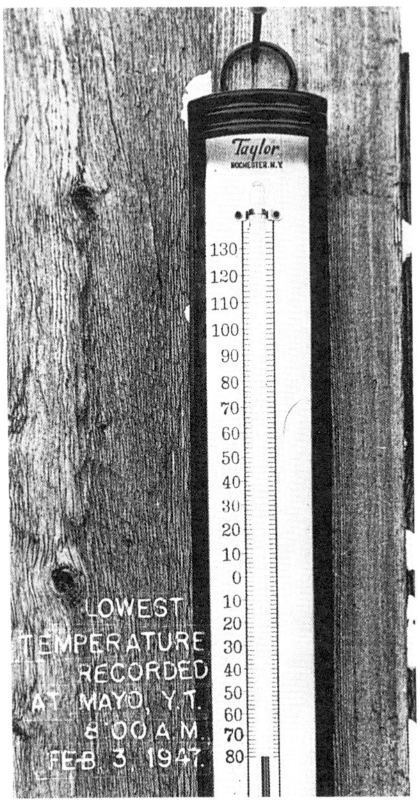

Record low of minus 80°F, recorded February 3, 1947. *G. A. McIntyre photograph*

Harry Ewing's first raft of logs from upstream Stewart River, circa 1948. Mayo is in the background. *G. A. McIntyre photograph, M. Rich Collection*

in Whitehorse. Son, John Grant was born in 1933 in St. Lambert, Quebec, and lives in Mayo. Their second daughter Elizabeth Suzanne was born in 1949 in Dawson because Mayo had no hospital at that time. She and her family moved from Mayo to Whitehorse in 1985.

After the posting in Dawson; Harry, Elizabeth and "Jo" (Esther Mary) moved to Camp Borden for a year; where Harry took further training. Then they moved to St. Hubert Airport in Montreal for two years. It is at this time that Harry had radio contact with engineers of the R100 dirigible that docked at St. Hubert in 1930. Then Harry, Elizabeth, Jo, and John moved to Fort Smith, N.W.T., in 1936 and then on to Ft. Resolution, N.W.T. in 1939; enjoying the Army Signal Corp. postings. The R.C.S.C. was responsible for all northern communications and their weather reports were invaluable to the southern regions of Canada. In late 1939 they went to Barryfield (Kingston, Ontario) and on to Camp Borden in 1941.

Harry and Dorrie Ewing.

In 1942, the family moved to Mayo where he ran the station until his retirement from the Royal Canadian Signal Corp. in 1947. Harry had his Meteorological Observer's Certificate and was in charge of Mayo Station when the record low of minus 80°F was recorded on February 3, 1947. Harry took the thermometer reading and Gordon McIntyre took the picture for a permanent record of the low temperature.

Harry also held an electricians trade certificate and enjoyed that phase of his life.

In 1947 they moved to Bedford, Quebec where Harry tried different work, one being in the needle factory, but that wasn't for him. So, in the spring of 1949, they bought a truck and moved back to Mayo where he lived until his death in September 1985.

Harry's family, back in Mayo, went into the firewood business and later into the sawmill business. Harry had taken sawyers training from an old time sawyer, Frank Cantin, so he did the sawing for the family company, Ewing Transport, for many years until at age 78 he relinquished his sawyers position to grandson Dick. Harry and his son John ran the Ewing Transport company for many years, having the sawmill, supplying mining timber, and operating cats and trucks. Harry was always interested in the mining industry, having staked mineral claims in the Dawson, Carmacks, and Mayo areas, with special interest in Galena and Keno Hills and Mount Haldane in the Mayo Mining District. He was also interested in placer mining and in 1962 he and son John did some placer mining on Minto Creek. Harry subscribed to the Northern Miner newspaper for over 50 years, keeping abreast of mining concerns.

In the fall of 1961, Harry married Doris Batty and they had a good life together. They built a retirement home in Bedford, Quebec, but they never did retire there as Harry spent his retirement years working in his work shop in Mayo.

Harry was acquainted with many noted persons such as: Sgt. Dempster, Pierre Berton's father, Lew Green who wrote "*The Gold Hustlers*," Dr. A. Aho, and Dr. Bostock to name a few.

Harry was a much loved patriarch.

[*Harry Ewing passed away in September 1985 and since the writing of this article Harry's son John Grant Ewing passed away on July 30, 1989.*]

DAVE & MARIE FOTHERINGHAM

The Historical Society has collected information about Dave and Marie Fotheringham from various sources. Little is known from local records about the Fotheringham's other than that they moved to Mayo in the 1920's and Marie published the *Mayo-Keno Bulletin*. Dave Fotheringham ran a small steamboat on the Stewart River.

Mrs. Lillian Cheatham, a long-time friend of Marie Fotheringham's wrote the following:

My memories of Marie are happy ones. I stayed with her frequently when I was in Dawson. If I remember correctly she came into Dawson as a reporter for a Toronto newspaper. Dave Fotheringham was a N.W.M.P., and he drove her around. They were later married and he had to leave the force to marry her. He went overseas with the Boyle bunch and came back after the war.

Marie was a very kind person and a true Christian. Not the kind that goes to church every Sunday and prays for others sins, but a person who believed in the Bible and its teachings. She was always helping someone in trouble and the working man. She did not like politicians and felt they were crooked. She fought for women's rights and against liquor. I feel she was fifty years ahead of her time.

Marie was a very brilliant woman. I have a book of poetry she wrote which was published in 1895. I also have most of her other poems. I do not remember when I first met her. It seems like she was always there so I must have been very young when we met.

Linda Johnson, in her paper *History of the First Through Eighteenth Wholly Elected Yukon Councils 1909–1961*, wrote the following about the Fotheringhams:

"Marie Fotheringham née Joussaye was married to D. H. Fotheringham, a former member of the North West Mounted Police in Yukon and veteran of the Boer War. They were involved in a dispute with federal lands officials in Dawson over a homestead they occupied at Indian River circa 1908–10. Marie Fotheringham sent a long and strongly worded letter to Frank Oliver, then Minister of the Interior, complaining of the unjust actions of various Dawson officials when their application for the land was refused. She claimed to have 'had a great deal to do in the past, in helping to mold world opinion. Ask W.F. Maclean, M.P. of the Toronto World, or your Secretary of State, Mr. Charles Murphy, whether they think that if Marie Joussaye told this story to the Canadian people they would not help her fight her battles even as they helped her fight the battles of others who were oppressed and ill treated. These two men have stood side by side with me on the public platform and helped me to secure justice for the workingmen and women of Canada. . .' Apparently Mrs. Fotheringham brought a wealth of experience in public speaking and lobbying to this new campaign. P.A.C., R.G. 85, Vol. 1854, file 140379, Fotheringham to Oliver, February 28, 1910."

ELMER & BESS FRIESEN

by Elizabeth "Bess" Friesen, 1989

Elmer Friesen was born in Nipawin, Saskatchewan, in 1927 and his wife Bess was born Elizabeth Profeit in Mayo. Elmer came to Mayo in 1950 to work for United Keno Hill Mines at Elsa and then worked on the dam construction project at Mayo Lake for two years. It was while he was employed at the dam that he met Elizabeth "Bessie" or "Bess" Profeit. They were married in Edmonton in 1955.

Elmer and Bess followed construction through Alberta and British Columbia until 1963 when the call of the Yukon brought them back to the Mayo area. They partnered with Steve Mosich on a placer mining operation at Thunder Gulch for a year. Elmer and Bess then leased the Chateau Mayo Cafe for three years and then bought the Pelly River Lodge at Pelly Crossing. They retired from the Pelly River Lodge in 1974 and returned to Mayo where they bought an acreage just outside of town and built a home.

Elmer and Bess placer mined on Mayo Lake and also leased the Variety Store, which was located in the Chateau Mayo Hotel, for approximately three years. Elmer carried out numerous carpentry jobs during the winter months when he was not mining.

Both Elmer and Bess were very active in the Curling Club and they played on the mixed team that won the Yukon Championship three times. Elmer was also very active in the Dawson Branch of the Yukon Order of Pioneers, and Bess in the Pioneer Auxiliary. Bessie curls competitively with the senior ladies and has been to the Nationals twice. Bess has worked as enumerator and returning officer during federal and territorial elections for the past twenty odd years.

Elmer passed away in 1985 and Bess moved to Whitehorse in 1986.

THE GERMAINE FAMILY

As told by Paul Germaine to his niece Catherine Germaine, 1989

My father, Jermie, originally came from Fort Good Hope on the Mackenzie River. It was there that he met my mother, Sophia Christine, who had come from Aklavik with her adoptive parents Et'tsì ts'ò and Mädlin. Jermie and Sophia, and Et'tsì ts'ò and Mädlin came to Lansing with other people from Fort Good Hope.

My father had a younger brother. People said he died of shock, while he was very young, when another child pretended to threaten him with a knife.

The Jermie family lived at Lansing with the rest of the people from Fort Good Hope. They travelled (up) the Stewart River, Nadaleen River and Swan Lake areas to dry meat and gather food for winter. During the winter they caught white fish, grayling and trout in Ladue Lake.

From left: Mary St. Pierre, Mary (née Kendi) Hagar and Ellen Germaine at the Old Village in 1925. *Dave Hager Collection*

ABOVE, Matt Butijer, Taja (Mädlin), and Et'tsì ts'ò. *Herb Foster photograph, O'Neill Collection*

BELOW, Three generations. Front, from left: Sophia Christine Germaine, baby Catherine Germaine and her mother Ellen Germaine Olin. Archie Martin is standing at the rear. *Thomas McKay Collection*

Left to right: Sophia Christine Germaine (mother of Colley, Paul, Ellen, Frank and Joe Germaine), Mary St. Pierre, Frank Germaine, "Big" Dave Hager and Ellen Germaine. *Dave Hager Collection*

When St. Pierre, the head man at Lansing, and his wife Mezie died, their daughter Mary came to live with our family. When Kätini's husband, Philie Unëbel died, Kätini stayed with our family when we were at Rackla and Ladue and then she went to stay with Mike Tomoff's friend at Keno City. Kätini was Mary St. Pierre's sister.

My father, Jermie, died before sickness came to Lansing. After he died the family moved to Rackla River. My brothers and I trapped around that area. We built a house at Ladue Lake and our family lived there as well.

There were at least eight children in the Jermie family, three died very young. The ones who grew to adulthood were: Colley, Paul, Ellen, Frank and Joe.

[*Paul Germaine passed away July 1, 1990, at the age of 84.*]

WILF & G. JEAN GORDON

Interview & transcription by Lynette Bleiler, 1988

I was born in Harris, Saskatchewan in 1912. The family moved to Saskatoon for a short while and then we moved to North Battleford, as my father had a job working on the boiler at the insane asylum. He got a job there because he had been working for the Liberal Party and they told him that they would give him employment. We didn't stay there long before we moved to Edmonton for a short while and then to Grand Prairie, Alberta in 1919, where I was more or less raised.

When I was only 16 years old I homesteaded on the banks of the Wapiti River. You were supposed to be 18 to homestead. I didn't fully prove up my homestead and couldn't make any money, so in 1929 I went placer mining for an uncle of mine in Tulameen, B.C.

From Tulameen I progressed to Barkerville in 1931 and 1932 where I did some placer mining. However, mostly I was working for a transportation company packing horses. Crawford Transfer had their headquarters in Stewart B.C., so when they pulled out of Barkerville I had to go with them to get the money they owed me. It was there in Stewart that I met my wife, Jean Matheson. When work at the Big Missouri Mine became scarce a fellow by the name of Bill Scott, who had been driving taxi in Stewart, and myself decided to look for work in Atlin, B.C. We had heard there was lots of work there. We bought our tickets in Prince Rupert B.C. to come up the coast to Skagway during the winter of 1933–34.

It was 4 p.m. in the afternoon when we got off the White Pass and Yukon Route train in Carcross, Yukon and to our disappointment we ran into couple of fellows on the platform who had just come from Atlin. They told us there wasn't any work in Atlin.

Well we didn't have any money to go any further on the train, so I caught up with a fellow who had travelled up on the same boat to Skagway with us and asked him to take our baggage and packsacks to Whitehorse. We were in a dilemma, we couldn't go back through American territory without any money so we decided to walk to Whitehorse. So with the January temperature reading 45°F below, Bill Scott and I left Carcross around 4 p.m. in the afternoon and walked all night to Whitehorse arriving at 1 o'clock the next day. It was a wonder we didn't freeze to death. Bill Scott was dressed in a blue serge suit, overshoes and oxfords and because I had been working at the Big Missouri Mine I had on Bone Dry pants and coat with a sweater underneath and hob nail boots. Guess we were just a couple of lucky fools.

We spent our last $2 on a moose steak dinner at a Whitehorse restaurant and afterwards decided to look for our packsacks. Our packsacks were at the Regina Hotel owned by Mr. Erickson who asked where had we been. I told him we had walked from Carcross to Whitehorse, and now we're looking for a job. "Well," he said, "I don't think you will find any work, but don't let that worry you. You can stay here."

"But we haven't any money to stay here," I replied.

"Don't let that worry you," he said, "you'll find a job. You can't go back and you have to have a place to stay."

The next morning I got up early, Mr. Erickson was on shift and he said he knew of somebody who wanted some wood cut. Did I know anything about cutting wood? Well I had worked in the bush and cut lots of wood on the homestead, so he gave me the name of this fellow (I can't recall his name now). I went down and found him just as he was getting ready to go to the bush at the back of where the Whitehorse General Hospital is now. He wanted the wood cut in 4 foot lengths and split and piled in two cord piles. Well the trees were about 100 feet apart and there wasn't any way I could make wood piles that big so I just had to go back into Whitehorse. When I got back Mr. Erickson asked me how I made out. I told Mr. Erickson, "Well there's wood to be cut there but there's no way at $3 a cord that I could make my board at it."

"Well," he replied "I thought that's the way it would be, but I thought I'd let you go and have a look at it. However," he said, "I have a big wood pile at the back of the hotel and I have to have it split and piled. If you fellows want to help pay for your room and board you could to that."

Then Mr. Erickson suggested that we go and talk to a Mr. Phelps who was a partner in the overland stage going to Dawson City. He thought if we talked to Mr. Phelps we could probably get down to Dawson C.O.D.

"You'll find work there this summer because Y.C.G.C. are building some new dredges. Once you get to Dawson you can get credit, because you won't be able to get out of there until you pay for everything you get," said Mr. Erickson.

A couple of days later we went and had a talk with Mr. Phelps. He indicated yes they would take us into Dawson, C.O.D. The fare was $50 apiece, and we rode on top of a truck that had racks on it up to the top of the cab full of mail. There were 18 passengers on the top of that mail going into Dawson. I can't recall for sure but I think the travelling time from Whitehorse to Dawson City was 39 hours. We stopped at Braeburn for a hot meal, and after that when they stopped to gas up from the caches they had along the road they would hand us frozen egg sandwiches. I can't look at an egg sandwich to this day after eating all those frozen egg sandwiches. There was a place called The Junction near where Stewart Crossing is now. That's where you turned off to go to Mayo. We met some passengers there from Mayo who were going to Dawson. There was Sam Wood and his wife, Rose, with their small daughter who was about a year and a half old, and a Mrs. Sullivan and her daughter and small baby. They were loaded on the top of the truck on top of the mail too, for the journey to Dawson. Well, we had quite a trip and it amazes me to this day that we ever made it because the trees were hanging over the road loaded with snow. We didn't have any heat on top of those mail sacks. All we had were some old dog skin robes, fur robes they called them—most were made of dog skin or cow hide and that was all.

It was five or six in the morning when we got in to Dawson City. I stopped a fella going up the street and asked him where Bill and I might find a cabin or room to rent. He said, "Come along with me and we'll talk about it." His name was Angus McIntyre and he was going to the barn to feed his horses, as he was a stage driver. Well, we went along with him. He advised us what to do and where to go to get a cabin but Bill Scott said. "No, I'm not going to stay in a cabin, I'm going to stay in a hotel if I can get credit."

"You'll be able to get credit to stay in a hotel," said Angus McIntyre, "but, it will cost you about $15 a month and you could possibly get a meal ticket at the restaurant. They will probably give you one meal a day until you go to work." I wanted a cabin so he told me to see Charlie Tennant who had cabins for rent.

I had to wake this fella up and he said yes he had cabins to rent and he asked me what I could do and so on. I told him I had no money right to start with. So he said he would let me have a cabin for $10 a month and I would have to get my own wood. However he also said he needed someone who could drive a horse because he had cut some wood and wanted to haul it but he didn't know how to drive a horse. He decided he would rent a horse if I would drive it and I would get wood that way too.

He took me down to the N.C. Store when they opened about 9 a.m. and I got $10 worth of groceries on credit there. Then I went to work for Charlie Tennant hauling wood from Swede Creek until the end of March. Following that I went to work for the Y.C.G.C. (Yukon Consolidated Gold Corporation). They had their head office at Bear Creek so I had to walk from town out there quite a few times. It was about a 10-mile walk to check at their office to see if there was any work coming up and then I think it was around the 1st or 2nd of April 1934 when I went to work for them driving a team of horses. They were having a difficult time getting experienced teamsters, as most of their drivers were getting too old to drive any more. They paid $150 a month, which was top wages at that time. I thought I was going to have to drive four or six horses; however, as it turned out, I was driving one single team.

Y.C.G.C. sent me to Quartz Creek with this team of horses. It took me two days to go from Bear Creek to Grand Forks where their main camp was. The first day I stayed overnight at Grand Forks and the next day I had to go to Quartz Creek over a real high summit called Calder Summit. Y.C.G.C. were building a small dredge on Quartz Creek, so I worked all summer there and then I went outside in the fall.

I had saved my money and I was able to go "outside" and spend the winter of 1934–35 in Vancouver. I should mention that Bill Scott didn't find work that summer until June 28 when he got a job on the dredges. He pretty well spent his life around Dawson working on the dredges. He got married in Dawson the next year after he came in from Vancouver where he stayed with his mother and sister. However, that fall, he didn't have any money. He got around with a bunch who liked to party so he spent all his money. He had to send to his sister for his fare out and that's the way he came back in again the next year.

The next spring, 1935, we both had to borrow money to return to Dawson City. The fare wasn't that much. I think it was around about $75 from Vancouver to Dawson at that time. It was pretty cheap. Of course being there the year before we had no problems. In fact, we were sitting in the restaurant having a meal when the Bank Manager came in and asked us where we had come from. When we replied Vancouver he asked if we were in Dawson last year. We said yes and he replied, "If you need any money or anything, just come over and see me."

So you see it was pretty easy to get by once you had been there and knew the ropes.

We didn't borrow any money from the bank. We arrived in Dawson in March when work was just starting. I thought I would work again at Quartz Creek for Y.C.G.C., but while I was bumming around town waiting for a job I met a fella by the name of Louis Brown who had come into the country from the back door. He came down the Mackenzie River over to Fort McPherson then he walked across country from Fort McPherson up the Porcupine River and then down river by canoe to Fort Yukon and up the Yukon River to Dawson. He was going to drive a team of horses for Y.C.G.C., too, that year. We were both sent to a place called Black Hills over on the Stewart River where Y.C.G.C. had a drilling program going. Louis was driving one team of horses and I was driving another team of horses, and so I spent the summer of 1936 on Black Hills Creek.

In the fall when I came back to Dawson I looked to see if I could find a place where I could go mining. The Company (Y.C.G.C.) had a piece of ground on Dominion Creek where they gave me a chance to do some work. I fixed up a cabin there and I bought $700 worth of groceries which I figured would do me pretty well for the winter. I was figuring on Bill Scott coming and working with me but it turned out that he decided to get married and didn't want to go mining. Well I couldn't find a partner so I loaded the groceries on the overland stage and let the stage driver take them in to Dawson. I told the driver to give the groceries to Bill Scott as my wedding present to them.

" The thermometer didn't show that it was very cold but I tell you when we got in the bush we found out how cold it was . . . "

I then went to work for wages again driving a team of horses for a fellow by the name of Andy Taddy. He had six or seven teamsters and teams of horses hauling wood that winter. It was a really cold winter. We had, as I remember, nearly a month of 50 and 60 below weather. Andy Taddy used to have the thermometer fixed so that the warmth from the building went out through a hole in the wall. The thermometer didn't show that it was very cold but I tell you when we got in the bush we found out how cold it was because the fellows who were cutting in the bush sometimes couldn't go to work because it was too cold, but here we were hauling wood in that kind of weather. I hauled wood all that winter until April 1937 when I went into Dawson and took a week's holiday. I came back and worked for Andy Taddy until I went to drive a team of horses at a place called the Granville Camp. There were about 250 to 300 men working in that camp as Y.C.G.C. were building another dredge there that year.

It was the fall of 1937 when I finished working at Granville and I was supposed to go trapping with Louis Brown, whom I mentioned earlier. He had left in July to trap in the Wind River. I didn't know what he had for dogs so I enquired around the country to see if I could buy some dogs to go trapping but I couldn't buy any dogs any place. I wouldn't go trapping without dogs so I walked from Granville to see some fellas on Quartz Creek where I had worked a couple of years before. They still owed me some money so I thought well I better go and collect the money if I could get it. That evening we heard on the radio that Grant McConachie was starting a new air route from Edmonton to Dawson City. His company was called Yukon Southern Air Transport Ltd. and his first flight as official carrier of the mail for the Post Office was on its way to Dawson City. I said to these fellows, "Well gee, I'm going to start walking to Dawson tonight and see if I can get out of here on that plane because I don't want to stay in the Yukon this winter."

So I started walking in the afternoon and got into Dawson 'round about 2 a.m. the next morning. I went to the hotel and asked them to wake up Grant McConachie. But the fellow on the night desk didn't want to wake him up. He said, "Oh you can't do that to Mr. McConachie; you can't wake him up in the night."

"Well," I said, "I've got to get out of here tomorrow and if I don't wake him up tonight and find out if they can take me, I won't know if I can go or not."

Anyway I did some fast talking and we went and woke Grant McConachie and he said he would be down in the restaurant at 7 in the morning as they were going to leave early. He also indicated that if I was there we would talk about me going on the trip with them.

So later that same morning I met Mr. McConachie at 7 o'clock and we had a talk and he asked when could I be ready to leave as they wanted to leave early.

"Well," I said, "I have to get to the bank and that bank doesn't open until 10 o'clock."

They had wanted to leave earlier. However after more discussion he decided to wait while I withdrew some money from my bank account. Then he wanted to know why I was so anxious to go on this trip. I replied that he would be flying over quite a bit of country that I had ridden saddle horses over. Which was all the way from north of Fort St. John to Sturgeon Lake. I told him I was raised in Grand Prairie. Grant McConachie was surprised and told me that his wife's name was MacLean and that she was also raised in Grand Prairie.

> **"I had an idea that I might get married so I went back to Stewart . . ."**

"I went to school with a Peggy MacLean, but we used to call her 'Pig Tail' MacLean" I said, "She had red hair, pig tails and freckles."

"That's my wife." replied Grant McConachie.

A few minutes later she came walking into the restaurant with her sister who was on the trip with them. Mr. McConachie said they were setting up stations in Dawson, Whitehorse and Watson Lake—Upper Liard they called it then—Lower Post and Fort Nelson. So some of the people he had brought with him were left as radio operators and managers on the way out. I was the first passenger from Dawson City to Edmonton on that trip.

I had a brother living in Edmonton at the time, who worked for Safeway Bakeries. I stayed a few days in Edmonton, but I got kinda of tired of it. There was nothing doing there. I had an idea that I might get married so I went back to Stewart, B.C. and that's where my wife, Jean, and I were married on September 20th, 1937.

I had spent the summer of 1937 on Clear Creek drilling for Fairbanks Exploration Company. They were interested in the ground on Clear Creek so they decided to carry out a drilling programme. I had a pretty good summer, didn't spend any money so I was pretty well fixed for money. After we got married we went to Vancouver and my uncle wanted me to work on a mine that he had in Hope, B.C. So I spent a couple of months at his mining camp. My wife, Jean, meanwhile had to live in Vancouver with my parents.

So, I said, "To heck with this, what the heck did I get married for? I was single when I was living in the camps. I have had enough of it." So I quit my job and we came back to the Yukon in 1938.

I got a job driving a team of horses and hauling wood that winter. Then part-time I was trapping a little bit. Then that next spring, I got blackballed with the Company (Y.C.G.C.) I couldn't get a job with them. However I also won the ice pool that year. I had two tickets and I won $2,700 so that was a pretty good thing for us. We needed the money. At the time I was cutting 16 foot lengths of wood for $3 a cord. Cutting and piling it and walking about 5 miles both ways to do that. So it was nice to have that little nest egg as our daughter, Betty, was born in the Dawson Hospital in August 1938.

Later that summer the company came to me and wanted me to go back to work for them driving horses. I refused and asked them to give me a job that I wanted. Y.C.G.C. asked me what type of work did I want. I told them that I wouldn't mind looking after that Australia Ditch. The company seemed surprised that I would want to go to Australia Ditch as the fellows they had there before couldn't stand being alone so the Company was happy to give me the job. We stayed there and worked there until they shut it down in 1944 I think. Then we stayed in the area and did a little wood cutting, and in 1945 we had to move some place so that our daughter, Betty, could go to school.

We moved to Mayo in 1945. We came via our trapline over to the Stewart River and caught the White Pass boat *Neecheah* on its way up in the first part of June. We came to Mayo on that barge and boat and when I got to Mayo I got a job working for White Pass as longshoreman and a year later a fella by the name of Rupe Steeves quit the job of longshore boss or wharfinger as they were called and I got that job. I worked for White Pass for a few years, then I did some guiding for Louis Brown who was in the big game guiding business. Then I went into the wood and timber business on my own. I bought a couple of trucks and Cats and sold timber to the mine in Elsa. However I was having a hard time getting help. You couldn't depend on people to work in the bush, so I finally gave that up and gave part of the wood business to my brother-in-law, Don Baker.

Don Baker worked for me for awhile. Then we had a fire and lost our garage, a truck and station wagon and all the tools and power saws—everything. So, I said to heck with it and gave up the timber business. I staked some ground on the lower end of Highet Creek and I started placer mining there in the early 1960's and worked there for three years. I think we got four ounces of gold. However, by the time I found out where the gold was coming from, we were broke again. So I went back in the wood business by myself and sold wood to the government and to the Department of Indian Affairs. Then in early 1970's I staked some ground on the upper end of Highet Creek as well as on Rudolph Gulch and started mining there. "Wilf" Tuck came and worked with me for a few years.

Jean and I are still mining on upper Highet Creek and just celebrated our 50th Wedding Anniversary (1987) so I guess we'll probably be in the Yukon for a long time to come.

DAVID "BIG DAVE" HAGER

by the Hager Family, 1986

Dave was born at McQuesten Village on September 27, 1905, to Old Jenny. He had one brother Charlie and two sisters Liza and Emma. He was raised at the old village in McQuesten, and later moved to Mayo where he remained.

David Hager's wife, Alice, who died in 1958. *Dave Hager Collection*

In 1929, he married Alice Bìter or Beeter. Alice was a sister to Caroline, Johnson Lucas's wife. They had six daughters: Martha, Rosie, Jenny, Irene, Laura, and Mary.

He had many stories to tell about his work on the S.S. *Keno*. Later he worked for the Yukon Territorial Government (YTG) Highway Maintenance in Mayo.

Dave Hager on the S.S. *Keno*. *Dave Hager Collection*

Dave enjoyed telling old stories about friends and about hunting trips. He liked to pass on his knowledge and wisdom to others. He had a good sense of humour and loved to tease. Dave Hager was a fine amateur photographer.

David Hager had 29 grandchildren and 20 great grandchildren (1986).

[Dave Hager died in Mayo on April 9, 1986.]

Alice and Dave Hager's daughters. From left: Laura, Irene, Mary, and Rosie, circa 1946. *G. Jean Gordon Collection*

MARY HAGER

Interview by Doris M. "Dorrie" Ewing, October 1987. Transcription by Linda E. T. MacDonald

My father, Julius Kendi, was born in the year 1881. My mother, Persis, was the daughter of Chief Robert of the Peel River Loucheux. She was born in 1890 at La Pierre House. I was born in 1907 at the Head of the Porcupine River and my brother, John, was born in the same place in 1909. My parents had other children but they did not live.

When Bishop Stringer asked my father to become a missionary he said, "No, I have never been to school and I don't know how to read." My mother asked him to try it for one year. Father agreed and remained to serve the church for the rest of his life.

From 1911 to 1913 we were stationed at the head of the Peel River. In the summer of 1915 we came to the Mayo district to do missionary work. We arrived at Fraser Falls where the Mayo people were at fish camp. My father and Rev. Frank Buck held church services in a tent. My father was the very first Indian minister in this district. It was at Fraser Falls that I met my friend, Lucy, for the very first time. She was about one year older than me.

My family didn't speak the same language as the Mayo people when we first came. After about a month and a half, my brother John and I could speak and understand the language perfectly. Lucy's mother, Eliza, and Chief Johnson could speak my father's language. My father never did learn to speak the Mayo language, but he could understand it, so I spoke his words to the Mayo people in their language. In this way they learned about God. My father also taught them things about child rearing and the cleaning of their homes.

In 1915 the Mayo people moved from Mayo to their own village on the Stewart River. When we got to the village only one house had been built and half of another one. As soon as we arrived the people built the church and a house for us.

On July 28, 1918, my father became a Deacon in the Anglican Church at a ceremony at Moosehide conducted by Bishop Isaac O. Stringer.

We stayed at the Mayo Indian Village for eight years and then the Minister in Dawson left and we went there. After two years in Dawson we went to Rampart. It was winter and we left for Rampart by dog team. In those days we travelled by dog team in winter and dog pack in summer. When we were small my father carried me on his back and my mother carried John.

Some trappers came over the trail and brought a letter to my father from the Bishop. The letter asked my father to return to Mayo because the Mayo people missed him and wanted him to come back. So we started back to Mayo. I was about 16 years old at the time. We arrived in Dawson in May. A road had been put through from Dawson to Mayo and we walked over the road to Mayo.

My mother and father worked very hard for the church; they devoted their lives to it. It is sad that no one ever mentions it any more except for Lucy and I, but it's alright, God knows.

When the school was established in Mayo, Indian children were not allowed to attend. In those early days, the Mayo school was only for white and half-native children. Edwin and Big Dave Hager were allowed to go to the white man's school and they stayed in town in order to attend. Lucy and I were not allowed to attend the school. My father eventually established a school for the children in our village.

In those days young people were supervised at all times and were not allowed to go out together alone. When I asked my father if I could go to a dance he said, "No." Finally he let me go but just for two hours and after that I had to go home. I liked to dance.

In 1923 I married Edwin Hager. He was one of the people at Fraser Falls when we arrived there for the first time, but I didn't know him then. I met Edwin at a dance at the village.

My first child, Peter, was born in the village in 1925. Two years later Robert was born. We named

The Kendi family, circa 1922–23. From Left: John Kendi, Rev. Julius Kendi and his wife Persis, and Mary Kendi. *William Hare Collection, Yukon Archives*

him for Edwin's father who drowned when the *Princess Sophia* went down. Our son Robert "Bob" died when he was seven years old. The doctor operated on Bob to try and save him but it was too late. We later named another son Robert in honour of Edwin's father. In the years that followed, Julius, Jimmy, Neil, Helen, Bella, Mabel, and Robert were born.

Mayo residents camped at the Mayo Airport during the flood of 1964. From left: Mary Hager, Dr. J. V. Clark, and Lucy Peter. *Lucy Peter Collection*

Lucy and Sam were married in 1922. As wives, Lucy and I worked hard, we worked like men. We cooked, cleaned, and maintained our homes and birthed our babies. We tanned moose skins and made moccasins to sell at the store in Mayo. We got $1.25 for plain moccasins and $3.50 for beaded ones. When our husbands were away hunting we looked after everything at home. If we ran out of wood while the men were away, we cut and carried it ourselves. In those days women and children rarely went to town. We made all of our own clothes and clothes for our children. We bought material at Gene Binet's store and sewed it on treadle sewing machines. Ella and I both had machines. When silver was discovered at Keno, Edwin and Sam worked there in the summer and made good money.

The flood in 1936 destroyed the original church at the village. The people rebuilt it a little further from the river using the logs from the original church.

A hunting party. From left: Peter Ellis and his wife Jessie, Mary and Edwin Hager. *Dave Hager Collection*

My mother became involved in the W.A. in 1912 in Old Crow. She continued her work with the group and later on Mrs. Stringer presented her with a gold pin to commemorate her dedication to the organization.

Christmas was a highly celebrated event in the village in those days. Feasting and dancing went on for a week. Just before New Year's a sleigh pulling a big canvas was prepared and it was pulled from house to house with a group of singers. They would sing at each home and people would put gifts in the canvass. Sometimes the canvass would be so full it was necessary to bring another one. On New Year's a feast was prepared from the gifts loaded on the canvasses. White people would come from Mayo town to the village for the nightly festivities, even if it was –50°F they would still come. Nobody drank back then, we just had fun. We had a phonograph to play music and sometimes Edwin and other people would play the fiddle.

In his last years, my father was serving in Old Crow (1947 and 1948). It was there that he fell ill and was flown to the hospital in Dawson City. I flew there to see him and he took my hand and said, "Don't cry, God will watch over us. Look after your children well and teach them about the Lord." My father died in the Dawson hospital in July of 1948.

We moved from the Old Village to Mayo in 1953.

In 1973 Edwin and I celebrated our 50th wedding anniversary. Our family and friends prepared a surprise party for us at the Revival Hall. Edwin and I knew nothing about it. Robert came and said to put our coats on and come out for supper. When we got there, the hall was full of people. We were so very surprised.

My husband died around the year 1983 and our son Julius died the following year. My other children all live in Mayo except Peter who is a minister of God in Vancouver. My youngest son, Robert, was Chief of the Na-Cho Nyäk Dun for ten years. My brother John still lives in Old Crow.

On July 1, 1987, I visited the Old Village. Most of the people who lived there have died now. Sam, Lucy and I are almost the only ones left. When I first came to Mayo I was eight years old and Lucy was nine. We became friends even before we could speak the same language. We have remained friends for all of these years, never once becoming angry at each other, and now we are old.

Edwin Hager and his dog at the Mayo Winter Carnival. This dog won the One Dog Pull at the Mayo Winter Carnival in Mayo and the Sourdough Rendezvous in Whitehorse in the middle 1960's. *Verna Klippert Collection*

PETER "SWEDE" HANSON

Kevin Shackell, Yukon News, January 31, 1983

A 40-year resident of the Yukon, Hanson went to the Mayo district in 1946. He hardly considers himself pioneer material, but then, he's not even sure of his real age.

Armed with a Grade Seven education and little more than the clothes on his back, Hanson left his strife-torn family when he was 12. During the Depression the school of hard knocks was a stern teacher.

He worked a string of mines across the country and when the Second World War broke out, joined the Black Watch Regiment, doing a tour overseas.

In his twenties after the war, Hanson was working in a central Manitoba mine when he was offered a job by Keno Hill which was reopening the Calumet deposit. He saw it as an opportunity to save enough money to return to England. "Well, the main reason I was up here was to save money to go back to England, I'd been over in wartime and liked it. At that time what I earned for a year could keep me in England for three years," Hanson commented. But Hanson never returned to Britain.

About six kilometres up the hill from Elsa, Calumet was a world of its own. When Hanson arrived to work as a drifter the site consisted of the mine works, two tar paper bunkhouses, a cook-house and outhouse. Mining men were used to the primitive facilities. It went along with the job. The draughty old bunkhouses were heated by large fuel drums converted into wood stoves. Water for the gravity feed showers was heated on top of the stoves. "These guys don't realize what it was like (today's miners), but it was no different than any of the mines across the country," Hanson explained.

Full-fledged miners were making $1.20 per hour and helpers earned $1.10. The company charged $1.25 per day for food and accommodation, Hanson said. About 80 men, most of them single, were employed by Keno Hill at Calumet. Women in a mining camp was unheard of even in the 40's. Miners worked eight hour shifts, but double shifts were common when extra manpower was needed to fix problems like frozen water lines. All the equipment was prewar vintage except for a mucking machine. "If you worked a double shift, your clothes were never dry," Hanson recalled of the underground work. "We used to work seven days a week too."

In the 40's, ore was still being placed in sacks for shipment to Whitehorse on the river boats. Ore mined at Calumet was transported to Elsa by a tramline and from there to Mayo it went by truck.

Recreation facilities at Calumet were also limited. "The recreation shack had one prewar radio and a record player with three records. One record was [the] last mail call," he chuckled.

Spare time was occupied by card games and drinking. Liquor, which was supplied by a bootlegger from Mayo, was in short supply that first winter, Hanson recalled. "We had a shortage of booze, nothing but Kings Plate rye. I haven't seen it since. It was terrible stuff. It burnt your throat out. We used to say it was good to get the dust out," he laughed.

Whenever the miners needed to blow off steam they'd slip into Mayo for a toot. Hanson said the men would either catch a ride to Elsa on a company vehicle or ride the tramline down, which was highly illegal. A taxi ride from Calumet to Mayo was $20 and from Elsa to Mayo it was $15. "We used to say there was one crew going, one crew coming and one working," he laughed.

In his younger days, Hanson's propensity for liquor got him into trouble more than once. If he'd been away too long, Hanson said the mine would get a hold of Elsa R.C.M.P. Constable Ivor Mast who would be instructed to round up the errant miner. "The mine would say, 'Look if Swede's not back he hasn't got a job,' " Hanson recalled.

When Hanson's job at the mine came to an end because of a wild party at the mine in the summer of 1947, he went to work for Kimbel's sawmill. He had to quit that job when he had trouble cashing his paycheck.

Hanson was then hired by White Pass as a deckhand on the S.S. *Keno*. Wages on the boats weren't as good as working in the mines, but Hanson said the clean air was a welcome change from dirt and grime of the stopes. The S.S. *Keno* plied the Stewart River from Mayo to Stewart City—the transfer point for passengers and ore destined for Whitehorse on the larger boats. Hanson liked the river life and enjoyed the people who made their living along the Stewart's banks; people like the Burians, Tony Besner and Ike Taggart. Hanson said he would have stayed with the S.S. *Keno* but he saw the river boat era was fast drawing to a close.

In 1949, Hanson ran the Silver Inn, a Mayo landmark long since destroyed by fire. Originally constructed on the Duncan Creek Road, the inn was later torn down and rebuilt in Mayo for owner Bob Palmer. It was a favourite watering hole for many of the miners.

During the 50's, Hanson spent his time prospecting in the summer and working for Keno Hill during the winter. While he was living in Keno during the winter of 1950–51, the Mayo district was hit with a flu epidemic. The bug had managed to travel to McQuesten Lake where a number of trappers were living. An old Indian woman in her nineties caught the virus and needed medical attention. Hanson, R.C.M.P. Constable Terry Carrol and Dan Golding mounted a rescue mission and went into the lake by dog sled to bring the woman and her husband out to the Mayo hospital. The woman was placed on the sled and hauled halfway to town, but the deep snow proved too tough for the dogs. Hanson had to snowshoe back into Elsa to round up a caterpillar tractor to haul the woman the remaining distance into Elsa. The old man, also in his nineties, snowshoed the whole way.

Although the Mayo district was changing, life was made more interesting due to the old-timers who were still around. The renowned "Hard Rock" MacDonald was one such personality. A prospector of incredible endurance, MacDonald walked one summer from Keno City to Norman Wells. He used to get a bath twice a year whether he needed it or not, Hanson recalled.

> **"He used to get a bath twice a year whether he needed it or not."**

Charlie Brefalt was another of the old prospectors who spent a lifetime on the creeks searching for gold and silver. Brefalt discovered, among others, the Elsa claim named after his sister and the No Cash deposit. Hanson recalled making a special trip out to Thunder Gulch to deliver a head of smelly Oka cheese to the old prospector because the postmaster couldn't convince anyone else to do the job.

Connected to the outside world by road in 1950, Mayo was exposed to changes brought on inevitably by the highway. The boats ceased running and postwar prosperity gave rise to 17 registered mining companies, said Hanson. United Keno Hill put in its own grocery store, beer parlour and recreation facilities.

In the late 50's, Hanson left Mayo for Edmonton. Married once before, he tied the knot again while outside, but it didn't last. He returned to Mayo to take up prospecting as a single man.

Throughout the 60's and early 70's he worked at a variety of jobs including several stints as town foreman and bus driver for Hutton Services.

Even before party politics arrived on the Territorial scene in 1978, Hanson worked as a Progressive Conservative supporter. Although he had no intention of entering politics, Hanson said he was prompted to run in the 1978 election by the party organization. He won the Mayo riding. In the territorial general election in June of 1982, Hanson lost the riding to NDP Piers McDonald.

[*Still considering Mayo his home, Hanson lived in Whitehorse until his death in September 1986.*]

LEONARD "LEO" HAVRANEK

by Astride Wenigerova-Noga

(Havranek means "A Little Raven")

Leonard Havranek was born in Slovakia on October 6, 1913. He was the son of a farmer. His childhood was full of hard work, dreams, and a love for hunting. The latter made him a poacher. He carried in his body a shotgun bullet—a souvenir of an outraged Slovakian gamekeeper. Of course Leo was proud of it because when Leo was a young man, poaching was considered a noble sport for adventurous young men.

While Leo was doing his army service, he met a cute little lady called Maria. She was very shy and unapproachable. Leo, being the kind of man who did not like an easy victory, set about to court Maria with particular stubbornness. Even though he preferred digging holes to writing letters, he was to send Maria so much mail over the following three years that she finally married him for better or worse.

Together, Leo and Maria, started a little business as wholesale and vegetable merchants. Life looked good after the war and it was during this time that a little boy was born to them.

Then the communist coup changed their lives drastically. One day Leo's brother got drunk and demolished a portrait of Stalin in a public place. In those days it would have meant a long imprisonment.

Leo's brother decided to leave the country immediately and Leo helped him to the border as he was more familiar with the well-guarded border than his brother was. Unfortunately they were both seen near the border by a fanatical party member from their village. Leo realized he would be accused of helping his brother and this would mean even more years in prison than it was for defacing a picture of Stalin. Leo's only chance was to leave the country as well. He returned to collect his wife and baby but left without them as they decided it was too risky to run with the baby. So Leo kissed his wife and little John, hardened his heart, and left. Tragedy struck again when their little boy died of pneumonia [at the age of 7 years]. Long years of loneliness lay ahead for Maria.

Leo made it safely over the border with his brother and eventually came to the Yukon in 1951. The Yukon captured his soul. He began working in the Elsa-Keno mine. Although learning English was difficult, he never doubted himself, his ability, or his dreams. His most important dream was to strike it rich so that his wife could join him. After making some money in the mines he became a prospector. With only a basic knowledge of prospecting, he ventured alone deep into the wilderness to prospect for silver. The money he invested in prospecting would have made a nice nest egg in the bank but Leo wasn't the type to play it safe. So year after year he tried anew.

He bought a trapline and a little cabin on Gulf Creek and many a winter trapped there for fur. Leo always enjoyed the harsh Yukon conditions because he enjoyed proving himself and rejoiced in overcoming the hardships. His diaries report in broken English; good take, Christmases with a radio, weather, and concerns for his dog. Years passed and still he did not strike it rich. A box of beer here, a night in a bar with the boys there; somehow money never stayed with Leo. He never felt rich enough to invite his wife to join him and was too proud to bring her to his tiny trapper's cabin. At one time he decided to offer his wife her freedom. For five years he never sent a word to her, giving her a chance to forget him and start a new life with somebody else. Apparently he did not know his little Maria, for she kept praying and hoping and writing and remained faithful to him.

When asked, Leo remembered only one serious situation in his lonely life as a trapper. With little money left after a summer of prospecting; he went to his trapper's cabin for the winter with only flour, matches, tobacco, and salt—hoping to make it through the winter on moose meat. He was so short of cash he could not even buy potatoes. Soon he became sick and lost some teeth. The winter was harsh and he did not have a snowmobile or any other means of transportation. Realizing he needed help, he attempted to walk to the highway on snowshoes. He couldn't make it more than a mile and returned to die in his cabin; or so he thought. The hand of God intervened; the previous owner of the cabin had left a box of bric-a-brac under the bed and Leo went through it looking for anything that might help. He found a bottle of pills marked "1-A-Day." Absolutely ignorant as to what they might be, he began to take them. To his surprise, they cured his illness. Later he found out that he had had scurvy and that the miraculous pills were vitamins.

Leo never bothered to read about how to keep healthy or about anything else for that matter. His reading ability wasn't great anyway. He certainly didn't fret about his health as many of us do. Once when he was in the hospital on a special bland diet, he ventured to the Mayo store in his pyjamas and bought himself a few pounds of good sausage. The nurses never suspected. Another time he had his leg in a cast and became quite upset about the "darn thing" while hobbling around his cabin. He took an axe and chopped the cast off. He looked very guilty when the doctor scolded him about it but never contested the doctor's medical authority. In fact he agreed that the doctor was right, but could he help it? The "darn thing," drove him crazy!

Leo had as great a respect for the R.C.M.P., as he had for doctors.

"Missus," he told me, "People are no good talking against policeman. We need somebody to fear, if people fear nobody, how would the world look? I love them. They put me, or my dog Lucky, sometimes behind bars; but what else can they do with me when I am drunk and start on some trouble. I am thankful they keep me from more trouble. And Lucky, well sometimes he goes for a walk without me and how would the streets look if all the dogs were fighting there. I also like Indians. They are my friends. I thank them for teaching me how to trap and hunt. I am their friend and they are my friends, I never had trouble with them except for once. An Indian boy was drowning in the river and I was called to help by an old man. I got the kid out of the river, was it cold! Next day I met his mother on the street, she slapped me on my face. Oh my, was I surprised, I was mad, but a man don't hit a woman so I had to take it. Later she came and apologized; her brat told her I pushed him into the river, but she found the truth. So we had a good laugh. Oh my, did she slap me a good one!"

Leo valued honest and straight people. With them he felt at ease.

"I hate lying, it is impractical because it is hard to remember what you said to whom; and Missus, never kick a dog. I once kicked my Lucky when ice broke under us on a lake and he was pulling a sleigh. I got mad and kicked him. He was innocent. He run away from me. After three days I found him in Keno. I apologized to him. He understood. Dogs are my friends."

It was the truth. He did not spoil them but treated them as dear friends.

In 1976 Leo got drunk one evening and came to visit with us. Somehow he had signed a promise to invite his wife to join him immediately and not wait any more for the luck. The next morning, scared, he wanted to back out of his promise.

"One more year, Missus, I for sure will sell my claims by that time and then Maria will come to a comfortable life." Then he overcame his pride and said, "A man has to keep his promises and this is my signature."

So he mailed the invitation the same day. When his wife finally arrived Leo had a house ready for her in Mayo, and even if his claims were not sold, he had many dreams for their life together. One month after their second honeymoon he told me.

"Missus, an ox was I. Why didn't I invite her to join me years ago. She is a jewel."

And so they lived happily ever after. Leo finally sold his silver claims and although he did not become rich, the money he invested in Yukon was partially returned to him. Leo died in January of 1983, leaving his Maria once more; until our Lord brings them together again.

Leo loved the Yukon. He would express it awkwardly with words (if ever); but his faith in the Yukon was unyielding. He loved the Yukon with every beat of his heart and he is part of the legend. Rough and yet gentle-hearted, without fear, and sure of himself; Leo lived his life fully, without bitterness, and enjoyed every minute of it.

A man so complete in his simplicity and so simple in his completeness.

[*Maria died in the spring of 1985*.]

HILDA HELLABY

by Hilda Hellaby, April 1982

I was born in England in 1898 and when my father died, mother brought my sister and I to Vancouver where my uncle lived.

In 1920, I joined the staff of the Board of Oriental Missions under the Anglican Church, and worked among the Chinese in British Columbia for 30 years. Then there was a vacancy at St. Pauls Hostel for a matron, and I and a young companion, [Felicity], came up to spend the summer in Dawson. That was in 1951 and we have been here ever since. Felicity is married [to Carl Skoreyko] and I am retired and we hope to stay in the north if all goes well.

When we had been in Dawson for a year, Mayo was left without a minister, owing to the death by drowning of the Rev. Wareham. Bishop Greenwood asked me to go to Mayo in 1952, to keep Sunday services going as well as any other pastoral work, and reopen the school in the Indian Village which was then located several miles down stream.

It was there that the following incident took place. There were two rivers to cross between Mayo and the village by the shortest route, and when I could get a lift from some of the Indians going to town I was glad to do so. On this occasion I walked with a group who said they had left a canoe on the second river in order to cross the Mayo River. As we drew near, the canoe was visible and I began to ask myself,

"Do I really see what I think I do?" Yes it was true; a perfect adaptation of modern material to traditional needs. The bent willow frame was the same, but covered in canvas instead of bark. Paint instead of spruce gum. Tin in the bottom of the boat so that no one could put their foot through it. And my vision was true—the pin holes where the paint had flaked off were patched with Band Aids!

This was too good to keep to myself so I wrote to the makers—Johnson & Johnson, and said:

"It says on the package that they are water proof and so indeed they are."

Three adults, four kids and a dog crossed the Mayo River with only Band Aids between us and the rushing river.

FATHER HENK HUIJBERS

by Kevin Shackell, Yukon News, May 26, 1982
with story additions by Father Huijbers

Father Huijbers was born and raised in Limburg, Holland. He studied in Wareghem, Belgium, and after college entered the order of the Oblates of Mary Immaculate. He made his first religious vows in 1937. He then studied Philosophy in Velaines, located on the French border, and was there when World War II broke out. The Belgium government urged all students to flee west to avoid being taken as war machine workers as was the case in 1914. The students arrived in Dunkirk and volunteers were asked to join the Red Cross. Father Huijbers volunteered as an ambulance driver for the International Red Cross. As a Red Cross ambulance driver he was able to return evacuees to their homes in Belgium and Holland from the French coast and these activities brought him into the Resistance. His indoctrination into the resistance movement came shortly after he successfully evacuated two injured Allied pilots who were shot down over Holland. He became chaplain for the Dutch Resistance, known as the L.O.K.P.; and the Belgium Resistance, known as the White Brigade. It was the beginning of a distinguished five years fighting the German occupation which saw him help hundreds of soldiers, airmen and civilians.

Bob, Father Henk Huijbers, Taja (Mädlin), and Et'tsì ts'ò, 1948. *O'Neill Collection, Herb Foster photographer*

In 1943, Father Huijbers was arrested by the Gestapo on suspicion of his involvement with the underground. The same night he was captured, several of his friends in the movement engineered a daring escape. The guards were plied with Schnapps laced with a drug. About 2:00 a.m. a rescue team burst through the window and held the intoxicated guards at gun point. Father Huijbers was hustled away in the dark and hidden for a number of months. His hair was dyed red, papers were changed and he was made to wear glasses.

Under the new identity, he spent the remainder of the German occupation continuing in his work with the sick, injured and needy, especially those who were in hiding or in danger of falling into the hands of the Gestapo. Twice he had to change his name so that the Germans wouldn't catch up to him.

When World War II drew to a close in 1944 in western Europe, he spent the next three years as chaplain for the Dutch troops attached to the American Army in Germany. He survived the Battle of the Bulge and re-entered Holland in Limburg and Germany in Rhineland.

Many of his friends did not survive the war, especially the people who died in the concentration camps after they were arrested. He has a list of 304 names, kept in the sacristy at Christ the King Mission in Mayo, of the people who did not make it through to see the Liberation. Special prayers are offered up to each of those individuals each day as the pages are turned revealing a new column of names, and the 56 villages where they came from.

In 1948 he came to Mayo to assist Bishop Jean-Louis Coudert for one year; and after more than 40 years, Father Huijbers is still here. Except for a five year stint in Cassiar, he has spent the rest of his time in the Mayo-Elsa area. He built St. Henry's church in Elsa and also constructed another church in the now vacant community of Calumet. Father Huijbers has received the Commissioner's Award for his work in the Yukon.

In May of 1982, he flew to Vancouver where he was united with fellow expatriates who he worked with from 1940–1945 supporting the Dutch Underground efforts. He was presented with the "Silver Resistance Remembrance Cross," for his work with the underground, by the Dutch Consul General in Vancouver, in memory of the Queen of Holland.

Reflecting on those times, Father Huijbers said, "We did simple ordinary things in an extraordinary way."

LENA HUMMEL

by Lena Hummel

My dad, Billy Malcolm, used to be chief at the Indian Village which was about three miles below Mayo.

In March of 1936, Mom and Dad went to Dawson by dog team and, in Dawson, I was born. In May of 1936 Mom and Dad walked all the way back from Dawson to Mayo. I believe that Dad packed me on his back all the way from Dawson City.

When I was 13 the children played basketball by the teacher's house [Sadie Dougall] at the Indian Village. Sometimes the children had picnics, just around the church, not going too far away.

Before 1940 the Indian people always went out to dry meat and also fish. Now today, the people are not doing that. I know that they have changed.

MILO & ANNIE JUROVICH

by Eve Jurovich

In 1926, Milo left his home and the high unemployment rate of Yugoslavia and made his way to Antwerp, Belgium. From there he was enticed to Canada with the promise of many available jobs.

He went to Ontario where his first job was digging sewers at 50¢ an hour. Milo then went to work on a farm in Saskatchewan where he earned 75¢ an hour; but found the sleeping quarters; beds of hay in boxcars, too cold. Continuing westward, Milo worked at various mining jobs in British Columbia, but was blacklisted for taking part in a strike at Anyox Mines in 1933. Losing his job was a mixed blessing because during that strike, he went to Prince Rupert where he met Annie.

Unable to find work near Prince Rupert, Milo decided to head north in March of 1935. He travelled to Skagway by boat then by train to Whitehorse. (Train fare at that time was $10, with a stop for a $1 meal at Lake Bennett.) In Whitehorse he bought a $50 ticket to Dawson on a huge snowmobile with a seating capacity of 18. The 400 mile trip usually took 98 hours.

Although arriving too early to work on the dredges, he and his friend, Bob Milanovich, were hired by Archie Fournier of the Yukon Gold Company to cut firewood. For a short time they lived at the company camp in Guggieville. In 16 months Milo and his partner had put up 900 cords of wood at $4.25 per cord. After 16 months they had saved $1,000 each. Milo used $200 to buy a small shack in Dawson.

In the spring of 1936, Milo went to work on the No. 3 Dredge for the Yukon Consolidated Gold Corporation. On the bull gang he earned $6 a day for a 12 hour shift before being promoted to deckhand. Each day he would ride a bicycle from his house in Dawson to the dredge on Bonanza.

Since leaving Prince Rupert he had kept his romance with Annie alive by constantly exchanging letters. Finally, having a house and steady job, Milo sent for her to join him. In January 1937, Annie travelled by boat and train to Whitehorse. She then took a seat on mail bags aboard a Northern Airways plane to Dawson where the temperature was –59°F.

The next day, January 5th, despite offers from the Catholic and Anglican priests, Milo and Annie were married by the R.C.M.P. Inspector. The ceremony was held in the R.C.M.P. barracks with the 16–member force attending in full dress uniform.

Soon there were three children and the expenses became a great burden. So great, that the family went back to Prince Rupert in 1943 to see Annie's family and to search for a better paying job. Milo returned to Dawson alone in 1946 and got a job as an oiler on No. 4 Dredge. Annie and the children joined him in the fall of that year, this time making the Whitehorse to Dawson journey aboard the river boat *Aksala*.

In 1947, Milo broke his shoulder and hand while working on the dredge. He was hospitalized for three months and wasn't well enough to work for five months after that. During his recovery, another son was added to the family.

One day Jack Gibbons, Commissioner, and Jimmy Johnson, Postmaster, came by to offer government assistance. Milo turned it down, and instead managed to get by with the help of his friends.

Harry Gleaves, owner of the Royal Alex Hotel; approached Milo in 1949 and asked him to lease the hotel, bar, and poolroom. Gleaves kept the cafe and backed Milo on a $3,000 bank loan. With Mike Franich as his partner, Milo leased the hotel. Annie provided laundry services for the hotel. This venture continued until 1951 when Gleaves sold the hotel and Milo again found himself unemployed.

Mayo, meanwhile, was bustling with activity as the Northern Canada Power Commission was building a power plant there. Milo decided to move to Mayo to either get a job or buy a business. It was then that he met Alex Arthur, owner of the Silver Inn Hotel. Arthur offered to sell the hotel for $30,000, half cash, and half on monthly payments. Milo and his partner, Mike Franich, took over the hotel on June 1, 1951.

Milo and Annie rented a cabin (where the North Star Motel now stands), then built their own house. They operated the Silver Inn Cafe for a few months until they found someone to lease it.

After Mike Franich passed away, Milo hired Pete Pavelich and later Gus Debastien to be bartender and watchman and for many years Isabel (Kimbel) Mosich was chambermaid. Milo added the Silver Inn Taxi service. The hotel was subsequently leased to a number of different people, among them Oliver Hutton, Alex Ruchin, Lino Battaja, Steve Van Bibber, and Drago Kokanoff. During this time Milo found employment as custodian at the Mayo General Hospital.

Milo once again began to run the Silver Inn himself, but government regulations required costly renovations. It was with reluctance that he decided to sell the hotel. The new owners, Alan McDiarmid and Brian Kates, converted the cafe to a beer parlour and in 1976 the hotel was destroyed by fire.

Annie remained a dedicated homemaker. While still in Dawson, she made most of the children's clothes and continued to amaze people with her beautiful knitted and crocheted articles. Equally amazing were her banquet-style tea parties. Annie kept active in the Church and belonged to the I.O.D.E., Rebecca Lodge, Anglican Women's Auxiliary (later known as the WIGS), and served on the Library Board. In 1969, their eldest son bought the Northern Commercial Company store where Annie worked until 1985.

Milo and Annie decided to sell their family home in 1975 and moved into the more modern government housing. During this time, Milo served on the Housing Board. Following a successful hip replacement operation in 1979, Milo enjoyed his retirement and spent time driving his car for pleasure, often stopping to talk to his friends in the park on the riverbank.

In May, 1985, Milo needed medical treatment in Whitehorse, then continued on to Langley, British Columbia. They took up residence in an apartment near their four children. They have fourteen grandchildren and four great-grandchildren, most of whom live in the Vancouver area. That summer Annie returned to pack up their belongings and leave the Yukon.

Milo's success as a businessman in Dawson and Mayo is a good example of how a combination of pride, determination, and perseverance can help achieve a goal.

Although Annie and Milo are now living far away, after fifty years in the north, the Yukon will always be "home."

[*Annie Jurovich passed away in Vancouver on July 7, 1987 and Milo passed away February 13, 1989.*]

"BERT" & VERNA KLIPPERT

by Verna Klippert

In October, 1959, we decided there had to be something better than trying to survive in the city. Previously, we had gone on a holiday to visit relatives in Mayo. We decided the country there was beautiful and that it would offer more than city life could for the boys and ourselves. So we packed all of our belongings in a 1953 three ton truck; and the two boys Calvin and Wayne, Bert and I, and a pup, headed up the highway. There were hardships on the way but we managed.

At that time, the bridge in Carmacks had recently been built but the ferry was still operating on the Stewart River.

When we arrived in Mayo we moved into a house on Centre Street owned by Irene and Oliver Hutton. Later we bought Annie and George Besner's house (now owned by Bob Adair). Then in 1968 we bought a house on 4th Avenue from "Kippy" Fisher where we lived until we moved to Whitehorse in 1985, and sold that house to Joan and Dick Ewing.

The school the kids first attended was held in the teacherage and Eve (Nichol) Jurovich was their first teacher. Calvin took Grade 11 in Dawson and one semester of Grade 12 in Whitehorse. He discovered the best math teacher, Mr. Montgomery, was in Mayo and so took one semester of Grade 12 in Mayo. Wayne attended the Mayo School from Grade 1 through 12. Philip was born in Mayo and graduated from Grade 12 at the Mayo School as well.

Calvin and Wayne Klippert on "Crocus Hill" with hats full of crocuses, 1960. *Verna Klippert Collection*

In our initial business undertaking, we hauled dry wood for customers and later we cut timber for Keno Hill. With the arrival of Bert's brother Con and his family [wife Ina-Mae, daughter Charmaine, sons Dan and Kim] in 1963, we went into the sawmill business. In 1978 we decided to take a different tack and formed our own construction company which we still operate with our three sons.

I well remember the flood in 1964. Many hours were spent filling sand bags; even the school children helped. We were all given typhoid shots.

During my 25 years in Mayo, curling was always a popular sport no matter how cold the weather got. In 1978, at age 15, Philip and three friends won the Territorial Playdowns and, with Wayne as their coach, went to Charlottetown, P.E.I., for the Junior Men's Playdowns.

The population of Mayo, more or less, remained unchanged during the years I was there but the face of the town changed tremendously with the construction of new buildings and the removal of many of the little cabins that were left from another era.

RECOLLECTIONS OF EDWIN "ED" KUNZE

as told to Lynette Bleiler, 1985

The Klondike Gold Rush was at its peak when Ed Kunze's parents arrived in Dawson City from San Francisco, California, coming over the White Pass and down the Yukon River to Dawson. They walked from Lake Bennett to Dawson in February 1900. Dawson was a boom town boasting an area population of around 40,000. It was a hastily assembled community of log cabins, some pretentious sawn lumber structures and tents. The currency was gold dust and every store was equipped with gold scales.

Ed, the eldest child, was born in the latter part of 1900 on Bonanza Creek. He, his brother Bonner (Bon) and sister Florence lived near the Ogilvie Bridge during their youth and Ed remembers meeting Robert W. Service (The Bard of the Yukon) on one of his walks out to Ogilvie Bridge on the Klondike River.

The Kunze children grew up there and went to school in Dawson and later on moved into Dawson itself to live.

One of Ed's first jobs during the time of World War I was to drive a Model T taxi from Dawson out to the creeks of Hunker, Dominion, Sulphur, Bonanza and Eldorado, and he even went over to Quartz Creek once or twice.

During the summer of 1919 he worked at the 12 Mile Power House where there were three little 1,000 horse power Pelton wheels connected to the small 600 kw generators. A few years after that he worked at the North Fork Power House on the Klondike River.

Up to 1925 Ed's employment had been of a seasonal nature, so when the opportunity arose for full time employment, he went by overland stage from Dawson to Mayo.

A minimum of 200 horses were strategically located to operate these mail, express and passenger stages. During the periods from the freeze-up of the Yukon River until the ice was solid and there was sufficient snow, wheel stages were used. Roadhouses and stables were located at intervals of about twenty miles. Food for passengers and drivers, and fodder for horses was maintained at each post. The roadhouses were miniature hotels of log construction, heated by wood fires. They provided comfortable beds with white flannelette sheets and good meals; sourdough pancakes being a breakfast staple.

During the winter months sleighs drawn by four- or six-horse teams were used, and carried up to 14 passengers besides mail, baggage, and freight. Carbon foot warmers were changed at each roadhouse and buffalo robes were supplied for all passengers. In the spring and fall the drivers had to contend with floods, washouts, mud and slides, as well as river crossings when ice conditions were dangerous. The route from Dawson to Mayo followed the "Overland Trail" through Flat Creek over to Clear Creek, down the McQuesten Valley and on to the junction at Stewart Crossing where the trail branched off to Mayo. The driver of the team on that arduous spring journey was Jack Stewart, recalled Ed.

On May 3rd, 1925, Ed started work for the Treadwell Yukon Company, a silver, lead and zinc mine in the Keno area under the management of Livingston Wernecke. Ed was a diesel powerhouse operator for a short while and later on he worked in the machine shop doing different kinds of mechanical repairs.

Nicknames were common in the North in those days, the most well known name being Klondike Kate, a dancer and actress in Dawson. The Mayo and Keno area had the MacDonalds; they were not related but each had earned a self-describing nickname. There was "Hard Rock" MacDonald, "Big Sandy" MacDonald, "Scarface" MacDonald and "Pretty Jack" MacDonald who came to the Yukon with the Royal North West Mounted Police. "Pretty Jack" married Ed's sister, Florence.

George Reynolds had some placer claims on Ledge Creek which runs into Mayo Lake. It was some time in early 1930 that George and Ed went out to the cabin on Ledge Creek to catch fish near the mouth of Nelson Creek which empties into Mayo Lake. They put out their net the first evening and pulled in more than enough fish for themselves and the dogs the next morning. Together they cleaned the fish smoking some and salting the others. The put them into 50 lb. butter kegs then transported the kegs the 60 miles back into Mayo and gave them to George Nagano for his restaurant in the Binet building on Main Street (now the Chateau Mayo).

Mayo was a nice, neat, clean little town with a stable population. Most of its economy was based on the mines in the area. The concentrates from the Treadwell Yukon Co. mill were brought in all winter and stockpiled on the riverbank until navigation opened up in the Spring. Often there would be eight or nine thousand tons of ore piled up on the flat along the river. The Company warehouses were also located there, they always held a tremendous quantity of supplies and spare parts.

Mail was very important to the townspeople and Mabel McIntyre (née Huffman), the Postmistress would sort mail till the early hours of the morning as a convenience to the public. No extra pay of course. "The Post Office at that time was an old log building, with a sod roof, and stood on Centre Street quite close to 4th Avenue," recollected Ed.

Bill Hare lived in the area at that time and with his love of photography, his camera captured many events and situations of the period.

About 25 miles from Mayo on the original road to Keno was Mrs. Agnes Kinsey's roadhouse at Field Creek. It was a two-storey frame building located just before the bend in the creek. There was also a building for the tractors and a bunkhouse for Treadwell Yukon Co. employees. Quite often on a Sunday, 8 or 10 of the young men from the Wernecke Camp would phone Mrs. Kinsey and make arrangements for dinner. They would drive down in the afternoon to enjoy her fine home cooking in the evening. "In fact it was because Mrs. Kinsey was so skilled in the kitchen that people planned on eating and staying at her roadhouse," mused Ed. (*See photo, page* 399.)

"They would drive down in the afternoon to enjoy her fine home cooking in the evening"

People made their own entertainment in those days and Ed remembers one occasion; a New Year's Eve Party held at the Y.O.O.P. Hall in Mayo. Part of the entertainment was a play, and a group from Wernecke Camp consisting of Jack Andison, Walter Murphy, Roy McDermid and Ed Kunze sang a song together the words of which were composed by Bill Langham, an office worker for the Treadwell Yukon Company.

The song went like this said Ed, crooning into the tape recorder:

How do you do, everybody, how do you do,
How do you do Mayo friends, how are you?
We've come 40 miles today just to help you with your play
Now we're here we'd like to say, 'how do you do?'
How do you do, how do you do,
How do you doodle, doodle, doodle, doodle do,
You love music, mirth and laughter,
And that's just what we are after,
How do you doodle, doodle, doodle, doodle do.

In the early part of June 1930 having left the Treadwell Yukon Co. Ed Kunze set off with Ed Barker, Norman Wightman and Frank Gillespie down the Stewart River while the water was low to mine bar gold. They returned to Mayo in the fall, and just before freeze-up in October Ed got a 16-foot boat from Ed Kimbel and added the 7 h.p. Elto engine that he and his partners had used in the summer, and drifted down to Dawson taking three and a half days.

"The night before leaving Mayo I had a talk with Mr. Binet and asked him of my chances of the ice giving me any trouble." Ed recalled. Together they went down to the riverbank where the water was as clear and as smooth as could be. "If its as smooth as this in the morning you can expect no trouble," said Mr. Binet. So Ed arranged to have some gasoline in cans left on the riverbank and early next morning set off for his first evening's stop at the mouth of the McQuesten River. There was a slight amount of ice around his boat the next morning and it was the same the following morning at Maisy May Ranch, and at the mouth of the Stewart River. He arrived in Dawson around 3 p.m. on the fourth day and the following morning when Ed looked out his mother's window the Yukon River was banked full of ice.

Ed again returned to Mayo in 1945 with his brother Bon and took over the power generating plant and distribution system from Lou Titus and Alex Nicol.

Electricity was supplied to the homes in Mayo from 8 a.m. till noon on Mondays to allow the weekly wash to be done and on Wednesdays, 8 a.m. till noon to allow the ironing to be completed. On other days the power was off except in the evenings when the plant would run till midnight. Of course if there was a special occasion or party, arrangements would be made and the hours would be extended.

A few years later when the Royal Canadian Corps of Signals came into Mayo they asked the Kunze brothers if they were interested in a contract to supply them with power; this meant an expansion of their equipment to provide a 24 hour service. They bought two Caterpillar generating units, one a 311 and the other a 318. They ran those units either singly or together if the demand warranted it, until the Northern Canada Power Commission bought them out. Ed continued to work for N.C.P.C. until his retirement in 1966.

Prior to his retirement in 1966, Ed along with John and Molly Donald projected the first "talkies" in Mayo using the Y.O.O.P. Hall once a week. However they could only use the stage at the Y.O.O.P. (Yukon Order of Pioneers) Hall so after the Wareham Hall was built over by the Anglican Church, they moved over there and showed movies every Wednesday evening. Later Wes Buyck took over and showed movies in the "new" Community Hall.

Ed left Mayo for Whitehorse to work for N.C.P.C. but after a few months returned to Mayo and worked for Oliver Hutton in the office as well as driving the minibus between Mayo and Stewart Crossing, and between Mayo and Elsa until early in 1968 when he married the Australian Nurse-in-Charge of the Mayo Hospital, "Betty" Podger; a second marriage for both of them.

Together they set off for Vancouver, B.C. (via Dawson City, Mt. McKinley, and Fairbanks Alaska), then down the Alaska Highway and the Hart Highway to Vancouver where they boarded the *Oriana* for Australia.

They have a lovely cottage in McCrae, Victoria, Australia with a beautiful vista that takes in Port Phillip Bay, and when they aren't travelling they can both be found tending their magnificent garden of fruit trees, dahlias and green lawns.

GORDON ALEXANDER MCINTYRE

by Gordon McIntyre, November 1983

Although I did not arrive in Mayo until 1932, my father, Angus McIntyre, left Dawson in 1920 to manage Greenfield and Pickering's branch in Mayo. Greenfield and Pickering had the contract to haul Keno Hill Ltd.'s (Yukon Gold Company's) silver-lead ore from the top of Keno Hill to the Mayo waterfront. My dad had had experience, before

Angus McIntyre wearing his Yukon Order of Pioneers banner on Discovery Day, 1921 or 1922. ***Ralph Rogers Collection***

coming to the Yukon in the Gold Rush, of hauling ore from the side hill mines in the Kootenay country of British Columbia so he had to set up the whole system of carrying the ore to Mayo. Since the road from the "Guggs" to Keno Hill was too steep for sleighs the ore was "rawhided" to that point, and from there hauled by horse-teams and sleighs to Mayo. The teams had to be changed every 10 to 15 miles so there were three roadhouses on the Duncan Road—Hoffman's, Van Cleaves and Minto Bridge. He left Mayo in 1924 when this contract finished.

Gordon A. McIntyre, circa 1932. McIntyre came to Mayo as the school teacher (1932–1941). He then went on to become the Mayo Mining Recorder from 1947 to 1965. *Middlecoff-Bleiler Collection*

I came to Mayo in August, 1932. This wasn't my first time in the Yukon as I was born in Dawson and lived there until 1920. I received most of my education in Vancouver, where I attended Britannia High School and Provincial Normal School, getting my First Class Teaching Certificate in 1930. I taught school in Mayo from 1932 to 1941 when I joined the Canadian Army, trained at Vernon and Petawawa, served in England with Personnel Selection and the 13th Canadian Field Regiment; in Italy with Headquarter First Canadian Armoured Brigade; in Northwest Europe with Headquarter 1st Canadian Corps Troops. I was discharged in February 1946 and returned to Mayo.

Instead of returning to teaching I worked for the Yukon Territorial Government for a year, then transferred to the Federal Government, as Mining Recorder and Crown Timber and Land Agent.

During my time in Mayo I was Secretary of the Hospital Board, Juvenile Court Judge for the Mayo District, a Justice of the Peace, Marriage Commissioner, Notary Public, Coroner, and Lieutenant in the Mayo Company of the Canadian Rangers.

In 1955 I married Ruth Batty who had come to Mayo to teach school in 1946. After teaching for three years she went to Edmonton for a Commercial course. On her return to Mayo in 1951 she opened a restaurant first in the Silver Inn and later in the Chateau Mayo. She also opened a variety store in the Chateau building. The restaurant was sold to Anne Kwasnicki and the variety store to Bessie Friesen. Ruth also taught school in Keno in the 1953–1954 school year.

We have three children, Gordon Angus, Mary Evelyn, and Norman William.

In 1965 I was promoted to Supervisor of Lands and Chief Mining Recorder and moved to Whitehorse. I was the first Regional Director of Resources for the Yukon and while in that position also served as Registrar of Land Titles and Administrator of Yukon (Assistant Commissioner). On retirement in 1974 I was elected to the Territorial Council as M.L.A. for the Mayo District and served in that capacity until 1978. I have been retired since that time. Ruth resumed her teacher career after we moved to Whitehorse, but is now also retired.

McLaren residence, Mayo, 1948. *L. Cheatham (McLaren) Collection*

MALCOLM & SARAH McLAREN

by Lillian Cheatham

My father, Malcolm McLaren, went to the Klondike in 1897. My mother, Sarah, and the three oldest children: Malcolm Earl, John and Mary; followed him as far as Skagway in 1900. They lived in Skagway until papa (Malcolm) came out from Dawson. Two more children, Dorothy and Lillian, were born in Skagway. In 1905 we all went to Dawson. Kathleen and Nora were born in Dawson, and we (the children) all went to school there.

When Dorothy accepted a teaching position in Mayo in 1918; mama (Sarah) and children, Kathleen and Nora, moved to Mayo with her. Lillian followed later that year. Papa visited Mayo several times and moved there in 1922 or 23. Dorothy taught school in Mayo for two years. My oldest brother, Malcolm Earl, never lived in Mayo; he lived in Dawson and left there in 1914. John lived in Mayo and worked as bookkeeper at Keno Hill for several months after World War I. John then went to Whitehorse where he was customs inspector for approximately ten years. He was married in Whitehorse and his two sons were born there. Mary never lived in Mayo but visited there. Dorothy, Lillian, and Kathleen lived in Mayo from 1918 until approximately 1924. Nora lived there from 1918–1926.

When my mother first went to Mayo in 1918, they lived in the house which later became Dave Hager's. It was the school house back then. I was there [Mayo] when Louis [Bouvette] struck silver.

When I was in Mayo they had no hospital, doctors, ministers or churches. Randy Maclennan took care of those he could. I cut the end off my finger with an axe and they had to go to the mine at Lookout Mountain to get Randy to patch it up.

We have all lived in Seattle since leaving the North except for John who lived in Victoria. (John passed away in 1972). Mama and papa both died in 1941 and are buried in Seattle. Dorothy, Nora and myself (Lillian) are living in Seattle [1989].

THE ALEX MACLENNANS OF MAYO 1925–1942

by Barbara "Betty" E. Taylor

Alexander Malcolm Maclennan was born on February 28, 1862, in Glengarry County, Ontario. While roving through Idaho, he heard and answered the call to the Klondike. He climbed the Chilkoot Pass in the early spring of '98 and arrived in Dawson in April. After a time of backpacking on the pass in 1899, he became a stage coach driver. At one of his stops, Silver City roadhouse, he met Alice Mills. She had recently arrived from Ireland and was working for her sister Mrs. E. Warne. Alex settled in Dawson where he found work as a winchman on one of the gold dredges. In October, Alice joined him in Dawson and they were married by Bishop Stringer on October 21, 1914. Margaret, Alex's sister and her husband, Richard Gillespie, were the attendants. Alice's gown was a beautiful heavy pale blue satin and was purchased from Madame Trembley's.

Three daughters were born of this marriage; Barbara Elizabeth "Betty" on August 18, 1915, Vera Margaret "Mickey" on November 19, 1917, and Alice Alexandra "Babe" on September 15, 1919.

Alex went to Mayo to work with Max Landreville on the building of the first Mayo hospital in 1922. The family moved to Mayo in 1926 having bought a log cabin from Jack McLean and Jack Carpenter. This is the present site [1989] of the Northwestel residence [Block 16, Lots 10 & 11]. Near neighbours were Joe Longtin, Joe Hoyne, Jim Fairborn, John Valentine Smith, Mr. and Mrs. Rupert Steeves, Dr. Ross, George Miller and daughter Hilma, and Mr. Dube. The Gillespies, Dempsters, Bellerbys, and Schellingers also lived near by.

Alex held the position of Fire Chief until his death in 1936. The fire hall consisted of an anchored barge with a large gasoline driven pump. This was lifted out in the fall when ice formed in the river. Years later a well was dug near the fire hall on Front Street, where a large electric pump was utilized.

The Maclennan sisters double wedding. From left: Alexander "Alex" Smith and his bride Alice "Babe" Maclennan, Vera "Mickey" Maclennan and her husband Sinclair Dunnett Jr. The couples were married by Rev. Wm. Valentine at St. Mary's Anglican Church on August 15, 1937. *Elizabeth "Betty" Taylor Collection*

The three girls attended the little one room school on the corner opposite the Catholic Church. The teacher in 1926 was Miss Laura Gillespie, who was followed by her sister Vera. Teachers who came later were Noel Manuel and Gordon McIntyre.

The family was active in community enterprises. Alice and later the girls were members of the Women's Auxiliary of St. Mary's Anglican Church, the I.O.D.E., and also supported the tennis court socials. Alex belonged to the Y.O.O.P., and was very active in Conservative politics. He always had a big garden and loved cutting and bringing in the winter wood. The family enjoyed the frequent dances, bazaars etc., held in the Pioneer Hall. Weekly bridge parties were well attended in the I.O.D.E. rooms. Another favourite haunt was the baseball game every summer evening. This was played at the school corner. Teams consisted of all who wished to join from age eight to eighty. Snowshoeing, hiking and tobogganing were winter activities before skiing became popular in 1935.

Barbara "Betty" married Charles Taylor on February 23, 1936. He was the manager of the Taylor & Drury store in Mayo from 1932 to 1942. Vera "Mickey" married Sinclair Dunnett Jr., mechanic and plant operator at the Silver King mine and Elsa; and Alice "Babe" married Alex Smith, a mechanic and power plant operator at the new mill at Elsa. Both of the younger girls were married at a double wedding in the Anglican Church on August 15, 1937.

Alex Maclennan died in hospital from pneumonia in May of 1936. Mrs. Maclennan then took over the cooking position at the hospital. Miss Page was the matron. She was followed by Miss Burns who married George Reynolds a few years later.

On the closing of the Treadwell Yukon mine in 1942, and the beginning of the Alaska Highway, the three families moved to Whitehorse. Mrs. Maclennan soon followed and passed away there on March 3, 1970.

The three "Maclennan girls" and husbands continue in good health. Each couple has achieved over fifty years of marriage. The offspring are Barbara (Taylor) Conway of Prince George, B.C.; Gordon Taylor deceased; Verna (Taylor) Hart of Whitehorse; Peggy (Dunnett) Thoreson, Red Deer, Alberta; Dwight and Warren Smith of Creston, B.C.; and Malcolm Smith, of Whitehorse.

SIMON "SAM" MASON-WOOD AND ROSE TURGEON

by Simon Mason-Wood

Sam arrived in the Yukon in the late 1920's and started work managing the Whitehorse Inn Hotel. He then worked for T.C. Richards driving a cat, hauling mail, freight and passengers. It was on one of these trips to Mayo that he met Rose Turgeon. They were subsequently married in Mayo by Sgt. Tidd of the R.C.M.P. They had three children; Eileen, Margaret and Simon.

Rose Wood (Mason-Wood), Mayo, 1948.
L. Cheatham (McLaren) Collection

Sam worked at several occupations—from miner at Mt. Freegold in the Carmacks area to Dawson City employed by the government. He was transferred to Mayo where his office duties included: Mining Recorder, Liquor Vendor, Territorial Agent, Forestry, Justice of the Peace and Coroner. After approximately 10 years he left the government to start a trucking and wood business.

Sam was a great gardener and enjoyed trying different varieties of vegetables and flowers not common to Yukon gardens. He brought the first Siberian May Day tree into the Yukon. He was also an avid outdoorsman and fly fisherman—most summers were spent with the family at their Mayo Lake cabin. He passed away in 1950.

Rose was born and raised in Mayo. She was an active lady who also enjoyed the outdoors and family life. She had a love of sports and was a participant in many a baseball, tennis and badminton tournament. She was one of the first members of the Mayo Curling Club. Rose was also actively involved in the I.O.D.E. and W.I.G.S. organizations. She passed away in 1980.

Eileen lives in Mayo, is married to Emery Shilleto and has three children, Roderick—deceased, Blair and Blaise.

Margaret lives in Whitehorse and is married to Dick Wallingham. They have three children, Elaine, Sam and Walter.

Simon lives in Mayo, is married to Beverley O'Connell and has two children from a previous marriage, Michael and Charyl. Bev has three children, Dawn, Cheryl and Darin.

ELMER & NELL MIDDLECOFF

by Doreen Doidge daughter of Evelyn Lee

Elmer Elsworth Middlecoff, born October 1, 1868 in Clinton, Missouri, U.S.A., ran away from home at the age of 17. He made his way to California, finding work at a citrus ranch supervising a crew of Chinese labourers planting and grafting orange trees. Several years later he returned to Missouri and worked for a time for his father, travelling the countryside purchasing calves to be fattened on grain stored on the family farm. During one such buying trip he met Nell Edith van Buskirk, (born February 3, 1874 in Appleton, Missouri), whose father owned a large farm—and many calves.

Middlecoff family. From left: Elmer Middlecoff Sr., Evelyn, Ethel (rear), and Nellie. *Middlecoff-Bleiler Collection*

On his 21st birthday Elmer received $2,000 from his father and promptly headed for Montana where he became a sheep rancher. He did well for a time but later lost everything when sheep farming became uneconomic.

In 1898 he was told he had tuberculosis and that his best chance of recovery was to stay in the fresh air as much as possible. He was advised to send his wife and infant son, Laney, to her family in Missouri to protect them from contracting the disease. (Laney later died of meningitis at the age of 10 months.) Elmer had decided he would either join the Army and go to the Philippines to fight in the Spanish-American War or join the stampede to the Klondike, leaving the decision to Nell, who felt he would be safer going North!

So Elmer headed west, eventually arriving in Victoria, B.C. where he acquired his outfit and probably, the advice to take the route he did. He boarded a steamer sailing for Wrangell, Alaska, took another boat from there up the Stikine River and travelled overland as far as Teslin Lake where he spent the winter prospecting (1898–99). Teslin consisted of a small community of prospectors, among them was George Black. In the spring of 1899 Elmer and George Black headed "outside" with just enough gold to pay their way as far as Los Angeles. George continued on, presumably to New Brunswick, while Elmer made his way to New Mexico where Nell was living with her parents who had sold their Missouri farm and bought a ranch in New Mexico.

Elmer only stayed a short while before heading north again, this time accompanied by Nell. They rented a small cabin in Skagway while Elmer was building a scow at Lake Bennett. The train line had been completed as far as the lake by this time so Elmer travelled to Lake Bennett each morning and, with the help of a couple of partners, whipsawed lumber and built a scow. When it was completed he, Nell and the partners set sail for Dawson, arriving October 6, 1899.

Shortly after their arrival in Dawson, Elmer along with a partner, took a lay on a claim on Bonanza Creek. Nell's first home in the Yukon was a small, one room log cabin with a pole floor and one window made of brown bottles chinked with moss. She often said she was happiest there and remembered those days as the best of her life. She spent her time baking bread and cakes which she sold to miners in the area, her profits keeping her, Elmer and the partner in groceries for the winter. Ethel May was born in a roadhouse on Bonanza Creek, April 14, 1900.

They worked the lay until the summer of 1900 when Elmer acquired a fraction on a bench above Lovett Gulch. During the summer when there was not much work done on the claims Elmer started a freighting business. There were no roads to the creeks so pack horses and mules were used. Elmer had had experience handling packhorses so this situation was ideal for him. He began with a few horses but soon needed more because of the demand for his services. This business proved more than profitable. However, Elmer only freighted for a summer or two before selling off the horses because the expense of keeping them over winter was too great. Nell and Ethel May left Dawson in the spring of 1902 for Raton, New Mexico where my mother, Evelyn Abilene was born August 29, 1902. They were joined by Elmer for the winter, returning to Dawson in the spring of 1903.

In 1903 Elmer heard of a stampede to Highet Creek in the Duncan Creek Mining District. He investigated and staked a few claims. During slack periods on Lovett Gulch he would return to Highet Creek to do representation work on the ground.

It should be mentioned that at this time there was an established community at Duncan Creek an area of placer mining activity since the Duncan Creek discovery in 1898. There was also a small settlement on the river below Mayo Lake. There was a trading post at Gordon Landing below Fraser Falls, a

Middlecoff family at Highet Creek, circa 1933. Left to right: Elmer Land Middlecoff Jr., Flora Middlecoff, Elmer Middlecoff Sr., Nellie Middlecoff, and Ethel Middlecoff. *Middlecoff-Bleiler Collection*

trail led from there to Janet Lake, Mayo Lake and Duncan Creek. The town of Mayo itself wasn't begun until around 1902–03. By the time Elmer returned to Highet in 1906, there was a good wagon road from Mayo so there was no difficulty bringing in supplies shipped by boat from Dawson.

Elmer sold his claim on Lovett Gulch in 1905 to one of the dredging companies believing that he had taken out all the gold there and that the company merely wanted the ground for a tailing dump—what a surprise to learn that a further quarter million dollars in gold was taken out by the dredging company when gold was $16.00 an ounce!

Following the sale of the claim the family travelled to Missouri, for the winter and returned to Dawson in the spring of 1906. Elmer returned to Mayo to mine on Highet Creek with the family joining him for the summer. They then went "outside" again that fall, visited Chicago and relatives in Missouri before proceeding to Seattle where Elmer bought a house and settled his family. Elmer wintered in Seattle, returning to Highet each spring; the family remained in Seattle until 1909, returning to Highet for the summer of that year. Elmer Land was born in Seattle, December 5, 1910. In 1911 the family again returned to Highet remaining a year. They spent the winter of 1912–13 in Seattle, returning a final time to Highet the summer of 1913. The girls had had no formal schooling in the Mayo area as there was no teacher. Their Mayo schooling commenced in 1915 with Emma Allen as their teacher.

Hunters Mary Ross and Evelyn Middlecoff, circa 1919. *L. Cheatham (McLaren) Collection*

Subsequent teachers were Florence McLaren, Edith Murphy, and Dorothy Hoggan. Until the girls attended school the family lived on Highet Creek. The George Miller family, the Mason family, Mr. and Mrs. Frank van Buskirk and quite a number of bachelors also lived along the creek. The Ross family lived about seven miles away at Minto Lake, so life wasn't lonely. There were many chores to be done with few idle moments but there was a lot of visiting back and forth plus the girls used to go hunting for rabbits and ptarmigan in the winter and grouse in the fall. Mary Ross and Evelyn bagged a moose when they were 17!

During the summer of 1916 Nell gave birth to their fifth child Flora Elizabeth on July 26 in Dawson City.

Many people along Highet Creek had gardens in which lettuce, onions, radishes were grown in abundance. Potatoes, swiss chard, cabbage, cauliflower also grew well and a profusion of flowers could be found in front of many homes. Supplies for the winter were transported from Dawson so there was rarely a need to go into Mayo. Later, when the family lived in Mayo, they had a farm located where the airstrip is now. On it they grew hay for the horses and their cow; also vegetables such as potatoes, carrots, rutabagas and beets which supplied the Middlecoff crew on Highet. The crew numbered up to 35 men in some years.

It was 1916 when the first dredge was constructed on Highet Creek by the Klasco Co. of Seattle with a Mr. Knott in charge of operations. The company brought a tractor in to haul the equipment to the creek; once assembled it wouldn't work so teamsters, Alex Nicol, Joe and Louis Cantin were hired to do the job. A building was erected below the Highet canyon and equipment set up to supply electrical power for the dredge. When the plant didn't work completion of the dredge was given up as only the scow portion had been constructed. Elmer bought the building using the lumber for sluice boxes.

The second dredge was brought in from Stewart City in 1920 by a group which included Lou Titus, Harry Jones, and E.O. Finlayson or (Finliason), the Manager of the Bank of Montreal in Dawson City. Ed Barker was the dredge master, Joe Walsh and Mr. Ross were engineers, and Rupert Steeves the winchman. The company paid a percentage to five claim owners on gold retrieved, each of whom did get $5,000 during the time of operation. The dredge began operation on Highet Creek below Lindquist's cabin. On the last shipment of gold to Dawson, the Bank Manager, Mr. Finlayson, recovered his total investment, leaving nothing with which to pay the crew or the expenses of the dredge. The company couldn't continue operating so went bankrupt. Various parts of this dredge and also parts of the Klasco Dredge went to start up the Wernecke Camp on Keno Hill.

Dances, concerts and plays were enjoyed by many in Mayo but the Middlecoff girls weren't allowed to attend dances until 1919 when their father eased up on his restrictions. Skating parties were a major pastime for the girls as well as tobogganing and sledding with dogteams in the winter months.

It is hard to say who influenced the life of Evelyn, or any of the other Middlecoff children, but certainly the following people left their mark: Mr. and Mrs. George Miller and children Hilma and Henry, the Mason family, Ole and Magnus Lindquist, Louis Bowden, Fred Swanson, Grant Huffman, Frank Cantin, Chas. Rockney, the Ross Family, Joe Walsh, Rodolph Rosmusen, Alex Nicol, Dora Holle, Steve Arbutina and uncle, George Ortell, Cora and son Gilbert Rich, Sam Blackmore, John Darbolo, Tom Keating, Billy Bramley, Andy Johnson, Paul Gust, the Churchwards (Annie and Phillip), the Hargraves (and children Mildred, Barbara, Paul and Dick), Hughie Monohan, Jewel Binet, Henri LeBlanc, Jack Smith, Bobbie Fisher, the Ferrells, Oscar Letourneau, Captain Hoggan, the Rev. Frank Buck, Randy Maclennan, Ernest Corp, W.D. Dempster and his wife, Jim Fairborn, Mrs. Sunderland and sons Ted and Joe, Joe McCaffery, the Cunningham family, Edith Murphy, Emma Allen, Florence McLaren, Harriet and Rupert Steeves, as well as Theos and Frank Whitney.

REGINALD THOMAS & VIOLA ANNIE MILLAR

by Viola Millar, November 1989

Reg worked on the Prairies until 1931 when the Depression was at its worst. That winter farmers were paying their hired men $5 a month and the government was paying $5. Reg decided it was time to move on because farmers couldn't even pay him his wages from past years. With the money he had saved, he went to Vancouver, B.C., to look for work but when he got there he discovered there were ten men for every job and so he decided to go north to either Atlin, B.C., or Mayo, Yukon. People he met while he was travelling north on the boat advised him to go to Mayo and following their advice, Reg arrived in Mayo on December 23, 1931.

During the first three years he spent in the Mayo district Reg worked for Elmer Middlecoff Senior's placer mine at Highet Creek in the summer and cut wood and worked at any job he could find during the winter months. He sacked ore at the Silver King mine one winter.

I was born Viola Annie Paul, youngest daughter of Jim and Nellie Paul of Manson, Manitoba. I arrived in Mayo as a bride-to-be on the S.S. *Keno* on September 21, 1933. Reg and I were married on September 22, 1933, by Rev. A.C. McCullum, Rector of Mayo's St. Mary's Anglican Church. We lived in Mayo during the first three years we were married. Our first home was a cabin belonging to George Miller and later we bought the old Alcroft house. Reg's brothers, Rodney and Donald, stayed with us after they finished school in Vancouver, until they found work.

When Treadwell Yukon decided to build the Elsa mill, Reg was the fifth man hired as a carpenter's helper for Rupert Steeves. Reg later worked a small Caterpillar tractor between the Silver King and the Elsa Camp. It was there, half way between the two mines, that we built a log cabin and the location became known as Millarville. We remained there until Treadwell was shut down in 1941. Reg helped board up the windows on the Elsa mine buildings and he was the third last man to leave. We, like everyone else, boarded up our home. We had planned to find a place in Whitehorse and remain in the Yukon but there wasn't a house to be found at that time so we moved to Richmond, B.C. Reg's father, Malcolm Edgar Millar, came out from Elkhorn, Manitoba, and joined us in B.C., where he worked with Reg doing carpentry work. Malcolm died of acute appendicitis at the age of 62.

Reg and I have two daughters, Dorothy Jean who was born in Mayo on March 10, 1938, and Norma Irene who was born in Grace Hospital in Vancouver on April 16, 1939. Dorothy and I left Mayo for Vancouver on February 28, 1939, and returned with baby Norma at the end of June, 1939. My brother Sterling McLelland Paul accompanied us on our trip back to the Yukon and when we got back to Mayo Sterling went to work for Treadwell and worked at both the Elsa and Calumet mines. Sterling remained in the district until the mine closed and we moved to Richmond. The weather was so cold when we were ready to move that we had to stay in a hotel until it warmed up enough for the plane to take off.

Reg's brother, Donald was married after he left the Yukon to Joyce Wells of Marpole, B.C. Donald and Joyce have two daughters, Donna and Rosemary.

Our eldest daughter, Dorothy married Duncan Roberts in 1954 and they have five daughters; Irene, Barbara, Elida, Donna and Noreen. Dorothy and Duncan's girls are all married with families now. Irene Isaacson lives at McLeese Lake, B.C., Barbara Lane lives in Calgary, Alberta, Elida Bruce lives in Mackenzie, B.C., Donna Sturgeon lives at Abbotsford, B.C. and Noreen Edwards lives in Calgary, Alberta.

Norma, our youngest daughter, married Dean Barlow in 1958 and they had one daughter, Leslie Ann who died in infancy, and three sons, Ross Barlow, Chris Barlow, and Dallas Barlow who all live in Prince George, B.C.

My cousin Cliff Greig also went to the Mayo district in the 1930's and lived there for a number of years.

We very much enjoyed our years in the Yukon and we have made a couple of trips back. Our last trip was in 1983 on our 50th Wedding Anniversary when we rented a motor home and visited all of our old friends.

[Reg Miller died on January 18, 1990.]

MAY GILLESPIE FIELDER MORRELL

by May Morrell, May 1984

My father, R.L. Gillespie, was sent to Mayo in 1922 to open a Mining Recorder's office, so my mother and I went along from Dawson. We lived in the quarters—an office, a large bedroom and kitchen—for about six months while my uncle, Frank Gillespie, and Tommy Morwick built us a huge log cabin.

The mines at Keno Hill were just opening up and high–grade silver was being mined. As I remember the concentrated ore assayed 420 ounces to the ton. Unbelievable. This was after it had been put through the flotation mill. The ore was then placed in sacks and trucked to Mayo, where it was piled on the bank of the Stewart River until the S.S. *Keno* arrived with its barge. The ore at that time went to Kellogg, Idaho. Later on some was sent to the smelter at Trail, B.C.

I continued high school in Mayo with the Rev. Frank Buck teaching four of us in the Anglican Church—Nora McLaren, Annie Churchward, Henry Miller, and myself. We all passed and the next year Lieutenant H.E. Taber (of the Royal Canadian Corps of Signals) took us on. He was in charge of the "Wireless Station"—the only way we could get a message or telegram to the "outside." The next year I went to Vancouver and took a commercial course, then went back to Mayo and helped my dad out in the office.

By this time Galena Hill had opened up and claims were selling for $150,000. Clem Sinyard sold the Hector Claim. Don Morrison [and partners Harry Colley and Tom McKay] sold the adjoining Calumet claim. I remember Clem Barlow received a huge amount. Bobbie Fisher, who couldn't count to ten, was offered $100,000 for claims on Keno Hill. Mr. Wernecke had the papers all ready to sign and Bobbie said, "I 'tink' I should get $150,000." Mr. Wernecke tore the papers up.

I remember the time when a big forest fire started about four miles out of Mayo. It was headed right for the town, the hospital right in its path. There was a cache of explosives, for the road crew about two miles out, which had to be rescued or Mayo would have blown up.

The men were all out digging ditches and making a fire break. Dr. Chipman was handling a hose trained on the hospital. Dr. Chipman was a wonderful doctor and surgeon. I must also mention the nurses, Miss Page and Miss Smith. They both served overseas during World War I in England and France. Whatever brought them to Mayo I'll never know. "Smithy" married Staff Sergeant (later Inspector) Dempster of the R.C.M.P. Inspector and Mrs. Dempster have passed away. However, their son Hugh teaches computers at U.B.C. I used to babysit Hugh and Sheila. "Pagey" lived in Vancouver and was well into her nineties before she died.

The Steamer *Aksala* tied up on the grass at the Anglican Church during the 1936 flood. *G. A. McIntyre photograph*

I knew Mr. and Mrs. Wernecke very well—he was the General Superintendent for the Treadwell Yukon Company Ltd.; also Alec Berry, and A.K. Schellinger, the geologist. Livingston Wernecke had a company airplane and pilot. The pilot, Mr. Gropstis, had flown planes to Burma during the war so was very experienced; but they met a tragic death. They flew out of Mayo in 1941 and stopped at Hyder, Alaska to examine a tungsten property. However coming down the coast north of Alert Bay they saw a plane in distress in the water and so were going down to help. Their plane hit a tree and they were found a week later in the water. They must have been killed instantly when they hit the tree. It was a freak accident.

My husband and I visited the Wernecke's at Berkeley, California on our honeymoon in 1937. My husband, Leo Fielder—R.C.M.P., came to Mayo in 1935. We were married in Vancouver after he was transferred there.

The Stewart and Mayo Rivers came up very high causing the flood of 1936. The Kimbel Brothers, Ed and Carl, had a launch running up and down the streets. They rescued a piano and floated down the street with Dr. Herkenrath, the dentist, playing it. *(See photo, page 108.)*

Dr. Nash warned everyone to boil their water and to have a drink of liquor every hour. We had eight feet of water in our house and lost all of our furniture. The steamer *Aksala* came to the rescue and we stayed on it until the flood was over.

One winter in Mayo it was –72° F and to go from our place to the office you practically had to feel your way through the mist.

Old-timers I remember are Isaac Mallette, a French Canadian; Sam Blackmore; Jimmy Johnstone, Postmaster; Dr. Wm. Chipman; Dr. Ross; Randy Maclennan, druggist; Joe McIntosh, liquor store; old Lesperance, undertaker. (One year nine miners died of pneumonia, they were laying around in the parlour with blocks of ice and sawdust around them).

Dan Matheson should never be forgotten. He and his son, Malcolm, put the telephone lines in. I worked for him after school or on weekends. I well remember the day we connected with Keno Hill—Mr. Wernecke had calls from San Francisco and I was sworn to secrecy.

You will never believe what the road to Keno was like. All gravel and very narrow. Anyone that saw Mr. Wernecke coming got off the road. He really stepped on the gas. Mrs. Wernecke told me years later that they were driving from Seattle, and she persuaded Livingston to pick up a hitch-hiker who was about 20 years old. Mr. Wernecke was suspicious of him and so he drove at a speed of 80 or 90 miles per hour so the fellow wouldn't have a chance to draw a gun and as soon as they arrived at a lighted city—Wernecke ordered him out.

Clyde Wann promoted the first airline service into Mayo. The pilot [and Wann's partner] Andy Cruickshank organized flights for $75 for one half hour. Mr. Lefebvre, who operated a sawmill in Mayo, paid my way and someone paid for Wilma and we had the thrill of a lifetime. But a few months later, the plane crashed while taking off.

My uncle, Frank Gillespie, built the new Binet's Hotel in 1930. It used to be all logs, and the restaurant was managed by a Japanese fellow. Mr. Nagano's dinners were tremendous, especially at Christmas, real gourmet.

I musn't overlook Joe Longtin. He delivered water by the buckets with a horse drawn sleigh. Also, Lou Titus who managed the power plant for our electric lights in the 30's (we used to have Coleman lights). My brother, Archie, who published *The Mayo Miner.* Also Everett Wasson, a pilot for Wernecke—he was a boyfriend of mine until he met a nurse in Whitehorse and married her. What a shock that was in Mayo! However, I did marry a veteran pilot [Thomas Morrell] in 1972.

We really had a lot of fun as teenagers; good skating with big bonfires, dances when the boat came in. Harry Whiteside played the piano, somebody had a saxophone. We played badminton in the Pioneer Hall; and my how we worked with bridge games and dances to build the tennis court.

DONALD MORRISON

by Mary Jean Morrison

Don Morrison was one of the best known pioneer mining men of the Mayo District. He was a native of Oban, Scotland and came to the Mayo silver district in 1922. He and his long time partner, Harry Colley were the original stakers of the Calumet claim, one of the richest claims ever discovered in the area. Don Morrison played a prominent and progressive role in the development of the silver mines in the Keno-Elsa district.

However Don did not live to see the results of his explorations. He died at the age of forty-six. He was survived by his wife, Jean, Louis Bouvette's niece and their daughter, Mary Jean.

MARY MOSES

Interview by Doris M. "Dorrie" Ewing, November 1987
Transcription by Linda E. T. MacDonald

My father was Rev. John Martin. He was a minister for a long time. My parents came from Fort McPherson. Mother's name was Bella Kunizzi before she was married. She had a number of brothers and sisters in Fort McPherson. Mom and Dad were married in the town of Moosehide.

When we were in Aklavik, Bishop Lucas arrived on the boat in the spring. He asked my father if he wanted to be a minister. All my Dad said was, "Sure" and he was ordained by the bishop right there.

After some years the bishop sent Dad to Moosehide. When we first came down from Aklavik we walked as far as Old Crow, resting a few days at the little town of Rampart House on the way. In Old Crow, Neil McDonald took us down to Fort Yukon in his boat. Neil McDonald grew up in La Pierre House. He was Archdeacon Robert McDonald's son and his mother was my Dad's aunt. In Fort Yukon we got on the steamboat and went down to Moosehide. My parents travelled back and forth from Moosehide to Fort McPherson many times.

We were in Moosehide for two years. My Dad's brother, Rev. Richard Martin, lived in Moosehide. He was blind but he could still preach. After that Bishop Stringer sent Dad to Ross River. It was winter when we left Moosehide and we took the stage to Pelly and from there we went by dog team to Ross River. We stayed there five years.

There were 11 children in our family and three of them were born in Ross River. Their names are Maggie, Mary, Effie—she died young and another girl born later was named Effie—Robert, Joseph, Rene, Betty, Benjamin, Julia, and Tippy.

I have lived in the bush all my life trapping and hunting. When we were in Fort McPherson and Aklavik we would come down to hunt rats in April. I used to go to day school in May, June and July when we were in Aklavik and Dawson. The rest of the time we were in the bush. I can read the Bible in Loucheux, my own language. That is the language spoken by Peel River Indians. When we were in Ross River I taught children how to read and write. Sgt. Tidd of the R.C.M.P., taught me how to type and play the organ in Ross River. I could type without looking at the keyboard. I can no longer remember how to type or play the organ. I used to play the organ in church.

While we were in Ross River my father tried to translate the Bible into the native language. He could read a bit but not very well because he never went to school.

After we left Ross River we went back to Dawson and then moved to the Old Village in Mayo around 1934. I taught school in the church there. I met and married my husband David "Little Dave" Moses at the Old Village. After that I quit teaching to raise my own children. We had 11 children altogether and I only lost one little boy who died when he was born. My children are Bernice, John, Albert, Julia, Richard, Norman, Benjamin, Liz, Frederick, and Effie.

Christmas at the Mayo Village was very much like Christmas in Fort McPherson. It was an exciting time for everyone. Today Christmas is nothing. In Fort McPherson the people decorated a Christmas tree and held a Christmas concert. We sang Christmas carols in English. We danced every night between Christmas and New Year's. Many people could play the fiddle and we danced square dances, jigs, and a number of other dances. People collected money to put on a feast. There were about 600 or 700 people in Fort McPherson then.

During Christmas at the Old Village in Mayo, the Chief collected groceries, meat, and other things

The Rev. John Martin family. Front, from left: Betty and Rene (Renie). Back row, from left: Effie, Mary, Mrs. John Martin (Bella), Rev. John Martin, Joseph, and Robert "Bob" Martin. *G. A. McIntyre Collection*

and made a big feast. We danced every night between Christmas and New Year's and lots of white people came down from Mayo to dance at the Village. We didn't have a hall so we danced in Chief Tom's big house. Little Dave played the accordion and some of the people from town could play music as well.

In 1937 my Dad died. The following year Mom went back to Aklavik where the rest of the family were. She died there. I have only two sisters left from my 11 brothers and sisters. My sister Betty is here in Whitehorse.

I still do porcupine quill and silk work embroidery, lots of knitting and bead work, and make gloves and jackets.

ANNIE (née CLOSE) NELSON

Interview and transcription by Linda E. T. MacDonald, 1985

To the best of my knowledge, my grandfather, John Alexander Ross, went to Dawson City to work in the spring or summer of 1897. He left that same year and came back the following year. My grandmother, Lena Elefsen, went to Dawson in 1900. They were married there in 1902. At the time, my grandfather was working a claim on Hunker Creek called "Last Chance." The claim was situated at a point where Last Chance Creek flows into Hunker. I always teased my grandmother about marrying my grandfather on "Last Chance."

From left, Annie Close (Nelson), Nancy Whitney (Moulton) and Doreen Lee (Doidge). *F. & T. Whitney Collection*

Lena and John Ross had seven children. Mary, Lena, Nora, Kate, Ida (my mother), and Thelma, were all born on Hunker Creek. After Thelma was born, they moved to Minto Lake where later their only son John was born on October 29, 1916. Mary married Gilbert Rich. Lena moved to Seattle. Nora married Bill Hare. Kate was paralysed and only lived to the age of 22—she is buried in the Mayo Cemetery. Ida Margaret Ross (my mother) married Archibald John Close (my father) on July 22, 1928, in Dawson City. Thelma married George Stevens. John lives in Whitehorse.

My parents, Ida and Archie Close, also had seven children; Annie born in 1930, Robert born in 1932, James born in 1934, Al born in 1936, John born in 1939, Muriel born in 1946, and Bill.

In 1922 my father worked underground at Keno Hill. Of course without the assistance of machinery and equipment available today, the work was much harder. Miners went underground using ladders and the ore was raised from the mine using a windlass. Once on the surface the ore was placed on steer hides and slid down the hill. Horses and sleighs brought it to the waterfront in Mayo where it was stockpiled and shipped out by boat during the summer months. Dad didn't think much of the underground work at Keno Hill. It was very dangerous. He had previously worked for several companies in Ontario including International Nickel where the working conditions and equipment were much better.

Having decided underground mining was too dangerous, he went into the freighting business. The only freighting company operating at that time was Greenfield & Pickering. He called his company A. J. Close. He went to Vancouver and purchased 28 horses and 6 new sleighs. He had to return to the Yukon immediately to organize his business; and so he arranged for his brother Hugh, an excellent teamster, to accompany the horses and sleighs travelling to Whitehorse by White Pass. White Pass would not allow horses to travel without a handler. My father gave Hugh the money to pay the freight when the shipment arrived in Whitehorse. Hugh was a ladies man and a bit of a gambler. He got into a card game on the boat and lost all of the money. When Hugh arrived in Whitehorse he wired my father that he had arrived in Whitehorse with the horses but White Pass wouldn't release them because he'd lost all of the money. My father received the wire on either Christmas Eve or Christmas Day 1926. He made immediate preparations to snowshoe to Whitehorse and left Mayo on Boxing Day. In a pack on his back he carried his sleeping bag, blanket, tin pot, and a bit of grub; and in his pocket the money for the freight. The temperature was –60°F. He walked from Mayo to Whitehorse in six days. He was so mad at Hugh the anger kept him warm all the way to Whitehorse.

After a time other freighting companies moved into the area and his business began to get a little further and a little further behind. Finally he quit the freighting business.

He decided that farming might be lucrative and took up a homestead. His original homestead was located where the Mayo Group Home is today. He cleared and broke all that land himself using horses and one big plough. He never did own a tractor. Clearing and breaking the land was a slow process and so when a man named Ringle died he bought the Ringle Ranch from the estate in order to acquire more cleared land.

There was a small property near Ringle Ranch owned by a man named Louis Bowden.We always called it the "Bowden Patch." When Mr. Bowden died, my father planted crops on that property as the land was not being used by anyone. I remember picking strawberries there when I was a kid. There were also beautiful strawberries at the edge of the field at the Huffman place—now owned by Simon Mason-Wood.

My father used to grow hay, clover, oats, sweetclover, alfalfa, potatoes, and vegetables. United Keno Hill Mines used to buy tons of potatoes from Dad. He used to experiment a lot with different crops. He tried pumpkins, corn, and other things. One time he got some hardy Northern Ontario apple trees and planted them. Later when the snow came the rabbits ate the apple trees. In the 1940's, the government started an experimental farm north of Whitehorse to experiment with different types of grains. Mr. Abbott was the manager and he used to come to Mayo every spring and give my father seed to experiment with in this area.

As the Ringle Ranch was up on a bench there was no water. The Mayo River flows in the valley below the ranch. Dad had a 45 gallon drum on a stoneboat. He would hitch the team to the stoneboat and we would go down the cliff to get water from the river. When the drum was full he would tie a tarp over the top and the horses would haul the water back up the cliff. He didn't water the garden of course because it was too big, but he did water his special plants such as the apple trees.

My mother seemed to be very much in tune with nature. One afternoon when I was six or seven we were picking strawberries at the Ringle Ranch. My mother was very quiet while she was picking. I had a five pound lard pail to put berries in. Suddenly I noticed a bear on the other side of the strawberry patch. I ran up to mother gasping, "Mom, Mom, there is a bear on the other side of the patch." She looked up and said very calmly, "Yes, it has been there quite a while and one of us is going to have to make a move pretty soon." Finally the bear just went away.

We lived in Mayo in the winter months and went to school and in the early summer we always moved to the home farm. When we moved to the home farm we walked the one and one half miles to school and back. When the crops were planted on the home farm we moved up to plant Ringle Ranch and so on. When we finished one place we would start on another. But most of the time we lived at the main camp at the home farm. We slept in a big tent with a board floor and sidewalls. The cook tent also had a board floor.

When we were kids we had to hoe four rows of potatoes before we could go and play. I was sure the rows went on to eternity and I always thought I would never finish.

The nurse who delivered me was Miss Veronica Page. "Pagey" is her nickname but you had to know her very, very well before you could call her "Pagey." She was Matron in Charge of the hospital when I was born in 1930. She was well into her nineties when she died in Vancouver, B.C. a few years ago.

My first school teacher was Gordon McIntyre. He taught me from 1936 to 1940 or 1941. The school was a one room log building with a cloakroom near the entrance and was heated with a wood stove. It was located where the North Star Motel is today. Gordon taught Grade 1 to 8. The attendance ranged from 30 to 40 children. Gordon maintained firm control over us. I thought he was grouchy in those days. I swear I got a "lickin'" every day but of course I didn't really. In the morning we lined up and marched in. The child with the furthest desk from the door was first in line. Standing by our desks, we said the Lord's Prayer and then we were seated. Instruction began with the younger grades and as they began working on assignments, he would instruct the higher grades. Every Friday there was a spelling bee. When Gordon went overseas we got a lady teacher Mrs. Carthum and later an Irish man named Hugh Hildidge.

To this day I have never gone to a Christmas concert that was better rehearsed and presented than the ones organized by Gordon McIntyre. The concerts were always held in the Pioneer Hall. I am a terrible singer but I just love to sing. When we were practising for the Christmas concert Gordon would always say "Now Annie, sing low." He was too polite to tell me I was not a good singer. The night of the concert I would always forget to sing low and Gordon would have to remind me from behind the curtain. Gordon played the piano for the concerts. (He was a good dancer too. He used to do cart wheels across the front of the classroom.) George Andison, who owned a butcher shop in Mayo, used to play Santa Claus. Everybody got a present—one apple, two oranges, and some peanuts.

"...Gordon would always say, 'Now Annie, sing low.' He was too polite to tell me I was not a good singer."

It was a big day when the steamboats came. We got the day off from school. The whistle would blow when the steamboat turned the bend and when we heard it everyone in town would go down to the waterfront. We would get fresh bananas from the boat and that was a real treat. The *Keno* came in most often. The *Aksala*, *Klondike*, and *Whitehorse* came only at high water. Transporting the Keno Hill ore made the back haul from Mayo most lucrative for White Pass (each ore sack contained 132 pounds of high-grade silver ore). I should think it was the most lucrative back haul in the entire Yukon. It was a sad day when the last boat left. After that any freight we got had to come overland. In the early days freighting was done with horses, as my father did in his business, and later with track vehicles.

Later on of course airplanes came in. Andy Cruickshank used to fly the *Queen of the Yukon*. On one trip the *Queen of the Yukon* got stuck on the ice after it had landed. I had a picture of my father pulling it out with horses. I believe my brother has the photograph now as he is in the flying business.

Joe Longtin and his horse Eli used to deliver water in Mayo. The water was 25¢ a bucket. Joe had a straw hat for the horse. If Eli was being difficult an argument would ensue and Joe would shout at the horse, "Who da boss, who buy da oats." Joe also had what seemed like thousands of cats. If you visited his house, cat eyes would stare at you from everywhere.

Treadwell Yukon had a big diesel power plant here. They used to start it up at 6:00 a.m. and shut it off at 5:00 p.m. After five you had to light the coal oil lamps or candles. My Dad used to be upset if the light bill was more than $4.50 for the month.

My brothers used to say Dad was the tightest man with a dollar they ever met. He was a character; he marched to a different drummer. He used to say, "There is a difference between needing and wanting," and "Close by name, close by nature." Another of his sayings I remember was, "He had the education of a Philadelphia lawyer and the common sense of a canary." My brother used that saying at the placer mining meeting with federal officials held in Dawson in 1983.

Bud Fisher had a car. My Dad had a truck. Charlie Taylor had a car and he liked to drive around Mayo in it. My Dad used to say "There goes 'Gasoline Charlie', up one street and down the other. No place to go and nothing to do when he gets there."

Seed orders were sent to Spences Seeds in Vancouver in February or March. Most of our clothing, household goods, etc., were ordered from T. Eaton Co., and Simpson-Sears. Each spring and fall we would send in a large order. If they were out of an item you had ordered, they would send a substitute—an item of similar quality and price. We seldom returned any of the substitutions. In those days you had to make do.

> ***"I never saw a neon light until I went to Whitehorse when I was sixteen years old."***

I never saw a neon light until I went to Whitehorse when I was sixteen years old. I flew from Mayo to Whitehorse. The air fare at that time was $48.70. Of course I was not allowed to find a place to live on my own. My Dad arranged for people we knew, who used to live in Mayo, to meet the plane and I stayed with them under their supervision.

Whitehorse sure looked big to me. It was just at the end of the war so there were still a lot of Canadian and American soldiers there. I believe there were about 30,000 people there then. My first job was a chambermaid at the Whitehorse Inn. It lasted about a month—I couldn't stand it. Next I worked for some Chinese people in a cafe called the S & A. It was located across the street from where the New North Hotel is today. I got $16 a week plus two meals and tips. Then I worked for White Pass at Bennett Lake. After that, another woman and I ran a laundry in Elsa camp for a time. Returning to Whitehorse, I went to work for the airbase where I met the man I was later to marry, Alf Nelson. (He was out of the army then and had come to Whitehorse to work.) When I came back to Mayo in 1952 Alf was working up at the Elsa mine. I kept running into him everywhere I went: Mayo, Elsa, Whitehorse; and finally we were married in 1956. I guess I knew him about nine years before that. I always tease him about that, I ran so fast and had so many things to do it took him nine years to catch me.

When my father was in his seventies he got so he couldn't harness the horses or manage the farm any more. He offered all of us a chance to take over the farm but we declined. We did not feel you could make a living on the farm any longer without having another job as well. So the parcels of land that comprised the Close farm were gradually sold. By this time us kids were all living in different places. Jimmy and Bill in Watson Lake; myself in Whitehorse; Bob and Al "Lugs" in Dawson; Muriel in Edmonton, Alberta; and John in Spruce Grove, Alberta.

It is really hard to explain to people that have been here a shorter time what it was like in 1936. You can tell them about it but it is impossible to recreate the atmosphere of that time with words.

ALEX NICOL

Aaro Aho[2]

Alex Nicol, a wiry young French Canadian, with looks that belied his strength and stamina, crossed Canada with one dog and $300 borrowed from his brother. He scaled the Chilkoot Pass in 1898, packed his supplies to Lower Laberge, and at the age of 18 had been in the first wave of boats to reach the Klondike.

He was also among the early stampeders into Duncan Creek and staked claim Number 35 below discovery. He built a cabin on his claim the next July; sank a few prospect holes, but got only a few colours.

2. Yukon Archives. Aaro Aho Collection, 82/161, "Keno Hill: An Era of Individualism in Yukon's Great Silver District," circa 1972.

In the winter of 1902–03 he worked on the government shaft, and heard that the government might build a road from Mayo Landing. Snowshoeing down the trail in March, he arrived at the river to build the first small cabin at Mayo Landing, finishing it by April 1. He returned to Duncan Creek before the trail over the ice of the Mayo River deteriorated with spring breakup.

Alexander J. "Alex" Nicol, one of the founding fathers of Mayo. *G. A. McIntyre Collection*

One day while Alex had been working on his cabin at Mayo Landing, surveyor Raoul Rinfret, Joseph E. Beliveau, and J.D. Bell arrived overland from Dawson with a mule, tent, stove, camp outfit and supplies. The beginning of the best trail to the Duncan Creek diggings, was a logical site for a town so they scouted the landing briefly and then staked and surveyed a town site of 185 acres, for this steamboat landing on the flat bank at the river confluence.

In 1905, Mrs. Potter, a ravishing redhead, wife of one of the miners at Duncan Creek became critically ill. The nearest doctor was in Dawson, some two hundred miles away. As Mrs. Potter's condition worsened, it was decided that she must be taken by dog team to Dawson. Alex Nicol had just returned from Dawson. He knew the trail well and had no qualms about his ability to make the journey quickly.

Alex, George and Lena Potter left on December 24. Mrs. Potter was placed in a large box that had been fitted to the dog sled, wrapped in quilts and surrounded with heated stones. The first night, after thirty-five miles of travel, they sheltered at Crooked Creek with Cap Williams. There they were met by Sergeant Bates of the Northwest Mounted Police from the McQuesten detachment. He accompanied them as far as Dominion Creek.

They arrived in Dawson on New Year's Eve, 1906. But all the perseverance of Nicol and Potter in this winter's ordeal were to no avail. Mrs. Potter died in the hospital twenty-four hours later.

From 1910–1915, Alex Nicol freighted mining supplies into Haggart Creek and Dublin Gulch and Highet Creek. Supplies were freighted in the spring before break-up.

In 1913 Nicol hauled ore for Jack Alverson and Grant Huffman from the Silver King mine for $50 per ton.

He was one of the early stakers in the Keno Hill discovery.

On the first day that Livingston Wernecke arrived in the area, Alex Nicol acted as guide and took him to see the Shamrock and Guggenheim workings.

With Treadwell Yukon's interest in Keno Hill and the great excitement over the Sadie, Ladue and No. 9 mines operating; Alex Nicol sold his 4-horse team and saddle horse to Harry Jones and joined the claim staking, selling, and prospecting. Nicol lived well but conservatively by holding and selling various claims on Keno Hill until his death in 1965.

GEORGE ORTELL

by Edward Ortell, January 1985

On a cold frosty December morning Maggie Ortell, looking out the west window of her home, heard the lonesome whistle of the train at Silvy Switch. It was the yearning call that had led her younger brother, George, off to the Klondike. Maggie was 27 now, 10 years earlier at the age of 17, her mom had died and left her with 5 younger brothers and sisters to raise. Their names were John, Katy Cockrell (who had no children), Louise Vogelsang, George, and Emma. Maggie had no children of her own. The

George Ortell, a prospector and trapper who lived in the Mayo area. *Middlecoff-Bleiler Collection*

siblings were pretty well raised by now, the farthest had gone to Chicago which was nearly two hundred miles east of this little Iowa town located between Dubuque and Davenport.

It was the time of the Klondike and brother, George Ortell, could not resist the call of the whistle. He had boarded the train heading for Bellevue, Iowa; then the river boat to St. Louis. Maggie kept and treasured the letters he wrote as he made his way westward. Perhaps the businessmen of Seattle made more on the gold rush than any of the sourdoughs, because Uncle George wrote Aunt Maggie he had very little left after he purchased his equipment for the Klondike.

It was on the river boat that George met an old gambler. He asked George where he was heading. George said he thought the Yukon. The gambler remarked, "It's a beautiful place. I've been there before." George was deeply moved by such a tribute as this, and he said when later times, the frigidity of the climate made him hollow inside, the beauty of the Yukon, the Northern lights running wild replaced the hollowness with an emotion he could not express.

George's nieces and nephews numbered many:

John Ortell had five children—the namesake George was first born, then came Harold, Margaret, Imogene and Edward.

Louise Vogelsang had three daughters—Mabel, Bertha and Clara.

Emma Quimby had one daughter, Janet.

He wrote to them all and often sent souvenirs varying from little nuggets of gold to grizzly bear skin rugs. Also, he often had a story to tell of life in the Yukon.

Maggie had married now. She and her husband Fred Harloff were living in La Grange, Illinois. George wrote her for permission to get married. George was now 50, but Maggie had always worn the mother role so he asked her for permission. The woman he wished to marry was a native woman. Maggie refused to give him permission so he did not marry.

Perhaps there was another reason. George had been in love with another man's wife in the Yukon. But those were the days of symbol—to honour a great moral code—the code of conduct and chivalry. The source of this information said that, as a child with a child's intuition, she could tell George was in love with her mother. But she was quite sure her mother was never aware of George's obsession. George did not attain great material wealth in the Yukon. Instead it taught him the wealth of character—to be proud and unbending in honest failure, but humble and gentle in success. He did not substitute words for actions or seek a path of comfort, but he faced the stress and spur of difficulty and challenge. In the winter of 1943 at 71 years of age George started down the trail for Mayo Landing. The temperature was –60°F. The webbing in George's snowshoe broke and while trying to fix it, his hands became frozen. He then had to go on without his snowshoe. His feet became frozen. So as not to fall over, he heaped snow around his waist and there, twenty hours later, a Mountie found him.

George was taken to the hospital in Dawson. There he had quadruple amputation. But his spirit was still undaunted, and from the writings of Archie Gillespie comes the quote, "Save what you can Doc—I've got to hobble around somehow. Figure on getting me a bunch of mighty fine beaver over on the McQuesten later this year."

George never acquired material wealth, but he left a legacy of legend to be added to that of the Yukon. There are only two mountains in the Yukon named for Americans: Mt. Kennedy and Mt. Ortell.

LUCY PETER

Interview by Doris M. "Dorrie" Ewing, November 1987
Transcription by Linda E. T. MacDonald

My mother's name was Eliza and my father's name was Lucas. I believe my father came from Selkirk and my mother's father came from Dawson people. Big Maisie was my mother's sister. Billy Malcolm's father was my father's brother. I was born in 1906 and I was the only girl in my family. Old Frank Cantin told me that I was the first child to be born in Mayo town. He said I was born two days before the New Year. My brothers are Johnson, Paul, John and Moses Lucas and they were all born before me.

In 1915, Julius Kendi and his family came to Fraser Falls. Everyone was at the falls catching and drying King salmon. I was about nine years old and Julius' daughter, Mary, was about eight. Julius and Rev. Frank Buck held church inside a tent. Many people were there and Rev. Buck baptized Sally Lucas, Ella Profeit, Maggie Mervyn, and Chief Tom's boy Albert, in the church tent.

Julius was originally from Aklavik. When he first came here we had a little trouble understanding him. He spoke the language of the Moosehide people. Chief Johnson, who came originally from Moosehide, could understand Julius. Julius' Bible, handbook, and prayer book were all written in his language. My mother, Eliza, could understand and converse with Julius as well.

Mayo Elder Lucy Peter. *L. R. Bleiler photo*

At that time we were living in Mayo. Our homes were located at the mouth of the Mayo River. When Julius came down from the falls to Mayo, he called the people together and asked us to find a place to build a village of our own. A policeman was at the gathering as well. My Dad wanted to move upstream on the other side of the old sawmill, but my Dad was not chief. Tom was chief and he wanted to move across the river and downstream. So, in 1915, we moved to the new village. Logs were brought from up river and we built Julius' mission, the chief's house, and our homes.

We enjoyed going to Julius' church very much. We walked to church carrying our babies and holding the little ones by the hand.

After a time Julius went back to Old Crow but the Bishop asked him to return because the Mayo people wanted him back. While Julius was away, Mary Moses' grandfather, Joseph Kunizzi, came to stay for two years. When Rev. Totty left, Julius went to Dawson and Moosehide for two years and then came back to Mayo. In 1947 Julius left for Old Crow. He died after that. Julius had taught us to take care of the old and the poor. After he died I looked after his wife Persis. She was blind by that time.

After Rev. Kendi left, John Martin came. Then Sadie Dougall came and taught school to the children; after her came Lillian Connor. Miss Hellaby came in 1952. She was the last one. In 1959 everybody moved to Mayo.

Sam and I were married at the Old Village in 1922. Rev. Kendi performed our wedding ceremony. My children were all born at the Old Village. Everyone's children were born there in those days. I had ten children altogether; five lived. My firstborn, Mary, died when she was five years old; my boy Lucas died as well. Another boy and two more girls died. My children are John, Walter, Johnson, Kathleen, and Ida. Kathleen finished school and then went outside. My other children live in Mayo.

I raised a number of other children as well as my own; among them, my brother John's son Jimmy "Shing-Shing" Lucas, Albert Peter, and Christopher Louis. My grandson Albert Peter is currently Chief of the Na-Cho Nyäk Dun. Simon Lucas was also my brother John's son, his wife was Gladys Ellis.

My brother Johnson Lucas married Caroline, a Mackenzie woman. Peter Lucas is Johnson and Caroline's son. My brother John married my husband's stepsister Sally Peter. After John died, Sally married Steven Louis. Charlie Jimmy, who came originally from Pelly, married Sally's daughter Alice. Charlie's sister, Lizzy, was Julius Hager's wife. Charlie and Alice's children are Sally, Annie, Betty, David and Douglas.

My father taught me how to tan skins when I was a little girl. In those days we tanned a great many hides. The skins were always soaked in rain water before they were smoked. Tanning moose hides is a two-person job. I used to tan many hides but now that I am old the work is too hard for me. I still make plenty of dried meat. When the hunters bring back moose fat, I render it into lard. You cut the fat into pieces and place it in a large pan in a slow oven. The moose lard (or grease) is used to make bannock, in the cooking of meat and other good things. Moose lard will soften dried meat so that old people with only a few teeth can eat it. My family and I eat very well.

Chief Tom was chief at the Old Village for a very long time. He always made a happy Christmas for everyone. Just before Christmas people would come down from the trapline. Taylor and Drury bought furs at that time and everyone had plenty of money for Christmas. Chief Tom hunted fresh meat and arranged a big party for everyone. There was always plenty of food prepared for the celebration. In addition to the fresh meat, there was dried meat, blueberries, dried fruit, bread, and other good things to eat. We were like one family then.

When they told us we had to move to Mayo I did not want to go and neither did Sarah. Finally Sarah moved to 17–Mile and just my husband and I were left in the Old Village. The Indian Agent told us we could not stay there alone. In the spring of 1959 we moved into Mrs. Asitsaw's house in Mayo and I looked after her until she died in 1960. After that I wanted to go back to my home at the Old Village but the Indian Agent said we could not return there alone. Joe Armashot, the Indian Agent, said we could have the house at the mouth of the Mayo River that Sam had started to build in 1950. So Sam finished the house and we moved into it. Right at that time the Mayo Indian Band was organized. Joe Armashot was the last Indian Agent. We moved into the house we are in now in 1980. The old house at the mouth of the Mayo River is still there. Andy Lucas and Sophie lived in it most recently.

In the summer of 1987 I went down to the Old Village. My house that Sam built for me in 1916 is broken down now. I would still like to live down there. You know, I liked Julius Kendi's church.

SAM PETER

Interview by Doris M. "Dorrie" Ewing, November 1987
Transcription by Linda E. T. MacDonald

It is possible that there were Indian people along the Stewart before the Selkirk Indians arrived. Sometimes people die off.

A long time ago, Selkirk Indians hunted and trapped through this area. They travelled from Selkirk up the McQuesten and Beaver Rivers, down the Stewart River and back to Selkirk. That's the way the Indians, and the Company, trapped in those days.

Later, when gold was discovered up the McQuesten River, Indian people from Selkirk moved to the mouth of the McQuesten where a trading post had been established to supply the miners working the McQuesten and it's tributaries. Miners poled boats up the river to mine on Haggart, Highet and Johnson Creeks.

Selkirk Indians lived at the McQuesten three or four years. They left to fish, hunt, trap, and dry meat; and would then return to McQuesten post. Indians did not trap in only one place. If traps were set and nothing was caught, they moved to another area. My old father moved over the whole area to trap. People trapped through the winter until spring came and then hunted beaver.

My father's name was Charlie. He had a brother named Joe Duncan. My mother, Maisie Johnson, was Chief Johnson's daughter. I was born at Ethel Lake. I am not certain of the year I was born but I believe it was about 1902. My father died around the year 1905 and my sister died soon after. They are buried together across the river on the high bank, across from the mouth of the Mayo River. My brother died at the age of 12 or 13 sometime after my father and sister died.

A long time ago, Rev. Totty would come to Mayo from Dawson once in a while. I think he came

From left: Lucy Peter, Persis Kendi, and Sam Peter. *Lucy Peter Collection*

every three years or something like that. Rev. Totty baptized me when I was about eight years old.

In 1916 I trapped up Nogold, Willow Rock, to Moose Creek and Big Kalzas, to Moose Lake, and way up to those creeks where that couple who trap today have built a house. It is still my trapline. In 1916 we trapped from Moose Lake up to what we called Russell Post. Art Zimmerlee and a man called Schofield had a store there. Armstrong was looking for gold at Russell Post at that time. He had obtained a concession from Duncan Gillies.

My son, Walter, works a trapline today that we started in 1912.

My stepfather, Peter, was a good carpenter. I think he came from Ross River. He showed me how to make mooseskin boats. To make one you must double the mooseskin and double sew it, that is, sew it and then fold it over and sew it again. A small spruce tree was bent to shape the frame. A big corner needle and sinew was used for sewing. We built one to four boats at a time. Each boat would hold two men.

When Binet built the store in Mayo in 1903 and the road was put through to Duncan and Haggart Creeks, people began to move from McQuesten post to Mayo. The store in McQuesten shut down in 1904. We settled at the mouth of the Mayo River in Mayo.

When Julius Kendi arrived in 1915, we moved to our own village, three miles down and across the Stewart River from Mayo. We built the church, minister's house, and our homes ourselves. After the church and rectory were built, I supplied them both with firewood. In the early morning I would go to the church and start the fire and clean.

One time, Lucy and I, Peter Hager, Johnson Lucas, Peter Lucas, and Johnson Peter were hunting moose and drying meat at Old Stewart Crossing. We loaded the meat in a boat and started back up the Stewart River. The boat we had was a big river boat, with two inch sides, built by Renny Burian. At Devil's Elbow, a tree, floating down from upstream, struck the boat. It went right through the boat and broke off the propeller. The boat immediately turned around in the middle of the river and then started to sink. I told the boys to upset it to get rid of the weight. All the fresh meat, dried meat, and guns were lost. We swam clinging to the boat which was constantly sinking and raising in the water. Ice was running in the river at that time. Finally the boat hit a tree close to shore. Johnson threw a blanket to the shore and he and I crawled in on the shore ice. The two Peters hung on to the tree and then made their way in. When we were all on the shore, I showed them how to dry the matches in their hair. Your hair has lots of spark you know. So Peter Hager did that and built a large fire and we slept. We had managed to save the eiderdowns because they float. Then we walked.

Another time I had a white man with me, Carl Kimbel, and we were on our way to get logs for Kimbel's sawmill. We were heading upstream and suddenly, right in the middle of the channel, the bow of the boat went under. Carl had the pole so I told him to push it back. He just looked down and got scared. I don't know why he was scared then, it had already happened. Anyway we managed to save the boat and didn't go down. This was around the time that John was a baby.

Lucy and I, Peter Lucas, Doug Whiteman—they called him Randy the White, John and Walter were on the river going through Five Mile Rapids. Randy went ahead and I was behind. He went too far out, hit the rocks, and the propeller broke off just like that. The boat started downstream, the engine working, and everyone prayed. I said, "Jesus where are you." God answered,"You called my name, I am going to do it;" and just like that a big wave came and took the boat to shore.

THE PORTLOCK, CANTIN FAMILY

as recalled by their grandson Franklin Portlock
March 1989

William Portlock (1876–1924) arrived in Canada in 1898 from Preston, England. He crossed Canada eventually reaching the Yukon after travelling over the Chilkoot Pass and settled in the Mayo area. He mined on Highet Creek, Dublin Gulch, and Haggart Creek in partnership with the Abbott Bros. His fiancée Alice Stevenson (1884–1955) arrived from Preston, England in 1914 and they were married in Dawson that year. Living on the creeks was an entirely new experience for her as she was a trained

The Portlocks Haggart Creek cabin, 1916. From left: unidentified, unidentified, William Portlock holding baby Thomas Portlock, Agnes Kinsey, Alice Portlock, and Olive Kinsey. *F. Portlock Collection*

seamstress and had never dealt with the other problems of household chores. Their son Thomas William Portlock (1915–1976) was born in Dawson City after Alice travelled by foot, dogsled, and stagecoach in the early stages of her pregnancy. She returned with her newborn son to their home on Haggart Creek the following March (1916). They lived on the creek until William became ill in 1922 and then they moved into Mayo. Alice cooked at the Mayo hospital and did seamstress work for local residents until William died in 1924.

Alice was then courted by an eligible bachelor by the name of Francois "Frank" Xaviar Cantin (1876–1958). Frank had mined on Highet Creek in 1915–16, and 1923–24 and later on Dublin Gulch and Haggart Creek. He eventually started a wood business in Mayo supplying firewood, mining timbers and lumber from his sawmill located at the mouth of the Mayo River on the Stewart River. Alice and Frank were married in Vancouver, B.C. in 1931 where Alice was living while Tom finished high school. After a trip to Quebec to visit Frank's family (he had 16 brothers and sisters) they returned to Mayo to live the rest of their lives.

Frank continued in the wood business. He also built and rented several log homes. In addition he rented the log school building (the present North Star Motel site) until 1941 when he built a frame school at the request of the Yukon Territorial Government on part of Joe Cantin's homestead and rented it to the government until they eventually purchased it. The favourite story told of Frank concerns his finding of a large gold nugget which is now part of a watch chain. He was walking near his sluice box after the day's work and saw something on the ground. In his broken French-English he exclaimed "I taut is was my watch, but it was tis big nugget."

Tom became involved in the business with Frank for a number of years. In 1936 a young nurse ar-

Mr. & Mrs. William Portlock and possibly Mr. Abbott with pipe. Probably taken at their Haggart Creek placer property. *M. Rich Collection*

rived in Mayo to work at the Mayo Hospital which was then located on the north side of town next to the Mayo River. Frances Elizabeth Waugh (1912) and her sister Millie were born and raised in Dawson City. Their mother Elizabeth Waugh(1885–1940) née Black operated roadhouses in the Carmacks area after the death of their father Aaron in 1929. Frances completed her high school in Whitehorse and then trained as a registered Nurse at the Vancouver General Hospital in Vancouver, B.C. She graduated in 1936 and worked at that hospital for a year. She applied for the nursing position at the Mayo Hospital and eventually arrived back in the Yukon.

Tom and Frances were married in 1937 in Mayo. A son Franklin Thomas was born in 1941. Tom secured employment with the British Yukon Navigation Company (White Pass) and the family moved to Whitehorse in 1942. Franklin married Carole Lani Kempster in 1956 and they have two children, Steven Joseph, and Suzan Elizabeth who married David Brillinger in 1987. The Portlock family now reside in Victoria, B.C.

The following prose is of unknown authorship about Alice and William Portlock's arrival on Haggart Creek after their wedding in Dawson:

So Many Memories—

We were excited on Haggart and in our little cabin and for hours watched the trail-like road for Will Portlock and his English bride whom he had gone to Dawson to meet—then by river boat to Mayo—and, I suppose, a rig of some kind to the road—so called—which took off to Haggart Creek. Anyway they had to walk the last 50 miles or so—what an introduction to married life. She had been a seamstress, they called it—made suits and dresses—and believe she had never worked a day in her life as she was to do the rest of her life.

Then they came along the path in front of our old log cabin—I remember how blond she was and how very pretty. Am sure she had long skirt on—but I recall that she looked such a lady! But they would not stop, even for tea—so anxious to get to their own home. And a comfortable one she made of it too. So precise, everything so clean. A beautiful garden of course.

Only remember one thing clearly about him—he would not swear at all—and when he had to refer to the "Go-Devil" which was a small sled with a gee-pole on it—he called it a "Get Thou Behind Me Satan."

Memories—
The "road" still takes off to Haggart—
We saw it a few years ago—still about impassable!

RETALLACK – THOSE WERE THE DAYS

by Athol Retallack, July, 1987

We couldn't consider ourselves as Mayo "old-timers" but we were certainly "great timers."

When we arrived at Mayo airport on a chilly, March morning in 1952, the town was just abustle, and our taxi driver, Tim Cole, was the most enthusiastic booster one could hope to encounter.

The dam for the hydro was being constructed, government survey crews were heading out in all directions, the road between Whitehorse and Dawson City was under construction, every possible hole in the hills around Elsa and Keno had a mining engineer and crew of men working, two hotels and their bars were filled to capacity, two general stores, gas stations, several taxis, the Royal Bank, liquor store, federal and territorial offices were busy, the White Pass freight shed a beehive, men coming and going from the construction offices, full staff at the hospital—it was a most exubcrant community in which we were to make our home, a striking contrast to the measured pace of a rather posh resort town of Minaki, Ontario.

The Northern Commercial Company store of which my husband, Tom, was manager, was an immense, two storey log structure, with a high, false front. It always appeared to me like a monstrous, stuffed elephant, one of the homeliest pieces of architecture one could imagine. It was about 40 feet wide and 60 feet long. Our son, Tommy, soon explored the attic and informed us that there were piles and piles of fluffy sawdust up there and old wires running in all directions. I thought about those wires, the sawdust and the large number of mice who had established residence before we did and I spent some uneasy nights, but the old place survived to be torn down many years later.

If the store area was almost 3,000 square feet, with a small area set aside for storage upstairs, our living quarters were the same dimensions. It was a big apartment and we soon found it was great for all sorts of gatherings—the Commissioner of the Yukon asked if he could use our living room for a meeting with townsfolk, the Whitehorse Choral Society held their rehearsals there, and we would lose count of how many people were guests at a party.

One of our larger gatherings happened soon after we arrived. One day Tom said he had had a letter from the Edmonton Chamber of Mines saying that a group of their members would be in Mayo to tour the mines in the district and he thought it would be nice if we served them coffee before they left for the mines. Of course I agreed and thought little more about it. A few days later he reminded me that the tour would be arriving two days hence, "How many should I prepare for?" Ten or twelve was what I had expected.

Mayo's Front Street in the early 1950's. The Chateau Mayo is front and left and further down on the right side of the street is the Silver Inn. *Athol Retallack Collection*

"I've just had a message that there would be 82 in the party, including Mayor Hawrelak."

I sent out an SOS to my friends and it was a delight to see the expressions on the faces of our guests as they came into the living quarters—I doubt that the setting and food could have been excelled anywhere—the delicious cakes and rolls set on the finest china and lace table cloths, with the silver serving pieces and spoons liberally set with gold nuggets. That was Mayo's style.

We did everything in style, it seems, now looking back through pictures—we wore formal gowns to the community dances or house parties before the dances, as we did when attending parties put on by the mining companies at Elsa and Calumet. The staff halls were at 4,000-foot elevation and to dance, or stand out on the decks at midnight in the twilight, looking across the broad valleys to Lookout and the Hungry Mountains were moments memories were made of.

We soon became involved in the activities of St. Mary's Anglican Church. The tragic drowning accident of the Reverend Wareham who had been a dynamo in the community, stunned us; then we had the good fortune to welcome Archdeacon Kirksey as our clergyman. He decided to renovate the interior of the church and he swept us along with his enthusiasm. What was really needed were pews to replace the rather primitive seats which had been installed when the church was first established. Tom had heard that one of the large churches in Dawson City had closed and there were pews for the taking, first come, first served. So off he went to Dawson and it was a toe-to-toe situation to see who would get the beautiful, hand carved, oak pews. He got them, and had them brought to Mayo. Kirksey was delighted but when the men tried to get the pews in to the church, they were too long. What a disappointment. Kirksey said something about ". . .there was another Man in my business who was a fair carpenter," and he labouriously sawed off the ends of the solid oak pews, extracted a few feet from each one, and replaced the carved end pieces; the hours and hours of manual labour he put in to adjust those pews for our church. He then had pale grey drapes placed behind the altar and our little church was all ready for Christmas Eve service.

While Archdeacon Kirksey was there, the rectory became a drop-in centre. The coffee pot was always on and Kirksey encouraged young men who didn't want to spend an evening in the bars, to drop by to have a game of chess or cribbage.

He was a man of many talents. Our twelve year old son wanted to know if we would mind teaching a couple of his friends how to dance so that they would go to the community dances with him. We cut out paper foot prints and laid them on the dining room floor, "a la Arthur Murray" method of dance instruction, got suitable records for the initial One, two three, One, two three, dance steps. The first session worked out fine and we were ready for the next session in a week's time. When we opened the door that evening, there were five or six eager would-be dancers. We sent a message to Kirksey and down he came and showed himself to be an excellent ballroom dance instructor. How that man could schottische.

A few years later Archdeacon Kirksey was killed in a traffic accident near his parish of Pouce Coupe. He is remembered with affection by many one time Mayo residents.

We were among the charter members of the Mayo Curling Club. I think the membership was $40 and we were soon curling on a rink whose walls and ceiling were plain, uninsulated, galvanized iron. When it was 40°F below zero outside, it was 40°F below zero inside and we became good strong curlers on that heavy ice. Of course in the spring we became equally good at sloshing the rocks down the wet ice too.

The first earthquake I ever experienced was when I was on the curling ice which rippled, the walls wavered, and I felt queasy. When someone yelled "Earthquake" all I could think about was that old NC building, all those logs piled up, one on another, and if the bottom ones rolled out, the whole thing would come down like a match stick house. When we got home the light fixtures were swinging and the pictures on the walls were askew, but the old monster was still in one piece.

Looking through old photographs, I came across a picture captioned "Coronation Day Parade in Mayo" and there was Dale Robertson driving a Model T truck with the royal entourage consisting of

Mary Jean Boyle, complete with crown and fur trimmed cloak, and her attendants, Lois Cole, Sybil Bleiler, Janet and Maureen Cripps. The 24th of May photographs were of a parade of decorated bicycles and tricycles, the Mayo baseball team in sharp, red uniforms—Steve Mosich, Dunc McGeachy, Ian Smith, Ian McLennan, Bill Bonner, Rusty Martin, Lorne Vance, Jim Davies, Ed Champion, and "Kippy" Fisher.

While our stay in Mayo was only for three years, we saw great changes. The hydro electric project was completed and with that came improvement in our life style, with unlimited domestic power for appliances and refrigeration. The all weather roads to Dawson and Whitehorse ended reliance on aircraft and steamboats and the so called "isolation." Actually, I never thought of being "isolated" surrounded by family, friends, and a good active social and community life.

We also had the good fortune to know some of those who had come to Mayo during the early days of settlement. Alex Nicol, for one, was a fellow passenger up the coast and over the rail to Whitehorse and we listened with great interest when he spoke of the early days in the district.

During those hectic days of development and construction, we all worked long hours at our respective businesses and occupations, yet found time to enjoy life to the full and form friendships which will last a lifetime. Mayo holds its own special place in our memories.

MARY HARLAND (née ROSS) RICH

Interview and transcription by Linda E. T. MacDonald, March 1987

My father, John Alexander "Jack" Ross, came over the Chilkoot Pass in 1897. He came to the Yukon with a peg-legged fellow named Tom Boswell. Boswell had come from the Yukon to Seattle with a heavy poke of gold. My father met Boswell when he tried to interest the mining engineer, that my father flunkied for during the summer, in investing in his gold prospects. The mining engineer asked my father to go with Boswell, and grubstaked them both, so that he could determine if it would be a worth-

This photograph of the Ross family was taken on July 13, 1913, at Minto Bridge by J. E. "Jim" Ferrell. Mrs. J. A. (Lena) Ross is holding Kate and clutching Margaret. Nora and Mary are standing behind Joe Cantin the wagon driver, and Lena is sitting next to her mother. Jim Ferrell had ridden the pony tied behind the wagon from Mayo to Minto Bridge to pick up the mail for Mayo. *Mary Rich Collection*

while investment. Boswell told the mining engineer that the gold had come from his mine on the Stewart River.

Several men from Seattle, on their way to Dawson, travelled with my father and Boswell. There was a doctor (O'Dell), a butcher and a blacksmith.

While on the trail, they encountered some school teachers who were also travelling to Dawson. These women had a number of goats with them. The goats all carried packs and at a place on the river had wandered out onto the rocks. Papa and his partner helped get them off the rocks and rounded up.

When my father and Boswell arrived at [the then unnamed] Chapman's Bar on the Stewart River, Boswell told my father there was no mine. He had acquired the gold by panning the gravel on the bar. It appears he deliberately misled the mining engineer in order to get a grubstake.

Grant Huffman with the Ross family at Minto Lake. *D. Walloch (McLaren) Collection*

My father only stayed for a few days as it was getting late in the season and he wanted to get back to Seattle. He planned to raft down the Stewart and Yukon Rivers to Dawson and catch a boat there. When he was ready to leave, Tom Boswell asked Papa if he would sell him his outfit. My father agreed because he did not need it as he was not planning to stay. Boswell said, "Thanks Jack, I am glad you said that because I had made up my mind to kill you and take the outfit if you had refused to sell it to me, and I don't like to do that."

Soon after my father arrived in Dawson, he contracted typhoid fever and did not get out of the hospital until the last boat had left. When he was well enough to get around he realized that there was only $14 left in his money belt and there had been $1,200 in it when he fell ill. He told Dr. O'Dell about it. Dr. O'Dell had travelled to Dawson from Seattle, hoping to practice medicine there but he was not allowed to (even though they needed doctors badly) because he was an American. Dr. O'Dell went to see the police about the wallet and they were able to recover all of Papa's money. Dr. O'Dell was then given permission to care for my father for the remainder of his illness.

When my father recovered from the fever, he went up the Klondike River to hunt. The butcher, who had travelled up from Seattle with him, went along as did Dr. O'Dell—for some of the time. The butcher was a master at his trade and my father said he could prepare a caribou almost as fast as he could shoot them. They hauled the meat to Dawson, which was about 125 miles from where they were hunting, and got $1 a pound for caribou and $1.25 or $1.50 a pound for moose meat.

Papa went back to Seattle after that with plenty of money in his pockets and bought fur coats and presents for all of the girls in his family.

He returned to Dawson in 1900 or 1901 and worked for wages on Hunker Creek. He had a second degree steam engineer's license and he ran hoists and engines and things at the creek. My father resented the fact that Americans were not allowed to have a first class steam engineer's license in Canada and he disliked the officialdom in Dawson very much.

Papa had a lay on some ground on Last Chance Creek and he boarded at the roadhouse there. It was at the roadhouse owned by Dave Cunningham, that he met my mother Lena Elefsen. She had come from Norway in 1900 and was cooking there at the time. We used to tease mother about meeting Papa at Last Chance.

I was born on Hunker Creek in 1903.

Some years later Papa went to Minto Lake and established a placer mining operation. He left Dawson on the first boat to Mayo one spring and we waited for some time in Dawson until he was settled. In 1913 we left Dawson on the *Pauline* to join Papa at Minto Lake. The *Pauline* and the *Vidette* were owned by the Barringtons and the *Pauline* was the smaller (only 145 tons) of the two boats. The boat was extremely crowded and was loaded with freight. The river was very low and I think it took us seven or eight days to come from Dawson to Mayo.

It was fascinating for me, but I don't know how my mother stood it. Can you imagine what it must have been like travelling in that crowded little boat with five children. One of my sisters was spastic and required constant care and the youngest, Margaret, was just walking and had to be watched continuously lest she fall off the boat into the water. I would have been yelling my head off but I never heard her complain or say an unkind word.

We slept on little canvass cots, I think they were about three high. Nora, Lena and I slept in one cot. Mrs. Betsy Cunningham, an enormously fat person, had the top bunk and when she got into bed the cot sagged down and touched the tip of my nose. I was very hot and uncomfortable.

Isaac Lusk, a teamster, was also travelling to Mayo on the boat. He used to barricade the door of his room and watch Margaret so that she could play out of harm's way.

When we arrived in Mayo I thought it was a lovely place. There were big cottonwoods (poplars) everywhere and stands of birch, green against the summer sky. Dewberries and strawberries grew in profusion in patches on the ground. I was thrilled. It was one of those magical moments a child shares with nature that lingers in one's memory for a lifetime.

Mrs. Cunningham and her husband had a hotel at the end of the street. I don't know that it had a

The John Ross family home at Minto Lake. *William Hare Collection, Yukon Archives*

name, it was just a rather low log building. We had supper there and then Mr. Ferrell took us over to Sam Blackmore's cabin (now known as Mabel McIntyre's cabin) where we stayed until we could get to Minto Lake. The cabin didn't have the lean-to built on it then.

After a few days, Joe Cantin took us out to Minto Lake in his wagon. I have a picture of us all in his wagon at Minto Bridge. The day after we got to Minto Lake was my tenth birthday.

That fall, we went to school in Mayo. Alex Profeit was the teacher. That was the first year they had a school in Mayo. There was supposed to be seven of us; that is, a total of seven students were required to open a school. There were the three Kelly children, the three Ross children and Olive Kinsey. Olive, however, didn't arrive until March or April and by that time our family had gone back to Minto Lake; so there never were seven students that year. The school was a new log cabin.

At Christmas time Mr. Profeit put grey construction paper around the walls of the classroom and we drew designs on it. I think Mr. Profeit taught the following year as well but we went back to Dawson in the fall because my mother was pregnant with my sister Thelma. That winter we went to school in Dawson. We had gone to school in Dawson other years as well, but I never did finish a grade with the other children because Mom always wanted to get out on the creek where Dad was. She didn't think finishing a school term was important and it probably wasn't.

We came back to Mayo in the spring on the *Vidette*.

We were well acquainted with Mr. and Mrs. Ferrell. They operated the post at Lansing for several years and then moved to Mayo where they ran a general store. Mrs. Ferrell had some medical training and when they first arrived in the Yukon she had hoped to work at the hospital in Dawson. As it turned out, the duties of a nurse in those days were more or less matters of housekeeping and she would not have been allowed to perform all of the duties she was capable of.

“ Our old cabin had been torn down and used for wood and the sand bar where we played was chained off. ”

During their time at Lansing, George Crissfield brought his partner, Jim Christie, to the post. Mr. Christie had been mauled by a grizzly bear. Mrs. Ferrell, being quite handy surgically, patched the man up and set his broken bones. She had the hide of the grizzly that mauled Mr. Christie for many years.

Gilbert Rich and I were married with three children: Ross, W. J. “Bud” and Mary Cecile, when Mr. Ferrell died. We were living at Flat Creek (near Elsa) at the time and received little news. I knew that Mr. Ferrell was not well. One morning I decided to go to Mayo and walked up on the road and caught a ride with the youngest of the Kimbel brothers. He had built a house at Flat Creek and married Isabel Profeit. After we had travelled some way down the road, he told me that an old-timer had died a couple of days before and there was to be a funeral in Mayo that day. I asked him who it was and he said he thought the man's name was Ferrell. I was just sick about it. Mr. Ferrell was as close as my own father. When we got to town the funeral was already in progress at the Yukon Order of Pioneers Hall. I stayed that night with Mrs. Ferrell. It was terribly sad. After that Mrs. Ferrell went to Oregon where, I believe, her brother lived.

As a child, I was curious about Indian people. The only Indian words I remember were “dé-tse” which means bad and “dook” which means go away, so I must have been a nuisance.

Sam Blackmore was a teamster. He either came from Nova Scotia or Maine. I believe he was nearly illiterate but the ridge pole on any building that he built is level to this day. He built the cabins and the cook house at Minto Lake. He was a very fine man.

About 20 years ago my daughter, Cecile, took me out to look at our Minto Lake home. Our old cabin had been torn down and used for wood and the sandbar where we played was chained off. There used to be a big tree right at the edge of the lake. I had a cream case nailed, like a seat, up in the branches and I used to climb the tree to hide from the children when I wanted to read. The tree had fallen down; it was lying there, quite rotten, with the box still in place.

CHARLES D. TAYLOR

by Kevin Shackell, Yukon News, November 15, 1982

Charlie Taylor, born and raised in Whitehorse, went to the Mayo country in 1932 to take over operation of the Taylor and Drury Store from his older brother Bill. He was well prepared for his managerial duties as he had learned the merchant's trade from the bottom up in his father's Whitehorse store.

The Mayo Taylor and Drury Store was established in 1919 after the [Keno Hill] Mine went into production. Throughout the Depression, it continued to supply the residents of Mayo, Elsa, and Keno City with goods and services.

Betty and Charlie Taylor, circa 1935. *Charles Taylor Collection, donated by Rick Cocks*

Shortly after Taylor arrived in Mayo, the Depression hit the town pretty hard. The local economy, at that time, revolved around the Wernecke Camp, the big silver-lead producer. But in 1933, the price of silver plunged to 25 cents an ounce and lead tumbled to four cents a pound and the company was forced to shut down operations.

"There was a general layoff and most of the families left the Yukon. Single men laid-in at Keno with supplementary rations," Taylor explains. Welfare and unemployment insurance hadn't been invented yet and even if it had, the territorial government wouldn't have had the financial resources to pay welfare to those thrown out of work by the mine shutdown. Commissioner George Jeckell came up to Mayo and "laid it on the line." The T & D store was instructed by the territorial government to give the unemployed men $20 per month in rations. A deal was struck whereby the company would carry the out-of-work miners over the winter and the government would repay the debt on April 1, the beginning of the next fiscal year.

"The winter of 1933, there were 100 people wintering at Keno City. The Wernecke camp had closed and there was no other place to stay."

Depressed metal markets spelled the end of Wernecke Camp in 1933 but in 1935, market recovery allowed [Treadwell Yukon] to start [mining] the [Hector-Calumet, Elsa] and Silver King deposits. Men who were laid off when the Wernecke Camp closed went "up the hill" to work on getting the new mine into production.

"Keno City was like the village below the mine," he says. "All the trappers and prospectors used it as their base."

Mining exploration in the old days wasn't conducted by specialized geologists using complicated survey equipment. Independent prospectors, people like Charlie Brefalt, picked and poked away at the ground and sunk exploratory shafts. Brefalt discovered the famous Elsa silver deposit which he high-graded before selling out to [Treadwell Yukon].

Most prospectors spent their lives searching for silver and gold during the summer, using the profits from their winter trapping activities to finance summer prospecting expeditions.

Mayo, on the banks of the Stewart River, was the service hub for mining activity in the area.

The T & D store was an important part of the trade and commerce of the area. The 40 x 60 foot building carried everything from soup to nuts to calico and lamp gas, recalls Taylor. Although the Northern Commercial Store also vied for a chunk of the retail market, according to Taylor, the Eatons mail order catalogue provided the stiffest competition in the 30's.

Life revolved around the Stewart River. A majority of T & D's supplies were brought in by river boat from June through to the end of September. "We brought all our supplies in by September because we had to wait until the following June when the boats came in again," Taylor comments. Some perishables like fruit, vegetables, and eggs were transported to Mayo by Cat train from Whitehorse, though, providing residents with a welcome relief in midwinter when freshgrown produce was running low.

The Cat train referred to was the early model Caterpillar tractor with a string of sleds in tow.

Methods of storing food stuffs were also considerably different. There wasn't the fancy packaging and labelling found on supermarket products today. Many of the items were preserved by packing them in barrels of brine.

Food retailers of today would appreciate the delicate cash balance that remote general stores faced in the good old days. With a large part of the work force seasonally employed, the stores, including T & D's, used to extend a hefty line of credit, or "jawbone" as the natives called it, to regular customers.

"One of our jobs was outfitting the trappers and prospectors," says Taylor, "there was an element of risk involved to it but it operated on trust." The store would supply the necessary goods to trappers, prospectors and families who had little winter revenue coming in on the understanding that the debts would be paid off in the spring. It was never the store's policy to grubstake. Grubstaking was the risky practice of outfitting a prospector for a chunk of the profit he made after discovering the big find. The trouble is, the motherlode often never materialized, and the store would be left with a large loss.

Life in Mayo was closely tied to the Stewart River. Spring breakup heralded the arrival of the river boats, and the start of a new employment season. Tons of ore bagged over the previous winter lay on the riverbank awaiting the arrival of the river boats to carry the concentrates to Whitehorse and then over the White Pass and Yukon Route to southern markets.

Taylor remembers the sight well. By spring there were 10,000 tons of sacked ore piled in front of the Chateau Mayo waiting for the ice to go out. The ore had been milled and dried at Elsa, placed in heavy jute sacks and then hauled down to Mayo over the winter by a cat train.

In early spring, when the water was still high, Taylor says the larger river boats could bull their way up the Stewart River to Mayo. But after a short time only the smaller steamers like the *Nasutlin* and the *Keno* could ply the river because of the large draw-down which took place.

Stewart City, at the confluence of the Stewart and Yukon Rivers, was an important transfer point. Ore shipments coming down from Mayo were loaded onto the bigger river boats for the upstream journey to Whitehorse, while diesel fuel and tons of other supplies were offloaded from the big boats at Stewart and placed into the shallower-draft vessels for the upstream voyage to Mayo.

It was a fascinating era of trade and commerce, Taylor fondly recalls. Life as a store manager drawing $150 a month during the Depression years wasn't easy, but Taylor says peoples' expectations were much lower then. "We didn't have the advantages of the social programs you have today, like housing, medicine and U.I.C. These things all had to be paid for by the individual."

Taylor met his wife Betty (Maclennan) in Mayo and the couple was married in 1936. Like many Mayo residents, a certain amount of the family sustenance was garnered from the land. "If you didn't have a good garden and weren't versatile with a gun, it was pretty slim pickin's," Taylor explains.

Between the store and other chores which included cutting a winter's supply of wood, there wasn't much time to sit around getting bored, says Taylor. "Those early days in Mayo were terrific," he says.

Times may have changed, but the cyclical nature of Yukon's economy has remained virtually the same and Mayo will still be there when the metal prices pick up again.

FRED TAYLOR

by Fred Taylor, 1989

I arrived on Dublin Gulch about 1936 to find Ted Bleiler mining by hand, and since the water was very low, Ted was enjoying his spare time nursing a sore back (those rocks were heavy and there was an inexhaustible supply), playing his violin (which he still does well into his eighties) and caring for a bountiful garden.

This garden had to be protected by a picket fence to keep out marauding rabbits. The garden was important to reduce the difficult chore of hiking to Mayo for supplies.

In winter it was a little easier because one could use snowshoes, skis or dog teams, although there were still problems with breaking trail, crossing "glaciers" and overflow ice on creeks, and cold, short winter days. It took about five to six hours (barring problems) to ski from Hughe's cabin at Half Way Lakes to Dublin Gulch, a distance of 26 miles. A good skier today with modern equipment and good trail could do that in about two hours. With a loaded toboggan and dogs it could easily take two days. Stopovers could be made at Bobby Burn's cabin at the McQuesten or one of several dilapidated cabins on Haggart Creek. No matter how rudimentary, these cabins were a welcome hostel for a poor, exhausted musher on a cold dark night.

I can still recall a blackened stump about five feet high that stood some six miles downstream from Dublin. Several times during darkness while I was in a semi-stupor from exhaustion, the image of that stump startled me and my first thought was, "What's that ahead on the trail? A man frozen to death?"

Ted Bleiler left for the "outside" in the fall of 1936 and I stayed on the creek and prospected and trapped "unenthusiastically" on the North Fork of the McQuesten River.

The Taylor's Dublin Gulch home, circa 1955. *Fred Taylor Collection*

Dreaming of the inevitable "old channel," Ted managed to get a shaft and drift to bedrock on the left limit of Dublin Gulch at the mouth of Suttles' Pup, thereby establishing that there was indeed pay dirt and a deeper channel than had previously been worked. Capital and modern machinery not being available, this was not practical to work until later years.

Frank Taylor. *Fred Taylor Collection*

The following spring (1937) Ted Bleiler returned with his bride, Almuth. He also brought a huge outfit of several hundred pounds of food for the summer which had to be hauled up Dublin with dogs and a "go-devil." Ted even brought a bathtub.

We lived high that summer, fresh bread, tasty meals prepared by Almuth's capable hands. Once she made huge cream puffs bursting with cream. I realized the joke when my full beard became imbedded with cream. The culinary variety was enhanced with grayling from Link Fork, moose from the North Fork and vegetables from Ted's garden.

Mining equipment consisted of wooden sluice boxes, picks, shovels, blasting powder, wheelbarrow and an automatic gate.

The sluice boxes were 12 feet long, with bottoms especially sawn by Lefebvre's sawmill in Mayo, 1.5 inches thick and tapered sides so the ends telescoped together. There were 2-inch by 4-inch "collars" to strengthen the sides. The centre collars were put just at the balance point. (This was so there was no way a man could carry a wet, dirty, heavy box uphill and over unstable rock piles without the collar digging into his shoulder or one end or the other bumping on the ground!)

Jim Taylor, Dublin Gulch, 1955. *Fred Taylor Collection*

Once Ed Barker and his partners were moving boxes and Ed heard a demoniacal screaming up ahead of him. On topping over the rock pile, there was Norman Wightman kicking and screaming at a poor innocent sluice box that happened to elude his grasp.

Both Ted Bleiler and I used a flatbed iron wheeled barrow to take boulders up an inclined whipsawn plank. (Ed Barker had also done this using a dog to help pull the wheelbarrow.) An alternate was to carry the boulders up in our hands. Those too heavy for one man were rolled on a stretcher made of canvas between two poles and carried by two men. One is reminded of this when ones sees photos of Chinese coolies.

We did not use powder for the bigger boulders unless absolutely necessary, because powder cost about $16 a box and had to be hauled in. Also we would rather struggle rolling one big rock up a plank, than carry a dozen or more sharp pieces.

On fine afternoons Almuth would come down to the cut with drinks and cakes etc. She would then reinforce our spirits by helping us carry out rocks. One problem with this was after passing each other on the plank a few dozen times, glancing at the distorted expressions on the loaded one's face, we got to giggling, considerably reducing the payload.

In later years I built a wooden derrick to take out the rocks. This was "powered" by a two ton Beebe winch. This worked so well that a better derrick was built powered by a small Wisconsin engine. We were in business, and eventually scrapped enough gold together to buy a D-7 Cat.

But that is another story . . .

CHARLES BLAISE "CHARLEY" TURGEON

by A.A. Gillespie, The Mayo Miner, October 21 & 28, 1938

Charles Turgeon was born in St. Isadore, Quebec on July 26, 1856. He and Michael Brandemore were partners in the logging game before they came to the Yukon. (Mr. Brandemore lived in Dawson and died there in 1937.) When Charley first came north in 1893, he went to Circle City. It is interesting to note that he hiked over the Chilkoot Pass several years before that famous trail became a byword to the thousands of early day gold seekers who thronged to the Klondike in 1897 and 1898. Charles Turgeon came to the Yukon with another grand old Sourdough from Quebec, John Tremblay. When they came it was Tremblay's second, or third trip into the country.

From Circle City Charley Turgeon, then in his prime, went to Dawson where for several years he

From left: Charles Turgeon, Ruth (MacDonald), and "Big" Maisie. *Ruth MacDonald Collection*

was engaged in the hotel business. Before coming north he had operated a hotel and logging business in Wisconsin.

A great deal of life, as it was lived in Dawson during those hectic, gold rush days, is reflected in a small, time-worn note book which the late Sourdough always kept. On one page for instance is a memo of a loan of one thousand dollars which he had made to Tex Rickard, the late world-renowned sports promoter. As that item stands today there is still a sum of $275 which has never been marked off. [Mr. Gillespie wrote this when Mr. Turgeon died in 1938. It is not known where the note book is today.] There are many other items listed—all of them highly interesting in that they throw a clear, concise light on various phases of the early day Yukon.

Suzzane Turgeon, left, and Dorothy McLaren, circa 1919. *D. Walloch (McLaren) Collection*

When the main stampede started Charley was at Ogilvie. He arrived too late to stake in on any of the rich claims, but from the pages of his note book, it can be seen that he became interested in quite a few claims. It is understood that he bought 8 above on Bonanza, property which he later sold for $65,000. He was also interested in a claim on 35 Eldorado in partnership with the famous "Swiftwater Bill" Gates. He owned claims at different times on Hunker, Dominion and other creeks in the Dawson area.

One undertaking turned out far from profitable for Pioneer Turgeon and that was that he bought a steamboat and began carrying supplies from Whitehorse to Dawson. On his second voyage down river, the steamboat, about the size of the *Keno* and named the *Golden Star*, ran into trouble coming through rough water and sank. Packed on board was a cargo valued at around $100,000 and Skipper Turgeon lost heavily after he had paid off this expensive venture.

After leaving Dawson he prospected on the McQuesten for six or seven years, coming to Dublin Gulch, in this district during the stampede of 1908. For many years he trapped and did various kinds of work around Mayo, including carpentry work. As a matter of fact he was still running a trap line up until 1931 at which time he was 77 years old.

In a ceremony performed by Rev. John Hawksley, Charley married Maisie, a Mayo Indian woman, on August 5, 1912. Maisie had three daughters; Ellen, Julia, and Suzzane. The couple raised Maisie's granddaughters Mrs. Rose Wood, Mrs. Mary McDiarmid, and Ruth MacDonald. The family lived upstream from Mayo on a large acreage in the 1930's and had a wonderful garden.

Of a genial, happy disposition, Charley was known and highly respected by hundreds of Yukoners throughout the country. One of the highest tributes that can be paid to Pioneer Turgeon is that he never said an ill word about anybody else. He had a keen sense of humour and was always an interesting conversationalist.

Charley loved the Yukon. On several occasions his niece in San Francisco, Mrs. Alexine Chadborn, tried to persuade him to make a trip outside but he never wanted to go. As he used to say, "even if he had been the possessor of much wealth," he still wouldn't want to go outside. Mr. Turgeon also had a nephew in Prince Rupert and another in Iowa,

Charley Turgeon died in Mayo on October 21, 1938, at the age of 82. Mrs. Turgeon died several years before that.

IRA & SON PAT VAN BIBBER

R. Coutts, Yukon: Places & Names
Yukon News, May 27, 1987
Jo Anne Gatey, Dan Sha News, May 1987

Pat was born and raised in the Pelly Crossing (Mica Creek) area. Son of Ira and Eliza Van Bibber, he was one of 14 children (Leta, May, Abe, Ben, Archie, Alex, Helen, "JJ" John James, Pat, Kathleen, George, Lucy, "Linch" Melinda Jane, and "Dode" Theodore).

> Ira Van Bibber with his two brothers Theodore and Patrick, came from West Virginia to the Klondike in 1898. Ira loved the country and married an Indian wife. They trapped for three years in the then almost totally unknown Nahanni country. He became one of the best big game guides in the Yukon. He made his home for most of these years at Pelly Crossing. Mount Van Bibber, in the McArthur Group about 35 miles south of Mayo, was in the midst of the region where he hunted and trapped. This name was requested by Mrs. D.J. Fulton of Cranbrook, B.C., in March 1967 to commemorate her father, Ira Van Bibber. After more than 60 years in the Yukon he died, leaving many descendants to carry on his traditions. (R. Coutts, *Yukon: Places & Names*)

Pat began a lifelong career in highways in 1949 when he was employed by Fred Mannix who had a contract building the Whitehorse to Keno Highway. He worked there until 1950 and then operated the ferry in Carmacks for the YTG Highways division. From 1951–1953 he operated the Van Bibber sawmill with his brothers in Pelly Crossing. In 1953 and 1954 he operated the YTG ferry at Stewart Crossing. In 1954 he went to work for United Keno Hill mine's road crew. In those days, the mine was responsible for the highway from Whitehorse to Keno. He operated a grader in the winter for two years, and ran the ferry in the summer. During his last two years in the employ of United Keno, he was road foreman at Mayo. Since that time Pat and his wife, Ada, and their family have made Mayo their home. In 1957 the YTG took over highway maintenance operations and Pat remained as YTG road foreman from 1957 until his retirement in 1987, at age 65.

Pat and Ada Van Bibber's children are: Debbie Van Bibber, Shirley Ellis, Pat Van Bibber Jr., Carol Taylor, and Karen Van Bibber.

WILKINSON FAMILY

Aaro Aho

The Wilkinson's were well known in this district. Mr. and Mrs. Wilkinson came to the Yukon in the early days from the Western U.S.A. They raised their three children Jared, Ethel, and Ed on the MacMillan River. In 1940 they bought Pelly Farm from George Fairclough. (Pelly Farm was established in 1901 by Edward Menard. Menard acquired partner George Grenier in 1906. Frank Chapman and Pete Olsen bought the farm in 1915 and May and George Fairclough bought Pelly Farm from Chapman in 1927.) At Pelly Farm the Wilkinson's trapped, raised beef, and had several horses. They eventually operated a big game guiding outfit. As a young man Jared Wilkinson was almost mauled to death by a grizzly bear and on another occasion his brother, Ed was attacked by a moose but was not badly injured. The senior Wilkinson's sold Pelly Farm to Hugh and Dick Bradley in 1953 and moved to B.C. Their sons, Jared and Ed, remained to guide and trap.

The following account about the disappearance of Ed Wilkinson was written by Mayo trapper Peter Beattie.

I came onto the long arm of the lake at noon of November 24, 1978. There was plenty of light and good visibility, but high clouds blocked the sun and coated the world in a cold, steely blue-grey. I saw the tracks from a good quarter mile off, coming at right angles to my line of progress along the shoreline. My first casual thoughts were "must be a moose down early." As I came upon the track I saw that it was no moose, but "ursus horribilis" himself, the grizzly bear. The tracks were made in about six inches of snow that fell four days before; the two inch snowfall of two days previous was in the tracks. A big grizzly walking in that much snow makes a track that looks about two feet wide. He'd marched with awesome regularity along the shoreline for about two miles, only once deviating into the bush to investigate. I remember thinking that it looked like some large bulldozer, that somehow ran on one track, had preceded me. For those two miles that track was a help, the dogs snorting and sniffing, pulling with that frenzy they have when they smell something "interesting," and the hardened tracks holding the toboggan up in the loose snow.

That night, as I sat in my tent, I thought back to the events of the previous year when Ed Wilkinson, a friend and one of the last of the "old breed," was killed by a grizzly bear.

In the fall of 1977 Ed (then aged 59) and Jared (then aged 62) Wilkinson were living at Lansing, site of a long abandoned trading post at the confluence of the Lansing and Stewart Rivers. This is about eighty air miles due east of the town of Mayo, the nearest point of civilization. Their uncle, Bill Thompson, had come into the country over the "Trail of '98" and their father, Jared Sr., followed shortly thereafter. These men had grown up and spent a lifetime in the Yukon wilderness—trapping, fishing, hunting, trading, freighting, ranching, and outfitting.

Together with my wife, Mary and daughter Moriah (then aged 3), we'd been living in the Yukon bush for the previous six years. That fall we had moved onto the Stewart River, some fifteen miles below Lansing. The Wilkinson brothers were our only neighbours and there was no one living between us and town. The first of December that year started a cold spell typical to these latitudes. The sky cleared and the temperature dropped to the –50°F to –65°F range for the next twelve days. If you are in a snug cabin this is a most beautiful time of year; crystal clear days, spectacular northern lights, and a chance to read a fat book or do some of those "little jobs" you've been too busy to do. After about a week of this particular cold snap we were surprised to see Jared walk up to the cabin on snowshoes. He'd been at his tent camp, about four miles upriver, and decided it was time to come down and see how cabin life was. The next day he went back to his tent camp and returned with his dog team.

During the next few days we naturally spent quite a bit of time "jawing" and drinking tea. One thing that Jared mentioned was that a bear had come out of the hills from the north, just a few miles upriver from us, and had headed up towards Lansing on his trail. It had been around Lansing for several days and he'd seen its track on some of his "side lines." At the time we thought nothing of this, as there are many grizzlies in this area and their tracks in the first snows of October and November are common. We concluded that this, the first real cold spell of the season, would probably "put him to sleep" for the season.

After three more days the weather broke. Jared headed back towards Lansing and I headed south towards Pleasant Creek, where I was trapping that year. I returned home on December 19th after a week on the trapline. One look at Mary's face told me something wasn't right. I was told that Eddie was dead.

Jared had gotten home to Lansing at dusk of December 15th. About two miles from the cabin the trail comes out onto the river. Within a few hundred yards of this point he came upon a curious hump only a few yards from the trail. Upon closer inspection he discovered it to be a dead grizzly bear, covered by a couple of inches of snow.

When he got to the cabin he found it just as it was on November 29th, when he left to come down river. The book Ed had been reading was on the table; the biscuits he'd made that day were still there; his toboggan, which he'd waxed in preparation to go on his own trapline, was there; the windows had no frost on them—which they would have had if there had been heat in the cabin during the cold weather. Jared even noticed some of the fire wood he'd hauled into the cabin just before leaving. Two out of five of Ed's dogs had gotten loose and had been dragging pieces of moosehide around the yard, so the other three had managed to get ahold of some. The dogs were all lean but had survived quite well.

Adding this all up Jared realized that most certainly Ed had somehow met his fate and was probably near the cabin. His parka and rifle still hung over his bunk. By this time it was full darkness and impossible to investigate. After a sleepless night and feeding Ed's dogs an extra large ration, Jared hooked up his own team and headed back to our place at the crack of dawn on December 16th. He spent the next day at our cabin, waiting for me to return, then headed back towards Lansing on the 18th as the dogs there needed care.

That was the story that awaited me when I returned home the next day. We immediately shuttered up the cabin and on the 20th we all travelled up to Lansing, arriving at dusk. After I got my dogs unharnessed and chained up, Jared led me some one hundred yards north of the cabin, through a dense patch of alders to a dry creek bed. Even in the fading daylight, and with two or three light snowfalls since November 29th, one could see evidence of a violent struggle. I returned to the spot the next day in full daylight. I didn't want to disturb anything, as this was obviously a place for the "proper authorities," but it looked as if the bear had sprung off the bank of the creek as Ed had passed by; one could at any rate see the general direction of the attack. There were a few scraps of clothing hanging from the branches of the alders and some visible blood. There were grizzly tracks (all with the snow from the month of December in them) throughout the alders. Also there were wolf tracks—I would say a group of six. There was less snow in these tracks indicating that they had been there after the bear.

That afternoon, I took my dog team downriver and hauled the carcass of the bear back to Lansing. The size of the head and feet indicated that this was a good sized bear, but he was so skinny that his back fit between the handlebars of my toboggan, which are only sixteen inches apart. For about a quarter of a mile along the riverbank one could see the bear's final tracks. He had repeatedly left the trail and crawled up under the shelter of the cutbank. All this certainly indicated that there was something seriously wrong with the animal. I noticed some blood on its back, but in its frozen condition it was impossible to tell if if was a wound.

As it turned out this was an old bear (over twenty years) and it had a broken scapula (shoulder bone). A splinter from this bone had come up and punctured the skin of the back, thus the spot of blood I noticed. He hadn't been shot so presumably the injury resulted from a fight or a fall. He only weighed around 230 pounds, and in good health he would have been double that.

"That afternoon, I took my dog team downriver and hauled the carcass of the bear back to Lansing."

At dawn on December 21st, the shortest day of the year, I left Lansing heading down the Stewart River towards Mayo, around 125 miles following the river. I spent the first night in our cabin, the next under the spruce trees, and got to a good cabin just below Fraser Falls on Christmas Day. Nearly all of this was without any trail so it was a pretty tough couple of days for my dogs who were then just pups with about six weeks of experience in harness. I was lucky to have an experienced lead dog so the dogs did all the trail breaking and I just had to come along behind. The weather was warm and cloudy, so the days were pretty brief and I travelled quite a bit in the dark. It certainly was one of the strangest Christmas Days I have ever spent—the simple comfort of a cabin with a hot stove, a pot of coffee, and some old magazines to read seemed like quite a bit to be thankful for.

The remaining forty miles from the falls to Mayo had a fresh ski-doo trail so the last day of my journey was easy. I left in the dark and was in town for the noon meal.

That afternoon, in the company of Tony Nette, Conservation Officer for Mayo, I went to the R.C.M.P. barracks and told my story to Corp. Ed Basaraba. That night I was talking to J.J. Van Bibber, a lifetime friend of Ed Wilkinson who spent his younger years hunting and trapping over much of the territory. He asked me if the bear was "iced up" and in what condition its feet were in. The bear wasn't iced up and its feet weren't frozen so it seemed safe to assume that it had expired before the onset of the extremely cold weather. The next morning at dawn we were in a helicopter heading for Lansing, retracing my steps of the previous five days in an hour. We spent the few hours of daylight searching the immediate area. The search turned up only a few bits and pieces of clothing, a boot, more frozen blood, bits of hair, and wolf and bear scat. Another search group flew out the next day, with about the same results.

In May we were living in Mayo, working for a sawmill, and I again flew to Lansing with Corp. Basaraba. We researched the area that was now without snow cover. Nothing of major importance was found at this time; a paddle, used for covering traps with snow, a handful of bear hair, a wool cap, and more of the same evidence found in December.

Trying to analyze this tragedy Jared and I concluded that Eddie was probably alerted by the dogs and was heading to a marten trap only about a hundred feet past the scene of attack. He had already caught several marten in that set that fall. As the only clothing found was of the light summer type and rubber gumboots, it was certain that Ed wasn't intending to travel very far, even in the warmish weather of November 29th of that year.

In June of 1978 an inquest was held in Mayo. All the evidence was considered by a group of men who were all experienced in the Yukon bush. No evidence of human flesh or bone was found in the stomach of the bear or in the recovered scat of the bear and wolves. Based on all the assembled evidence it was concluded that on or about November 29, 1977, Ed Wilkinson was attacked and killed by a grizzly bear about 100 yards north of his cabin at Lansing. Thus ended the life of a man who was born, raised, and lived his life in the wilds of the Yukon.

In that fall of 1977 some canoeists had stopped by and mentioned that Ed (who was at Lansing alone then) hadn't got a moose yet. I'd been lucky enough to get one the day before so I headed up to Lansing with a little fresh meat; a good excuse to spend the day talking with Eddie. Of course he had just got done hanging one up. He heard it snorting up the Lansing River, went up there and called it out to the bank and shot it only a few feet from the riverbank. The conversation that day ranged from tunnel boats to wolf sets and of course got around to bear stories. Here are a couple of them:

"I was heading up the Pelly early in the spring and saw a big grizzly on the shore just below Needlerock (this is at the mouth of Granite Canyon on the Pelly River). I took a shot at him and missed so I tied up the boat and started tracking him. He ran for a few hundred yards but the tracks began getting closer together so I knew he was slowing down. I decided that following a bear who knew I was there and wasn't afraid could get to be a real bad policy, so I went back to the boat and went about my business. I came down river a week later and decided to pick up that track just to see what that bear had done. He'd gone up the side hill and doubled back around to a ledge just overlooking the trail he'd just made . . . looked like he spent awhile there, too."

"I decided that following a bear who knew I was there and wasn't afraid could get to be a real bad policy. . . "

"I was on a rat and beaver hunt that spring. I was standing on a beaver dam with a muskrat in one hand and a .22 rifle in the other and my 30/30 over my shoulder. Now people say a grizzly will always stand up or give some indication before he'll charge, but when I saw this one he was on all fours coming at me just as hard as he could, maybe a hundred feet away. I dropped the .22 and the rat and by the time I got the 30/30 around I just shot twice from the hip and the bear dropped two rifle lengths from my feet. The first shot was a solid (intended for a beaver) and went right into the skull. The other was a regular hunting bullet and glanced off the top of the skull."

I'll always remember what Eddie said at this point in the conversation concerning run-ins with bears, "If you live in the bush, sooner or later it happens to everyone—and always just when you least expect it!"

MARGARET MAY "MAGGIE" WOOD

My father was James Hastings Mervyn. He came from Stratford, Ontario. He left Stratford and went to the United States where he helped build the first wash house in Salt Lake City, Utah. He moved on to Washington and built a sawmill in Walla Walla. He had planned to become an American citizen but the Klondike gold rush brought him back to Canada. He walked in over the Chilkoot Pass to Dawson City. When he got there all the ground had been staked and so he became a carpenter.

He left Dawson and went to Duncan Creek when gold was discovered there. He prospected on Duncan Creek for a while and then bought Lansing Post from a man named Ferrell. My father traded using metal tokens with his name on them instead of money.

My mother, Julia, was a Lansing woman who had originally come from the Mackenzie River in the N.W.T. She had two sisters and two brothers. Senoah, the oldest (Norman Niddery's wife), lived in the Mayo district; Liza stayed in the N.W.T.; Pete, who was older than my mother, went back to the N.W.T.; and Fred, who was blind, lived in Carcross.

I was born up in the Black Sands country at Margaret Lake. It was during winter and Dad put Mom in a toboggan and took her up there where a group of trappers lived. There was a woman there who helped Mom with my birth and later helped with the births of almost all of her 11 children. The children's names are Maggie, Florence, William "Kaiser," Jerry, Nora, Alice, David "Dick," Gordon, Carl, Norman "Dinky," and Alexander.

There are only four of us left: Norman, David, Carl and I. David lives in Edmonton, Alberta; Norman is in Prince George, B.C.; I believe Carl lives in Aklavik, and I am in Whitehorse. My brothers Jerry and Kaiser drowned in the Stewart River [Jerry in the rapids above Fraser Falls on September 29, 1935]. Gordon was killed in a hit-and-run accident at Prince George and Alec died at a young age (17 or 18) of meningitis.

I went to school for a year in Dawson and two years in Kaslo, B.C. I came back to Mayo and took Gordon McIntyre's commercial course. Dad built a house in Mayo while he was still running the post at Lansing (the house is now owned by Margaret and Dick Wallingham). Florence and I stayed there with the kids so they could go to school in Mayo.

I met my husband James Edward "Jim" Wood around 1933 when he came to Mayo with the Royal Canadian Corp of Signals. Signals were in charge of all communications in Mayo in 1935. Jimmy was not of high enough rank in the signals to be permitted to marry so he had to buy his way out. We were married in 1935.

In 1937 Dad left Lansing to go outside on business. Jimmy and I went up to look after the post while he was away and by the time we came down to Mayo again Dad had bought the Chateau Mayo hotel and store.

One other time when Dad went out on business he brought back quite a number of Manitoba maple trees and planted them all over Mayo. There used to be one between the old Northern Commercial store and the butcher shop.

Jimmy and I lived in Elsa for about three years. We moved back to Mayo and Jimmy worked for Bud Fisher for a while. In 1939 or 1940 Jim went to Whitehorse to work and the children and I followed in 1941. My husband worked for Yukon Electric for 30 years. I worked as a cashier for Charlie Taylor at Taylor & Drury's store and after that I worked the courtesy counter at Tourist Services. I held a number of jobs in Whitehorse over the years including government work in Motor Vehicles and Legislation.

My father returned to Ontario and died there in 1957. After Dad died Mom moved to Whitehorse and married Baptiste St. Louis. Mom died in Whitehorse and is buried in Mayo. Our son Doug lives in Whitehorse and our daughter Bobby is in California. Three of Mervyn's children (Louis, George and Roy), live in Whitehorse and her son Norman lives in Mayo.

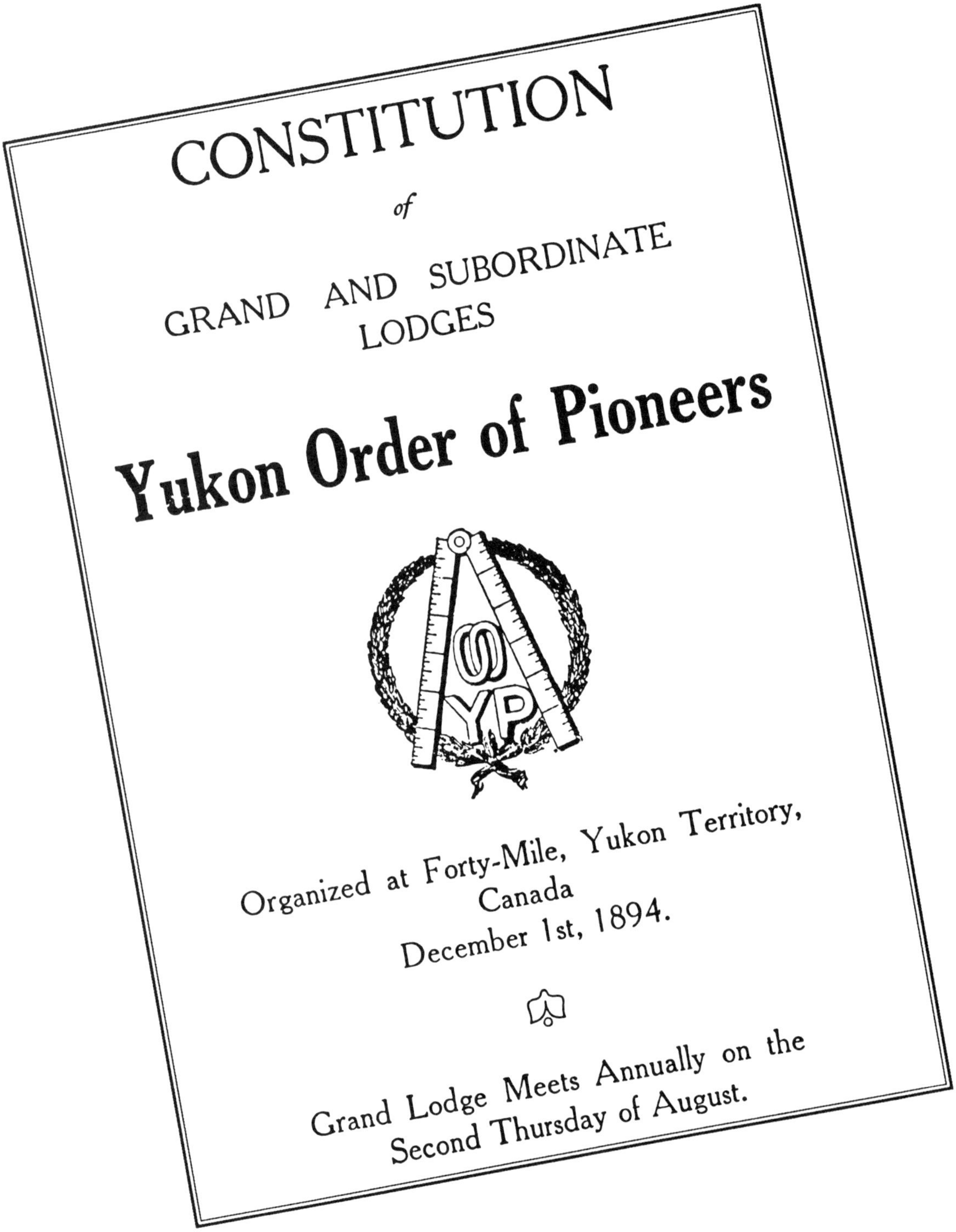
CONSTITUTION

of

GRAND AND SUBORDINATE
LODGES

Yukon Order of Pioneers

Organized at Forty-Mile, Yukon Territory,
Canada
December 1st, 1894.

Grand Lodge Meets Annually on the
Second Thursday of August.

Cover page of the Yukon Order of Pioneers *Constitution* 1917

PART THREE PIONEER LIST

PIONEERS AND EARLY SETTLERS

The names for the *Pioneer List* stem from Polk's *Alaska-Yukon Directory and Gazetteer* for 1915–1916 and 1923–1924, the only two editions with listings for Mayo, as well as from Robert Coutts' *Yukon Places and Names* and *The Mayo Miner*. It has been augmented by pioneers of the area and their descendants. No doubt, memories being less than perfect, some names have been missed and some facts omitted. While every endeavor has been made to cross check each piece of information, verbal accounts tend to differ over the span of ninety years as the narratives pass down through the generations.

Mayo district pioneers came from all over the globe, many to establish new lives and new identities. To add to the difficulty of tracing their roots, there was an unwritten code in the early days of the settlement that if a person did not volunteer background information it was not to be requested.

Men predominate this list, as wives and families were often left behind when they travelled to the harsh conditions of the mining camps. Women that accompanied their men were generally known by their husband's names, not their own Christian names. If a woman's name was known it has been included in the list.

ABBOTT, Nathan

Nathan Abbott was a partner in a mining company that operated on Haggart Creek from 1910. Portlock and Abbott Bros. comprised William Portlock, and the brothers Nathan and William Abbott.

ABBOTT, Mr. & Mrs. William

William Abbott was a partner in Portlock and Abbott Bros., a mining company operating on Haggart Creek from 1910, and his wife cooked for the crew. William was Nathan Abbott's brother.

ABRAHAM, Jack

– Archie Gillespie, *The Mayo Miner*, February 26, 1937

Jack Abraham was born in Iowa but moved to Ontario at an early age with his parents, and spent most of his boyhood years there. He came to the Yukon in about 1907, and worked on the dredges in the Dawson district for many years. He came to the Mayo district in 1922 to work for the Treadwell Yukon Company, in the Mayo camp. He was hoist man at the Silver King Mine, and a member of the Mayo Fraternal Order of Eagles. He died of the "flu" at the Mayo hospital on February 20, 1937 at age 55.

ADAIR, John "Jack"

Jack Adair staked the Punch claim in the 1920's and optioned it to the Treadwell Yukon Company. He later got the claim back when no ore was found on it.

ADAMI, Frank

Frank Adami was a long-time Yukoner who trapped in the Stewart River area from 1936 to 1957. Frank held degrees in chemistry and physics from the university in Milwaukee, [Wisconsin], and the Sorbonne [in France], but he preferred the solitary life of a trapper to any other. He died in Vancouver, B.C. in 1958.[1] Adami had a wife and at least one child.

AIMES, Ruster S.

Ruster (or Rutser) Aimes mined on Dublin Gulch from approximately 1905 to 1914. His claims were acquired by John Suttles and later by Frank Cantin.

AITKEN, Thomas P.

Originally from Nova Scotia, Thomas Aitken was a successful gold dredge owner in Fairbanks, Alaska. He financed McWhorter's work on the Silver King (after McWhorter had cancelled Alverson and Huffman's lease) with an option to buy the claim if it proved to be good.

In 1915 Aitken and his partner Monroe bought the Silver King from McWhorter. He worked the property until 1916 and then returned to gold mining in Alaska. Aitken eventually retired to rural Washington where he died.

ALCROFT, Bill

Bill Alcroft mined for a time on Duncan Creek in the early 1900's (circa 1903–1906). He and his wife lived in Mayo in the 1930's, and their son Ian attended school in Mayo circa 1932–33. Alcroft became the diesel operator of Treadwell Yukon's Mayo generator. The family moved to Dawson in the late 1930's.

ALGUIRE, Daniel

Daniel Alguire was a teamster in Mayo in the 1920's. The Alguires built a home in Mayo in 1922.

ALLAN Dorothy

See *Pioneer Accounts*

ALLEN, Margaret Emma

Emma Allen was the school teacher in Mayo in 1915–16. (Her father was R. L. "Bob" Allen, a Yukon pioneer and World War I veteran. Born in County Bruce, Ontario, he came to the Yukon during the gold rush, and for many years was the government timber agent in Dawson.) Emma had graduated from high school in Dawson, and she later returned to Dawson. There, she worked for the Yukon Consolidated Gold Corporation, eventually becoming an executive. On May 19, 1939 she married John Francis "Jack" Sealey in Dawson.

ALVERSON, John "Jack"

Jack Alverson was born in Le Grand, California, in 1878. He acquired extensive prospecting and mining experience in Colorado. He came to the Yukon in 1899. He lived and prospected on a small tributary of Scougale Creek in the Davidson Range (later named Alverson Gulch). In 1912 he teamed up with Henry McWhorter and they leased the Midas and Ophir Lode gold claims on Dublin Gulch. Finding nothing worthwhile, they moved to the location where Jake Davidson had found rich galena float in 1903. On February 23, 1913, Alverson staked the Webfoot claim, McWhorter staked the Silver King, Huffman staked the Mabel and Mark Evans staked the Adam.

In 1914 Alverson and partner Grant Huffman acquired a 100% lay on the Silver King. They mined and shipped the first high-grade, silver-lead ore in the district. (See *Silver History*) In 1915 he staked a galena vein on Rambler Hill and he and James Scougale staked veins on Mount Cameron. He eventually dealt away his interest in the Webfoot claim.

From left: ? Smith, Raoul Binet (nephew), Lillian McLaren, Jack Adair, Kathleen McLaren, Louis Bouvette, Dorothy McLaren. *D. Walloch (McLaren) Collection*

He prospected for a time in the Klondike with Jerry Drapeau and George Potter. In 1928, Alverson prospected with Cecil Poli on Arizona Creek. In the late 1930's he, Poli and Potter found and mined placer gold on Gem Creek. Alverson died in 1955.

AMERSON, George

George Amerson is listed at Keno Hill in Polk's 1923–24 directory.

ANAM, Olaf

Olaf Anam, a Norwegian, operated a boot and shoe repair and harness-making business in Mayo from the 1920's to the late 1940's. He worked George Andison's and R. G. Lee's claim on Bunker Hill for them in 1938. Alex Boris bought the boot and shoe repair business from the Public Administrator after Anam's death.

ANDERSON, Emil

Emil Anderson was a trapper and prospector who lived in Mayo, and later in Keno, from the 1920's until his death in 1959. He came to the Mayo district in 1914 with his partner Matt Butijer.

Olaf Anam seated outside his boot repair and harness-making business in Mayo, circa 1935. *G. A. McIntyre photograph*

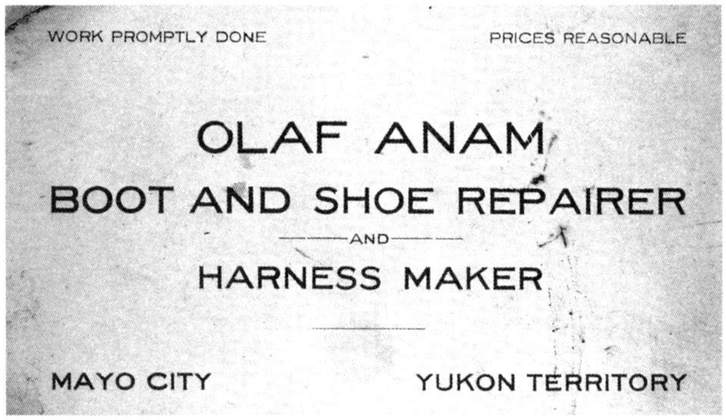

Olaf Anam's business card. *Verna Klippert Collection, Mayo Historical Society*

ANDERSON, "Fish"

Doreen Doidge (née Lee) recalls that "Fish" Anderson was a gentle, kindly Scandinavian gentleman who lived outside Mayo, beyond the Mervyn residence at the Mayo-Keno Road junction. She said that during the bitterly cold winter of 1942–43 while her father was in Whitehorse, "Fish" would walk into Mayo every day to make sure that her family had a good supply of wood stacked by the back door and bring with him a freshly-caught rabbit or whitefish.

ANDERSON, Harry Roland "Toby"

Toby Anderson was born in 1921 in Inwood, Manitoba where he spent his childhood.

While working in Edmonton, Alberta, he heard United Keno Hill Mines in Elsa was hiring. He came to the Mayo district in 1950 to work for the mine as a carpenter. He worked for the company for six months and then quit to start his own timber and carpentry business and eventually bought a truck for the wood haul.

In 1966 he went to work as fire boss for the Forest Service in Mayo and remained in their employ for 20 years. Anderson retired from Forestry in 1986 and continues to live in his home in Mayo.

He is completely fluent in Swedish.

ANDERSON, James Alexander

James Anderson prospected in the Mayo and Keno Hill districts in the late 1900's (circa 1916–1919), 1920's, and 1930's. He was one of the partners in the Yukon Silver-Lead Mining

Company situated on the Lookout Mountain property.

Louis Bouvette staked the first claim on Keno Hill which he called the Roulette, and Anderson staked the second on July 21, 1919, naming it the Rico. Anderson worked with Louis on the Roulette during the summer of 1919, and he later sold the Rico to Guggenheim's Keno Hill Ltd.

Anderson died in 1935 and at the time of his death he held claims on Highet Creek, Dominion Creek, and had interests in the Ladue Fraction, Liberty Bell, Matt, Ripp, Jake, and Olive No. 2 claims.

ANDERSON, Peter

Peter Anderson is listed in Polk's 1915–16 Alaska-Yukon Gazetteer and Business Directory as being a miner in Mayo.

ANDISON, George

George Andison arrived in Mayo in 1929. He was the Mayo agent for Klondike Airways Ltd., and manager for Burns and Co. Ltd., in Mayo in the late 1920's and the 1930's. He and his first wife Ethel Andison had one son, Jack. After Ethel's death in 1931, Andison married Phyllis Marshall, Fred Marshall's mother. Following the death of his second wife in 1940, he married Christena Matheson. Christena's daughter, G. Jean Matheson, married Wilfred Gordon. Andison was born in 1882 and died in 1969. George, Ethel, Phyllis and Christena are buried in the Mayo cemetery.

ANDISON, Jack & Agnes

See *Pioneer Accounts*.

ARBUTINA, Steven "Steve"

"Steve" Arbutina, an Austrian and chef by profession, arrived in the district circa 1910 with his uncle. Elmer Middlecoff's Time Book shows that "Steve" Arbutina worked for him on Highet Creek in 1913 and 1917.

He lived at Minto Bridge, trapping and prospecting in the Mayo Lake area, and also cutting wood for a living. He was a loner who rarely made trips to town, but he had a cabin in the Nelson Creek area and another one at the foot of Mayo Lake. He also had a cabin and worked a small prospect on the Lookout Mountain Road (Mt. Haldane) near the summit. He died alone in the summer at a remote cabin at the foot of Mayo Lake circa 1967.

ARMSTRONG, Lt. Col. Neville Alexander Drummond

Lt. Col. Armstrong was variously described as a 98'er, prospector, miner, geographer, big-game hunter and conservationist, soldier and gentleman.

Born in England in 1874, Armstrong at the age of 17 spent three years in Texas and New Mexico as a day labourer. In 1898 he went to the Klondike via St. Michael as assistant manager of the Yukon Goldfields Company, acquired rich claims for the company on Bonanza Creek and Cheechako Hill, and became manager of the company. In 1901 he made a deal with Duncan Gillies, the discoverer, for a placer concession on Russell Creek on the MacMillan River. On August 3, 1901, together with his crew of six, he took the steamboat *Prospector* up the MacMillan River to Russell Creek. For this journey he was made a Fellow of the Royal Geographical Society.

In 1904 and 1905 he was appointed Game Warden for the MacMillan River country. Until 1914 he spent almost every summer and several winters on Russell Creek prospecting for gold. His wife, Marion Armstrong, spent 18 months on Russell Creek in 1905–06 with him and their friends George and Jessica Leith. Neither woman had ever spent any time outside England, and they were completely isolated during their time on Russell Creek.[2]

ARNOLD, Alfred "Fred"

Fred Arnold was from Dawson where he had a barber shop. He came to Mayo for the 1920's mining boom, then returned to Dawson.

ARTHUR, Alexander & Peggy

Alex Arthur, wife Peggy and their two children, Henry and Margaret came to Mayo in 1947. Arthur bought Palmer's Hotel in Keno and the Silver Inn in Mayo from Bob Palmer. He later sold the Keno Hotel to United Keno Hill Mines and the Silver Inn to Milo Jurovich and opened Porky's Menswear in Mayo, Keno and later Whitehorse. He named the store Porky's using the nickname of "Porky" Levine, a former N.H.L. goaltender, whom Arthur employed to run his clothing stores in Mayo, Keno and later

Whitehorse.

In 1954 he closed these businesses and reopened Porky's Menswear in Whitehorse which was bought by Rolf Hougen in 1958. He died in 1961 in Whitehorse at age 45 and is buried in the Pioneer Cemetery.

ATWATER, Ben

Ben Atwater mined on the McQuesten River with Ray Stewart and Charlie Dickson in 1892 and on a bar a few miles below the mouth of Johnson Creek. He left the district to join the rush to the Fortymile.

AYLWIN, George S. & Emelia

See *Pioneer Accounts*.

BACON, Frederick & Elsie

Frederick Bacon was a cook at Elsa in the 1930's. Later on he was Chief Steward on the riverboats out of Whitehorse. He and his wife, Elsie had three children, Eileen, Denis and Terance who attended school at Elsa. Terance "Terry" joined the Diplomatic Service and has served as an ambassador for Canada in several countries overseas. He married Katherine Lee and they have four sons.

BACON, George

George Bacon was Treadwell Yukon's mine foreman at the Hector-Calumet and Elsa Mines between 1936 and 1941. The Bacons and their children came from Elsa to Mayo in 1938 and moved into Bud Fisher's house, formerly the Major home, on Fourth Avenue next to Whitney's. Evelyn and Beverly Wilms were George Bacon's nieces.

After Treadwell Yukon ceased operation he obtained a 180-day option on all of the Treadwell Yukon properties, but was unable to interest any mining companies.

BAKER, Don & Hazel

See *Pioneer Accounts*.

BALLENTINE, David W.

See *Pioneer Accounts*.

BAMFORD, William Edward "Bill"

Bill Bamford was a purser on one of White Pass's main steamboats circa 1930. Later in the 1930's he owned and operated the Stewart roadhouse in Stewart City. He built a frame building for a trading post. In July, 1937, he sold the roadhouse to Rudy Burien and Rudy's sister Mrs. L. Lechner. Bamford continued to operate the store until he sold it to the Hudson's Bay Company in 1939.

BANISTER, Garnet Fitz-William

Garnet Banister, born in Buenos Aires, Argentina in 1875, worked for Thomas Aitken at the Silver King in 1915. He was killed in a circular saw accident at the mine on March 2, 1916 and is buried in the old cemetery at the Mayo-Elsa road junction.

BARCHEN, Hans & Regina "Gina"

Hans and Gina Barchen originally came from West Germany. Hans arrived in the district in the 1950's and went to work for United Keno Hill Mines. He left U.K.H.M. in 1959 to mine placer gold with John Sandanger and Klaus Djukastein on Johnson Creek and the outfit was named Bardusan Placers when Hans joined the partnership. Hans bought out his partners and now operates Bardusan Placers on Thunder Gulch (1989). Sons Claus and Ralph are also placer miners.

BARKER, Edward Harmon "Ed" & Matilda "Tillie"

Ed Barker was a dredgemaster in the Klondike (in 1907 at age 23) and Mayo districts as well as for a period in Siberia. His wife "Tillie" accompanied him to Siberia. He was a dredgemaster on Highet Creek for the Titus Dredge Company (also known as the Highet Creek Dredge Company) in 1921 (McVicar, Finlaison, Jones and Titus). Barker was master mechanic for Treadwell Yukon Co. Ltd. at Wernecke Camp from approximately 1923 to 1931. Treadwell's General Manager, Livingston Wernecke, dismissed him following a disagreement, in 1931.

Barker tried bar mining on the Stewart River and in 1932 he made a prospecting trip with Jim Gibson and staked the ground below Dublin Gulch. He worked the ground with his partner Kit Watters that season and cleaned up $5,000. A few years later he bought a small gas Cat with scraper but found it was too small to work the ground, and in the late 1930's he and his partner Irvin Ray bought a D-4 Cat and formed Haggart Mining Company. He later obtained George Potter's ground and by 1940 held 22 claims and

12 miles of prospecting leases along Haggart Creek. The partners closed the operation down in 1942, though they later mined together again in about 1946.

In 1952 he bought the Chateau Mayo Hotel from Jim Mervyn and in 1953 he bought another D-7 Cat to prospect for hardrock in his spare time. He also had a large log cabin on the bench across Haggart Creek. This cabin was owned previously by Jahnke and Gill.

By this time, Barker held some 100 hardrock claims. From 1955 to 1959 he leased his placer holdings to Waddco (Maynard Wilson, Jack Acheson, Bill Drury, and Clyde Day; also known as Spruce Creek Placers) who had successfully mined at Spruce Creek in the Atlin district for several years. In 1959 when Waddco moved to Dominion Creek in the Dawson district, he again mined for a season near his cabin.

He and his partner Irvin Ray later bought the property in Whitehorse occupied by the Yukon Inn [1989] and former Hudson's Bay Store. They operated it as Tourist Services—one of the Yukon's first supermarkets.

In his later years, he maintained his fascination with hardrock prospecting, staking claims and acquiring the large group O'Neill had staked in the early 1950's. In June 1961 he took a Cat and crew to Cecil Poli's Peso property. The day after returning, Barker suffered a stroke in his cabin on Haggart Creek and died enroute to the hospital in Mayo. He is buried in the Mayo cemetery.

BARKER, F. M.

On November 8, 1898, F. M. Barker staked a Discovery claim on a tributary of the Stewart River (20 miles above its mouth) and named the creek Barker Creek. However, he found very little gold, so all claims were soon abandoned. In 1903 Louis Marret found good prospects and staked a new Discovery claim on Barker Creek. There was another stampede later on this creek in 1906, but it never lived up to its early promise.[3]

BARR, John

John Barr came to the Yukon in 1898 and ranged the north MacMillan River country for many years, trapping and prospecting. Tall, rugged, and a supreme bushman, he was known as the "King of the MacMillan Country." Always courteous and helpful, he was well-liked and respected by all. A tributary of the North MacMillan River was even named Barr Creek because he had a cabin at the mouth of the stream.[4]

BARTON, James

James Barton was a fireman for Keno Hill Ltd., at Keno Hill in the 1920's.

BARTON, Thomas Henry

Thomas Barton was one of the original party of eight Scottish prospectors who found some success prospecting a stream they called Thistle Creek (a tributary of the Yukon River just above the mouth of the Stewart River). The same party also prospected and named several other streams in the area. Barton died in Dawson in 1916.

BATTY SISTERS

The Batty sisters were from Assiniboia, Saskatchewan. They were all graduates of the Provincial Normal School. Ruth was the first one to come to the Yukon. After her discharge from the RCAF (Reserve) she attended the University of British Columbia then came to Mayo to teach school in 1946. She went to Edmonton in 1949 to take a commercial course and returned in 1950 to open a cafe in the Silver Inn, moving later to the Chateau Mayo. She also opened Ruth's Dress and Novelty Shop. One summer she cooked for Waddco Placers on Haggart Creek. The year before her marriage to Gordon McIntyre she taught school in Keno as well as running the two businesses in Mayo. She sold the restaurant to Anne Kwasnicki and the variety store to Bessie Friesen.

Doris's ("Dorrie" Ewing) account appears in the *Pioneer Accounts* section.

Margaret came to the Yukon to teach school at Destruction Bay. She then came to Mayo where she taught one term and worked for the Royal Bank. She left the Yukon to attend U.B.C.

BEAUDET, G.

G. Beaudet and Denis Parent prospected Parent Creek (a tributary of Duncan Creek) circa 1902 and found placer gold.

BEAUMONT, R. C. "Chuck" & Florence

Chuck Beaumont first came to the Yukon in the spring of 1932 as a deckhand on the steamer *Keno*. By 1936 he had worked his way up to the position of purser on the *Casca*, and by 1940 he was working in the White Pass office in Dawson.

In Dawson he married Florence Troberg. (The Trobergs are a pioneer Dawson family.) He was transferred to Whitehorse during the war years and posted to Mayo as agent from 1946 to 1950.

He was later transferred back to Whitehorse and promoted to a management position. In 1963 he was transferred to Vancouver B.C. where he eventually retired. The Beaumonts had two sons, Ralph and Douglas. Chuck Beaumont died in Vancouver in the mid 1980's, and Florence died there on May 31, 1989.

BEAUPRE, Joseph & Marie

Joseph Beaupre came to Mayo from Dawson. He was a clerk at Binet's hotel and store in the 1920's. Marie Beaupre, Joseph's wife, is believed to have originally come from Louisiana.

The *Dawson Daily News* credits Beaupre with building and owning a two story business block on the corner of First and Centre Streets in Mayo in 1922.

The Beaupres are buried in the old cemetery at the Mayo-Elsa Road junction.

BEAUVETTE, Louis

See Bouvette, Louis

BECK, Charles

Charles Beck staked the Black Hawk claim on June 29, 1922. He continued to mine, prospect and stake claims in the district for many years. In 1936 he and Sandy McPherson leased Hans Formo's Rocket claim, on the north end of Galena Hill, and put down 40 holes before striking a vein.

BECKMAN, John

John Beckman was at Keno Hill in the 1920's. He later became a carpenter in Mayo and built the log cabin that Alex and Daz McCarter lived in during the 1930's. (Alex Smith later lived in the cabin.)

BELIVEAU, Alphonse

Alphonse Beliveau arrived in the Mayo district around 1901. Alphonse, Victor Savage, and Elphege Desautels staked claims on Duncan Creek in 1903. He staked claims on Beliveau Creek and staked No. 25 above Discovery on Edmonton Creek in June 1904. He sold his claim on Beliveau Creek to Victor Savage and Elphege Desautels in July, 1903.

BELIVEAU, Ernest

Ernest Beliveau was from Quebec, and became an active prospector connected with many of the early finds in the Mayo district. He discovered good placer gold on Beliveau Creek (a tributary to Duncan Creek) and staked a Discovery claim on January 7, 1902. Ernest staked on Steep Creek in 1903. Mount Beliveau is named for him.

BELL, Henry

Henry Bell was a barber by trade on ocean-going ships. He trapped at the mouth of Clear Creek and when he was in Mayo he lived in a cabin on the road to the sawmill. He operated a barber shop in Mayo in the 1930's and 1940's.

BELL, Jim

Jim Bell was a native of Montenegro, Yugoslavia. He came to the Yukon in 1906 and remained in Yukon and Northern British Columbia for the rest of his life. He worked in the Mayo district as an employee of Treadwell Yukon for many years. He died following a lengthy illness at the Mayo hospital on May 16, 1939, at age 59. Treadwell employee Nick Sutolovic and Jim Bell were born in the same place in Yugoslavia.

BELLERBY, John "Jack" "Pop" & Rose

"Jack," also affectionately known as "Pop" Bellerby was the wharfinger for White Pass from 1923 to the 1940's. He transferred from Nenana, Alaska to Mayo in 1923.

"Ma" and "Pa" or "Pop" Bellerby were an institution in Mayo—kind and generous to all the children. "Ma" spent every December baking and decorating fancy cookies and gingerbread men which "Pa" delivered by the bagful to all the kids in town. He was also the umpire for baseball games.

The grounds of their home were the most expansive and attractive in Mayo with numerous flower beds as well as a greenhouse. The Bellerbys were also the only people in Mayo to have the trees on their property lit up for Christmas.

The Bellerbys had a son, Russell, who was an officer in the U.S. Navy. Rose and Jack Bellerby left Mayo and retired to Seattle, Washington in the 1940's.

BENNETT, Charles E.

Charles Bennett found placer gold and staked the Discovery claim on Bennett Creek (a tributary of Minto Creek) on May 25, 1903.[6]

BERG, Conrad "Connie" & Kathy

"Connie" was born in 1931 near Munich, Germany, and Kathy was born in Czechoslovakia. They apprenticed and worked in Munich, he in the shoemaker trade and she as a machinist for Siemen's Electric.

In 1954 Connie came to Canada and worked on construction in Edmonton, Alberta and then in Hamilton, Ontario. In the spring of 1957 he came to the Mayo district to work for United Keno Hill Mines in Elsa as a timberman. In 1958, he married his sweetheart from Munich in Edmonton, Alberta. He returned to Elsa and Kathy arrived three months later. They set up their first household in Keno City and later lived in Calumet, Flat Creek, and Elsa. Their son, Eric, was born in Mayo.

During his years with United Keno Hill Berg was a timberman, underground miner, shift boss, safety supervisor, and from 1972 until his retirement in November 1987 he was United Keno's personnel officer. He was active in the Mount Haldane Lion's Club and was also Notary Public for the community of Elsa.

Their love of the outdoors greatly influenced the Bergs' decision to retire to Marsh Lake near Whitehorse.

(Information supplied by Claire Briand)

BERG, Ole

Ole Berg owned a restaurant at Keno Hill in the 1930's.

BERMINGHAM, Claude Hubert

In 1921 Claude Hubert Bermingham and Charles Reed Settlemier acquired the Arctic and Mastiff claims which during the summer of 1925 were not only the most important but also the only claims being mined on Galena Hill. Bermingham died in 1958 leaving a legacy to the Vancouver Yukoners Association so that they might continue to hold their monthly meetings and "uphold the spirit of the Yukon."[7]

BERRY, Alec F.

See *Pioneer Accounts*.

BERTON, Frank G.

Polk's *Alaska-Yukon Gazetteer* shows that Frank Berton was the Mining Recorder in Mayo from 1923 to 1924. However, he was only in Mayo temporarily to reopen the Mining Recorder's Office for the district when Louis Bouvette struck silver in 1919.

He was the father of author Pierre Berton.

BESNER, Ferdinand & Adelia

The Besner family came to Mayo from Dawson. Besner mined in the Mayo area in the early 1920's, and his wife, Adelia, ran a laundry in Mayo during the 1920's and 1930's. Their children are Tony, Rose (Dubois), Josie (Pelland), Emelia (Aylwin), Yvonne (Farr), George and Gus.

BESNER, George & Annie

Before the Besners moved to Mayo, George and his father, Ferdinand, hiked over the Conservative Trail from Dawson to Mayo in 1921. The pair worked for two months yarding wood with horses for Joe Longtin and Morisette. Eleven people died during a flu epidemic while the Besners were in Mayo. When the job was over, father and son returned via the Liberal Trail to their home in Bear Creek where Mrs. Besner (Adelia) ran a hand laundry for the Boyle Company dredges.

When George was 15 he returned to Mayo (to work at Oscar Letourneau's hotel) on the boat which Lord Byng (Governor General of Canada, 1921–26) had chartered from Dawson to visit this silver district. At age 17 George went to work for Treadwell Yukon, under mine foreman Jack Maloney. During his years with Treadwell he was a mucker, driller and timberman, and while he was hoistman at the Lucky Queen he rode the tram to work from the Ladue. Besner also partnered on the Black Cap with claim owner Tony Hollenbeck.

George built the house on Block 12, Lots 23 and 26 (see *Buildings and Artifacts*), as well as the house on Block 16, Lots 13, 14 and 15 (later the Earl Graham home) which was constructed of materials salvaged from the old hospital.

Besner married Annie Churchward and they have one daughter, Adelia. The Besners retired to B.C. in 1965. They celebrated their 58th wedding anniversary and George's 84th birthday in 1990.

From left: George and Annie Besner with Annie's father Guy Churchward, Mayo, 1948. *L. Cheatham (McLaren) Collection*

BICKERTON, Ed

Ed Bickerton was a Treadwell Yukon employee at the Hector in the 1930's.

BIDLAKE, Geoffrey "Geoff" & Mary

The Bidlakes lived in Mayo from 1939–41. He worked for the White Pass & Yukon Route. They had two children, Larry and Diane.

BINDLE, H.

H. Bindle was a miner at Keno Hill in the 1920's.

BINET, Joseph Eugene "Gene" & Jewel

Gene Binet was born in Charlebourg, Quebec. In 1895, he climbed the Chilkoot Pass, and travelled that summer with two others to both the Fortymile and Circle City. He returned the following season and was on the Yukon River between St. Michael and Circle City when the Klondike was struck. He arrived in the Klondike in late December 1896 and began to buy claims. His brothers Joe and Phil joined him, and they mined on several of the creeks. They also operated hotels, saloons, and mercantile businesses not only in Dawson but in Nome, Alaska.

When Binet heard that the townsite of Mayo had been surveyed and that a road would be built, he arrived late in the winter of 1903 to buy lots and build a hotel. He brought building tools and a crew of six men for the hotel construction, and he personally felled the first spruce from the hotel site. By the time the ice went out on the Stewart River, Binet's 30x60 foot, two-storey log hotel was assembled, and by July 1903 the rooms were ready and the dining room operating.

Binet soon sold a number of his lots, then built a store on to his hotel. The establishment became known as J. E. Binet's Chateau Mayo Hotel and Store (during the time Jim Mervyn owned the hotel it was known locally as Mervyn's Hotel although the official name remained Chateau Mayo). The following year he enlarged the hotel to 60x60 feet with 26 rooms.

He also grew potatoes in commercial quantities, and kept a cow, chickens, and several pigs. By 1915 he owned a stable and had invested in a sawmill with Fred Lefebvre as partner.

During the 1920's, Binet married Jewel and built a home in Mayo. Jewel Binet was born in 1886. It is said that he imported the wallpaper for his house from France. This house still stands in

Gene Binet (one of the "Fathers of Mayo") on the left and prospector Jack Alverson on the right, 1937. (Alex Nicol was the other "Father of Mayo.") *G. A. McIntyre Collection*

Mayo, and it is located on Block 8, Lot 12. The Binets had one son, Eugene Jr., who with Jewel lived in California during the winter months and in Mayo during the summer.

On July 13, 1930 Binet's log hotel burned down. He immediately set to work erecting a new frame construction hotel.

In 1938, Binet sold the hotel and store to Mervyn, and his interest in the Binet and Lefebvre sawmill to Bob Palmer. The "Father of Mayo" retired and left for California in the fall of 1938. He died on December 26, 1943, in Los Angeles. When Gene Binet died he had a brother living in Riviere-du-Loup, Quebec.

Binet's potato patch, Mayo, 1919. From left: Raoul Binet, Lillian McLaren, and Gene Binet. *D. Walloch (McLaren) Collection*

BINET, Phillip

– Archie Gillespie, *The Mayo Miner*, March 12, 1937

Phillip Binet came North in 1897 during the Klondike Gold Rush and lived in Mayo from 1906 until 1909. During his stay in Mayo he was involved in the hotel business with his brother Gene Binet. He was well-known in both the Mayo and Dawson districts. He died in February 1937 in Montreal, at the age of 74.

BINET, Raoul

Raoul Binet was Gene Binet's nephew. In 1920 Joel Sunderland gave Raoul the Friendship claim in exchange for the recording fee, and Raoul gave Dick Mercure a half interest in the Friendship for doing the assessment work. A rich vein was discovered on the claim and the following year Binet and Mercure sold the Friendship to the Guggenheims' Keno Hill Ltd. for $40,000. Binet then returned to Montreal, Quebec.

BISEAK, Joseph

Joseph Biseak is listed as a miner in the Mayo area in Polk's 1923–24 directory.

Binet's Hotel, Mayo. *A. K. Schellinger Collection*

BJONNES, Ellef "Nels"

"Nels" Bjonnes obtained a 3-year lease on the No Cash claim from Charlie Brefalt in the 1930's. He spent many years high-grading on the Ruby Fraction and the No Cash and did relatively well. He spent his final years in the late 1950's in Vancouver, B.C., where he lived at the Hudson Hotel.

BLACKMORE, Samuel "Sam"

Sam Blackmore was a prospector in the district from 1902 and a teamster in the 1920's. Upon his death he willed the cabin currently known as "Mabel McIntyre's Cabin" to Mabel McIntyre. He also owned the old Mining Recorder's Office which was located next to "Mabel's Cabin" (to the south). At one time, the Mining Recorder's Office was used as a high school classroom, then as an annex to Ruth's Novelty Shop. It is currently Dick Ewings' shed.

BLAIS, Raymond

Raymond Blais staked claims on Duncan Creek, Parent Creek, and Williams Creek in 1901. He acquired a claim on Highet Creek in 1906.

BLEILER, Clarence A. & Helene

In 1938, at the age of 21, Clarence Bleiler came to Mayo over the winter trail by snowmobile to join his older brothers "Ed" and "Ted." He moved into Ted's cabin at Minto Bridge and got his first job in the Yukon loading wood onto the Treadwell Yukon trucks. This job paid six dollars a day. He recalls that two of the truck drivers were Alex McCarter and Roy Thomas, an ex police officer.

The following summer, Bleiler's brother "Ed" arranged for him to work on Highet Creek which he followed by a winter of trapping. He spent the next few years placer mining in the summer and trapping in the winter. One of his trapping partners was an experienced trapper and miner, Louis Bouvette.

There was little work available after Treadwell Yukon closed down in 1942, so he moved to Whitehorse and was employed by White Pass in the automotive shop at the airfield. After working there for a while he got the opportunity to rent a small Caterpillar with a saw mandrill to go into the wood business. From 1943 to 1944, Bleiler, along with partners "Ed" Bleiler and George Besner cut firewood to supply the camps along the Canol Road.

While in Whitehorse, he became reacquainted with his childhood sweetheart Helene McKinnon from Edmonton. They were married in June 1944 and settled in Whitehorse, and in November 1945 their first son, Dwight was born. Not long after that, they sold their house and bought a house in Mayo that Ed Kimbel had built so that Bleiler could go into a placer mining partnership with his brother, "Ed" on Highet Creek. After a summer of placer gold mining, he did some underground exploration work on a silver prospect on Bunker Hill without any success.

Bleiler spent the next summer sacking silver-lead ore from the old dumps with Claude Bermingham, and with mining still in his blood, he took a lay on the Bermingham property. He hired "Big" Ellis Johnson, who was known for his strength, to help him develop the ore body which was visible on the surface. Meanwhile, his wife, Helene went to Edmonton to have their second son, George.

He invested his savings on developing a mine, but met with very little success. He then took his equipment and got a contract to de-water and de-ice the Brefalt Mine near the tramline angle station with Cliff Greig as a partner. After that, he and "Ed" decided to go into the wood business again, cutting and hauling timber to United Keno Hill Mines at Elsa. He would then get a backhaul of six or seven tons of ore. Not long after this, he sold out to his brother "Ed" and in 1949, moved to Lund, B.C. and later to nearby Powell River, B.C. where a daughter, Teresa, was born.

Powell River, is still home to the Bleiler family. Clarence, Helene their children and grandchildren and his brother, "Ted" with his wife Almuth all make their homes there. [1989]

BLEILER, Edwin "Ed"Charles F. & Flora

See *Pioneer Accounts*.

BLEILER, Theodore "Ted" Roosevelt & Almuth

See *Pioneer Accounts*.

BOERNER, C. A. "Kippy" & Enid

See *Pioneer Accounts*.

BOICH, Michael

Michael Boich is listed as a miner at Keno Hill in Polk's 1923–24 directory.

BOLD, R. J.

R. J. Bold was a World War II veteran who worked at Elsa for United Keno Hill Mines in the 1950's. He was a charter member of The Royal Canadian Legion, Keno Hill Branch No. 3. Bold is buried in the Mayo cemetery.

BOND, Jack

Jack Bond was a bartender for J. E. Binet in the early 1920's. In the 1930's he was Treadwell Yukon's foreman at the Hector.

BOND, Samuel

Samuel Bond staked a claim on Johnson Creek in 1916. He is listed as a miner in the Mayo-Keno Hill areas in Polk's 1915–16 and 1923–24 directories.

BOON, Jack "Doc"

Jack "Doc" Boon was originally a schooner sailor, and is remembered as a big, powerful man who spoke with a British accent. He was considered the best Cornish single-jack miner in the district, and he worked for Harry Colley and Don Morrison on the Calumet claim in 1935. As a result of his efforts, several rich veins were located. It is said he acquired the name "Doc" after assisting in the delivery of a child in the district in the early days.

BOOTH, H. H.

H. H. Booth is believed to have been a recluse who trapped and prospected in the Beaver River Country. While alone in the bush in 1934, he suffered a very painful illness and when he could stand the pain no longer he shot himself. Booth is buried in the Mayo cemetery.

BORIS, Alex & Blanche

See *Pioneer Accounts*

BOSSET, Leo OMI

Father Bosset was the resident priest at Christ the King Mission in Mayo from 1940 to 1942 following Father Monnet's departure. He trapped during the winter to eke out a living.

BOSTOCK, Hugh S.

See *Pioneer Accounts*. (Although Dr. Bostock never lived in Mayo he was involved in several exploration and mapping excursions in this district as a member of the Geological Survey of Canada.)

BOSWELL, George & Thomas

The Boswells were from Peterborough, Ontario, and they were among the first prospectors in the Yukon watershed. Records show that Thomas came over the Chilkoot Pass in the spring of 1882 and George came either the same year or the year after. They were active men and among the first to find rich bar gold on the Stewart River at Chapman's Bar. In 1885 Thomas Boswell, John Fraser, and a man named Chapman, and others explored the upper reaches of the Stewart River and named Fraser Falls.

They ranged widely across the Yukon and Alaska for years, always in the forefront of pioneer activity. Thomas lost a leg to a grizzly bear in Alaska in 1891, but he and George still joined in the Klondike stampede. Thomas Boswell (and perhaps his brother George) renamed the Nadaleen River the Boswell River. The name of the river was changed back to Nadaleen in 1905.[8]

BOULYER, Matthew

Matthew Boulyer is listed as being at Keno Hill in Polk's 1923–24 directory.

BOUVETTE, Louis

Louis Bouvette (commonly misspelled Beauvette) discovered silver-lead ore in 1919 while hunting mountain sheep on Sheep Mountain (later renamed Keno Hill). A pioneer prospector of the district, he and Charles Johnson staked the first placer gold claim on Carlson Creek (a tributary of Bear Creek) on March 10, 1911 and staked on Highet Creek in November 1913. Bouvette's story, prepared by his niece, Mary Jean Morrison, appears in the *Pioneer Accounts*.

BOWDEN, Lewis Franklin

According to Elmer Middlecoff's Time Book 1910–1918, Lewis Bowden was employed on Highet Creek as a miner from April 1910 to September 1918. He may have continued working with Middlecoff, but there are no records to support this. He had a small farm on the road that went to Archie Close's farm. Albert and Neoma

Mayo, 1948. From left, back row: Jack Hawthorne and Jean Boyle. Front row: Mary Jean Morrison Boyle, Frances Morberg, and Baxter Wareham.
L. Cheatham (McLaren) Collection

Pelland lived in the log cabin on the property in the 1950's. Bowden was an associate of Irvin Ray. Lewis Bowden was born in 1868. He died in 1940 and is buried in the Mayo Cemetery.

BOWER, Fred

Fred Bower is listed as a miner at Keno Hill in Polk's 1923–24 directory.

BOYCE, John & Kay

John Boyce was a Northern Commercial Co., employee who came to Mayo from Dawson circa 1958–59 to manage the Northern Commercial store, an old two-storey log building. He then moved the business into the new building (now the Mayo General Store [1989]). He left the company circa 1960 to build and operate the Tim-O-Lou Motel with his wife Kay. (The motel is now known as the North Star Motel [1989].)

BOYD, R. J. "Bob"

"Bob" Boyd was Treadwell Yukon's General Superintendent at the Elsa Mine from 1936 to 1938.

BOYLE, Jean

Jean Boyle, née Nixon, came to this district in 1922 as Jean Wolsey to visit her uncle Louis Bouvette. She returned to Keno in 1930 to keep house for her uncle. She moved to Mayo in the 1930's and lived there almost continuously until her move to Whitehorse in 1980. Charlie Nixon was her brother. Jean met her husband, Don Morrison, in Keno and they had one daughter Mary Jean Morrison. After Don died, she married William Boyle. She is remembered by all who knew her for her warm and charismatic personality. She died in Whitehorse in1982.

BOYLE, William "Billy"

"Billy" Boyle was a teamster for Keno Hill Ltd. at Keno Hill in the early 1920's, and later became a Cat operator. He was married to Jean Boyle. He served in World War I and lied about his age to serve in World War II. He worked mostly as a bulldozer operator on airport construction jobs in England. Boyle was U.K.H.M.'s foreman in charge of loading ore on the sternwheelers during the late 1940's and early 1950's. He died in Mayo in 1958.

BRADLEY, Fred

Fred Bradley, who never actually lived in Mayo, was President of the Treadwell Yukon Company based in San Francisco. He graduated from the University of California as a mining engineer and successfully operated the Spanish Mine in California on very low grade ore. He became president of Bunker Hill and did considerable work on the lode gold deposits of California. Fred also received the Saunders Medal for bringing the Alaska Juneau mine into successful production on $1 per ton ore.

In 1921 he sent Livingston Wernecke to the Mayo district to examine Keno Hill. This transaction began the Treadwell Yukon era in this district. Fred Bradley died in 1933 and his brother Phil Bradley became President.

BRADY, Patrick

Patrick Brady was mining on Blackhills Creek

circa 1910. He is listed in Polk's 1923–24 business directory as a labourer at Minto Bridge.

BRAINE, Frank

Frank Braine had been a trader in northern Saskatchewan, but was drawn to the Yukon in the 1890's as were many other men from around the world. In 1897–98 he and Percival Nash travelled through this district on their way to the Klondike from Edmonton, but were disappointed in their search for gold, so they returned to the vicinity of Fort Norman in the Northwest Territories to trap and trade. In 1902, he asked a number of Indian people from Fort Good Hope to guide him across the Mackenzie Mountains. He built a trading post, known as Lansing Post, on the Stewart River at the mouth of the Lansing River and supplied it from Dawson City.

BRAME, Albert Leonard & Bertha Sarah

Albert and Bertha Brame came to Canada from England and settled in Peterborough, Ontario, where their son Albert Leonard Bramley Brame was born. They later moved to Thorold, Ontario, and remained there until 1923 when they moved to the Yukon. The couple settled on Duncan Creek in the Mayo district where Mrs. Brame's brother William "Bill" Bramley was mining. Mr. Brame worked a couple of summers on the creek with Bill Bramley and then went to work for Treadwell Yukon on the Friendship claim where he worked in the assay office and as a surveyor. Albert Brame died in 1948. His widow remained in her home on Duncan Creek until approximately 1955 and then moved to Mayo.

There is an unknown grave on a hill near where the Brame's cabin was located on Duncan Creek. It is believed the man buried there was of French descent and that he died before 1920.

BRAME, Albert Leonard Bramley & Martha

When Albert Brame Jr. was approximately 17 years old, his parents, Albert and Bertha Brame, left their home in Thorold, Ontario, to move to the Yukon and Albert Jr., who had finished school by that time, left to travel the world.

He arrived in the Mayo district in 1927 and

James H. "Spot Cash" Breaden standing near the United Keno Hill fuel storage area in Mayo. *G. A. McIntyre Collection*

worked for a time on Duncan Creek and then went to work for Treadwell Yukon. He remained until the Depression caused the mine to close temporarily in 1932. He left for the U.S.A. and was married in Tacoma, Washington, where the couple's sons were later born. He returned to the Yukon with his wife and sons and worked for Treadwell Yukon from 1937 until the mine closed permanently in 1941, when the family moved to Whitehorse where Albert and Martha lived until 1971. The Brame children are James Patrick Brame, Albert Leonard Brame (who was killed during construction of the Mayo Road), Mary Elice Brame who was born in Mayo, and Brenda Lee Brame, born in Whitehorse.

BRAMLEY, William Leonard

– Archie Gillespie, *The Mayo Miner*, September 24, 1937

William Bramley was born in 1867 in Warwickshire, England, and came to the Yukon during the Klondike Gold Rush of 1898 via the Edmonton Trail. From the time he arrived until 1907 he mined and prospected in various parts of the Yukon, and from 1907 to 1937 he mined and prospected in the silver district of Mayo, particularly on Duncan Creek where he operated a placer mine.

Bramley died on September 22, 1937 in the Mayo General Hospital. His relatives included Mrs. A. L. (Bertha) Brame of Duncan Creek, and Mrs. A. E. Child of Humborstone, Ontario.

BREADEN, H. James & Kitty

Jimmy Breaden was born in Vancouver, B.C. and was known as "Spot Cash" or the "Spot Cash Kid." He earned this nickname when he ran a "jitney" between Dawson and Bear Creek after World War I. While overseas with the Yukon Infantry Company he met and married Kitty in England.

A first class mechanic, he worked at that trade during his years in Mayo. The Breadens' three children are: Audrey, Vera, and Henry. Jimmy and Kitty later moved from Mayo to Whitehorse.

BREADEN, Henry & Alice

Breaden was First Mate aboard the SS *Keno* when the boat made it's last historic trip down the Yukon River to Dawson, arriving on August 29, 1960. See *Pioneer Accounts*

BREFALT, Charles "Charlie"

Charlie Brefalt was born in Amotfors, Sweden, on September 21, 1886. He came to America at age 23 in 1909 and arrived in Denver, Colorado where he learned hardrock mining. He prospected through Colorado, Arizona, Nevada and Wyoming, worked for the C.N.R. mill in Hedley, B.C. for a time, then mined in the Coeur d'Alene district of Idaho until 1916.

In the spring of 1916 he came to the Yukon and worked at the Pueblo copper mine in Whitehorse, but became convinced that the Pueblo was unsafe, so he quit and went to Dawson City. He cut wood that first winter in Dawson and then mined unsuccessfully on Henry Gulch.

In 1920, he came to the Mayo district and went to work for the Yukon Gold Company at Keno Hill. Following that he staked claims on Stand-to Hill with his partner Joe Michaud, prospected the property for eight months, then returned to Keno and took a lease on Bert Lamb's Gambler claim in March of 1922. Later, he and Lamb worked as partners to mine some ore from the Gambler.

In 1923 he prospected the Beaver River and beyond with Matt Butijer (Butyer), then returned to Keno and took a short term option on Joe Walsh's Croesus claim with his partner Joe Michaud. He also had a lease on the Bunny claim with Sam Swallow as partner.

In 1924 he sank a shaft for Charles Settlemier and Claude Bermingham and struck a good ore shoot. On September 13, 1924, Brefalt staked the Elsa claim, naming it after his sister, and in May of 1925 he staked the Lucky Strike claim. In October of 1925 Charlie, Dan Tolmie and a crew of 12 men sacked and shipped 50 tons of ore from the Elsa claim. By the end of the winter they had shipped 192 tons of ore. Also in 1925, he staked the No Cash claim.

In the spring of 1928, Treadwell Yukon optioned the Elsa for $150,000. The Elsa and the Lucky Queen were Treadwell's richest mines.

In 1933–34 he worked the No Cash with his partner Dan Tolmie, with little success. While working on the No Cash, he also acquired a one year lay on the Bermingham-Settlemier property and worked the ground with Elmer Gustaveson. In 1938 Brefalt and Theodore Erickson made a prospecting trip east of Keno and staked and worked a claim on Mount Hinton. In March 1941 Treadwell Yukon leased the Elsa to Brefalt and Elmer Gustaveson who worked the property for four years before United Keno Hill Mines took over. He continued to work the area in the 1940's when the major companies had all shut down. He helped Frank Buckle sample the Hector-Calumet.

In the 1940's he bought two cabins at Thunder Gulch and staked abandoned claims, and he leased Alex Nicol's Vanguard claim on Keno Hill. In 1949 he sold nine claims to "Rees" Powelson and together with other claims he had recommended, Mount Keno Mines was assembled.

In 1950, he went to Toronto to sell his Ajax fraction to Newkirk and while there, suffered a stroke. He returned to Keno in 1954 and worked for a time on the Vanguard claim. He retired at 68 to a cottage near the ocean at Roberts Creek north of Gibson's Landing, B.C. In 1963 he returned to his home in Amotfors, Sweden. He died there in October, 1970.

Charlie was known locally as the "Lucky Swede"

BRENNAN, James

James Brennan is listed at Keno Hill in Polk's 1923–24 directory.

BREWSTER, James Alex

James Brewster mined near Lake Creek circa 1913. He is listed at Keno Hill in Polk's 1923–24 directory.

BRIBA, Pete

Pete Briba was Master Mechanic at the Wernecke Camp circa 1930–1932.

BROADFOOT, Mrs. Sophie

Mrs. Sophie Broadfoot and daughters Mary, Kate, and Esther (Mrs. C. G. May, wife of C. G. May, officer in charge of the Royal Canadian Corp. of Signals staff in Mayo) lived in Mayo in the 1930's. Mary Broadfoot worked at the Silver Inn in the late 1930's and was very active in Mayo's amateur theatre. Mary left Mayo for Vancouver in November, 1938.

Kate Broadfoot cooked for the Middlecoff

Charles "Charlie" Brefalt came to Mayo in 1920. He was a successful prospector and staked the Elsa and No Cash claims. *G. A. McIntyre Collection*

placer camp in the summer of 1938, and frequently cooked for Ed Barker's placer camp.

BROSSEAU, Malvina

Malvina Brosseau lived on Field Creek in the 1920's.

BROWN, Louis John & Dolores

Louis Brown came from Stettler, Alberta into the Yukon via one of the Edmonton trails in the early 1930's. He had a ranch at 17-Mile on the Stewart River, and operated a big game outfit for many years with his wife, Dolores. Dolores Cline Brown wrote a book about their guiding experiences entitled "*Yukon Trophy Trails.*" Brown travelled areas such as Clear Creek, Dublin Gulch, the Hanson Lakes Road, and the Bonnet Plume country. He found rich copper mineralization near Fairchild Lake in the early sixties, and Bonnet Plume Copper Mines Ltd. was organized to explore the deposit. He also found copper prospects on the Bonnet Plume River which were explored by Leo Proctor in 1967. Often guiding with him were the best Indian woodsmen from the area; men such as Lonnie Johnny[9], Paul Germaine, Edwin Hager, Freddie Alfred, Harry

Baum and Johnson Edwards. Brown died in Whitehorse in 1987.

BUCHARD, Theophilo

Theophilo Buchard is listed at Keno Hill in Polk's 1923–24 directory.

BUCK, Reverend Frank Hepworth

Reverend Frank Buck was a camp mission worker in the Klondike and Mayo areas from 1916 to 1924, and in 1922 he became the first rector of St. Mary's Anglican Church in Mayo. He taught the first high school classes in Mayo, but later resigned his post in 1925 to study at the University of British Columbia. In his later years he taught high school in Victoria, B.C. Reverend Buck was born in Truro, Nova Scotia.

BUCK, William "Bill"

Bill Buck was a prospector at Christal Lake in the 1920's.

BUCKWAY, Edward & Edna

Edward Buckway came to Mayo in 1941 and worked that year in the Elsa and Calumet mines. He joined the Army in communications and was posted to England, Belgium and Germany. He and his wife Edna live in Vancouver, B.C.

BUCKWAY, Frank & Sadie

Frank Buckway came to Whitehorse with his sister, Daisy Keobke and her two children in 1935. He drove the overland stage for Klondike Airways Ltd. with Donald Keobke. He worked in Dawson, then drove ore trucks from Keno Hill.

He joined the Army in 1942, went overseas, saw fighting in Holland, Belgium and Germany as a sniper. He and his wife Sadie were killed in a car accident on the Hart Highway in 1977.

BUCKWAY, Peter

Peter Buckway worked with his brother-in-law Neil Keobke as a heavy equipment operator from 1937 to 1939. After leaving Mayo, he was employed for 36 years in Montreal with the Federal Government. He now lives in Abbotsford, B.C.

BUNTING, Samuel "Sam"

Sam Bunting was mining on Scroggie Creek in 1912. In his senior years he lived and cut wood on Duncan Creek just above where the Jack Ross family was located. Bunting is buried in the Mayo cemetery.

BURIAN, Alfred "Alf"

"Alf" Burian had a wood camp at Independence Creek in the 1930's and another at Stewart Crossing (located 30 miles below the present Stewart Crossing). His brother Renny took over the wood camps after he drowned in the

The Burian family. From left: Renny, Louis Lechner, Rudy, Alfred, Amanda, Martha and Carl in the centre with Martha's daughter, Wilda. *R. Burian Collection*

Stewart River in 1939.

BURIAN, Carl & Rosie

Carl Burian had a wood camp at the mouth of the Stewart on the Yukon side for a number of years. Alfred, Carl and Renny Burian are brothers. Carl and his wife, Rosie from Pioneer, Alberta, were married in the spring of 1937.

BURIAN, Renny & Mary

See *Pioneer Accounts.*

BURNELL, Edward W. "Ed"

Ed Burnell was a member of the R.C.M.P. in Mayo in the early days. He joined the army in World War I and returned to Mayo in 1917 to trap on the Stewart River. He had a cabin across the Stewart River from the White Pass sheds and another at 17-Mile as well as one at 26-Mile, and would often hike to town or travel down the Stewart on his raft, *Nogo Verifast*. Burnell died in Mayo in 1942.

BURNS, Bobby

Bobby Burns, an Australian, mined and prospected in the district prior to the 1920's. During the 1930's and 1940's he lived in a cabin near the log jam on the McQuesten River, on the old Lookout Mountain (Mt. Haldane) Road to Haggart Creek. He mined briefly on Haggart Creek, including the summer of 1934. Bobby hunted wild game commercially, and was also a trapper. Ed Bleiler recalls that Bobby's bread-baking skills were the envy of all the women in the district.

Bobby Burns died in Mayo, circa 1942–43.

BURNS, J. T. "Tommy"

Tommy Burns was a trapper and prospector, utilizing a dog team, in this district during the 1920's, 1930's, and early 1940's. Burns was always a "loner" although other people considered him a very interesting character. Burns was born in 1860 and died in Mayo in 1944. He is buried in the Mayo cemetery.

BURRELL, Louis

Louis Burrell arrived in Mayo in the late 1940's. He operated a cafe in the Chateau Mayo and when the road to Dawson was completed, he opened a small eating place called the "Little Brown Jug" near the McQuesten River.

BURRELL, William Anderson "Billy"

Billy Burrell staked claims on Duncan Creek during the stampede to that creek in 1902 and records show that he was still working his Duncan Creek claims in 1913. He staked claim #62 below Discovery on Minto Creek in October, 1903. Burrell prospected Keno Hill following it's discovery.

BURRETT, A.

A. Burrett is listed as a miner in Mayo in Polk's 1923–24 directory.

BUTIJER, Matthew "Matt" (also spelled Butyer)

Matt Butijer was born in Dalmatia, Yugoslavia in 1886, and went to Dawson in 1913 by dog team from Fairbanks, Alaska with Emil Anderson. He mined placer gold on several of the creeks flowing into Mayo Lake, and was mining on Gull Creek with George Crisfield and Arthur Hester, circa 1915–16.

In 1919 he staked the Stone claim on Keno Hill and managed to drive a 700-foot tunnel into the permafrost with hand steel. Another feat was digging a 1,200-foot tunnel into hard rock by himself with nothing but hand steel and a wooden wheeled wheelbarrow. During the boom in the 1950's he was paid $25,000 down for his claim on a $250,000 option. He also once mined on Duncan Creek. In the early 1960's he was mauled by a bear near his cabin. He lived his last years in Keno, where he died in the 1960's. (See *Pioneer Accounts.*)

BUYCK, Helen

See *Pioneer Accounts*. Helen Buyck (née Hager) was married to Don Buyck.

BUYCK, Peter & Evelyn

Buyck is remembered as a tall, handsome man who originally came from Belgium. He was a long-time resident of the Mayo area, and employed by the Treadwell Yukon Company at the Silver King. He had a house near the Silver King, and he and his wife, Evelyn, had one daughter, Yukona or "Connie," and two sons, Wesley and Don.

He died of silicosis while his children were still quite young. Evelyn Buyck lives in Edmonton with her daughter. Don died in Mayo, and Wesley lives at Stewart Crossing.

CAMERON, Benjamin "Ben"

"Ben" Cameron was a logger in Mayo in the 1920's and 1930's who lived in a cabin upstream from the Anglican Church–the first cabin past Isaac Mallette's cabin. He also worked for Alex Nicol looking after the barn and the horses and doing odd jobs. Margaret Wallingham (née Mason-Wood) recalls that "Ben" was a fine old man who was good with children. Cameron died circa 1956.

CAMERON, Charles

Charles Cameron, who came from Ontario, was a labourer for Keno Hill Ltd. at Keno Hill in the 1920's.

CAMERON, Ewen

Ewen Cameron held claims on Dublin Gulch circa 1907.

CAMERON, G. I. & Martha

See *Pioneer Accounts*.

CAMERON, James

James Cameron is listed at Keno Hill in Polk's 1923–24 directory.

CANTIN, Francois Xavier "Frank" & Alice

Frank Cantin operated Cantin Bros. with his cousins Louis and Phileas. The Cantins mined on Haggart Creek, Highet Creek and Dublin Gulch from 1909 to the 1920's. He sold firewood, provided Lefebvre's sawmill with sawlogs, and held timber rights on both limits of the Stewart River from Mayo River to Fraser Falls. He owned the log school building (North Star Motel site) and rented it to the Yukon Territorial Government until 1941, when he built a frame school at the request of the government. This new school was built on Joe Cantin's homestead, and Frank continued to rent it to the Yukon Territorial Government until the government purchased it.

Cantin married William Portlock's widow, Alice on June 27th, 1931. (See Alice Cantin in *Pioneer Accounts*.) She died on September 3, 1955, at age 75. Frank (born 1876) died on June 19, 1957. Alice and Frank Cantin are buried in the Mayo Cemetary.

The Cantins (Frank, Joseph, Louis, and Phileas) were French Canadians from Trois Rivieres, Quebec.

Evelyn and Peter Buyck with children. From left: Connie, Don & Wes at Mayo Lake, circa 1930. *Buyck Collection*

CANTIN, Joseph "Joe" & Jessie

At various times, Joe Cantin was a miner, a freighter and a labourer in Mayo from the early 1900's (circa 1903) to the 1920's. He owned the 160 acre homestead adjacent to Mayo townsite on the north, which is now occupied by the Mayo hospital, school, etc. Jessie and Joe Cantin grew market vegetables on their homestead.

When the Yukon Territorial Government was looking for a site on which to build the new hospital, Commissioner Collins asked Gordon McIntyre to acquire five acres (present hospital site), but Cantin wanted to dispose of the entire homestead (less a few parcels). A deal was eventually struck, and McIntyre obtained the entire property for $1,000.

He was married to Maggie Hunter's sister Jessie, and the couple had two daughters, Josephine and Annette. After Mrs. Cantin died in 1917, Persis Kendi looked after the girls until a woman from Quebec arrived and took the girls to that province to attend school. Josephine and Annette were about the same age as Mary Hager and Lucy Peter. Maggie and Jessie's mother was a sister of Sam Peter's father.

Joe Cantin (on wagon) and Frank Cantin, Katherine Lee (left), and Leila Palmer (right). *G. & E. Lee Collection*

Mrs. Cantin is buried in the old cemetery located at the junction of the Mayo-Elsa roads. Joseph Cantin died in Mayo in 1953 and is buried in the Mayo cemetery.

CANTIN, Louis

Louis Cantin was a partner in Cantin Bros., and he mined on Haggart Creek from 1909 to the 1920's. He remained in Mayo from the 1920's until he died in 1959 and is buried in the Mayo cemetery.

CANTIN, Phileas

Phileas Cantin was a partner in Cantin Bros., and he mined on Haggart Creek in the 1900's (circa 1900–1919) and 1920's. He eventually returned to his home province of Quebec.

CARD, Stephen

A trapper on the Stewart River in the 1920's.

CARLSON, Mattson "Matt"

– Archie Gillespie, *The Mayo Miner*, April 16, 1937

Matt Carlson was born in Finland in 1867 and became a famous figure in the Canadian northland. He was held in high esteem by his many friends, and it is understood that he served as the model for the "The Big Swede" character depicted in Rex Beach's famous northland story, "*The Spoilers*." Matt had gone to the U.S. as a young man and worked in the Arizona and California mines.

He went to Dawson in 1899, then moved on to Nome, Alaska after the Klondike Gold Rush subsided. In 1922, he came to the Mayo district, where he remained for the rest of his life. He worked in the Mayo silver mines, and was a hard rock miner of outstanding ability.

According to reports, many years before he had ventured north, he and a partner had won a world championship drilling contest. He died of pneumonia at the Mayo General Hospital on April 11,1937 at the age of 70. When Carlson died, his nephew Karl Alvan Carlson was living in Mayo and he had a sister living "outside."

CARMICHAEL, A. C.

A. C. Carmichael was Chief Geologist at United Keno Hill Mines in the early 1950's.

CARPENTER, John H. "Jack"

Jack Carpenter was a pioneer prospector of

the Mayo and Carmacks districts where he held considerable property, and was considered an excellent bushman and prospector. He mined on Blackhills Creek with Peter Brady circa 1910. "He found the first rich silver-lead ore in the Beaver River district near Carpenter Creek, and the find sparked a sizeable rush to the area in 1920–23. However, the mineral finds were not economic at the time, so the rush faded and the district was deserted again. The big ridge to the northwest of the creek was later named Carpenter Ridge."[10] In 1925, he accompanied Emil Anderson, Ray and Don Stewart on a prospecting trip to the Beaver River country from Keno City, and in the late 1920's he prospected and staked property in the Clear Creek and Black Hills area. Carpenter was also one of the prospectors who participated in Livingston Wernecke's air reconnaissance prospecting from 1928 to 1931.

In 1939, the General A. D. McRae interests secured an option on the extensive placer holdings of Carpenter and Marsh on Clear Creek and Black Hills. W. A. O'Neill, formerly of Fairbanks, Alaska, was in charge of work on both properties for the McRae interests. The McRae party sank shafts and drilled on the Clear Creek option on ground extending for 9-1/2 miles along the left fork of the creek. The tests indicated that the ground averaged about eight feet to bedrock. After completing work on Clear Creek they moved to test Black Hills in September. Both of the creeks were first located by "Big Alec" McDonald, the Klondike King.

CARSCALLEN, Frank

Frank Carscallen mined and prospected in the Galena Creek area from the 1900's (circa 1900–1919) to the 1920's, and held hard-rock claims on Dublin Gulch. He lived in a cabin on the point across the Stewart River from where the White Pass and Yukon Route dock used to be. He was reputed to have been a lawyer. He was elected to the Yukon Council in 1931. Carscallen died in April 1936.

CARTER, Arthur

– Archie Gillespie, *The Mayo Miner*, January 21, 1938

Arthur Carter, a longtime Mayo district resident, was born at Wortley, Leeds, in England, in 1875. He was the first man from the Yukon to join the British forces when the Great War broke out in 1914. To do this, Arthur left Highet Creek, where he had been working [since 1911 for Elmer Middlecoff,[11]] and paid his own way to Vancouver. He became a reservist with the Imperial Army.

He later returned to Mayo, and was the orderly and janitor at the old Mayo hospital circa 1930. Carter died at the Royal Jubilee Hospital in Victoria, B.C., on December 4, 1937, at the age of 62. He had a brother (John) and a sister (Mrs. W. Barnett, in England).

CARTER, Blair B.

Blair Carter is listed at Keno Hill in Polk's 1923–24 directory.

CARTHUM, C. C. "Con" & Alice

"Con" Carthum was a well-known employee of the Treadwell Yukon Company in the 1930's. He married Alice Fisher (sister of Bud Fisher) on September 29, 1937, and they had a daughter, Genevieve. In 1938 he leased the Shamrock claim on Keno Hill with Ellis Johnson, Hans Formo, and Antonne Finnes.

CASE, John

John Case is listed at Keno Hill in Polk's 1923–24 directory.

CASTENGUAY, J.

J. Castenguay was on the McQuesten River in 1938.

CATHRO, William "Bill"

Bill Cathro came from Scotland to the Yukon in the early days to work for Burns and Co. in Whitehorse. He later worked for Taylor and Drury in Mayo and during the 1940's managed the Northern Commercial Co. store in Mayo.

CATTO, William

William Catto never actually lived in the Mayo area, but was a Dawson physician who grubstaked many prospectors in the 1900's (circa 1900–1919), including John S. Stewart who held quartz claims on Dublin Gulch near the Olive claim.

CESARONI, Venanzio

Venanzio Cesaroni is listed at Keno Hill in Polk's 1923–24 directory.

CHAMBERS, Art

Art Chambers prospected across the McQuesten Valley in the late 1920's for galena. In 1928 he staked silver claims which later became known as the Shanghai Property when restaked by Cecil Poli. He drove a tunnel on Gull Creek at Mayo Lake, but spent most of his time mining on Barker Creek down the Stewart River, using a monitor.

CHANTELOIS, Joe

Joe Chantelois had a small farm at the west end of the Mayo Airport (two acres of land two miles from Mayo on the west side of the Mayo-Minto Road) from the 1920's to the 1940's. His main crop was potatoes which he sold to the Treadwell Yukon camp. He was also known as "Bullcon Joe."

CHAPMAN, Charles H. "Chappy" & Mattie

See *Pioneer Accounts*.

CHAPMAN, Ernest

Ernest Chapman was a trader, trapper and prospector for many years in the Chappie Lake area (near the Bonnet Plume River, below its junction with the Peel River). He had a cabin and trading post on Chappie Lake in the 1930's and 1940's, [but his home was in Dawson City where he owned a dry goods store]. Chapman was killed in an airplane crash at Dawson City on January 31, 1941.[12]

CHASNI, Frank

Frank Chasni staked the lower Discovery on Steep Creek on August 16, 1903. He operated a roadhouse with his partner Gus Schogren at Minto Bridge in the 1920's. Chasni is buried in the Mayo cemetery.

CHAUVIN, Archie

Archie Chauvin operated a general merchandise store at Keno Hill in the 1920's.

CHERNUTA, Tom

Tom Chernuta prospected and mined in the Mayo district in the 1900's (circa 1900–1919). He and his wife owned a small parcel of land near the old Mayo hospital site and they also had a cabin on Bear Creek. They moved to Port Angeles, Washington in the 1920's.

CHIPMAN, Dr. William W.

– Archie Gillespie, *The Mayo Miner*, April 12, 1937

Dr. Chipman was born in Ottawa and was a graduate of McGill University in Montreal. He served with great distinction overseas during World War I. After the war he went to the Dawson district where he practised medicine for several years.

From 1922 to July 1926 he practised in Mayo and subsequently maintained a contract with the Treadwell Yukon Company to refer suitable doctors to Mayo during the remainder of the 1920's and through the 1930's. Dr. Chipman eventually left Mayo to practise in Vancouver B.C., but during his years in Vancouver he remained in contact with scores of patients and clients from the North. He died on March 30, 1937 in Vancouver.

CHISHOLM, James H. "Jim"

Jim Chisholm was a sourdough trapper who trapped for many years in the Mayo district, and was listed at Keno Hill in Polk's 1923–24 directory. In the early days he and Walter Johnson ran log rafts down the Klondike River to Dawson City. In the winter of 1938 he was trapping in the vicinity of the old Liberal Trail.

In his later years Chisholm lived in a cabin on the bank of the Stewart River past Sam and Rose Mason-Wood's place near where the airplane hanger was located. Chisholm's cabin was dug down about four feet so that he had to step down into it. The walls were only three feet high. Martin Forestberg later lived in this cabin.

CHRISTAL, Charles

Charles Christal found and staked the first gold claims on Christal Creek (a small tributary of the South McQuesten River) on November 14, 1902.[13]

CHRISTIE, James Murdoch "Jim"

Jim Christie was born on October 22, 1874 in Perth, Scotland. He came to the Canadian prairies as a young man, and in early 1898, he left Carman, Manitoba to take the Edmonton Trail to the Klondike. On this journey he found and made use of what became known as Christie Pass. He prospected the Yukon and Northwest Territories for years, but was mainly interested in a vast stretch of territory centering on the Rogue River.

It was near Christie Pass and mountain in late October 1909, that a huge grizzly bear attacked him. His skull and jaw were fractured, his right arm was broken and his right thigh was badly bitten, but he courageously made his way, in sub-zero temperatures, seven miles to a temporary camp. His partner George Crisfield tended his wounds with the only medicine they had—some Scotch whisky. Crisfield obtained the help of some nearby Indian people and their dogs, and after four long and terrible days over rugged mountain country, got him to the small settlement of Lansing on the Stewart River.[14]

There, Mrs. Ferrell (wife of the trader J. E. Ferrell) tended his injuries and set his arm and jaw. Mrs. Ferrell had received medical training and was familiar with surgical practises.

She and her husband nursed Christie for two months. On New Year's Day, he was taken by dogsled to Dawson, a journey of 17 days. He later went to Victoria, B.C. where his jaw was re-set and his other injuries given further attention. Just a few months later, in the summer of 1910, he was back prospecting.[15]

He later served in World War I.

The hide of the grizzly bear that attacked Christie is buried somewhere in Mayo with Ferrell's photographs.

CHURCHWARD, Guy S. & Edna Mae

Guy Churchward operated the first tinsmith in Dawson in 1898, and designed and constructed the Yukon Airtight Heater stove. (See Philip Churchward's story for further information on Guy Churchward in *Pioneer Accounts*)

In 1937 he built a huge furnace for the bunkhouse which was constructed at the Silver King. The furnace combined 2 tank units and was 7 feet long and 7 feet high.

Churchward and his wife, Edna Mae, moved to Mayo in the 1920's. Their children are Annie (Besner) and Philip. Churchward is buried in the Mayo Cemetery.

CHURCHWARD, Philip E.

See *Pioneer Accounts*.

CLARK, Dr. James V.

Dr. Clark spent his childhood in St. George, New Brunswick. In 1943, during World War II, he left the University of New Brunswick, in Fredericton, to join the R.C.A.F. as an airgunner. He was engaged in teaching gunnery and coastal patrol. In 1945, following his war service, he returned to the University of New Brunswick, graduated in 1946, and went on to study medicine at McGill University in Montreal. He spent six years in Montreal and then came north in 1952 to practise medicine at Elsa. In 1954 he moved to Mayo, where he still continues his practise. The J. V. Clark School in Mayo is named for him. Dr. Clark is an opera aficionado and also enjoys producing and directing amateur theatre.

Dr. J. V. Clark, circa 1954.
J. & A. Boyes Collection

CLARK, Jim

Jim Clark worked at Binet's Hotel in the 1900's (circa 1903–1919). He and a partner staked the Sadie claim on the northwest slope of Keno Hill. George Besner later optioned it but found very little. In 1921 Guggenheim's manager, Frank Short, optioned it for $30,000. The Sadie was one of Keno Hill Ltd.'s best high-grade producers.

CLARK, William.

William Clark is listed at Keno Hill in Polk's 1923–24 directory.

CLARKE, Walter E.

Walter Clarke was General Superintendent for United Keno Hill Mines in the 1950's.

CLEMONT, Floyd

Floyd Clemont was a bulldozer operator at Mayo and Keno Hill in the 1920's.

CLIFTON, John A. & Margaret

John Clifton worked for Treadwell Yukon in Mayo in the late 1930's. His wife Margaret worked for a time as a stenographer in Mayo's Treadwell Yukon office in 1938. Clifton was

Joseph W. Clifton's nephew. Both John and Joseph Clifton had wives named Margaret.

CLIFTON, Joseph Wellington "Joe" & Margaret

Joe Clifton was a former member of the N.W.M.P., H Division, in Whitehorse. From 1909 to 1914 he was the bookkeeper for Burns and Co. in Whitehorse, and from 1915 to 1920 he was a clerk in the Whitehorse Post Office. In about 1920 he came to Mayo to clerk in the Taylor and Drury Ltd. store, and in 1925 he went to work for Treadwell Yukon in Keno.

He was the original occupant of a house in Mayo built for him by the Kimbel brothers. The same house was later occupied by the Gordon Lee family, the Bidlake family, the Ed Bleiler family, and the Cliff Greig family, among others. His wife, Margaret, died in 1935 and is buried in the Mayo Cemetery. After her death, he moved to British Columbia where he eventually remarried. His daughter from his second marriage is Mrs. Isabelle May Poirier of Victoria, B.C.

CODY, Andrew

Andrew Cody is listed as a miner at Keno Hill in Polk's 1923–24 directory.

CODY, James "Jim" & Margaret

James Cody was the agent for the Northern Commercial Co. Ltd. in Mayo in the 1920's. His wife's name was Margaret.

COLE, Mike & Tim

Mike and Rosalie Cole, his brother Tim and wife Norma, operated a garage and service station in Keno in the 1950's. Mike and Rosalie lived in Mayo for a while before they moved to Keno to operate the garage there.

Mike and Tim later ran a garage in Mayo which they sold to Joe Wuerr. Tim and Norma, lived in Mayo with their two sons during the time they owned the garage in Mayo. Both families moved to Dawson City where they mined on Eureka Creek and worked in road construction until they left the Yukon and moved to Victoria, B.C.

COLLEY, Howard "Harry"

Harry Colley was born in Richmond, New South Wales, Australia in 1888. He studied mining engineering at the University of Stellenbosch

Joe and Margaret Clifton's house in Mayo, circa 1930. *F. & T. Whitney Collection*

in South Africa, worked as a reporter in San Francisco, then taught school for a while.

He came to the Yukon in 1911 and prospected, mined and studied the country for many years. He was fluent in Dutch, and for that reason served with British Intelligence in World War I under Viscount Alexander.

On April 27, 1920, Colley, Joe Walsh, and Harry McCrimmon staked the Calumet claim, naming it after the great native copper mines in Michigan. In the winter of 1921, Colley made a dog-team journey from Whitehorse to Norman Wells to stake oil lands, then returned via the Mackenzie, Bell, Porcupine and Yukon Rivers. Little work was done on the Calumet until 1932 when he began to prospect it further with Tom McKay. In the spring of 1934 Don Morrison was added to the partnership. The three men continued to work the property, and eventually struck a rich ore shoot which they mined out. In 1935 they sold their Calumet claim to Treadwell Yukon.

In the summer of 1938, he staked placer ground on Britannia Creek (60 miles above Stewart City) and on its tributary, Canadian Creek. He continued to prospect in the Mayo and Dawson districts into the 1950's. He died in 1956.

COLLINS, Thomas

Thomas Collins was a Mayo carpenter in the 1920's. During this time he built the house that Dorothy Allan lived in during the 1950's. (See *Thomas Collins House* in *Buildings and Artifacts)*

COMSTOCK, Charles M.

Charles Comstock staked a claim on Duncan Creek in 1902. Records show that he mined on Highet Creek in 1907.

CONNOR/S, Roger

Roger Connor/s was a cook in Mayo circa 1915, and owned property in the town.

CONNORS, David

David Connors is listed as a miner in Mayo in Polk's 1915–16 directory.

COOLEY, John S. "Jack"

Jack Cooley came into this district right after World War I. He prospected the silver district through the 1920's and 1930's, and operated the diesels at the Hector in 1938.

COOMBS, Lee

Lee Coombs (also spelled Combs) was a teamster in Mayo circa 1915. In the 1920's he was a carpenter in Mayo and at Keno Hill.

Harry Colley, Malcolm and Kathleen McLaren, and "Deep Hole" Thompson. *L. Cheatham (McLaren) Collection*

COOPER, George

George Cooper was a Treadwell Yukon mill employee and diesel operator in the 1930's. He was coach of the Galena-Mayo girls team that won the Yukon softball championship in Dawson in 1939. He went to Vancouver in the summer of 1940.

COOPER (McDONALD), Shanon

See *Pioneer Accounts*.

CORBEIL, Camille

Camille Corbeil came from Dawson and he and his partner Gervais P. Godbout mined on Highet Creek in 1904. He continued to prospect and mine in the Mayo district for many years.

CORBET, Xheusune Fiorenza "Madam Zoom"

Xheusune Corbet was born in Picardy, France. She was well educated, well travelled, an accomplished pianist and seamstress. She had studied dressmaking in Paris and was always clothed in fine Paris fashions. It is believed that she was the daughter of a British diplomat.

In her younger days she was a dancehall girl in Dawson City and when activity in Dawson slowed she came to the Mayo silver district. "Madam Zoom" was, undoubtedly, one of Keno's most exotic and flamboyant residents. In 1940 she met a Canadian Army Reserve engineer, named Corbet, in Dawson City and they were married immediately. Corbet was killed at Gibraltar, while assisting in the war effort, shortly after the couple were married. He was awarded the George Cross posthumously and Xheusune kept the medal hanging above her bed for the remainder of her days. (The George Cross was a British award, also given to Canadians, for heroism in nonmilitary circumstances.)

"Madam Zoom" returned to Dawson in the 1950's. She was a very athletic woman and could perform numerous acrobatic dances. Even in her senior years she was still able to do a jette. When she was approximately 70 years old she became ill and returned to Mayo to be treated by Dr. J. V. Clark. She died of cancer in Vancouver, B.C., in the 1960's.

CORDERY, George

George Cordery came to Mayo in the 1920's, and was one of "Pop" Bellerby's longshoreman during the 1930's. He also worked for Treadwell Yukon, driving bulldozers in their oil storage yard in Mayo.

CORP, Ernest Joseph

Ernest J. Corp came to the Klondike gold rush in 1897–98 via the overland route (one of the Edmonton trails) down the Mackenzie River and over the Keele and Hess Rivers to the Stewart River. His is the only account of the continental divide route to the Klondike.

He made a fortune on the Anderson concession at Hunker Creek in the Klondike, and also did well on the Onek vein at Keno. He and his partner Dave Ryan prospected various properties in the Keno area, and also ran the post office in Keno City for several years. He was a member of Yukon Council for the Mayo district from 1934 to 1939. Corp retired to White Rock, B.C., in the early 1960's where he died.

COUTTS, Charles J. "Charlie"

Charlie Coutts came from Dawson circa 1921 with capital and bought into a number of claims on Keno Hill and adjacent areas. He ultimately returned to Dawson.

CRISFIELD/CRISSFIELD, George

George Crisfield was a prospecting partner of Jim Christie, 1908–09. He mined on Gull Creek, circa 1915, with Arthur Hester and Matt Butijer (Butyer).

CUNNINGHAM, David A.

David Cunningham, Ray Stewart's partner, was a former livery stable owner from Dawson. He had teamed up with Ray on a gold mining venture on the Fortymile, and the two prospected the Keno Hill and Galena Hill areas extensively. They staked the Ladue fraction on September 11, 1920 and later sold it to Treadwell Yukon. They also staked the Lake Group and some claims on Galena Hill.

Cunningham had a cabin on Duncan Creek. His two nieces, Viola and Effie Cunningham from Pittsburgh, visited him in Mayo in 1923. Viola was a secretary and Effie a schoolteacher.

CUNNINGHAM, George & Betsy

– Archie Gillespie, *The Mayo Miner*, July 16, 1937

George and Betsy Cunningham arrived in Mayo as newlyweds (circa 1900–1903), and for many years located in various parts of the Mayo district. Their eldest daughter, Elizabeth Gordon Stewart, was born in 1903 at a point on the Stewart River 40 miles above Mayo. Elizabeth was the first non-native child to be born in the Mayo district.

The Cunninghams operated a hotel and store in Mayo for many years during the 1900's, and George also mined and prospected on Galena Creek circa 1915–16.

Betsy was a native of Scotland and never quite lost her broad, Scottish brogue. She was renowned throughout the North for her sparkling wit and hearty good humour. George died of lockjaw on Lookout Mountain in 1918, and the couple's eight-year-old son "Jock" passed away suddenly from a strange and sudden stomach ailment. Betsy never recovered from the shock and tragedy of Jock's death.

It appears that all the Cunningham daughters eventually moved away from the Yukon. Elizabeth married J. W. Gellatly in 1936. Gellatly was a tea plantation owner in India and the couple went there to live. Florence (Cleary) settled in California, Mary in B.C., Margaret married Dr. Kells and lives in Wenatchee, Washington, and another daughter lived in Scotland. Mrs. Betsy Cunningham died on July 3, 1937 at St. Mary's Hospital in Dawson City.

CUNNINGHAM, Peggy

Peggy Cunningham was the postmistress in Keno in the early 1950's and was later the Mayo Mining Recorder's secretary. She died in Watson Lake.

CURRIE, Archie & Violet

Archie Currie worked at Taylor and Drury's Store from 1930 to 1934. He was an ex-R.C.M.P. officer and was married to Violet née Palmer. They had one child, Joan, who attended school in Mayo in the 1930's.

Currie operated a store and post office at Tagish after leaving Mayo. He died in Tagish in his senior years.

CURRY, Don & Cecile

Don Curry arrived in Elsa from Calgary on March 16, 1949 where he joined his father, Lorne Curry and brother "Bob." He later began work as a diesel mechanic for United Keno Hill Mines and has continued in that position until the present time (1989). Don met Mayo-born Cecile Rich (daughter of Mary and Gilbert Rich) in Elsa and they were married. The Curry's have five children.

CURRY, Lorne

Lorne Curry came to the Yukon during World War II. He took his discharge in Whitehorse when the war was over and was hired by United Keno Hill Mines in Elsa in 1947.

His son, Robert "Bob" Curry, came north in 1948. Robert married Linch Van Bibber and the couple have two sons. "Bob" was followed by another son Don who came to Elsa from Calgary as a teenager.

DAGLISH, Hamilton

Hamilton Daglish was born in England. He mined on lower Dominion circa 1909. He went overseas with the Yukon Infantry Company during World War I. He is listed as being at Keno Hill in Polk's 1923–24 directory.

DAHENECKE, Bill

Bill Dahenecke was an old-time Klondike placer miner who had mined in other countries including Korea. He mined and prospected in the Keno area in the 1920's.

DAHL, Ole

Ole Dahl mined in the Mayo area from 1915 onward. Records show that he worked for Elmer Middlecoff on Highet Creek in 1917.[16] He spent his later years in Essondale Asylum in Vancouver, B.C.

DAHLIN, John

John Dahlin was located on the Mayo-Minto Bridge Road in the 1920's.

DALTON, Mrs. Hazel

Hazel Dalton (née Palmer) cooked for the Middlecoff placer camp in the late 1930's, then left Mayo in February 1938 to work for Mr. and Mrs. Ernie Sommerton at the Bluebird Cafe in Dawson. Only a year later, in March 1939, she returned to Mayo, and along with her brother George Palmer, purchased the Club Cafe from Mrs. A. W. Wootton.

DANKER, Joseph

Joseph Danker staked the Black Wolf mineral claim in 1920.

DARBOLO, John

– Archie Gillespie, *The Mayo Miner*, October 1, 1937 and October 8, 1937

John Darbolo was born in Pogliano, Italy, in 1857 and came to the Yukon during the Klondike Gold Rush. In the early 1900's he stampeded to Duncan Creek in the Mayo district. He worked for Elmer Middlecoff as a miner on Highet Creek in 1910 and 1911, and he held claims on Highet Creek in 1913. He also mined on Highet Creek with his partner Louis Bouvette circa 1915–16.

After he broke his leg in an accident on his claims on Highet Creek, his mining activities came to a temporary halt. At one point he left Mayo for several years before returning in 1920. During the 1920's, he mined in the Keno Hill area where he also made his home. Darbolo died in the Mayo General Hospital on September 29, 1937 after a lengthy illness. At the time of his death he had no known relatives.

DARLING, Pete

Pete Darling worked as a logger for Binet and Lefebvre in Mayo in the 1920's and at the Palmer-Lefebvre sawmill in the 1930's.

DAVIDSON, Arthur

Arthur Davidson is listed at Minto Bridge in Polk's 1923–24 directory.

DAVIDSON, Jacob A. "Jake"

Jake Davidson was one of the four original stakers on Duncan Creek in 1901. He was also the first man to find galena and stake a silver claim in the Duncan Creek Mining District.

On July 20, 1903 he staked the Hell's Gate claim and recorded it on August 3, 1903 at the Mining Recorder's Office on Duncan Creek. He later let the Hell's Gate claim lapse when he moved on to prospect the Beaver River area.

In 1905, when he heard about the rich silver discovery in Cobalt, Ontario, he decided to go there. He and Weldy Young formed Young-Davidson Mines on a discovery in the Cobalt Camp. The company later acquired a successful gold producer and is still active today.

In 1913 Henry McWhorter staked the Silver King on what had been Davidson's Hell's Gate claim. (See *Silver Mining*.)

Mrs. Catherine Mulloy of Ontario is Jake Davidson's daughter.

Old Village people at the graveside of one of Billy Davis' children. Left to right: Chief "Albert Tom" Moses, John Lucas, Billy Davis, Alice Tom, Lucy Peter, Sally Lucas, Alice Jimmy, Sophie Davis, child, "Big" Maisie, Emma Johnson. *Dave Hager Collection*

DAVIDSON, James S.

James Davidson was an employee of Galkeno Mines Ltd. He was killed in a motor vehicle accident on May 19, 1957, and is buried in the Mayo cemetery.

DAVIS, John M.

John Davis is listed as a blacksmith in Mayo in Polk's 1915–16 directory.

DAVIS, Mrs. M.

Mrs. Davis was a cook for Keno Hill Ltd. at Keno Hill in the 1920's.

DAVIS, Sgt. J. W.

Sgt. J. W. Davis was one of Captain Steele's men of the N.W.M.P. He ran the post at the mouth of the McQuesten River from June 1901 until June 1903.

DAVIS, Billy & Sophie

Sophie and Billy Davis lived at the Old Village on the Stewart River. She was the daughter of Emma and John Johnson. The Davis children are Charlie Davis, Sarah Johnny (née Davis), Jimmy Davis, Arthur Davis, and Lena Hoogland (née Davis) who lives in Vancouver, B.C.

DAY, Hugh & Al H.

The Day brothers were among the first miners to enter the Yukon. They came over the Chilkoot Pass in the spring of 1884 and mined successfully on the Stewart River Bars in 1884 and 1885. On July 4, 1893 they discovered fine gold on Independence Creek (a tributary of the Stewart River).[17]

DEAN, Const. H.

Const. Dean was a member of the Mayo R.C.M.P. in 1922 and 1923.

DeCHAMPLEIN, Ernest

Ernest DeChamplein is listed as a Mayo engineer in Polk's 1923–24 directory.

DeCLARK, C.

DeClark was an employee of the Northern Commercial Co. in Mayo in 1937.

DeCOSTA, Mr. & Mrs.

DeCosta was a barber in Mayo. When his wife, Carrie, died of cancer in the 1930's, he left Mayo. Carrie DeCosta is buried in the Mayo cemetery.

DELAPORTE, Don R.

Don Delaporte was a mine superintendent for United Keno Hill Mines in the late 1950's.

DEMPSTER, W. J. D. "Jack"

See *Pioneer Accounts*.

DESAUTELS, Albert Edward

Albert Desautels arrived in this district during the Duncan Creek stampede and staked claims on that creek in 1902. He and his partner, Ernest Beliveau, discovered gold on Edmonton Creek (which flows into the southeast side of Mayo Lake) on June 13, 1904. He also staked on Parent Creek.

DESAUTELS, Elphege

Elphege Desautels staked claims on Duncan Creek during the gold rush that occurred on that creek in 1902. He staked the upper Discovery claim on Edmonton Creek in 1904 and later transferred the claim to Albert Desautels.

DESJARDIN, Alsed

Alsed Desjardin was a Mayo carpenter in the 1920's, and owned property at the junction of Haggart Creek and Galena Creek Roads. He lived in the Mayo district from 1901 to the 1920's and was an associate of Alex Nicol.

DICKIE, Crandell R.

Crandell Dickie lived in Mayo in the 1930's, and worked for Elmer Middlecoff's placer operation on Highet Creek.

DICKIE, Frank

Frank Dickie was a Treadwell Yukon employee at the Elsa in the late 1930's.

DICKSON, Charlie "Baggage"

Charlie Dickson mined on the McQuesten River near the mouth of Johnson Creek in 1892 with Ray Stewart and Ben Atwater.

DILLON, John

John Dillon is listed as a miner at Keno Hill in Polk's 1923–24 directory.

DINEEN, Mr. & Mrs. Jack

The Dineens lived in Mayo in the 1930's, then left and opened a roadhouse at Carmacks in the spring of 1938.

DIVINE, Peter

Peter Divine worked Parent Creek without success in 1907, but did find a $14 nugget on the creek. He was of French descent.

DJUKASTEIN, Klaus & Thordis

Klaus Djukastein mined on Johnson Creek in the 1950's with John Sandanger. This operation became known as Bardusan Placers Ltd. when young Hans Barchen joined the partnership in the 1950's.

Thordis, Djukastein's wife, managed the Silver Inn Cafe in Mayo for a short time. For a while, the couple managed the Stewart Crossing lodge. He then resumed mining while she operated a variety store in Mayo and a shop in Whitehorse.

Djukastein mined on Brewer Creek for five seasons and then, in the late 1960's, took over where Waddco had been working on Haggart Creek. After he left Haggart Creek, he mined on Revenue Creek near Carmacks for several years.

It was Djukastein who developed the "Derocker," a gravel screening device used by many placer operators.

The Djukasteins have twin sons, Erik and Torfim.

DOBSON, Geordie

Geordie Dobson originally came from England where, at the age of 15, he tried to join the Royal Navy. Being too young, he signed on with the Merchant Marine instead. His first post was deck boy on a Norwegian freighter. By the time he was 17, he had been around the world and visited numerous countries, among them Australia, New Zealand, South Africa, Brazil and Venezuela. He humorously recalls sharing a jail cell in Douala, Cameroon Africa, with a cannibal convicted of devouring his aunt.

In 1949, he was quartermaster on a ship that sailed to Toronto. When the ship reached its destination Dobson bought his way out of the Merchant Marine and went to work for Canadian Comstock on construction of the de Havilland airport.

In 1951 he came north to work for United Keno Hill Mines in Elsa. In 1954 he quit United Keno to work in Mackeno's machine shop where he remained until 1959 and then went back to work for United Keno. In April 1963 he bought the Keno City Hotel which he continues to operate today (1989).

Dobson owns the amazing "bottle house" in Keno City. The idea for his bottle house occurred at a time when the government had no bottle return policy. He decided to cover his house with the thousands of empty beer bottles generated by the tavern in his hotel and so he coated his house with cement and pushed a total of 32,000 bottles, neck first, into the cement, leaving the bottom end of the bottles exposed.

Geordie Dobson is a most entertaining personality. If a visitor to Keno City disappears while touring the town, there is a very good chance he will be found engrossed in lively conversation with Dobson.

DONALD, John F. & Molly

John Donald was a Northern Commercial Company employee from Dawson who managed the Mayo store in the 1950's. He returned to Dawson to manage that store and was later transferred to Fairbanks, Alaska.

DORAN, Ed

Ed Doran came to the Yukon in 1936 and was a Treadwell Yukon employee who worked at the Silver King and also at the Calumet until he joined the service and went overseas during World War II. He returned to Bruce, Alberta where he farmed and subsequently died.

DORR, Stillman M.

Stillman Dorr was in Mayo during the 1920's boom.

DOUGLAS, Frank P.

Frank Douglas was a clerk for J. E. Binet's hotel and store in the 1920's.

DOUGLAS, Mrs. Hannah

Hannah Douglas is listed at Mayo in Polk's 1923–24 directory.

DOUGLAS, Walter Scott "Hot Stove" Douglas

Walter "Hot Stove" Douglas was a Liverpool bricklayer who went to work at his trade in San Francisco after the earthquake. He later made his way up to Dawson City where he became a clerk in the B and F Store (an ice cream parlour and

English Football Pool enthusiasts. From left: R. Gordon Lee, "Shorty" Carter, and Walter "Hot Stove" Douglas. *W. S. Hare photograph, Lee Collection*

confectionery.) He may have got his nickname "Hot Stove" Douglas while working in the B and F in Dawson—he had the place so hot that everything melted!

He served in the first World War but not overseas, was at Keno Hill in the 1920's, and worked in the Treadwell Yukon Mill as an ore sorter.

He was also known as a top-notch soccer player.

Douglas lived in a small cabin (about 12 feet x 14 feet) on the waterfront. However, when Treadwell Yukon obtained a lease on the waterfront reserve for ore storage, Livingston Wernecke ordered "Hot Stove" to move his cabin. When he refused, Wernecke had one of his cat skinners put a cable around the cabin and towed it to a site right in the middle of the surveyed road (Front Street), perhaps one hundred feet beyond Alex Nicol's cabin and "Ma" Fisher's boarding house.

In 1938 he worked for the government road crew. He also worked in Whitehorse for the Americans during World War II and purchased Hugh Monohan's log cabin (upstream from the Anglican Church) from the Monohan Estate on his return to Mayo. Josie Pelland later lived in this cabin.

DOUGLAS, William Lawrence

William Lawrence Douglas was born in Maitland, Hants County in Nova Scotia. He joined the Black Contingent in Dawson in 1916, and returned to be the government vendor in Mayo in the 1920's. William Douglas is listed at Keno Hill in Polk's 1923–24 directory.

DOYEA, Frank

Frank Doyea is listed as having a greenhouse in Mayo in Polk's 1923–24 directory.

DRAPEAU, Gerrard "Jerry"

Jerry Drapeau prospected in the Mayo district and at Drapeau Creek (a short tributary of Hobo Creek) for many years. He also had claims at Keno Hill in the Silver Basin area, and did some prospecting in the Klondike during the late 1920's with Jack Alverson and George Potter.

DRAPER, Peter

Peter Draper, born in New Brunswick, went overseas with the Yukon Infantry Company during World War I. He was a cook for Keno Hill Ltd. at Keno Hill in the 1920's.

DRISCOLL, Barney

Barney Driscoll, a Welsh prize fighter and coal miner, worked for Treadwell Yukon in the 1930's at Keno Hill and was considered one of the best hardrock miners at the time.

DRISCOLL, Cornelius

Cornelius Driscoll used a rocker to mine gold at Steamboat Bar on the Stewart River in 1918.

DUBE, Fortuna

Fortuna Dube was an orderly at the Mayo General Hospital in the 1920's. He was renowned for his pea soup.

DUBOIS, Henry

Henry Dubois was a prospector and miner in the Mayo district. Henry, Amedé (who married Rose Besner) and Joseph Dubois were brothers.

Henry worked for Charles Settlemier and Claude Bermingham on the Arctic and Mastiff claims doing assessment work in the early 1930's.

Dubois is buried in the Mayo cemetery.

DUBOIS, Joseph & Clara

Joseph Dubois was a barber in Mayo from 1922 to 1927. His wife was Clara Dubois.

DUNCAN, Dr. Allan C.

Dr. Duncan came to practise medicine in Mayo in 1933, after graduating from the University of Manitoba in 1931. He was instrumental in the introduction of compulsory legislation and compensation for silicosis in 1940. Dr. Duncan married Miss Winona Spence who was a nurse at the Mayo Hospital. He later practised medicine in Dawson City and later again in Vancouver, B.C.

In 1989 he published a book titled *Medicine Madams and Mounties: Stories of a Yukon Doctor* about his experiences in the 1930's and 1940's.

DUNNETT, Sinclair "Big Sinc"

"Big Sinc" was a blacksmith in the 1920's and 1930's for the Treadwell Yukon Company, and he was the uncle of "Little Sinc" Dunnett who came to Mayo from Scotland in the 1930's.

DUNNETT, Sinclair "Little Sinc" & Vera "Mickey"

"Little Sinc," nephew of "Big Sinc" Dunnett, came from Scotland in the 1930's and worked for the Treadwell Yukon Company as a mechanic and plant operator at the Elsa and Silver King mines. He married Vera "Mickey" Maclennan in 1937 in a double wedding ceremony with "Mickey's" sister Alice "Babe" and Alex Smith.

Mrs. Andison, Mrs. D. R. Maclennan (née Kinney), Dr. D. R. (Randy) Maclennan, May Gillespie, Fred Marshall, Dr. A. Duncan and Winona Spence, Flora Middlecoff, and Annie Close. *May Gillespie Fielder Morrell Collection*

DURAND, John

John Durand arrived in this district during the Duncan Creek stampede and staked a claim on that creek in 1902. Durand also staked claims on Minto Creek and on May 30, 1904, he transferred 50% of claim No. 60 on Minto to Triffle Paris.

Dr. Allan C. Duncan sitting on a log by the riverbank, circa 1934. *Middlecoff-Bleiler Collection*

DURIE, A.

Durie was a well-known resident of the Atlin district. He moved to Mayo in September 1939 and opened a cleaning and pressing business in Isaac Mallette's cabin.

EDWARDS, Fred

Fred Edwards staked claims on Highet Creek in 1903.

EHLEN, Henry

Henry Ehlen operated a confectionery in Mayo from 1922 to 1927, selling candy, cigars, and ice cream. He also owned property on the Mayo-Minto Bridge Road.

ELLIOTT, Frederick

Frederick Elliott prospected the Beaver River country in the Wernecke Mountains during the 1920's. He was associated with Charles Settlemier another well-known prospector.[18]

ELLIOTT, William James "Rocky Mountain Bill"

William Elliott was born in St. John, New Brunswick. He joined the Black Contingent in Dawson in 1916, and was listed at Keno Hill in Polk's 1923–24 directory.

ELLIS, "Colonel"

"Colonel" Ellis was a woodcutter in the Mayo district during the 1930's. He cut wood for the Cantins on the Stewart River and later on the McQuesten River. Lloyd Ellis was his son.

ELLIS, Lloyd

Lloyd Ellis was a wood contractor on the McQuesten in the 1930's.

ELLIS, Peter & Jesse

Jesse and Peter Ellis were highly respected members of the Mayo Indian community who lived during the Old Village era. Their daughter, Gladys, married prospector Cecil Poli.

ELWELL, Edgar H.

Edgar Elwell is listed at Mayo in Polk's 1923–24 directory.

Inside Tommy McKay's cabin at the Shamrock. The man on left is either Paddy McIvor or Jack McHugh, and the man on right is Axel Erickson. *T. McKay Collection*

ERICKSON, Axel & Caroline

Axel Erickson came to the district circa 1919 and prospected on Christal Creek. He and his partner Tom McKay were among the first stakers on Keno Hill.

On August 4, 1919, McKay staked the Nabob and the Shamrock and Erickson staked Solo No. 1 and the Reno. The partners also had claims in the Calumet group.

Erickson's wife was Caroline "Ma" Erickson who ran a boarding house in Keno for many years. She entertained lavishly in her Keno home with feasts and liquor for the senior mine staff. She left in the late 1950's to live with her son, Theodore, in Seattle after her husband died. Theodore Erickson was in the U.S. Navy.

ERICKSON, Leonard. B.

Leonard Erickson was a pioneer miner in the Mayo area. He made one of the first discoveries of silver-lead ore on Erickson Gulch (a small tributary of Christal Creek on Galena Hill), and worked at the Calumet for Treadwell Yukon in 1940. Erickson had a sister in Alaska—Mrs. Alyce Anderson, who was a U.S. government teacher at Seldovia.

ERVIN, E.

E. Ervin found silver-lead ore on the banks of Ervin Creek (which flows into Carpenter Creek in the Wernecke Mountains) in 1923 during the rush to the Grey Copper Hill finds.[19]

ERWIN, Edward

Edward Erwin is listed as a miner at Keno Hill in Polk's 1923–24 directory.

EVANS, "Curly"

"Curly" Evans drove Cat trains into Mayo in the late 1930's.

EVANS, Mr. & Mrs. Mark

Mark Evans, a mining engineer from the U.S., and his wife lived in Keno for some time. Evans staked the Adam claim on February 23, 1913, which was the same day that Henry McWhorter staked the Silver King, Jack Alverson staked the Webfoot and Grant Huffman staked the Mabel. He also assisted Alverson and Huffman with their initial development on the Silver King option. He was a mining engineer for Treadwell Yukon in

Mr. & Mrs. Mark Evans (centre), celebrate Gordon and Evelyn Lee's first wedding anniversary, March 23, 1925, in Keno. Mark Evans was a mining engineer and both men worked for Treadwell Yukon. *G. & E. Lee Collection*

Keno in the 1920's. Evans and his wife later made their home in Idaho.

EVOY, Alex

Evoy was a miner at Elsa in the 1930's. He left the Mayo area to join the Army in World War II.

EWING, Doris M. "Dorrie"

See *Pioneer Accounts*.

EWING, John Grant & Dorothy

John Grant Ewing, son of Elizabeth and Harry Ewing, was born in St. Lambert, Quebec on February 19, 1933, where he lived until the family was posted to Fort Smith and Fort Resolution, Northwest Territories. They lived there for three years before moving to Kingston, Ontario for two years. Then, with great rejoicing, the family was posted north again to Mayo.

Ewing attended high school in Mayo with a final year in Vancouver, B.C. where he also took flying lessons.

He met his wife-to-be, Dorothy Boss, in Mayo. They were later married in Edmonton and returned to Mayo where he worked with his father, Harry, to establish the family business, Ewing Transport.

He worked with his father and sons in the business until late in life, when he went to work for the Village of Mayo, leaving the business in the capable hands of his sons.

Together, the Ewings led a full and rewarding life centered around their family of seven children: Judy, Linda, Dick, Carol, John Matt, Heather and Michael.

John passed away in Mayo on July 30, 1989, and is survived by his wife, Dorothy, their seven children, twelve grandchildren, two sisters and many other relatives.

From left: Jack Faulkner (Victoria Faulkner's father), Charlie Taylor and Leo Fielder, 1935. *Charles Taylor Collection, donated by Rick Cocks*

EWING, Matthew Henry "Harry"

See *Pioneer Accounts*

FAIRBORN, James A. "Jim"

James Fairborn was the agent for White Pass and Yukon Route in Mayo from 1922 to 1940. He had been with the White Pass in Dawson and Whitehorse since 1908 and came to Mayo from Fairbanks, Alaska. He was very active in community affairs and local theatre. He retired to the coast of British Columbia in 1940.

FAIRCLOUGH, May

May Fairclough is the second oldest child of Eliza and Ira Van Bibber. She and her husband, George, bought Pelly Farm from Frank Chapman in 1927 and operated it until 1940. Their children are James Sinclair Fairclough, Grace Joy Tyerman, George Ira Raymond Fairclough, and Robert Donald Fairclough. Fairclough lived in British Columbia for 25 years after she left Pelly Farm and then retired to Mayo in 1969.

FARRELL, J. H.

Farrell was a mining geologist from Los Angeles who arrived in the district in June 1920 to map the top of Keno Hill for the Guggenheim's.

FARRELL, Martin Edward

Martin Farrell staked No. 1 above Discovery on Steep Creek on June 7, 1905.

FAULKNER, John T. "Jack"

Faulkner was a labourer in Mayo in the 1920's. When Treadwell Yukon closed the Keno operation in 1932, he was left as watchman at Wernecke Camp on Keno Hill. He continued as watchman until 1938. He was Victoria Faulkner's father.

FAWCETT, John

John Fawcett is listed as a prospector in the Mayo area in Polk's 1923–24 directory. Records in the Mayo Mining Recorder's office show that Fawcett and partner Robert Henderson owned some 12 claims on Keno Hill in the 1920's and early 1930's.

FERGUSON, Dan

Dan Ferguson, originally from Prince Edward Island, was a sourdough miner, prospector and trapper in the Mayo district. During the Depression he and Henri LeBlanc mined ore on Alex Nicol's ground. He lived and mined on Davidson Creek in the Mayo Lake area in the late 1930's. Ferguson lived on Duncan Creek in the early 1950's before he returned to Prince Edward Island. He died there in the 1950's.

FERGUSON, Ruth (also spelled Fergusson)

Ruth Ferguson married a barber in 1914 and lived in Blairmore, Alberta for six years where she ran a boarding house. She later travelled north and arrived in Mayo on June 15, 1924. She settled in Keno and bought a dog team to make supply runs back and forth to Mayo.

In 1927, she decided to finance prospectors. She obtained a 10% lay on Sourdough Hill from Andy Johnson, grubstaked Jack Hawthorne to work the Sourdough Hill property, and backed Jack Alverson on his Webfoot claim for 1/6 interest. She later bought a half interest in the Webfoot, and when Alverson and Hawthorne became partners on the Webfoot, Ferguson often freighted up to 400 pounds of supplies over the 20 miles from Keno to the claim. Despite all her hard work, in the early years she made little from her investment.

Alverson eventually dealt away his share of the Webfoot, but Ferguson and Hawthorne located the Silver King vein on the Webfoot claim. They eventually sold the Webfoot to Treadwell Yukon, and she invested her earnings from the Webfoot in a new 17-suite apartment building just east of Granville Street on 12th Avenue in Vancouver. She left the Mayo district for good in 1940 to run her apartment block and remained there until her death circa 1960.

FERRELL, James Emmett "Jim" & Helen

Jim Ferrell ran the trading post at Lansing from 1908 to 1915, then operated a general store in Mayo under the name of Alverson and Ferrell (Jack Alverson and J. E. Ferrell) in 1915. He also doubled as mining recorder, had an interest in quartz mining on Galena Creek with Alverson,

"A toast." From left: Ruth Ferguson, Charlie Brefalt, Bob Greaves, Dora Nicol, Alex Nicol, and Emil Forrest.
A. Aho Collection

From left: Mrs. Jim (Helen) Ferrell, Dorothy McLaren and Victoria Faulkner photographed in Mayo, circa 1919. *D. Walloch (McLaren) Collection*

and even operated a real estate business at the same time. He was the Mining Recorder when Louis Bouvette staked the Roulette claim in 1919. Ferrell was a member of the Yukon Council for the district from 1922 to 1928. His wife, Helen Baird Ferrell, had extensive medical training.

In the winter of 1938, already in his 70's, Ferrell still found the time and energy to assist in Taylor and Drury's store. He died in his early 70's in Mayo on July 17, 1941. Mrs. Ferrell left soon after and it is believed that she went to Oregon where her brother lived. (The Ferrells are also mentioned in Mary Rich's account in *Pioneer Accounts* and in *Buildings and Artifacts* under *Jim Ferrell's Photographs*.)

FIELD, Fred

Fred Field of Gordon Landing went out hunting before the close of navigation in the fall of 1902 and did not return when he was expected. In February of 1903 he was reported lost, and search parties went out. On May 25, 1903, his body, surrounded by all of his belongings, was found on a raft in the ice on the Stewart River near Lansing Creek. He had apparently died of exposure and starvation. Field was one of the first men to prospect in the Mayo region. It is believed that he found gold prospects on Field Creek and also staked the first claim there.[20]

FINNES, Antonne

Antonne Finnes mined on Haggart Creek with F. B. Towsley in 1938.

FISHER, Alice

Alice Fisher was the first teacher in Keno in 1928 and succeeded Gordon McIntyre as a teacher in Mayo from 1941 to 1943. She married Con Carthum and had a daughter, Genevieve. Bud Fisher was Alice's brother.

FISHER, Clifford E. "Bud"

In the winter of 1930, Bud Fisher, his wife, Ila, and their children, Clifford E. Jr. "Kippy" and Doris, arrived in Mayo from McKeesport Pennsylvania. Children Ila and Patsy were born

Ila Fisher and Rose Wood (Mason-Wood). *Middlecoff-Bleiler Collection*

several years later. Ila died in 1935 at the age of 30.

During the 1930's Fisher owned and operated Fisher Service Company, which offered transportation between Mayo, Keno, and Galena Hill. He acquired transportation contracts and offered day and night car service and repairs. The company had three trucks, a touring sedan, a snowmobile, and a Caterpillar.

He acquired a fleet of 18 vehicles and he did most of the contract trucking, bulldozing, and timber supplying for the silver mines from 1947 to 1949. He Cat-hauled concentrates to the Yukon River at Minto and back-hauled oil and supplies.

He was also known for owning an 8-mm movie camera, with which he filmed various events in the district during the 1930's.

He married his second wife Catherine "Katey" McDougall of Mayo (formerly of Calgary) on February 17, 1938 in Vancouver, B.C. Their daughter, Marie was born in the early 1940's.

His mother was Mary Alice "Ma" Fisher.

"Ma" Fisher's Boarding House, 1936. Van Cleaves' home in later years. *G. A. McIntyre Collection*

Bud Fisher is in his 89th year [1989] and resides in Penticton, B.C., with his wife Sally (Harris) whom he married in Vancouver, B.C. in 1969.

FISHER, Mary Alice

From 1925 to 1936, Mary Fisher owned and operated a rooming house and restaurant in Mayo

"Ma" Fisher and her boarders. From left: Gordon McIntyre, Joe Longtin, Jim Fairborn, William A. "Bill" Hutchings, Dr. Allan Duncan, and "Randy" Maclennan. *G. A. McIntyre Collection*

known as "Ma Fisher's Boarding House." The boarding house was located near the Anglican Church. When Fisher left the Yukon in 1936, Laura Roche operated the boarding house for a while. Fisher's two children were Alice and "Bud."

FISHER, Robert "Bobbie"

– Archie Gillespie, *The Mayo Miner*, August 11, 1939 and August 18, 1939

Bobbie Fisher was born in Bonavista, Newfoundland and came to the Yukon by way of the Chilkoot Pass during the Klondike gold rush. He came north with Art Rouse, another well-known northern pioneer. In Skagway, Bobbie broke a team of 25 dogs for his partner.

For many years Bobbie Fisher mined on Dominion Creek and at Granville in the Dawson district. He came to Mayo in 1906 and remained in this silver district for the rest of his life. In the Mayo district he mined on Dublin Gulch where he staked the Olive claim which he recorded on July 1, 1908. After silver was found on Keno Hill, he spent 20 years mining on Keno and Galena Hills, and other parts of the camp.

He staked many silver prospects and he and his partner Archie Martin dug numerous holes at the head of Porcupine Creek on the Coral and Wigwam claims.

Although he was a hardworking prospector, and found many fine leads for others, Bobbie never struck it rich himself. At the time of his death, he was interested in a considerable number of mining properties in the Mayo area.

"Fred" Fluhmann at his claim on Duncan Creek, circa 1910. *R. M. Clark Collection*

He had two sisters in Mayo (Mrs. Agnes. J. Kinsey and Miss Maria Fisher), a younger brother and sister in Bonavista, Newfoundland; and a fourth sister in Gambol, Newfoundland. [His sister, Virtue also came to the Yukon about 1902 where she met and married John Lochore. They made their home in Penticton B.C. in 1908.]

His niece, Olive Powers (Kinsey), of Olympia, Washington, was well known in Mayo because she lived here with her mother for several years. Robert Fisher died at age 73 on August 9, 1939 while holidaying with his sister Maria Fisher at Steamboat Bar, on the lower Stewart River.

FLETT, Andrew "Bonnet Plume"

Andrew Flett, or "Bonnet Plume," was a Loucheux Indian chief. He had worked for many years as an interpreter for the Hudson's Bay Company, and his band made their home on the Bonnet Plume River. During the Klondike Gold Rush, many of the stampeders from Edmonton made their way through that part of the country. When some of these unfortunates were caught by winter on the trail to Dawson, Bonnet Plume gave his assistance to them. The Loucheux called the Bonnet Plume River "The Black Sands" River, because of the large amounts of black magnetite sand (iron ore) found in its bed.[21]

FLETT, Robert G. "Bob"

Bob Flett was a Scotsman from the Orkney Islands who came into the district during the Duncan Creek stampede. He trapped in the McQuesten River area for a long time, and lived alone for many years. Flett later became a game warden in one of Canada's national parks.

FLUHMANN, Fritz "Fred" & Emily

Fred Fluhmann was originally a Swiss watch maker. He operated a placer mine on Duncan Creek in the early 1910's and mined on Highet Creek in the early 1920's. He is listed as a Mayo resident in Polk's 1923–24 directory. Emily Fluhmann was his wife. Later in the 1920's, while working at the Cox, Snell and Cox sawmill in Mayo, Fluhmann lost a leg when his clothing became tangled in the machinery. He was taken to the Mayo hospital where he died of his injuries.

Fluhmann's grandson, the late R.M. "Bob" Clark, of Edmonton, Alberta, visited Mayo in 1986 to see where his grandfather had lived. The Fluhmann's granddaughter (Bob's sister) Betty Reynolds resides in Vancouver, B.C.

FLYNN, E. Miles

Miles Flynn assisted J. R. O'Neill with his silver interests in the Mayo district. The two were long-time associates. They worked together in the famous Porcupine Gold Camp and the Sudbury areas of Northern Ontario. Miles was from Mexico, where his father owned silver mines, and he graduated from the Colorado School of Mines as a mining engineer.

In addition to working extensively in Northern Ontario and the Northwest Territories, Miles and his brother, Milton, also a mining engineer, played a part in "pouring" the first gold brick produced in the "Porcupine Camp" at a mine known as the "Vipond Gold Mines Ltd."[22]

FLYNN, Michael

Michael Flynn is listed at Mayo in Polk's 1923–24 directory.

FOLEY, Jack

Jack Foley worked in Mayo in the 1950's as an expeditor for United Keno Hill Mines in the building which was formerly Treadwell Yukon's Mayo offices. He formed Foley Silver company in 1964 and he and partners Oliver Hutton and Wilf Tuck explored and mined claims on Stand-To Hill.

FOLEY, Matthew Levi

Matthew Foley arrived in the Mayo district during the Duncan Creek stampede and staked claims on that creek in 1902.

FOLEY, Victor "Vic"

Vic Foley was a product of Vancouver's "East End." He boxed as a bantam weight in the 1920's. He was Canadian Champ but lost the title bout with Tony Canzoneri. He retired from boxing and opened a night club called "The Narrows" at the north end of the Second Narrows Bridge in Vancouver. Unfortunately, the bridge was closed when hit by a freighter and Foley went bankrupt.

During the Depression, he went to Dawson where he worked as a cook in the Y.C.G.C. camps. He came to the Mayo area during the silver boom in the late 1940's and tended bar for Matthews and Mercure at the Chateau Mayo and at the Northern Lights Hotel in Keno for owner Robert "Bobbie" Greaves in the 1950's.

He came to Mayo from Dawson as an old man and had worked with Dr. Aaro Aho while Aaro was in the district and also worked at Elsa doing lighter jobs assigned to older employees. It is believed he became blind and lived at the C.N.I.B. residence in Vancouver, B.C. where he died at the age of 90. Vic Foley was a sign painter by trade.

FORBES, Alex

Alex Forbes is listed as a miner at Keno Hill in Polk's 1923–24 directory.

FORBES, Kenneth A. & Hilda

Kenneth Forbes was an old-timer from Dawson who became a mill worker at Keno Hill in the 1920's. He later left Mayo and bought a sawmill at Port Kells, B.C., but returned in 1934 or 1935 to supervise the construction of the Elsa mill.

He was General Superintendent for Treadwell Yukon at the Elsa in 1936. He was married to Hilda Potter (sister of George Ferguson Potter), and their daughter Mary attended school in Mayo in 1934–35.

FORBES, William H. "Bill"

William Forbes is listed at Keno Hill in Polk's 1923–24 directory.

FORREST, Emil Joseph

Originally from Quebec, Emil Forrest's father went to Dawson City during the Klondike Gold Rush and it was here that his sons Emil and Albert played hockey while still young men. Albert was on the Dawson City hockey team that competed for the Stanley Cup in 1904.

Emil made his way to Mayo and staked the Tango claim in 1919 and held it for over 40 years. He worked as the mechanic for Treadwell Yukon's airplanes in the 1920's and 1930's, worked as a mechanic for the White Pass in the 1920's, and at one time also operated the power plant in Mayo.

He was a favourite with all the children in Mayo, as every winter he made an ice slide on the banks of the Stewart River. He was not popular

Wernecke Camp, circa 1929. From left: J. B. "Bon" Kunze & Emil Forrest (power plant operators), Wally Jones (mill operator), Ed Runacres (mill oiler), Bert Walters (electrician), Roy McDermid (hoist operator),and Ed Kunze (mechanic). *Ed Kunze Collection*

with their parents, however, as the ice slide wore holes in the children's pants too quickly.

Forrest went to Whitehorse during the war years where he worked for the Whitehorse power plant as well as for White Pass and Yukon Route on the steamboats.

He was to have been the captain and pilot of the S.S. *Keno* on its last voyage to Dawson City in 1960 before it became a National Historic Site. He was very excited about this honour, but died very suddenly just prior to the S.S. *Keno's* departure from Whitehorse for Dawson City.

FORRY, George L. (sometimes spelled Forey or Forrey)

George Forry was at the Lansing Creek Post in November 1909 according to a letter he wrote to The Bank of British North America in Dawson. In 1910–11 and 1917 he worked for Elmer Middlecoff on Highet Creek. In Polk's 1915–16 directory, he is listed as quartz mining on Galena Creek, and in Polk's 1923–24 directory he is listed as a miner in Mayo. He also staked claims on Rambler Hill in the Davidson Mountains.

FOSS, Fred

Fred Foss is listed at Mayo in Polk's 1923–24 directory.

FOTHERINGHAM, Marie & David Hetherington

Marie Fotheringham (née Joussaye) was born Mary Josey in Belleville, Ontario circa 1864. Her name, as a writer in Toronto, was Marie Joussaye. She came to Dawson in the early days as a reporter for a Toronto newspaper (believed to be the Toronto World).

In Dawson she met David H. Fotheringham, a three-year member of the N.W.M.P. and a veteran of the Boer War. "Dave" was born in Merriton, Ontario circa 1875.

In those days, members of the N.W.M.P. were required to resign in order to marry. The Fotheringhams were married in Dawson on November 16, 1903, and moved to Mayo circa 1922 where she published the *Mayo-Keno Bulletin* in the 1920's while he ran a small steamboat on the Stewart River. David Fotheringham died on March 29, 1936 and is buried in the Mayo cemetery. Marie died in Vancouver, B.C. on March 24, 1949. They had no children. (See *Pioneer Accounts*.)

FOURNIER, "Joe" & Marguerite

Marguerite Fournier operated a restaurant in Mayo in the late 1920's. The Fourniers owned the "post office" cabin located on what is today the site of the Northwestel exchange. She died in 1933 and both she and her husband are buried in the Mayo cemetery.

FRANICH, Matha "Mike"

Mike was a partner and brother of Pete Franich who ran the "F and F" Cafe in Dawson. After Pete died in Dawson, Mike came to Mayo with Milo Jurovich and "backed" Milo in the purchase of the Silver Inn. Franich died in 1958 and is buried in the Mayo cemetery.

FRASER, Alexander

– Archie Gillespie, *The Mayo Miner*, March 25, 1938 and April 1, 1938

Alex Fraser was born in Inverness, Scotland in 1887. He was a member of the 54th Battalion from Kelowna and saw several years of active ser-

Dave Fotheringham's steamer, *Klondyke*. *Middlecoff-Bleiler Collection*

Dave and Marie Fotheringham, taken in Dawson. *McLaren-Swensen Collection*

vice overseas in World War I. After the war he came north to Dawson where he lived until coming to Mayo in 1925. For many years he worked for Treadwell Yukon on Keno and Galena Hills.

On March 25, 1938, Alex died of silicosis, aged 51, at the Mayo hospital after a two-year illness. His brother John "Buck" Fraser (not the "Buck" Fraser listed below) also lived in the Mayo district.

FRASER, John "Buck"

In 1885 a party comprised of John Fraser, J. Chapman, Thomas Boswell and others explored the upper reaches of the Stewart River and named Fraser Falls.

FRIEBERG, John A.

John Frieberg is listed at Stewart settlement in Polk's 1923–24 directory.

FRIESEN, Elmer & Bess

See *Pioneer Accounts*.

FRY, Samuel

Samuel Fry is listed as a Mayo miner in Polk's 1915–16 directory.

FURRER, Jacob

Jacob Furrer staked the upper Discovery on Cascade Creek on October 3, 1903.

GABOURY, Alexander, "Alex"

Alex Gaboury was a prospector and trapper on Carlson Creek in the 1930's. He also owned property on the Mayo-Keno Road.

GALBRAITH, Findley

Findley Galbraith is listed as a labourer at Mayo in Polk's 1923–24 directory.

GALLAGHER, J. D. "Jack"

Jack Gallagher was the government agent and liquor vendor in Mayo in the 1930's.

GAMMON, Albert

Albert Gammon is listed as a labourer at Mayo and at Keno Hill in Polk's 1923–24 directory.

GASPAR, Matt

Matt Gaspar staked No. 1 below Discovery on Steep Creek on August 16, 1903, and staked on Ledge Creek with Mike Pavisich on August 18, 1903.

GATEY, John & Nora

John Gatey worked for Fred Taylor on Fred's placer mining property at Dublin Gulch in 1938–39.

In the late 1940's, when the Keno road was closed one winter, Gatey hauled mail from Mayo to Calumet and Keno by dog team.

He married Nora Mervyn, daughter of Julia and Jim Mervyn and they had three children Alan, Marie and Johnny.

GAUNDROUE, Elmer & Margretta

Elmer Gaundroue worked at the Wernecke mill and the Elsa mill. He also worked on the dredges in Dawson. He was married to Margretta, daughter of Elizabeth and David Ballentine (see *Pioneer Accounts*). The Gaundroues had two sons and made their home in Elsa.

GAUVIN, Wilfred W.

In 1908 Wilfred Gauvin operated a roadhouse at Barlow which offered French cuisine. Gauvin and Hector Morrison mined on Barlow Creek circa 1913–14. He operated a roadhouse on Clear Creek circa 1915–16. He was listed as a miner at Keno Hill in Polk's 1923–24 directory.

GENERT, O.

O. Genert is listed as a carpenter in Mayo in Polk's 1923–24 directory.

GERGICH, Joseph

Joseph Gergich mined in the Mayo area in the 1910's, and is listed as a miner in Polk's 1915–16 directory. In the 1930's he held placer ground on Clear Creek. He also lived on this property, but would come to Mayo each year for his outfit. Sometime in the 1930's he sold his property to the Clear Creek Dredging Company.

GERLITSKY, Jack & Hazel

Jack Gerlitsky was Treadwell Yukon's mechanic at both Wernecke and Elsa in the 1930's. He had a taxi and in the evening he drove "boys," so inclined, to see "girls" at Keno. He was also a prospector and hard rock miner. He left the district and then returned in the 1950's to sell claims near the Silver King. He and his wife, Hazel, had one daughter. Gerlitsky was later involved in mining promotion in Vancouver and had his own company.

GERMAINE, Colley (Yezá)

– Archie Gillespie, *The Mayo Miner*, January

Colley Germaine, Mary St. Pierre and Ellen Germaine at the Old Village. *D. Hager Collection*

8, 1938 and February 11, 1938

Colley Germaine (written as Jermie in the old records) was born at Fort Good Hope in the Mackenzie district in about 1898. He had three brothers and one sister, and he lived and trapped in the Ladue Lake district.

In January 1938 Colley collapsed and died while out on his trapline. He had not been well, and after he had been gone for about ten hours on this particular day, his brother Paul set out to search for him. Paul found Colley on the trail about 10 miles out of Keno City.

GERMAINE FAMILY

(The name was originally Jermie.) See *Pioneer Accounts*.

GERRARD, James W.

James Gerrard did not live in Mayo but formed the Reserve Mining Company in the 1920's and held the Onek property in Keno. The Onek was later acquired by United Keno Hill Mines.

Gerrard had been the U.S. Ambassador in Berlin at the outbreak of World War I.

GILL, Fred R.

John Maynard, Albert Jahnke, and Fred Gill mined on Haggart Creek from 1908 to the 1920's. Gill eventually left the Yukon for the U.S.

GILLESPIE, Archie & Ada

Archie Gillespie was the son of R. L. Gillespie. He came to Mayo in 1932 and published the weekly *Mayo Miner* for seven years. All of the businesses in the area advertised in the *Mayo Miner* and he had a circulation of approximately $250 a month. He had an avid interest in airplanes and pilots and wrote glowing reports of their accomplishments.

In 1941 he moved to Dawson City and published the *Yukon Miner* through the 1940's. He retired to Whitehorse and occasionally wrote articles for the *Yukon News*. Archie Gillespie died an accidental death in Whitehorse when he was in his eighties.

He and his wife Ada (Poole), had a daughter, Ruth, and sons: Dick, Keith and Don.

GILLESPIE, Frank

Frank Gillespie was R. L. Gillespie's brother. He lived in Mayo from the 1920's to the 1940's. He was a carpenter and timberman and rebuilt the Chateau Mayo for Gene Binet after it burned down in 1930.

Laura (left) and Margaret Gillespie with their uncle Frank Gillespie in the Keno area, circa 1920's. *T. McKay Collection*

He also built the Galena Creek bridge at the Silver King for the Yukon Territorial Government.

Gillespie lived at Minto Lake for many years where he had a small placer mining operation. He was also an avid fisherman in his spare time.

GILLESPIE, Laura

Laura Gillespie was the daughter of R. L. Gillespie. Born on Bonanza Creek, she received her early education in the Dawson public school and the Dawson high school.

After graduating from Edmonton Normal School she returned to Mayo where she taught at the Mayo public school for two years.

In the late 1930's, she worked as Private Secretary to the General Manager of the Canadian General Electric Company in Vancouver. On April 11, 1938 she married Rio Elgar of Vancouver.

GILLESPIE, May

See *May Gillespie Fielder Morrell* in *Pioneer Accounts*.

GILLESPIE, Richard L. "Dick" & Margaret

R. L. Gillespie was the Mining Recorder for the Mayo district in the 1920's. He and his wife Margaret (née Maclennan) had four daughters: Margaret, Laura, Vera, and May and one son, Archie.

GILLESPIE, Vera

Vera Gillespie was the daughter of R. L. Gillespie. She taught school in Mayo following her sister Laura's teaching term. She is now Mrs. Herb Lintner of Vancouver, British Columbia.

GILROY, Jack

Jack Gilroy was at Lake Creek on the Stewart River in 1918.

GLEESON, Daniel

Daniel Gleeson, born in Ireland, went overseas with the Yukon Infantry Company during World War I. He is listed at Keno Hill in Polk's 1923–24 directory.

GLENDENNING, Horace

Horace Glendenning and his partner Sandy McPherson made important discoveries along the Silver King vein. Glendenning died "outside" in 1936.

GODBOUT, Gervais P.

Gervais Godbout was a partner in the "Little Guggs" Mining Company (R. Rosmusen, G. Miller, C. Rockney, M. Lindquist and G. P. Godbout) which operated on Highet Creek from 1910 to the 1920's. Godbout died in 1934 and is buried in the Mayo cemetery.

GONZALES, Manuel

– Archie Gillespie, *The Mayo Miner*, October 1, 1937

Manuel Gonzales was a native of Rendin, Orenso Province, Spain. He was a hard rock miner of many years experience having worked for years in the Premier and Big Missouri mines in Stewart, B.C. He came to the Mayo district in June of 1937, and was employed by Treadwell Yukon Co. Ltd., on Galena Hill in July, 1937. Tragically, less than three months later, he was struck and buried by a rock slide in the Elsa mine. The accident occurred on the morning of September 23. Gonzales was working the 100-foot stope above the main tunnel level when a huge quantity of rock fell on him from above, completely trapping him. Lars Burke and Jack

McCrimmon, working in the same stope, were about 75 feet away from Manuel when the accident occurred. Hearing the noise caused by the falling rock they rushed to the spot. While one ran for aid, the other began digging the injured miner from the slide. After releasing him from the mass of rock which covered him, it was necessary to lower the injured man on a rope to the tunnel level. R. J. Boyd, General Superintendent, rushed Manuel to Mayo, being met at Field Creek by Dr. E. T. W. Nash, who had been driven out from town. Gonzales died of his injuries the following day, September 24, 1937, in the Mayo hospital. His friend, Tony Lopez, said he had never heard Manuel mention any relatives. Gonzales was about 50 years old at the time of his death.

GOODING, Richard "Dick"

Dick Gooding was the government telegraph operator at Stewart City in the 1930's. The telegraph station was located atop the ridge across the river from Stewart City. He was considered to have one of the greatest collections of books and periodicals in the Yukon. His collection entertained him during the long winters at his secluded post. He was well known in Mayo and was dubbed the "Mayor of Stewart Island."

GORDON, Alex

Alex Gordon came to Keno in the 1920's from the Dawson area where he was a successful placer miner.

He owned claims on Galena Hill which he sold during the 1950's boom, and he later bought Alex Nicol's house in Mayo. Gordon died in Mayo.

GORDON, Andrew

Andrew Gordon is listed as a miner on Keno Hill in Polk's 1923–24 directory.

GORDON, George

The settlement of Gordon Landing was named for George Gordon, who homesteaded at Gordon Landing. He acquired the 160 acres of land on June 17, 1902, and it was located at the junction of the outlet of Janet Lake and Stewart River on the right limit of the Stewart River about 12 miles below Fraser Falls. Gordon Landing was abandoned when Mayo was established in 1903. Gordon mined for a time on Barker Creek, circa 1908, and moved to Mayo circa 1915.

GORDON, J.

J. Gordon is listed as a miner on Keno Hill in Polk's 1923–24 directory.

GORDON, Wilfred "Wilf" & G. Jean

See *Pioneer Accounts*.

GOUGH, Frank

Frank Gough was mining on Blackhills Creek circa 1910. He is listed at Stewart River in Polk's 1923–24 directory.

GOULET, Octave

Octave Goulet is listed in Mayo and the Black Hills area in Polk's 1923–24 directory.

GOULTER, Anne

Anne Goulter came from Carmacks and lived in Mayo during the late 1930's. She represented Mayo in the Miss Yukon contest in Dawson City in 1939.

GRAHAM, George

George Graham is listed as quartz mining on Dublin Gulch in Polk's 1915–16 directory.

GRAMLEY, James

James Gramley is listed at Keno Hill in Polk's 1923–24 directory.

GRAND, Oscar

Oscar Grand was from Finland, and worked as a wood contractor in the Mayo silver district. At one time he had a woodcamp on Glacier Hill, and in the 1930's he had a wood camp near Mayo Canyon.

GRANT, Victor Garibaldi "Vic"

– from "Tragedy on Keno Hill" by A. A. Gillespie, *The Alaska Sportsman*, August, 1944

Vic Grant was born in Eastern Canada in 1860. He and his partner, Jim Bungate, stampeded into the Yukon during the Klondike Gold Rush. They took a lay on a piece of ground on Eldorado Creek and named it the Victoria claim. They sank a 50 foot shaft on the claim and one day while Vic was going down the shaft, the rope that held him broke and he fell to the bottom. It took him three years to recover and the accident prevented him from further physical involvement in the mining game. Following his recovery, Vic went to work in the Mining Recorder's office in Dawson City. In 1912, he left Dawson to pursue business interests in Victoria, B.C., and other parts of Canada.

In 1922, Vic returned to the Yukon and, shortly thereafter, was appointed Agent for the Mayo Mining Recorder in Keno City. During the 1920's silver boom, "there was a steady stream of men coming into Vic's office to record new claims, to swear in representation work or to have him attend to their various business affairs." Vic remained Keno City's Government Agent until World War II and the drop in silver prices caused a reduction in mining activity and the Keno sub-branch was closed.

A couple of years later, and at age 83, Vic decided to fix up his cabin in Keno with the help of old-timers Charley Johnson and Alex Gordon. He began to dismantle an old cabin near Keno in order to obtain building materials for his cabin. While he was removing the last plank from the roof, the building collapsed. He managed to claw his way out of the rubble and start a fire to keep warm. Unfortunately, the wind whipped the fire into some old rags and the pile of cabin material and soon the site was a swirl of flames. Vic managed to crawl away from the fire but suffered severe burns to his hands and feet. When Grant did not appear in Keno by late afternoon, his friends Charley and Alex began to search for him. They found him unconscious near the burned-out cabin. Tommy McKay brought his pick-up truck, with a mattress in the back, to the scene and the three brought Vic back to his cabin in Keno. Mayo nurse Pheobe Reynolds was sent for and she nursed Grant in his cabin for four days until he was well enough to withstand the bumpy ride over the road to Mayo. Vic had sustained severe back injuries in the accident and he was flown from Mayo to St. Mary's Hospital in Dawson. Grant remained in the hospital for a year but did not recover. He died at St. Mary's Hospital in 1944.

GRAY, George

George Gray operated a woodsaw business in Mayo in 1925 and was the Fire Chief before Alex Maclennan.

GRAY, William "Bill"

Bill Gray was a member of the R.C.M.P. stationed in Mayo in 1937. He took his discharge in

The Keno cabin of Victor Grant, the liquor vendor, which Ted Bleiler rented for two years at $25 per month, in 1930. *T. R. Bleiler Collection*

Carmacks in June, 1938, and then came back to the Mayo district to prospect. In 1940, he worked for a mining company on Dominion Creek in the Dawson district.

GREAVES, Robert Bruce "Bobby" & Ada

"Bobby" Greaves was a former saloon keeper and gambler from Dawson. He was the proprietor at the Red Feather Saloon in the early days, and his wife Ada "Tiny" was a former Dawson dance-hall girl.

She cooked for the crew working on the McIntosh, Hamilton, Patterson, Davidson operation at Duncan Creek in 1902 and later cooked at Keno Hill Ltd.'s powerhouse in Keno. She was also the first postal clerk in Keno, in 1921.

"Bobby" went overseas with the Black Contingent during World War I. He held claims on Keno Hill, and operated the Keno Hotel in the 1920's and the 1950's. He was born in 1878 in Ontario and died in Vancouver at age 95.

GREENAN, James O.

James Greenan was sent by the Guggenheims to replace Frank Short as manager of Keno Hill, Ltd., when he left in the 1920's.

GREENAWAY, Thomas

Thomas Greenaway was born in Lewisham, Kent, England. He signed on with the Black Contingent in Dawson and went overseas during World War I. He owned a store and was postmaster in Keno City in the 1920's.

GREENFIELD, James E.

James Greenfield operated a freighting company from 1915 to the 1930's with J. E. Pickering called Greenfield and Pickering. The company was originally based in Dawson City, but it came to Mayo to freight ore from Keno Hill to the Mayo waterfront.

The partners also operated a property on the side of Lookout Mountain (Mt. Haldane) in 1919. In the early 1920's Angus McIntyre ran their ore haul contract from Keno Hill to Mayo. Greenfield and Pickering later held the mail contract from Whitehorse to Mayo and Dawson (1925 to 1929) and McIntyre became their Dawson manager.

Greenfield staked the Solo No. 2 claim on Keno Hill on August 15, 1919. Greenfield's daughter is Mrs. Lillian Baigent of Gibson, B.C.

Robert "Bobby" Bruce Greaves, the one time owner of the "Red Feather Saloon" in Dawson City and the Keno Hotel. *G. A. McIntyre Collection*

GREER, Leonard T.

Leonard Greer, born in Winfield, New Brunswick in 1868, came to the Yukon in June 1898. He was listed on the Stewart River in Polk's 1923–24 directory, and during the final years of the White Pass's overland stage service, he was employed to keep the roads in repair. Leonard lived on Scroggie Creek for the last twenty years of his life prospecting and trapping, and died on Scroggie Creek in October 1938.

GREIG, Clifford W. "Cliff" & Maureen

Cliff Greig worked for Treadwell Yukon on Galena Hill in the 1930's. He went overseas during World War II. He purchased the upper half of Fred Taylor's Dublin Gulch claim in 1955. He and his wife, Maureen, lived in Mayo with their son, Ieuan. He worked for United Keno Hill Mines in his later years. Greig retired to Creston B.C. with his son, Ieuan, and died there in the 1980's.

GRENIER, Gene T.

Gene Grenier trapped on the McQuesten River with Dick Mercure in the late 1930's and worked as a miner at the Wernecke and Elsa Camps. He also owned property on the Mayo-Minto Bridge Road.

GRITZUK, Nick

Nick Gritzuk was a transportation superintendent for United Keno Hill Mines in 1953.

GROPSTIS, Charles "Chuck"

Chuck Gropstis was Livingston Wernecke's airplane pilot, draftsman, and mill designer. He was piloting the plane when he and Wernecke were killed in a crash in Millbank Sound on the coast of B.C. in 1941.

GRUGAN, Frank

Frank Grugan was a clerk at J. E. Binet's hotel and store in the 1920's.

GRUNDMANIS, A. "Butch" & Virginia

Butch Grundmanis arrived in Elsa in 1951 to work for United Keno Hill Mines. He later became the mill superintendent. He and his wife, Virginia, have four children. They left the district in the early 1980's.

GUERIN, Napoleon

Napoleon Guerin staked a claim on Steep Creek on June 5, 1905 and the claim was later acquired by John Hamilton Baker. Guerin was at Telford Creek in 1915. He is listed in the Stewart River area in Polk's 1923–24 directory.

GUILBAULT, Wilfred "Fred"

The Guilbaults were Mayo pioneers. He was the blacksmith in Mayo from the 1920's to 1938, but left Mayo in March 1938 to work in the Dawson area.

GUNN, Thomas D. "Tom"

Thomas Gunn was a veteran of World War I and was a miner employed by Treadwell Yukon at Wernecke. He was killed in a mining accident at the Lucky Queen in 1932 and is buried in the Mayo cemetery.

GUSTAVESON, Elmer

Elmer Gustaveson prospected and worked silver claims in the Keno district with Ole Ostenstead in the 1920's. He continued to live in the district through the 1930's and 1940's and died in Mayo in the late 1940's or early 1950's.

GUTHRIE, John "Jack"

John Guthrie, originally from England, was a diesel mechanic at Wernecke and Elsa. When the mine closed in 1932, he moved to Mayo. Guthrie died in Mayo.

GWYNE, Thomas

Thomas Gwyne is listed as being in the Stewart River area in Polk's 1923–24 directory.

HAGER, David "Big Dave"

See *Pioneer Accounts*.

HAGER, Edwin

See Helen Buyck's story in *Pioneer Accounts*.

HAGER, Mary

See *Pioneer Accounts*.

HAGER, Robert & Liza

Robert Hager arrived in Dawson during the Klondike Gold Rush. He was a Special Constable for the N.W.M.P. at McQuesten Post from November 1905 to May 1906. He married Liza Jimmy (daughter of Jenny Jimmy) and they had one son, Edwin. Liza Hager died when Edwin was still a baby.

Hager worked as a carpenter in Mayo circa 1915, but was also involved in mining. He worked his own properties as well as working for wages, and in 1918 he rocked on the bars of the Stewart River at Johnson's Point (near the mouth of Moose Creek) where he made from $6.50 to $9.60 per day in placer gold.

In the fall of 1918, he left the Yukon and perished in the *Princess Sophia* disaster on October 25, 1918 off the coast of Alaska. The *Sophia* struck the Vanderbilt reef in Lynn Canal in a blinding snowstorm, and all 343 passengers and crew were lost.

George Black, who had become acquainted with Robert Hager in 1902 or 1903, testified on his behalf during the hearings that were held following the *Sophia* disaster. He described Hager as a man of good habits, who saved his money and invested most of it in mining.

Black also testified that Robert Hager was approximately 45 years of age at the time of his death, and that his son Edwin would have been about eight years old in 1918. After his father's death, Edwin Hager was raised by his grandmother, Jenny Jimmy.

HAGGART, Peter Findlay

In 1898 Peter Haggart and Warren Hiatt reached Nelson Creek, and after staking Discovery renamed the stream Haggart Creek. He also staked the old river bed at Minto Lake on May 19, 1903 and Highet Creek on June 3, 1903 with his partner Jimmy Scrivener. Peter and Thomas Haggart were brothers.

HAGGART, Thomas

Thomas Haggart was one of the area's first prospectors, finding coarse gold at Dublin Gulch (a small valley off Haggart Creek) in 1895. In 1896 he built the first cabin there and gave Dublin Gulch its name. With the influx of Klondikers, Haggart and others decided to record their claims for protection against claim-jumpers.[23]

HALDANE, possibly Robert

Haldane was an early prospector who lived and mined on Haldane Creek at the foot of Mount Haldane.[24]

Mount Haldane was officially named for this man on March 31, 1924, as part of the 18th Report of the Geographic Board of Canada (GBC). The name [Haldane] first appeared on a list of names from the report and plan of an exploratory survey of the Stewart River by J. J. McArthur, Dominion Land Survey (DLS), in 1898. Before that time and well into the 1940's, the mountain was referred to locally as Lookout Mountain. Neither the GBC nor the DLS report offer a first name for Haldane.

It is believed that his first name was Robert and that he is buried in the old cemetery at the Mayo-Elsa road junction, where there is a marker bearing the name "Robert Hald...." The last part of the name cannot be read at all and the marker has fallen down.

HALL, Arnold Theodore

Arnold Hall was born in England, and was the manager of the Bank of Montreal in Mayo from May 1934 until it closed in May 1942. His kindness and compassion towards the underprivileged, as well as his financial acumen, gradually earned him the respect of the Mayo community. Prospectors, woodcutters and trappers were especially fond of him, and he sometimes authorized substantial loans to men he judged to be of good

July 1934 on Keno Hill. From left: Mr. Parmeter, Bank Clerk; Mr. Hall, Bank Manager, Alec Berry, prospector; Mr. Robinson, Engineer; Wm. Sime, Govt. Assayer. *A. K. Schellinger Collection*

character, even if they held no security or financial assets. Hall's trust in such men resulted in the establishment of some successful operations which employed many people during the lean Depression years, and these unsecured loans were always repaid.[25]

Hall donated the lots beside the North Star Motel to the community to be used as a children's playground. After leaving Mayo, he worked in Dawson City. Hall was a World War I veteran.

HALL, Oliver

Oliver Hall was a mechanic for Treadwell Yukon at the Elsa in the 1930's.

HALLIDAY, M. R.

M. R. Halliday was a shift boss for Keno Hill Ltd. at Keno Hill in the 1920's.

HAMILTON, Colin

Colin Hamilton was one of the initial stakers of the three Swedes Discovery on Duncan Creek. He later sold his interest to Commissioner Mackenzie, then worked in other areas of the district. He is said to have died in Alaska.

HAMMOND, S.

S. Hammond was an engineer for Keno Hill Ltd. at Keno Hill in the 1920's.

HANNA, Sam

Sam Hanna worked at Elsa in the late 1930's. His uncle was Archie Close. He married Louise Myers of Swift Current, Saskatchewan on October 5, 1938 and returned to Mayo with her. His brother Art Hanna, an assayer by trade, arrived in Mayo in February 1939, looking for work.

HANSON, Peter "Swede"

See *Pioneer Accounts*.

HARDY, Sam

Sam Hardy was an old-timer of the Keno district. He died in 1940 and is buried in the Mayo cemetery.

HARDY, Thomas

Thomas Hardy is listed at Mayo in Polk's 1923–24 directory.

HARE, W. S. "Bill" & Nora

Bill Hare was a well-known photographer in the district in the 1930's. He arrived in Mayo in 1922 from South Africa and went to work at the Binet Bros. store. In 1923 he started grocery hauling for McCown up the hill to the "Guggs," as the Keno Hill Ltd. mine was known, and in 1925 he began working shifts as a mill operator at Wernecke Camp for Treadwell Yukon.

Bill produced top quality photographs which documented life in the silver camps as well as daily events, and in his sideline as a filmmaker produced a collection of 8-mm movie reels of this district. His photographs and film reels are now in the Yukon Archives. Bill married Nora Ross, daughter of J. A. "Jack" Ross, at the Ross family home at Minto Lake. The Hares had a home in Elsa in the late 1930's.

The family moved to Dawson City at the start of World War II where their two children, Alex and Bette, finished high school. In 1965, Bill and Nora retired to Salmon Arm, B.C., where Bill continued his interest in photography. Bill Hare died in Salmon Arm on September 12, 1983, and Nora died in 1989.

HARGRAVES, William Burton & Maud

W. B. Hargraves was an engineer for the Treadwell Yukon Company in the early 1920's. He was also a mining engineer and superintendent at Wernecke Camp, and in charge of the Elsa when it commenced operation in 1928. The Hargraves lived in Keno where their children (Mildred, Barbara, and Dick) attended school.

HARPER, Arthur

For his early history, see *McQuesten, Mayo and Harper* and *Early Bar Mining*.

By 1893 Arthur Harper and Joe Ladue had become partners and were prospecting in the Yukon. Men drifted from Fortymile to the Sixtymile and in the summer of 1894 Harper and Ladue set up a trading post on a large island in the Yukon River opposite the mouth of the Sixtymile River. It was initially called Sixtymile Post but they renamed it Ogilvie after William Ogilvie. Harper later split up with Ladue and in 1896, landed on the site of the original Fort Selkirk (on the east bank) with his wife and several children. He had a large stock of trade goods and started a post independent of the Alaska Commercial Company.

Although Harper had pioneered most of the major goldfields that were found in the Yukon and his experience and advice led other men to fortune, he never found the riches he sought. He left the Yukon in 1897, a poor man, broken in health from his privations and hardships, and died of tuberculosis in Yuma, Arizona, in November 1898.[26]

HARRISON, "Chuck" & Mildred

Chuck Harrison was a teller at the Bank of Montreal in the 1930's. He married Mildred Waugh who was a nurse at the Mayo hospital.

HARTNELL, Joyce

Joyce Hartnell was the daughter of Mrs. Eddie Kimbel, and the sister of Norman Hartnell.

HARTNELL, Mrs. Lou. B.

Mrs. Hartnell operated a restaurant in Mayo in 1930. She later became Mrs. Eddie Kimbel. See also *KIMBEL, Edward.*

HARTNELL, Norman

Norman Hartnell was a truck driver for Kimbel Bros. in the 1930's. He left Mayo to join the RCAF during World War II. Joyce Hartnell was his sister, and Mrs. Eddie Kimbel was his mother. See also *KIMBEL, Edward.*

HAVRANEK, Leonard "Leo" & Maria

See *Pioneer Accounts.*

HAWTHORNE, John "Jack"

Jack Hawthorne, from County Armagh, Ireland, came to the Yukon circa 1920 to work as a teamster on the Keno-Mayo ore haul. He mined, prospected, and held claims in this district from the 1920's right up until 1981.

Ruth Ferguson grubstaked him to work her interest in Sourdough Hill for a time and then Alverson, Hawthorne, and Ferguson partnered on the Webfoot. (Jack Alverson was the original staker of the Webfoot.) Ferguson provided the finances and expediting, while Alverson and Hawthorne did the work. Amos Noyd also worked for Hawthorne on the Webfoot, and in March of 1929 they struck a big ore shoot, an extension of the Silver King vein. Hawthorne and Ferguson struck a deal with Treadwell Yukon on the Webfoot for $300,000 with $30,000 down, but when the price of silver later bottomed the deal was re-evaluated with a lesser settlement.

Other claims staked by Hawthorne included the Armagh, Blueberry, Mountain King and Mountain Queen. Jack Hawthorne died in Whitehorse in 1981.

Four hardrock prospectors. From left: unknown, Jack Hawthorne, Tommy McKay and Oscar Letourneau in the Keno area, circa 1930. *Bleiler Collection*

HAYDEN, Raymond

Raymond Hayden is listed as a miner in the Mayo area in Polk' s 1923–24 directory.

HAYES, A. R.

Listed as a miner on Keno Hill in Polk's 1923–24 directory.

Robert Henderson of Klondike Gold Rush fame prospecting in the Keno Hill area. *T. McKay Collection*

HEFFRING, Henry "Harry" (also sometimes spelled Heffering)

Harry Heffring operated a sawmill at the mouth of Clear Creek in 1918 and into the early 1920's. He rafted the lumber to Dawson and sold it for $60 per thousand board feet. H. Norman and Carl Leoffler worked for him.

HELLABY, Hilda

See *Pioneer Accounts*.

HENDERSON, Harvey

Harvey Henderson, born in Kansas, went overseas with the Yukon Infantry Company in World War I. He returned to work for Treadwell Yukon at the Silver King in the 1930's, and is buried in the Mayo cemetery.

HENDERSON, Robert "Bob"

Bob Henderson of Klondike fame prospected the Mayo-Stewart River district for several years and acquired an interest in various properties. In July 1897 Henderson staked a 40-acre townsite on the banks of the Yukon opposite the mouth of the Stewart but it never succeeded.

He acquired a 50% interest in McIver's Discovery claim on Thistle Creek (a tributary of the Yukon River just above the mouth of the Stewart River). In the 1920's (and up until his death) Henderson and his partner John Fawcett owned 12 claims on Keno Hill, among them the King, Groundhog, Moose and the Vera.

He lived in Mayo from approximately 1921 to 1924 and owned a cabin and lot. His cabin and lot are believed to have been on the lots presently owned by Bob Adair (1988). He was a relative of the Ewing family who live in Mayo (an uncle of Harry Ewing's first wife, Elizabeth Grant Ewing). Henderson died in Vancouver.

HENEY, Thomas Edward

Thomas Heney and a party of three other Australians, were the first to discover gold on Cascade Creek where he staked the lower Discovery on October 7, 1903. He also staked a claim on Duncan Creek in 1903. Heney continued to prospect and mine in the Mayo Lake district until approximately 1920.

HENRY, Dutch

See *Henry Vorthin*.

HENRY, Samuel

In 1897 Samuel Henry took up a homestead at the mouth of Maisie Mae Creek (about 35 miles upstream from the mouth of the Stewart) and started Maisie Mae Ranch. He grew 26 tons of hay that year and continued farming successfully for many years afterward. He specialized in feed for the horses in Dawson, and the mines. In 1906 he had over 100 acres under cultivation.[27] By 1918 Joseph Roy had become his partner and Bill Stewart was working for them. They grew hay for horse feed, which they sold to the White Pass and Yukon Route. More recently, the ranch was owned by Rudy Burian.

HERBSTER, F. J.

F. J. Herbster was a miner on Keno Hill in the 1920's and well-known in Mayo in the 1930's. He prospected the Coal Creek district on the lower Yukon in 1937.

HESTER, Arthur

Arthur Hester mined on Gull Creek circa 1905 with George Crisfield and Matt Butijer (Butyer). He staked on Davidson Creek in 1913 and in the early 1920's staked the Vancouver claim. In the 1920's he lived at Mayo Lake where he was still mining on his Discovery claim on Gull Creek and where he manufactured some of the best moonshine in the district. He was well educated and artistic.

HICKS, Henry Brodie

Brodie Hicks was United Keno Hill's mine manager from the time the company was reorganized in 1947 until approximately 1953. He was granted the title of postmaster in 1949 so that mail could be delivered to Elsa.

HINTON, Thomas Hyde

Thomas Hinton was the Mining Recorder in Dawson circa 1900, and was the Mining Recorder at Minto Bridge from 1904 to 1911 for the Duncan Creek Mining district. Hinton and partner Weldon C. Young had claims on Duncan and Beliveau Creeks in 1902. He staked a claim on Corkery Creek on July 12, 1916. He is listed as quartz mining on Galena Creek in the Mayo district in Polk's 1915–16 directory. He staked claims in the Homestake group on Bunker Hill in April, 1920. Hinton is buried in the Mayo cemetery.

HOBBIS, Antonette "Tona"

Antonette Hobbis was a nurse at the Mayo Hospital circa 1941. She married Corp. William Sutherland of the R.C.M.P. who was also stationed in Mayo.

HODER, James

James Hoder was located in Mayo in the 1920's, and the records show that he operated a woodsaw business in 1924. He cut large quantities of wood for the Klasco Dredge which was being built on Highet Creek but never completed. Piles of wood can still be seen rotting on McIntyre Creek (a tributary of Minto Creek). Hoder died prospecting on the Pelly River.

HODER, Joseph

Joseph Hoder is listed as a miner in the Mayo-Keno Hill areas in Polk's 1923–24 directory.

HODINOTT, Charles

Charles Hodinott was a camp cook. He cooked for Colonel Armstrong on Russell Creek, then came to the Mayo area after that operation closed down. He lived at Minto Bridge for many years.

Dorothy Hoggan fishing at Minto Bridge in 1922.
Middlecoff-Bleiler Collection

HOFFMAN, Fred

Fred Hoffman was a pioneer prospector and trapper in the south Wernecke Mountains and a partner of Frank Rae.[28] The first roadhouse out of Keno was his.

HOFFMAN, Miss Hetwig

Hetwig Hoffman owned a claim on Parent Creek in 1903. She also acquired one half interest in claim No. 2 on Parent Creek and Nos. 58 and 59 on Duncan Creek in 1903.

HOGAN, Paul Shakespeare

Paul Hogan came to the Mayo area from Whitehorse in about 1920 and held property at the head of Silver Basin on Keno Hill. From 1920 to 1922 he was a member of the Yukon Council for Klondike which included the Mayo district.

HOGGAN, Dorothy

Dorothy Hoggan was the school teacher in Mayo from 1920 to 1924, having received her teaching certificate in B.C. She was also the first non-native born in Whitehorse, and grew up in Dawson City. She was a daughter of Captain John Hoggan, a riverboat pilot, and sister of Gregg and Johnny Hoggan.

HOGGAN, George

George Hoggan worked at the Elsa camp. He later became a tug boat operator out of Vancouver, B.C. where he eventually retired.

HOGGAN, Robert McGregor "Gregg"

Gregg Hoggan was a Treadwell Yukon employee at the Elsa in the 1930's. After leaving Elsa, he worked on dredges in Columbia, South America as well as in the Klondike.

HOLLE, Dora

Dora Holle worked for J. E. Binet's operations in the early 1920's. She married Alex Nicol.

HOLLENBECK, Anthony A. "Tony"

"Tony" Hollenbeck was one of the early prospector in the Ogilvie Mountains. He helped stake the Lake Group of claims in the Keno Hill area in the 1920's. Hollenbeck was also a blacksmith. He was born in 1869 and died in Mayo in 1947. (See also *John Hollenbeck.*)

HOLLENBECK, John E.

John Hollenbeck staked the Travice claim on

March 12, 1921. John and Anthony "Tony" Hollenbeck were part of the Hollenbeck family who ran the roadhouse at the Klondike River on the Mayo-Dawson trail, circa 1905 to 1929. (The name has also been incorrectly written Hollenback and Hollenbach.)

HOLMBERG, Edward

Edward Holmberg is listed as a miner in the Mayo area in Polk's 1915–16 directory.

HOLMSTROM, John "Windy John"

John Holmstrom was born on October 20, 1909, on Äland Island, Province of Finland. He went to sea in his younger days and was a dance instructor on ocean going ships. He came to Canada in March of 1927 with the idea of learning languages and then returning to sea.

He came to the Yukon Territory in March, 1938, and went to Dawson City where he worked on the Yukon Gold Company's dredges for a time. It was there he acquired the nickname "Windy John." He later trapped in the winter, did carpentry work in the summer and also did some mining in the Dawson area.

He moved to Keno in August of 1948 and worked for United Keno Hill Mines as a carpenter at Calumet, Wernecke, Onek, and Mackeno. He bought a cabin and four lots from the Thomas McKay estate.

In 1951, he moved to Whitehorse and worked for the Army and Airforce as a carpenter until he suffered a fall in 1962. It took a year for him to recover from his injuries and he was left with a permanent hearing loss in one ear.

In 1963 he retired and returned to Keno to prospect and do exploration work on the Sourdough. John started with eight hardrock claims and now he and his partners have 20 claims.

John Holmstrom, or "Windy John" as he is affectionately known, continues to work his claims in his 80th year.

(Information supplied by Claire Briand)

HOMER, Geoffrey

Homer was the company doctor in Mayo in the late 1930's. He later became a surgeon in Victoria, B.C.

HORROBIN, Bill

Bob Horrobin was Treadwell Yukon's Master Mechanic at the Elsa from 1936 to 1940.

HOSFALL, Joseph S. & Mrs.

Joseph Hosfall was an Englishman who came to the district during the Klondike stampede via

Grant Huffman's second cabin on Mayo side of the Silver King. Bill Tormey lived here for 15 years. *M. Rich Collection*

one of the Edmonton Trails. The country and the lifestyle suited him so well that he married one of Jack McQuesten's daughters and stayed. He earned a living trapping and prospecting. He also accompanied noted big-game hunters such as F. C. Selous and Charles Sheldon into the MacMillan River country, and spent a good deal of his time searching for the Lost McHenry Gold Mine. Surprisingly, he was a poor bushman and often became lost.

The Hosfalls started a farm on the east bank of the Yukon about five miles below Fort Selkirk and grew good crops of vegetables for several years. He died on or near this stream in about 1935.

Joseph Hosfall's wife was invariably spoken of in superlatives. She was well-liked and respected by everyone who knew her. She was educated at the mission school at Fort MacPherson and 20 years old when she married Joseph. She evidently inherited the best characteristics of both her Indian mother and her New Englander father. She could hunt and shoot, handle boats and canoes, trap, and build cabins as well as most men, and her knowledge of woodcraft and animal life was superior to that of most people. In addition, she was well educated, articulate, modest, and graceful. She did all the traditional tasks of Indian women, such as tanning skins, catching and drying fish, making clothes, and doing beadwork, and she was an excellent cook. The Hosfalls had four daughters.[29] (See also *Gerald Kelly*.)

HOUSTON, A.

Records show that A. Houston purchased a jeweller's business license from the Town of Mayo in 1925.

HOYNE, Martin J. "Joe"

Martin Hoyne is listed at Mayo in Polk's 1923–24 directory. (See *Buildings and Artifacts*)

HUFFMAN, Grant

– Archie Gillespie, *The Mayo Miner*, October 21, 1938

Grant Huffman was born in Glasgow, Kentucky in 1872. He and his brother arrived in the Mayo district in 1900 and located on Duncan Creek. In 1903 Grant owned a roadhouse midway up Duncan Creek. On February 23, 1913, Grant staked the "Mabel" claim, naming it for his daughter, Mabel Huffman. Grant, his partner Jack Alverson, and Grant's nephew Gilbert Rich, were also responsible for developing the Silver King Mine. They put up the first buildings on the claim and cut out adjoining roads.

From left: Cora Rich, Mrs. George Miller (Ellen) with daughter Hilma, Grant Huffman, and Henry Miller. *M. Rich Collection*

Grant Huffman and Alverson mined 59 tons of high-grade, silver-lead ore from the Silver King, the first ever mined in the Mayo district. They shipped the ore in 1914 on the steamboat *Vidette*, and when the ore was smelted they shared the $10,000 profit.

In 1915–16 Grant was still mining and prospecting on Galena Creek near Keno City, but in 1918 he took up a homestead near the Mayo Canyon as well as a pre-emption on the Mayo-Keno road four miles out of Mayo. In the 1920's he became a market hunter for the Treadwell Yukon Company, supplying them with moose-meat, and vegetables harvested from his homestead such as cabbages, carrots and potatoes. In 1930, he left Mayo and returned to Glasgow, Kentucky. He came back to Mayo in the summer of 1936 shortly after the Stewart River flooded the townsite of Mayo, then returned to Kentucky in October. He died at his home in Kentucky on September 18, 1938. Grant Huffman was Mabel McIntyre's father.

HUFFMAN, Virgil "Jim"

Jim Huffman came to the Mayo district with his brother, Grant, in 1900. He built a cabin on

Duncan Creek just above Van Cleaves. Jim left the district probably before 1913 and, according to some accounts, went to Alaska where he married an Alaskan Indian woman and raised a family. It is believed his cabin on Duncan Creek is still standing.

HUGHES, Louis

Louis Hughes was on Independence Creek in the late 1930's.

HUIJBERS, Father Henk

See *Pioneer Accounts*

HUMMEL, Lena

See *Pioneer Accounts.*

HUMPHREY, R. H.

R. H. Humphrey was an independent mining engineer from New York, hired by J. J. McCarthy of the Yukon Gold Company, to assess the Keno Hill properties in 1919.

HUNTER, Alfred & Maggie

Alfred Hunter was the Special Constable at the Old Village in the late 1920's and the 1930's. The Hunters had no children.

HUNTER, Mrs. Lucile

Lucile Hunter is listed as living at Keno Hill in Polk's 1923–24 directory. She had claims on Keno Hill, and prospected them herself. She was a black woman, and her son "Buster" was a Cat skinner. She died in Whitehorse on June 10, 1972, at age 94.

HUTCHINGS, William A. "Bill"

William Hutchings was the Acting Manager and then Manager of the Bank of Montreal in

Alfred Hunter and his wife Maggie in 1930. Alfred was a Special Constable at the "Old Village." *William Hare Collection, Yukon Archives*

Mayo from July 1932 to May 1934. He was transferred from Mayo to Bralorne, British Columbia. In 1937 he became manager of the Bank of Montreal in Dawson City when E. O. Finlaison retired.

HUTTON, Oliver & Irene

Oliver Hutton came to Mayo in 1953 and operated a taxi business. Following that he ran a wood business with Don Buyck Sr., then built a large garage (Lots 1–4, Block 6) and became the White Pass agent. He also made regular trips to Stewart Crossing to meet the Coachway's bus and pick up passengers, freight and dry-cleaning. Mechanics who worked in his garage include Wilf Tuck, Don Curry, William McIntosh, and Jack Besplug. He and his wife Irene continue as Mayo's White Pass agents today (1989). The Hutton children are Joyce, Don, Terry, Darlene, and Dale.

INKSTER, William

He was located in Mayo in the 1920's and was an associate of T. A. Dickson.

IVERSON, Iver

Iver Iverson is listed as a miner in the Mayo area in Polk's 1915–16 directory.

IVES, E. G.

E. G. Ives is listed at the Silver King Mine in Polk's 1923–24 directory.

JACKSON, Thomas Wesley "Tommy"

Tommy Jackson once owned the Royal Alexander Hotel in Dawson, and came to Mayo from Dawson circa 1920. He ran Jackson's Club in Keno in the 1920's and 1930's. This building is now the Keno Mining Museum.

JAHNKE, Albert

Albert Jahnke mined on Haggart Creek from 1908 to the 1920's with John Maynard and Fred R. Gill. He later left the Yukon for Germany.

JEFFREY, Nancy

Nancy Jeffrey was the daughter of William Jeffrey Sr. She married Ray Bird, who was a Non-Commissioned Officer for the Royal Canadian Corp. of Signals in Mayo during the 1940's and early 1950's.

JEFFREY, William Jr. & Thelma

William Jeffrey Jr. was the son of William Jeffrey Sr. He and his wife Thelma lived in Mayo in the 1930's and moved to Whitehorse in the early 1940's. They had a daughter Thelma and a son "Billy."

JEFFREY, William Sr.

William Jeffrey was licensed as a jitney operator in 1928 by the Town of Mayo. He then became the manager of Taylor and Drury's Mayo store from 1929 to 1943. He and his wife had a son, William and a daughter, Nancy.

JENSEN, Charles

– Archie Gillespie, *The Mayo Miner*, December 24, 1937

Charles Jensen was a native of northern Norway who had worked at the Copper King Mine near Whitehorse for several years before coming to Mayo in 1922. He was a well-known figure in the district. Jensen died at the Mayo General Hospital on December 16, 1937, at the age of 54. He was survived by a brother in Alaska.

JENSEN, Pete

Pete Jensen was in the Keno Hill and Mayo areas, and worked for the Yukon Territorial Government road crew circa 1950. He was the father of Johnnie Jensen.

JEWSEU, P.

P. Jewseu operated a jitney in Mayo in 1935.

JIMMY, Charlie & Alice

Charlie Jimmy originally came from the Pelly area. He taught school for a time at the Church in the Old Village. He married Alice Lucas, and the couple's children are Sally, Annie, Betty, David, and Douglas.

Mayo, circa 1920. From left: "Old" Jenny, Dorothy McLaren, Louis Bouvette, and Kathleen McLaren. *D. Walloch (McLaren) Collection*

Lonnie Johnny. *Lucy Peter Collection*

JIMMY, Jenny

Jenny Jimmy was affectionately known as "Old" Jenny. She originally came from Fort Selkirk and settled at McQuesten Village, then later lived at the Old Village on the Stewart River where she was an Elder. Jenny Jimmy had three children: Emma Jimmy who married Grant Huffman, Liza Jimmy who married Robert Hager, and David Jimmy Hager (known as "Big Dave.") Following the death of her daughter Liza and Liza's husband Robert Hager, Jenny raised her grandson Edwin Hager.

JOHANSON, Joseph

Joseph Johanson is listed in Mayo in Polk's 1923–24 directory.

JOHNNY, Doc & Sarah

Doc Johnny came from Carmacks to hunt beaver. While he was in Mayo he met his wife Sarah Davis, daughter of Sophie and Billy Davis, and thereafter remained in the Mayo district.

When the Old Village was abandoned the Johnny, Davis and Malcolm families decided to live at 17-Mile on the Stewart River where, at one time or another they all worked for Louis Brown. In the late fifties when Louis Brown moved across the river to the north side Doc Johnny and his family also moved, leaving the Malcolms and Davis' on the south side.

Sarah and Doc had ten children: Mabel (she died in her late thirties), Jimmy, Dorothy, Alec, Marie, Irene and Susan. The other three children died as infants. Doc Johnny died in 1988.

Dorothy married Carl Charet of Montreal and lives at 17-Mile with their two children, Marcel and Cecelia. Irene lives in Montreal and Marie is in Whitehorse. Susie, Alec and Jimmy live in Mayo.

JOHNNY, Lonnie

Lonnie Johnny's family came from Fort Good Hope in the Mackenzie River district and lived at Lansing Post. His father's name was Johnì, and Johnì had a brother named Lucas. His uncle was Alsän Mases, and Alsän had an older sister named Cecil. He had two sisters: the younger sister was Alphonsi, and the older one was Eliza Farr, who now lives in Dawson.

Both his parents died of gunshot wounds. His father accidentally shot himself with a .22 rifle while hunting rabbits, and Lonnie's mother was accidentally shot in the chest by a man from Pelly while he was cleaning his .33 rifle in a cabin at Lansing.[30]

JOHNSON, Albert

Albert Johnson came to be known as the "Mad Trapper of Rat River." He came by raft down the Peel River to Fort McPherson to Lansing, and then to Mayo. He trapped in Lansing part of one winter. James Mervyn, who operated the Lansing post, had known him to be quiet and kindly.

JOHNSON, Andrew "Andy"

– Archie Gillespie, *The Mayo Miner*, October 30, 1936

Andy Johnson, a Swede, was born in 1860. He went to sea in his younger days, and jumped ship in Canada to work in the coal mines in Nanaimo, B.C. He prospected the interior of B.C., and came to the Stewart River bar gold diggings about 1893.

He prospected the Mayo district in 1896 and found galena on Lookout Mountain (Mount Haldane). This was the first known discovery of galena in the area even though Johnson did not know what it was at the time.

Johnson left the area for Cassiar, B. C. but returned to join the stampede to Dawson in 1898. Johnson returned to the Mayo area and prospected Lookout Mountain (Mount Haldane) in 1905 finding galena at Bighorn Gulch.

He optioned the Lookout Mountain property to Greenfield, Pickering, Cale, Forrest, Anderson, and Maclennan who formed the Yukon Silver Lead Company. He mined one of his prospects on Sourdough Hill on a small scale. He took out approximately 3 tons of ore from Lookout Mountain, rawhided it 2,000 feet to the bottom of Bighorn Gulch, took it by dog team to the McQuesten River and by raft to the Stewart.

He was a good prospector and a fine, earnest, upright man. Andrew Johnson was very well known in the district. He lived in a cabin on Lookout Mountain for some time and spent his last years in a cabin at Minto Bridge. He died at the Mayo hospital on October 23, 1936, at age 76.

JOHNSON, Charles

Charles Johnson and Louis Bouvette staked the first placer gold claim on Carlson Creek (a tributary to Bear Creek) on March 10, 1911. Johnson was mining on Charles Creek in the Mayo area circa 1915–16, and records show that he held a wood dealer's business license in Mayo in 1923.

JOHNSON, Chief

Chief Johnson was the leader of the Stewart River people from the McQuesten Village era until 1925. Chief Johnson was assisted in his senior years by "Tom" "Albert" Alfred Moses who became chief after Chief Johnson's death.

Chief Johnson died at a very old age in the summer of 1925. His daughter, Maisie, was Sam Peter's mother.

JOHNSON, John & Emma

Emma and John Johnson lived at the Old Village on the Stewart River. Sophie Davis was their daughter. Emma Johnson was Chief "Tom's" (Chief Alfred Moses) sister.

JOHNSON, Frederick John "Fred"

"In 1894 the two Garrison brothers found gold on what later became known as Johnson Creek (which enters the McQuesten River just below Ortell's Crossing) but did not mine it. In the autumn of 1898, Fred Johnson, who had been unlucky thus far in the Klondike and had

From left: Jack Pickering and Andy Johnson on top of Lookout Mountain, July 12, 1931. Mayo Lake is in the background. *A. K. Schellinger Collection*

prospected this stream previously, returned and staked a Discovery claim, naming the creek."[31]

Johnson was still mining the creek in 1915, and that year he recovered 47.5 ounces. He also held property on Haggart Creek from 1917 to the mid 1940's.

Johnson was born in England and went overseas with the Yukon Infantry Company during World War I.

JOHNSON, Hans Rudolph

Hans Johnson is listed as quartz mining on Galena Creek in Polk's 1915–16 directory and as a miner in the Mayo area in Polk's 1923–24 directory. The Mayo Mining Recorder's Reports for 1922 and 1929 show that Johnson prospected and mined at the mouth of Ledge Creek during those years.

JOHNSON, Helmar

Helmar Johnson was mining on Quartz creek in the Klondike district circa 1910. He prospected in the Keno area in the 1930's and in 1935 Helmar operated a jitney in Mayo.

JOHNSON, Ole

Ole Johnson is listed in the Mayo area in Polk's 1923–24 directory.

JOHNSON, Peard

Peard Johnson was a Treadwell Yukon employee at the Silver King in the 1930's.

JOHNSTON, A. W. "Ash"

"Ash" Johnston was a consultant for United Keno Hill on the Hector-Calumet circa 1949.

JOHNSTONE, Alfred

Alfred Johnstone is listed at Keno Hill in Polk's 1923–24 directory.

JOHNSTONE, James Kennedy

James Johnstone, born in Ottawa, Ontario, played on the Dawson City hockey team that competed for the 1904 Stanley Cup. He went overseas with the Black Contingent during World War I.

He returned to become the Mayo Postmaster in the early 1920's and continued in that position until approximately 1930. His post office was the small log cabin that stood on the corner across the street from where the North Star Motel is today, near the Stewart Valley Enterprises building.

Klondike River Camp 1918. From left: Otto Kastner, Mr. Dever, Hazel Kastner, Mrs. Dever, Ethel Kastner, Roma Morisette. *A. K. Schellinger Collection*

JONES, Alma B.

Alma Jones was the matron at the Mayo General Hospital in the 1930's.

JONES, Edward K.

Edward Jones staked the Wigwam claim on May 4, 1921, and the Mastiff claim on July 10, 1921. It is believed Jones also staked the Huxley and the Midway and possibly other hard rock claims in the district.

JONES, Mrs. Nora

Nora Jones was a nurse at the Mayo General Hospital circa 1923.

JONES, Richard W.

Richard Jones is listed at Mayo in Polk's 1923–24 directory.

JOYAL, Eli

Eli Joyal is listed as a miner in the Mayo area in Polk's 1915–16 directory and in Polk's 1923–24 directory.

JUROVICH, Milo & Annie

See *Pioneer Accounts*.

KALDAL, George

George Kaldal was a Treadwell Yukon employee at the Elsa in the late 1930's.

KARDANOFF, F. M.

F. M. Kardanoff operated a barber shop (one chair) in Mayo in 1929.

KASTNER, Otto F.

Otto Kastner mined on Minto Creek with James A. Scougale, circa 1910.

KAZINSKY, George & Pearl

George Kazinsky was a Treadwell Yukon employee at the Silver King in the 1930's. His wife Pearl (née Kimbel) played the piano at dances.

KAZINSKY, Mrs. J. P.

Mrs. Kazinsky was a pioneer Dawson woman. She was the mother of Johnnie, Louis and George Kazinsky.

KAZINSKY, Johnnie

Johnnie Kazinsky worked at Wernecke Camp and ran the Mayo power plant in 1930's. He later worked at Elsa and Calumet as an electrician and power plant operator. He and his wife retired to Sidney, B.C.

KAZINSKY, Louis

Louis Kazinsky was born of pioneer parents in Dawson City in 1908. In 1926, he and Bon Kunze walked from Dawson to Mayo and went to work for Treadwell Yukon. He had a motorbike with a sidecar which he called the *Flaming Flamingo.*

He joined the army in 1939 or 1940 and served in both the Italian and European campaigns. After the war, he worked for Keno Hill when those mines were reopened. From 1948 until the mid 1950's, Kazinsky worked for Fisher Service in Mayo, and then for N.C.P.C. as a hydro operator until 1973. During his retirement, he spent his summers in Yukon and winters in Calgary, where he pased away in 1976.

KEATING, Thomas

– Archie Gillespie, *The Mayo Miner*, January 28, 1938 and February 11, 1938

Thomas Keating was a pioneer of the Klondike gold rush. He was born in Boston, Massachusetts and was of Irish-American parentage. He came from Dawson City to the Mayo district where he worked on Highet Creek for Elmer Middlecoff from 1916 to 1918.

He made his home in Keno, but spent his entire life in the northland hiking, prospecting, mining, and hunting. During his years in the North, he was always noted for his powerful physique and extraordinary strength. Once, in his younger days, he swam across the Yukon River near Dawson, a feat accomplished by very few men.

In the summer of 1933, this sourdough was also the main player in one of the most harrowing experiences recorded in this part of the country. In the company of Corp. McAskill, formerly of the R.C.M.P. Keno post, Thomas Keating went on a prospecting trip to McNeill Gulch, about 12 miles from Keno. While McAskill was a considerable distance away from Keating and hiking up the opposite side of the creek, Keating was suddenly attacked by a giant grizzly bear. In the struggle that followed, Keating was almost killed. McAskill quickly struck out for Keno to procure aid in packing the injured man out, while Keating lay unconscious until the rescue party returned several hours later.

Dr. Duncan, Mayo physician at that time, rushed by car to Keno and then to a point five miles down the trail where by this time, the injured prospector had been made as comfortable as possible in a cabin there. He was then rushed to Mayo General Hospital by Alex McCarter where an emergency operation was performed. It was thought that he would not recover, but his hardiness stood him in good stead. He gradually did recover, although it was almost a year before he was able to leave the hospital to return home to Keno City.

Even after this horrifying experience and despite his crippled condition, Thomas Keating never lost his thirst for adventure. The lure of gold and silver still beckoned him, and right up until the time of his death he often talked about the "finds" he was determined to make from further prospecting journeys. Thomas Keating died in his mid 60's in his cabin at Keno City on January 26, 1938. He had a sister in Lawrence, Massachusetts, and a nephew.

KELLY, Brian

Brian Kelly was born on Highet Creek. He was one of Phillip Kelly's sons and a well-known Mayo character. He cut wood, trapped extensively and worked at the mines. His career also took him to Dawson when work was scarce in Mayo. He married Alex Profeit's widow, Ellen. The couple later moved to Whitehorse. He moved to Vancouver after Ellen died and eventually remarried. Kelly and his wife Dorothy went back to Mayo in 1966 and then worked at Carcross before returning to Vancouver. Kelly died in Vancouver circa 1972.

KELLY, Gerald "Gerry" & Rose

"Gerry" Kelly was one of Phillip Kelly's sons. He married Rose Hosfall, granddaughter of Jack McQuesten. Their children are: Joy Wheeler (deceased); Eileen Scanlon and Vivian Harris who live in Carmacks with their husbands; Gloria Seeley (deceased); and Ken, the only son (deceased).

He trapped, cut wood, operated the ferry at Pelly, and was the operator for Yukon Telegraph at Stewart River Station. Kelly died of cancer in the fall of 1954 and is buried in the Mayo cemetery. Rose later married Mr. Nielson and after his death married Joe Page. They lived in Whitehorse. Page died and Rose was killed in a car accident in the 1980's.

KELLY, Phillip M. & Kathryn

Phillip Kelly stampeded into the Klondike Gold Rush but missed out on any big finds. He came to the Mayo district and mined for a time on Highet Creek near the head of the canyon. Phillip became Mining Recorder at Minto Bridge and remained in that position from 1913 to 1915. The Kellys moved to Mayo when the Mining Recorder's office was moved there in 1914. In Mayo, the duties of postmaster were added to his responsibilities as Mining Recorder. The Kelly's three children, Phyllis, Gerald and Brian, attended the Mayo Landing School in 1913 which was the first year the school was opened. They moved back to Dawson City circa 1919. The Kellys lived in British Columbia in their senior years.

KENDI, Reverend Julius & Persis

Reverend Kendi was an Anglican minister and the first Indian minister in the Mayo district. He was an Anglican catechist at the head of the Peel River from 1911 to 1913, and at the Old Village in Mayo from 1915 to 1918. He was ordained Deacon at Moosehide, a short distance down river from Dawson City, by Bishop Stringer in 1918 and continued as minister at the Old Village until 1928. From 1929 to 1940 he served at Rampart House, and in 1941 he returned to the Old Village at Mayo where he remained until 1946. In 1947 he returned to Rampart House, but in 1948 he fell ill there. He was flown to the hospital in Dawson, where he died in July of 1948. Reverend Kendi is buried at Moosehide. (See his daughter Mary Hager's story in *Pioneer Accounts*).

KEOBKE, Donald & Carrie

Donald Keobke lived in Mayo, Dawson and Whitehorse from 1937 to 1941. He worked for Klondike Airways Ltd. on the overland stage with Frank Buckway and at Mt. Freegold. He and his wife Carrie live in White Rock, B.C.

KEOBKE, Errol F. "Slim"

Errol "Slim" Keobke came to Whitehorse in 1927. He was a master mechanic and worked for the copper mines, then worked as an overland stage driver for T. C. Richards of Klondike Airways Ltd.

He and T. C. Richards were partners in

Laforma mines at Mt. Freegold in the Carmacks area in 1938, where they established a 20-man camp. Keob-Rich Alluvials made and took out the first gold brick from that area. Mayo prospectors A. J. "Hard Rock" MacDonald, Malcolm Ross and Bill Gray stampeded into the Mt. Freegold area also at this time. "Slim" married Helen Murich of Whitehorse and later Thelma Jeffrey.

He left the North in 1953 and died in Victoria in 1983.

KEOBKE, Neil & Daisy

Neil Keobke and Daisy Buckway were married December 29, 1928. Neil operated grain elevators in several Alberta towns. When the Depression hit, only senior men remained with the elevators. When his brother, Errol "Slim" Keobke invited Neil to come to Whitehorse to work for T. C. Richards as an overland stage driver he was not long in accepting. He arrived in Whitehorse October 23, 1934. Daisy, her brother Frank and two children, Neille (Bucky) and Mary Joanne joined him in April 1935. The family, accompanied by David McLean, moved to Mayo in March 1936. They moved via Klondike Airways, Neil driving the last stage of the season. Mr. and Mrs. Andison Sr. and Cpl. and Mrs. Tidd were also passengers. The trip took three days.

Upon arriving in Mayo, Keobke immediately started work, leaving his family in Binet's Hotel. Binet's two-storey house was vacant, sadly in need of repair, but they moved in. Binet supplied the material from his store, while the Keobkes supplied the manpower. He came to call and after looking at their handiwork, said, "You've got a nice place here, I'll have to raise your rent." And he did! The clothes closet in the bedroom upstairs was just large enough to hold one hanger. It was rumoured that when asked about it, Binet said, "I've only one suit." The Keobke's fourth child, Donald Francis was born in the Binet house with the help of "Ma" Miller.

Keobke worked as Government road foreman from 1936 to 1942. He supervised the building of the road from the Mayo cut-off to the Silver King, built summer roads to Haggart Creek and Dublin Gulch, and was involved in the construction of the South McQuesten bridge. In winter he maintained the road to Keno Hill mines, often in all hours of the night, rushing to thaw out rampant glaciers creeping over the road.

When RCMP Bill Sutherland made his last

Kimbel's logging truck, circa 1925. Standing in front of the truck, from left, Ed Kimbel, George Besner and Jack Sullivan. *Besner Collection*

patrol from Mayo by dog team, Keobkes huge white, part husky dog "Rip" was in the team.

Keobke went to Whitehorse to work for the White Pass. The family followed later, flying to Dawson and from there up river on the steamer *Keno* to Whitehorse. The river was very low, as it was September, and they took five days to make the trip. A fifth child, Linda Irene, was born in Whitehorse in October, 1950. Neil and Daisy left the Yukon in October 1974 and now live in Sardis, B.C.

Daisy Keobke recalls that when George Ortell came to Mayo from his trap line, he would stop at their house with an offering of beaver meat which he referred to as, "The finest meat you can find," and gum for the children. The Keobke's daughter, Joanne, had two black warts on her cheek and every time George saw her he would say that he could painlessly remove them. One day, when about nine school children were going to see George to have their warts removed, Joanne went along. In describing George's method, Joanne remembers that he rubbed something on each offending growth and chanted in a strange tongue. The children's warts all disappeared at about the same time.

Ed Kimbel. *L. Cheatham (McLaren) Collection*

KERR, George

George Kerr is listed as a miner in the Keno Hill area in Polk's 1923–24 directory.

KERR, John "Jock"

Jock Kerr was a retired R.C.M.P. sergeant who came to Mayo in the 1950's as Territorial Agent and Justice of the Peace.

KETALA, Gus

Gus Ketala was the official greeter and clerk at Binet's hotel for many years. He retained his duties after the hotel was bought by Jim Mervyn.

KIMBEL, Edward Elgin "Eddie"

The Kimbel brothers Eddie, Carl and half-brother Dick operated the Kimbel Bros. sawmill in Mayo in the 1920's. The Kimbel brothers were sons of a Mississippi steamboat captain. Eddie Kimbel went to Seattle during World War I to work in the shipyards as a caulker, then went to Iditarod country in Alaska before coming to Mayo in 1920 to caulk the dredge at Highet Creek for the Highet Dredging Company.

His other skills included boat building and carpentry, and in addition to managing the Kimbel Bros. sawmill business, he hauled wood to the silver camps, supplied a transportation service between Mayo, Keno and Galena Hill, and had the mail haul between Mayo and Keno in the 1930's. He operated the sawmill until the 1950's when he sold it to United Keno Hill Mines Ltd.

His brother Carl drove a truck for a time for Treadwell Yukon in the 1930's. Carl Kimbel married Wlodga Kondroski. Dick Kimbel married Isabel Profeit and Ed Kimbel married Mrs. Lou Hartnell and they had a daughter, Gail. Lou

Hartnell was a Palmer from Saskatchewan. During the Depression the Palmer brothers George, Wesley and Elmer, came from Melfort, Saskatchewan to Mayo to work for Eddie Kimbel. With the Palmer brothers came Glen Rear (who married Ella Profeit), and the Morberg brothers, Raydor and Roy. "Raider" Morberg married Maisie Profeit.

The Moens, Linns, Megers and many more came to Mayo from Saskatchewan during the Depression and worked for Kimbel's Sawmill.

When Ed Kimbel retired he went to live with his daughter, Gail in Prince George, B.C., where he eventually died. Gail (Pelland) continues to live in Prince George.

KING, Harold

Harold King was a bull cook at Northern Constructions camp during the building of the Mayo dam. He also worked at the Elsa mill. King is buried in the Mayo cemetery.

KINMAN, John August

– Archie Gillespie, *The Mayo Miner*, April 16, 1937

John Kinman was born in Stockholm, Sweden in 1857. He came to America as a cadet in the Swedish navy and came North with the Klondike gold rush. He prospected and mined all over the Mayo silver district long before Keno Hill was struck.

Known as the "Mayor of Keno," this old-timer was a colourful figure in the silver camp. Short and stocky, he always wore a Van Dyke beard and one of his pet expressions was, "Hello, neighbour."

Along with his partner "Rocky Mountain Bill" Elliott, Kinman was the original staker of the Keno townsite. He was also among the first to sell silver property on Keno Hill when Keno Hill Ltd. began operations there. He died in his cabin at Keno on April 11, 1937.

KINSEY, Mrs. Agnes J.

Agnes Kinsey was married to Clarence E. Kinsey, the well known Klondike gold rush photographer. Olive Kinsey was their daughter. Her brother Bobbie Fisher and sister Maria Fisher also lived in Mayo. Kinsey owned a property on Keno Hill with her brother, and by selling this property during the 1950's boom, she was able to retire to the State of Washington where she died. Olive Kinsey Powers still lives in Olympia, Washington.

KINSEY, Clarence E. & Agnes

Just after the Klondike gold rush Clarence Kinsey and his brother Clarke, excellent industri-

Field Creek Roadhouse dinner party, 1927 or 1928. From left: Bon Kunze, Fred ?, "Sandy" Jack MacDonald, Elmer Gaundroue, Ed Runacres, George Kazinsky, Mrs. Agnes Kinsey, Ed Kimbel, Ed Kunze, Walter Teare, Clarence Kinsey, Carl Kimbel, and Bill West. *Ed Kunze Collection*

Clarence Kinsey on a motorbike, 1928 or 1929. *Ed Kunze Collection*

al and logging photographers, left Seattle for Grand Forks, Yukon. There, the brothers had a photographic studio as well as several mining interests.

In 1906 their partnership came to an end when Clarke, his wife Mary and their two sons left for Seattle to continue their photographic business. Clarence, his wife Agnes and their daughter Olive stayed and mined throughout the Territory—living in Dawson, Mayo and various outlying creeks.

Clarence Kinsey mined on Haggart Creek circa 1915 with partners, Charles E. Merriman and John Mawhinney.

The classic photographs taken by the Kinsey brothers of the Yukon's Klondike years have become invaluable as historical records. Some of these photographs have appeared in Norman Bolotin's books *Klondike Lost* and *Klondike Scrapbook*.

KITAGAWA, Dick

Dick Kitagawa and Vic Foley ran the Northern Lights Hotel in Keno for owner Bobby Greaves in the 1950's.

KITZER, Lloyd E.

Lloyd Kitzer is listed as a miner in the Mayo area in Polk's 1915–16 directory.

KLIPPERT, "Bert" & Verna

See *Pioneer Accounts*.

KNORR, A. N. & Margaret

A. N. Knorr came to the Klondike in 1898. He raised money in Dawson to prospect the Bonnet Plume River country, and Knorr Creek (a tributary to the Bonnet Plume River) was named for him. He later married Margaret, a woman from Old Crow. The Knorrs lived in the Bonnet Plume River country in complete isolation between 1905 and 1910, but they disappeared without a trace after that. It is believed that the cabin on Margaret Lake (near the junction of Knorr Creek and the Bonnet Plume River) was built by the Knorrs in about 1907.[32]

KNUTSON, George

George Knutson was a blacksmith who came from Atlin, B.C. with his wife and family to work for Treadwell Yukon at the Elsa in 1938.

KOENIG, Bernice

Bernice Koenig was a Mayo beauty who participated in the 1937 Miss Yukon contest held in Dawson City. She returned to her home in Melfort, Saskatchewan in 1938.

KONDROSKI. Mrs. Anthony

Mrs. Kondroski was a Polish widow whose

husband, Anthony, was killed in an accident at the Sadie on August 5, 1927. He was working on the 200 level when he was hit by a fall of rock from the back of the drift.

Mrs. Kondroski lived in a house in Keno, supplied by Keno Hill management, for many years. She grew excellent vegetables and when the Elsa commenced operation she started a laundry business.

In the 1930's she lived in the two-storey house next to the Roman Catholic Church in Mayo. Her daughter, Wlodga, who was a talented pianist, married Carl Kimbel.

KRANTZ (KRAINTZ), John

John Krantz mined and prospected in the Mayo district circa 1900. He left Mayo circa 1918–19 and in 1922 was working in a coal mine at or near Ladysmith, B.C. John owned a cabin near the old Mayo hospital site.

KUNZE, Edwin "Ed"

See *Pioneer Accounts*.

LADD, Thomas H. & Eva

Thomas Ladd was a butcher in Mayo in the 1920's. His wife was Eva Ladd.

LADUE, Joseph

Joseph Ladue was a pioneer Yukoner who prospected the lower reaches of the Keno Ladue River (which flows north of Keno Hill to the Stewart River) in the 1880's. He bar mined on the Stewart River in 1885, and he was a partner of Arthur Harper for a time. (See *Early Bar Mining.*)

Joseph Ladue also staked the original 160 acre townsite of Dawson City.[33]

From left: Eva Ladd, Theos Whitney, and teacher Dorothy Hoggan, 1923. *Ralph Rogers Collection*

Picnic on Keno Hill. From left: Hilma Miller (rear), Wlodga Kondroski, and Evelyn (Middlecoff) Lee. *G. & E. Lee Collection*

LAKE, John

John Lake is listed as a prospector in the Mayo area in Polk's 1915–16 directory.

LAKNESS, Conrad C. "Connie" & Violet

"Connie" Lakness was a road foreman in Mayo in the 1950's. He was originally from

Saskatchewan and married Violet Somerton, a member of the Whitehorse Somerton family.

LAMB, Albert E. "Bert"

Bert Lamb was a pioneer Dawson lawyer who came to Mayo circa 1920. He lived in Keno, where he continued as a lawyer, and also held claims on Keno Hill. He worked the Gambler claim on Faro Gulch with his son in 1938. His son William worked for Treadwell Yukon in the 1930's, then left Mayo in 1939 to join the Air Force. The Lambs also had a daughter, Alberta.

LANDERS, George

George Landers was a baker for Treadwell Yukon in the 1930's. He became chef for Otter Creek Mines in Atlin, B.C., in 1938.

LANG, Bruce

Bruce Lang was mine superintendent for United Keno Hill Mines Ltd. in the 1950's, then later became production manager.

LANGHAM, W. J. "Bill"

"Bill" Langham, originally from England, was a prospector and miner on Keno Hill in the 1920's. He either staked, or was associated with, the Wolf, Lynx, Rita, Hard to Tell, and Alpha claims. Langham also worked in the office for Treadwell Yukon in the 1930's and later lived in Carmacks.

LANGRAM, Joseph & Lydia

Joseph Langram is listed as mining on Highet Creek in Polk's 1915–16 directory and he and his wife Lydia were still mining on Highet Creek in the 1920's.

LANING, O. S. & Laura

O. S. Laning operated a general merchandise store at Keno Hill in the 1920's. He also operated a jitney in Mayo in 1925. He and his wife, Laura B. Laning, owned a number of lots in Mayo. He was an employee of the Northern Commercial Co. Mrs. Laning died circa 1926. O. S. Laning died in 1930 and is buried in the Mayo cemetery.

LANSING, Samuel

"Samuel Lansing was one of the first prospectors in the Yukon Territory and mined the bars of the Stewart River in 1883. He prospected the Lansing River in 1883 when it received his name. He prospected widely and was one of the first at the Fortymile in 1886. He was moderately successful in the Klondike and was, for a number of years, the manager of the Harper and Ladue Co.'s operations there. He retired in 1926 to Sumner, Washington."[34]

LANYON, Art

Art Lanyon was the first teacher at the Elsa school when it opened on September 6, 1938. Lanyon was originally from Vancouver and attended Normal School there at the same time as did Hilma Miller of Mayo.

LARSEN, John

John Larsen is listed as a miner at Keno Hill in Polk's 1923–24 directory.

LARSEN, Louis

Louis Larsen worked for Treadwell Yukon and also trapped in the McQuesten Valley. Larsen died in 1942 and is buried in the Mayo cemetery.

LASKA, Joseph

Joseph Laska mined in the Mayo and Minto Lake area from circa 1901 to the 1920's. E. Middlecoff's Time Book shows that he worked for him on Highet Creek in 1918. In the 1930's he had a cabin at the summit of Lookout Mountain (Mount Haldane) on the old road, as well as at Minto Bridge. In 1943 he mined and prospected at the head of Highet Creek. He also prospected on Johnson Creek.

LAWRENCE, John C.

John Lawrence is listed as operating a general store at Stewart River in Polk's 1923–24 directory.

LeBLANC, Henri

Henri LeBlanc was a trapper on the McQuesten River, who from the 1920's to the 1940's lived in a cabin on the east side of Mayo townsite. During the Depression he and Dan Ferguson worked out 14 tons of ore on Alex Nicol's ground on Keno Hill.

LeBOEUF, Albert

Upon finding paying gold in 1912 on the left fork of Scroggie Creek in the Yukon, Albert LeBoeuf renamed this creek Walhalla Creek after the gold fields of Walhalla in the State of Victoria, Australia, where it is believed he once prospected. Walhalla Creek flows into Scroggie Creek which is a tributary of the Stewart River.[35]

He mined on Sharpe Creek circa 1914.

LeBoeuf was born in Canada and went overseas with the Yukon Infantry Company during World War I. It is also believed that this is the same Albert LeBeaut who is listed as being at Stewart River in Polk's 1923–24 directory.

LEE, Charles Alexander

– Archie Gillespie, *The Mayo Miner*, December 10, 1937

"Charlie" Lee was born in Boston, Massachusetts. He came to Mayo during the early days of the silver camp. In his early years he was a well-known logger.

He worked for the Treadwell Yukon Co. on Keno and Galena Hills for many years and was a highly respected member of the community. "Charlie" Lee died at St. Mary's Hospital in Dawson City on December 6, 1937, at age 83.

LEE, R. Gordon "The Duke" & Evelyn

Gordon Lee was born in England and as a young man followed the footsteps of his uncle Jack Lee to Dawson City. He came to Mayo where he was an accountant for Treadwell Yukon at Wernecke Camp and also at Elsa. Lee married Evelyn Middlecoff in 1924 and they had two children, Doreen and Katherine, both born in Mayo.

Lee was very active in amateur theatre and was frequently in demand for musical entertainments and weddings because of his trained tenor voice.

He was a partner in the Homestake property on Bunker Hill. During World War II Gordon joined Canada Customs and later became a member of the Yukon Council for the Mayo district from 1940 to 1943. He was very active in civic affairs and the Conservative Party.

His full name was Richard Gordon Lee, however his friends also knew him as "The Duke." Gordon and his wife, Evelyn, moved to Whitehorse where they owned and operated a store known as the Kee Bird. After they left the Yukon, they retired to Brentwood Bay on Vancouver Island, B.C.

Gordon died in the late sixties and Evelyn lives in Victoria, B.C., as does Doreen Doidge. "Katy" Bacon is at present [1989] in Yugoslavia where her husband, Terry Bacon, is the Canadian Ambassador.

ABOVE, Gordon and Evelyn Lee with their firstborn, Doreen, circa 1929. *G. & E. Lee Collection*

BELOW, group picture outside St. Mary's residence, Mayo. From left: Evelyn Lee, Gordon Lee, Jim Fairborn, and the Rev. Valentine with his dog "Maggie". Front row, from left: Katherine and Doreen Lee. *G & E. Lee Collection*

LEFEBVRE, Narcisse Alfred "Fred"

Fred Lefebvre and Isaac Mallette found gold on Secret Creek on September 7, 1911, and they mined their claims until August 1914. Fred operated a sawmill in Mayo from 1900 until the 1940's with his partner Gene Binet.

Bob Palmer bought Binet's interest in the sawmill in 1938 and the name of the business was changed from Binet and Lefebvre to Mayo Lumber Mills. The sawmill was located on the river between the townsite and the mouth of the Mayo River.

LEFEVRE, Jean Baptiste "John"

John Lefevre mined on Dominion Creek in the Dawson district in the early days (circa 1908). He acquired land near Gravel Lake in the 1920's. He is listed in the Stewart River area in Polk's 1923–24 directory. Lefevre was the grandfather of Virginia, Doris, Alice, Adele, Evelyn, Olive, Frank, and Bernice Burkhard of Dawson City.

LEIVEILLIER, Louis

Louis Leiveillier is listed as a miner in the Mayo area in Polk's 1915–16 directory.

LELIEVRE, Mr. & Mrs. Joe

Joe Lelievre was a pioneer rancher and wood contractor in Mayo. Joe was known in Mayo as "Joe Rabbit." ("Lelievre" is a French word for rabbit.) He lived on a parcel of land at the northwest corner of the Mayo-Keno Junction and had another parcel on the north side of the airport. He grew potatoes for sale in town and also made homebrew. Mrs. Lelievre was a pioneer women of the North land, particularly the Dawson district.

Their son "Bunny," Charles F. Lelievre, was a member of the Government Telegraph Service staff in Dawson. "Bunny" married Alice Burkhard, third eldest daughter of Mr. and Mrs. C. F. Burkhard of Dawson City, on October 27, 1938, in Dawson's St. Mary's Church.

Mrs. Joe Lelievre died suddenly, enroute to Whitehorse, aboard the *Keno* just shortly before the steamer arrived in Whitehorse on June 8, 1937. (Archie Gillespie, *The Mayo Miner*, June 12, 1937)

LESPERANCE, N. A. & Ruth

N. A. Lesperance was a welder in Mayo in the 1920's. Ruth Lesperance was his wife.

LESPERANCE, Alex

Alex Lesperance, a French Canadian, was born on January 21, 1856 and came to the Yukon in 1898. During the period 1900–1920 he ran the

Hunting party. From left: Three Cantin cousins, Oscar Letourneau, Evelyn Middlecoff, Cora Rich, and Alex Nicol. Front row: Elmer Middlecoff Jr. and unknown. *M. Rich Collection*

The Hotel Royal Alexandra, Mayo. *L. Cheatham (McLaren) Collection*

inn at Glacier Creek in the Boundary placer camp and during the 1920's and 1930's he was the undertaker in Mayo. He was still living in Mayo when he celebrated his 83rd birthday in 1939.

LETNES, Lorns

Lorns Letnes is listed at Stewart River in Polk's 1923–24 directory.

LETOURNEAU, Gideon

Gideon Letourneau had mining interests on the Sixtymile River circa 1914. In the 1920's, Gideon was doing carpentry work and living in Mayo. He was Oscar Letourneau's brother.

LETOURNEAU, Oscar

From the *Dawson Daily News—Yukon Silver and Gold Fields Edition, 1922*

"Oscar Letourneau was a pioneer of the Dawson, Ruby, Mayo and other northern camps, and always operated on a large scale. For a long time he was engaged in mining on Hunker Creek, near Dawson, and also was in business in the gold metropolis before coming to Mayo. In this district he mined on Duncan Creek extensively, and when he entered the hotel business in Mayo he built one of the largest structures in the city, and gradually enlarged it, until he had one of the largest blocks in Yukon. The Hotel Royal Alexandra stood three stories high, 100 feet deep and had a frontage of 60 feet. The building was higher than any other in Mayo and was a conspicuous landmark from far up and down the river and from inland points. The building was finished and furnished throughout in modern style, had steam heat in all parts, including the cafe which was conducted in connection with the hotel, and the general store in the same building. A large cellar for storage of perishables and freezeables was under the block, and a fine electric plant in the basement supplied the entire block with electric lights, while water tanks in the attic provided fire protection and water for all purposes, and a sewer to the river gave first class drainage."

"Eddie" McKenzie was Letourneau's partner in the general store in 1922. In addition to dry goods the store had a billiard table and they sold cigars and tobacco.

Letourneau was one of the founding members of the Mayo Chapter of the Yukon Order of Pioneers.

He always led the Grand March at New Year's Eve dances. He delighted people with his gifts of maple syrup which he had shipped from Quebec. The Hotel Royal Alexandra eventually burned down but he remained in Mayo until the 1950's when he returned to Quebec.

LEUILLIER, Rene

Rene Leuillier is listed as quartz mining on Galena Creek in Polk's 1915–16 directory.

LEVAC, Raoul "Bob" (also spelled LEVACE)

Raoul "Bob" Levac operated a general store and trading post at Fraser Falls from about 1915 to 1931. Steamboats delivered supplies to the Fraser Falls post after first stopping in Mayo.

In 1937 Fred Swanson took over the post and Levac moved to Mayo and lived in one of the waterfront cabins upstream from the Anglican Church. During the 1937 season he prospected the Hess River with Ted Skonseng and Arthur Turmil from a base camp at Kalzas Lake.

LEVY, Joseph

Joseph Levy mined on Dublin Gulch circa 1903 to 1909.

LINDAHL, Mrs. Augusta

Augusta Lindahl is listed in Mayo in Polk's 1923–24 directory.

LINDQUIST, Magnus

Magnus Lindquist was a partner in the Little Guggs Mining Co. which comprised Rodolph Rosmusen, George Miller, Gervais Godbout and Charles Rockney. They operated on Highet Creek from 1910 to the early 1920's and it is estimated that they took out $106,000 in 1910–11.

Their partnership was nicknamed "Little Guggs" because of the large number of men they employed. (The Guggenheims whom they were named after were big operators in the Klondike at the time).

The "Little Guggs" tried steam equipment on Highet Creek but when it didn't work out they ground-sluiced the top dirt and then used pick and shovels to dig the pay gravels which they transported to the sluice box in wooden wheelbarrows. The Highet Creek Dredge began operation on their property in 1919.

Magnus and his brother Ole came to the Yukon in search of their father whom they never found. They remained in the Yukon and prospected on Morrison Creek and mined on Highet Creek in the Mayo area. Ole Lindquist also worked for Elmer Middlecoff on Highet Creek in 1918.

LINDSTROM, Adolph

Adolph Lindstrom prospected the McQuesten River area in 1898 with Rodolph Rosmusen and mined placer gold near the mouth of Rodin Creek for a time. He continued to prospect the Stewart River area in the 1920's, and in 1938 he was on

"Bob" Levac's Trading Post at Fraser Falls, 1922. Part of this building is still standing (1989). *Wm. Hare photograph, Ralph Rogers Collection*

Rosebud Slough.

LINDSTRUM, Berger

– Archie Gillespie, *The Mayo Miner*, October 9, 1936

Lindstrum arrived in Mayo on July 29, 1936. He was 35 years old. He drowned in a boating accident at Fraser Rapids on October 3, 1936. Berger Lindstrum, Ted Skonseng and Syd Palmer were on their way up river to the Kimbel wood camp in a launch. About four and one half miles above the falls, there is a small island in the rapids some 15 feet from shore. The water hits this island with terrific force and forms a solid wall of water. It is a treacherous part of the rapids. The three men were lining their launch over this bad stretch and were nearly at the end of the bad water when the launch began to fill with water and swamp. Despite the frantic calls of Skonseng to "stay with the boat," Lindstrum, who was a good swimmer, jumped off the far side of the launch and was instantly swept under by the terrific current. Meanwhile the line was steadily loosening as the boat careened downstream. Palmer and Skonseng were able to leap from the near side of the launch and make it to the shore.

LINN, Roy Matson & Hazel

Roy Linn was from Melfort, Saskatchewan and worked for the Kimbel's sawmill in the 1930's. On March 25, 1939 he married Hazel Jaklyn Moen of Melfort, Sask., in Mayo, at a double ceremony with Marion Lindsay and Adolph Meger.

LISKE, Berthold "Bert"

Bert Liske came to Canada from Germany in 1951 and worked his way across the country. He arrived in the Mayo district in 1956 to work as an underground miner for United Keno Hill Mines. In 1958 he spent a year in Ontario and then came back to the Mayo district and never left again. He began placer mining with Bill McCoomb on Ledge Creek in 1963.

In 1967 he contacted George Reynolds, who was living in Penticton, B.C. at the time, and bought George's three claims on Ledge Creek. Liske has placer mined on Ledge Creek since that time.

He spent two years in France during his early years and speaks French fluently.

Joseph "Joe" Longtin, Mayo, Yukon. *G. A. McIntyre Collection*

LONGTIN, Joseph Arsene "Joe"

"Joe" Longtin was originally from Quebec, and is said to have come to Mayo from the Fort Selkirk-Pelly Crossing area. He staked a claim on Partridge Creek (five miles below the mouth of the McQuesten River) on November 13, 1908.

Beginning in the early 1920's, he and his horse, Eli, delivered water to homes in Mayo and continued to do so for many years. As Eli got older he had days when he did not want to pull the water wagon and as Longtin would not use force, the town went without water.

He also picked up garbage and ran a transfer business which meant he had the contract to deliver the boat mail from the dock to the post office. He was also the fire chief in Mayo for a number of years and was the school janitor in the 1940's–succeeded by George Miller.

Longtin was also very fond of cats and had dozens of them in his cabin. He even had a cat stairway up to a hole with a swinging door in the top of the cabin wall over his bed so that his cats could come and go as they pleased.[36]

He was a strong supporter of the Mayo Tennis Club which he joined at the organizational meeting in 1926 and stayed a member until 1938.

Longtin left Mayo and retired to Vancouver, B.C. where he lived at the Ambassador Hotel.

LOVDAHL, Herb

Herb Lovdahl was a shift boss for Treadwell Yukon in the 1930's.

LOVELAND, Daniel

Daniel Loveland is listed in the Stewart River area in Polk's 1923–24 directory.

LUDE, John "Jack"

Jack Lude was born in Stuttgart, Germany. He was a mechanic and came to this district from the U.S. Lude bossed the 30-man crew at Wernecke Camp while master mechanic Ed Barker was away in 1925. A few days after Barker returned, Lude hiked up to the old workings on top of Keno Hill with F. Herbster, on September 13, 1925, where he suffered heart failure and died at the Shamrock camp. He is buried in the Mayo cemetery. At the time of his death, Jack Lude was 43 years old and had a wife, Augusta ("Gussie"), and two sons, Edward and Clarence, in San Francisco, California.

LUSK, Isaac

Isaac Lusk was a freighter who carried the mail between Dawson and Minto Bridge for many years. He was a Yukon Councillor for South Dawson from 1912 to 1915.[37]

LYNCH, Thomas

Thomas Lynch is listed at Keno Hill in Polk's 1923–24 directory.

LYONS, James D.

James Lyons is listed at Keno Hill in Polk's 1923–24 directory.

MACAULAY, Norman D.

Norman Macaulay is listed in Polk's 1915–16 directory as owning and operating a hotel and saloon in Mayo. He also owned (or applied to purchase) Lots 10, 11, and 14 in Block 12 in Mayo in 1915. Macaulay's involvement in Mayo is uncertain as the Yukon Archives have no record of Macaulay ever being in Mayo, and his life in Yukon is fairly well documented.

MacDONALD, Alexander

Alexander MacDonald came to the Yukon from New Brunswick. He was one of the first men to prospect the Mayo region. In 1887 he found and named Mayo Lake and Mayo River after Alfred H. Mayo of the McQuesten, Mayo and Harper partnership. A solitary traveller, he was the first non-native to ascend the Beaver River and enter the watersheds of the Wind and Peel Rivers from the south. He found gold showings on and named Rosebud Creek. It is believed he took

Gordon Dickson and Allan John "Hard Rock" MacDonald just back from a prospecting trip to Kathleen Lakes. *G. A. McIntyre Collection*

Jim MacDonald meeting H.R.H. Prince Philip, Mayo Airport, 1959. *Ruth MacDonald Collection*

part in the Battle of the Rosebud River in the Sioux Wars in the American west during the 1870's. He died of natural causes, alone, at his camp on the banks of the Yukon River in 1894.[38]

MacDONALD, Allan John "Hard Rock"

"Hard Rock" MacDonald was a well-known prospector who explored much of the northern territory from 1920 to 1960. He came to Mayo circa 1920 as a teamster on the ore haul and was one of the few who graduated to catskinner on the Treadwell Yukon ore haul.

"Hard Rock" was a great trekker. When he heard about the Norman Wells oil discovery, he set off from Keno and walked across the divide to Norman Wells. He was the first to draw attention to the silver-lead deposits in the Kathleen Lakes area.

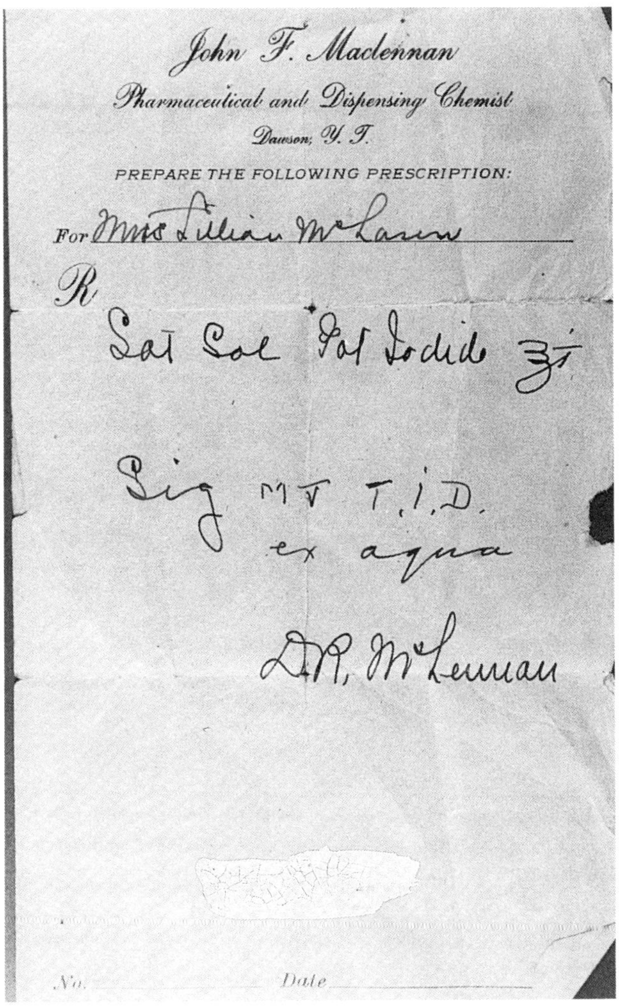

John F. Maclennan
Pharmaceutical and Dispensing Chemist
Dawson, Y. T.
PREPARE THE FOLLOWING PRESCRIPTION:
For Mrs Lillian McLaren
R
Sat Sol Pot Iodide 3i
Sig M T T.I.D.
ex aqua
D.R. McLennan
No. Date

A prescription written by Dr. "Randy" Maclennan of Mayo. *L. (McLaren) Cheatham Collection*

In the 1930's, he arranged to be flown out to a lake he had seen east of Mayo, to prospect. He did not know exactly where he was, but he prospected the area, then finally rafted down a river. To his surprise, he arrived at Fort McPherson on the Peel River. He returned over the Bell and Porcupine Rivers to Dawson, then up the Yukon and the Stewart Rivers to Mayo. It was a mistake similar to "Deep Hole" Thomson's in 1916.

MacDONALD, James "Jim" & Suzzane

"Jim" MacDonald was originally from New Zealand. He prospected and mined in Cobalt, Ontario, before coming to the Yukon. He was a Dawson pioneer, operating restaurants in Dawson, Whitehorse, and Mayo.

He was an alderman in Dawson circa 1903. In 1922 he found low-grade silver-lead ore on Cobalt Hill, just east of Keno Hill. He was married to Suzzane Turgeon. Ruth MacDonald of Mayo is their daughter.

MacDONALD, Johnnie

Johnnie MacDonald worked for Treadwell Yukon at the Silver King in the 1930's. He was also a saxophone player.

MacDONALD, "Red"

"Red" was a Cat skinner and taxi driver in the district in the 1950's.

MACLENNAN, Alexander Malcolm

See *Pioneer Accounts*.

MACLENNAN, Donald Randolph "Randy"

Randy Maclennan was a Doctor of Medicine graduate from McGill University. He played on the famous Dawson hockey team that competed for the Stanley Cup in 1904. He was the mining recorder at Stewart City in 1903, and came to Mayo circa 1904.

He refused to practice medicine on a regular basis and did so only when no other doctor was available in the district. He operated a small drugstore in Mayo from the early 1920's until 1935.

Maclennan was a principal in Yukon Silver-Lead Mining Company and worked to develop their claims on Lookout Mountain. He staked the Frog and the Faro claims on Keno Hill in 1919.

He was the postmaster in Mayo during the 1930's, following J. K. Johnstone's term, and was the mainstay of the Mayo Tennis Club. He married Margaret A. Kenney, a nurse at the Mayo hospital. Alex Maclennan was his cousin. Dr. Maclennan died in 1935 and is buried in the Mayo cemetery.

MACPARLON, "Red"

"Red" Macparlon was a miner for Treadwell Yukon at the Calumet in the 1930's. He later enlisted in the Canadian Army, and won the Military Medal, but was killed in action at Ortona, Italy, in 1943.

MAHER, Thomas

Thomas Maher staked claim No. 3 on Steep Creek on September 16, 1903.

The famous Dawson City "Klondike" Hockey Team that competed for the Stanley Cup in 1904. This photograph was taken at Deys Rink in Ottawa on January 14, 1905. Col. Joe Boyle was the team's backer and Weldy Young (who later formed Young-Davidson Mines with Jake Davidson) was captain and coach. (The Ottawa Silver Seven won the Stanley Cup that year.) Front row, from left: Albert Forrest, Col. Joe Boyle, Dr. Randy Maclennan. Back row, from left: Hector Smith, George Kennedy, Lorne Hannay, J. K. "Jim" Johnstone, and Norm Watt. Substitute players Dave Fairbairn and Archie Martin are not in the photograph. In later years Dr. D. R. "Randy" Maclennan became mining recorder and postmaster in the Mayo district, J. K. Johnstone became a postmaster in Mayo, and Archie Martin was a well-known prospector who lived in Keno from 1920 to 1950. Albert Forrest was a brother of Mayo's Emil Forrest. *Forrest Collection, Yukon Archives*

MAJOR, Richard "Dick"

Dick Major ran the poolroom, barber shop, and store at Wernecke Camp in the 1920's and at Elsa in the 1930's. He sold the business to Alex Smith when Elsa reopened after World War II and later moved to Alaska.

MAKELA, Elmer

Elmer Makela mined on Duncan Creek around 1913–1914. He built a ditch by hand and sluiced with water delivered through 1,500 feet of 10 inch pipe to a monitor however he earned little money. A mastodon tusk was found on his property. It was the first discovered on Duncan Creek.

MALCOLM, Billy "Ghechuck" & Jessie

Billy Malcolm was the Traditional Chief of the Mayo Indian Band from 1947 to 1949. He was known as Billy "Dechuck" or "Ghechuck" which means "porcupine." The word is pronounced "Dechuck" by Mayo elders.

He married Jessie Lackney, and they had six daughters: Jenny (married Frank Germaine), Sophie (married Paul Germaine), Mary (married Johnny Silverfox) Lena (married Hans Hummel) Alma (married Toby Besner) and Margaret (married Cathro Peter — no relation to the Mayo Peter family). The Malcolms lived at 17-mile on the Stewart River for many years. His father and Lucy Peter's father were brothers. (See his daughter, Lena Hummel's story in *Pioneer Accounts*.)

MALESICH, Martin

Martin Malesich mined on Thunder Gulch from 1908 to the 1940's.

MALLETTE, Isaac

Isaac Mallette mined on Secret Creek with his partner Narcisse "Fred" Lefebvre from about 1911 to 1930 and was also a partner in Lefebvre's sawmill in Mayo. During the 1930's and early 1940's he was the school janitor in Mayo.

He owned and rented out the cabin next to the Swanson/Barker cabin but he lived in the frame house next door to the Cal Harris cabin on the alley. In the garden behind the house he grew cabbages which he made into sauerkraut. He used an overhead watering system in his garden. Mallette died in 1939 and is buried in the Mayo cemetery.

From left: Jenny Hager, Martha Hager, Helen Hager and Sophie Malcolm, circa 1946. *G. J. Gordon Collection*

MALONEY, Jack

Jack Maloney came from Butte, Montana. He was mine foreman for Treadwell Yukon at Wernecke Camp circa 1925 and mine superintendent at the Silver King from 1935 until approximately 1939.

MANIFOLD, A. H. "Moose"

"Moose" Manifold was United Keno Hill Mines' chief geologist in 1957.

MARKLUND, Frederick

Fred Marklund was a miner. He lived in Mayo upstream from Ben Cameron. Marklund died in Mayo.

MARLOWE, Helen

Helen Marlowe lived in Mayo during the 1920's boom. She was a relative of Gene Binet's wife, Jewel. It is believed that she operated a small restaurant or shop during her time in Mayo.

MARSHALL, Frederick "Fred"

Fred Marshall was Phyllis Andison's son by a former marriage. Phyllis (Marshall) Andison originally came from England. She was very active in community and Anglican Church work. Phyllis died in 1940 and is buried in the Mayo cemetery.

Fred was a Cat operator for Klondike Airways Ltd., in the 1930's, and in the late 1930's he worked for Treadwell Yukon at the Silver King.

H. P. Peterson, cook (on left), and Archie Martin on top of Keno Hill overlooking Faro Gulch, 1948. *Herb Foster photograph, O'Neill Collection*

He was a top badminton player and helped organize the Mayo Badminton Club.

MARSHALL, John & Phyllis

John Marshall was Burn's butcher in Mayo during the 1920's. He was Phyllis Andison's first husband and they had one son, Fred Marshall. Marshall died in 1930 and is buried in the Mayo cemetery.

MARTIN, Archibald A. "Archie"

Archie Martin was a Dawson pioneer from Aylmer, Quebec. He played on the Dawson City Stanley Cup Hockey team of 1904, and served on the Yukon Council from 1912 to 1917. He came into the Mayo-Keno Hill areas in the 1920's to mine and prospect. He and his partner, Agnes Kinsey, had claims on Keno Hill.

MARTIN, Reverend John & Bella

– Archie Gillespie, *The Mayo Miner*, March 12, 1937

Reverend John Martin was born at Fort McPherson in 1883, and as a young man served as a guide to the Royal North West Mounted Police on their yearly patrols between Dawson and Fort McPherson. Mrs. Martin was the former Miss Bella Kunizzi of Fort McPherson.

In 1925 Reverend Martin was ordained as a Deacon in the Anglican Church by Bishop Lucas at Aklavik, Northwest Territories. He ministered among his people at the little Arctic outpost for four years, and then came to Moosehide near Dawson City in 1929. His brother Reverend Richard Martin also lived at Moosehide.

Reverend Martin spent the years from 1930 to 1934 doing missionary work among the Indian people at Ross River. Reverend Martin came to Mayo with his family during the summer of 1935 and was in charge of St. Mark's Church at the "Old Village."

He was also an able trapper and hunter. Reverend John Martin became ill at his home on February 20, 1937 and was taken to the Mayo General Hospital suffering from influenza which developed into pneumonia. He died on March 8, 1937 at the age of 54. (See his daughter Mary Moses' story in *Pioneer Accounts*.)

MASON, John

John Mason is listed as mining on Highet Creek in Polk's 1915–16 directory.

MASON-WOOD, Simon Stanley "Sam"

See *Pioneer Accounts*. See also *Wood, Simon Stanley "Sam"*

MATHENY, Cicero D. "Charlie"or "Ole"

C. D. Matheny was in the Fortymile district, and staked claims there, in 1909. He was a sawyer for Lefebvre's sawmill in Mayo in the 1920's and is listed as a millwright in Mayo in Polk's 1923–24 directory. He lived in the frame house that was later occupied by the Lovdahls and then Fred Taylor.

MATHESON, Charles

Charles Matheson is listed at Keno Hill in Polk's 1923–24 directory.

MATHESON, Christena "Chrissie"

Christena Matheson was a widow who came to Mayo to cook for Elmer Middlecoff. She later cooked at Lou Titus's boarding house in Mayo.

Christena Matheson was G. Jean Gordon's mother.

She married George Andison after the death of his second wife, Phyllis.

After his death, Christena sold the house in Mayo and went to live at Macaulay Lodge in Whitehorse where she died on January 16, 1976, at the age of almost 92 years. Christena (Matheson) Andison is buried in the Mayo cemetery.

MATHESON, D. A. "Dan" & Adolphine

Dan Matheson, also known by the nickname "Slippery Dan" Matheson, mined on Duncan Creek in 1907 with his partner Sam Thurber.

He and his wife, Adolphine (née Schoch), whose nickname was "Foxy Phinie," moved to Mayo circa 1920 and operated the telephone service (Mayo Utilities Co. Ltd.), as well as a confectionery, and dress shop on Front Street.

Matheson was the president and manager of Mayo Utilities Co. Ltd., and Adolphine was the secretary-treasurer. She had been the telephone operator for the Yukon Telephone Syndicate at Bonanza (Grand Forks, Yukon) circa 1903. Matheson sold Mayo Utilities to Stan Barker in 1950.

MATHESON, Malcolm & Mary

Malcolm Matheson was the son of Adolphine and Dan Matheson. He graduated in Electrical Engineering from the Milwaukee Institute of Technology. He came from Dawson City in 1938 to manage the Mayo Utilities Co. Ltd. for his parents who went to Vancouver because Mrs. Matheson was not well.

He had previously been employed in Dawson with the Dawson Electric Light and Power Co. Ltd., a subsidiary of Yukon Consolidated Gold Corporation (Y.C.G.C.).

On January 4, 1938 he married Mary Wyness in Dawson City. Mary was the daughter of Mr. and Mrs. John Wyness of Dawson City. Both had attended the Dawson public and high schools.

Mary worked with the stenographic department in the Dawson office of the Yukon Consolidated Gold Corp., before she was married. She ran the Art Store in Mayo. He was a pianist and gave piano lessons in Mayo.

In 1944, Mayo Utilities Co., started a telephone system in Whitehorse which was eventually bought by C.N.T. (Canadian National Telecommunications).

MATLOCK, George H.

George Matlock was born on September 22, 1860 in Iowa. He left home at 17 and crossed the west to British Columbia. He pulled stumps with a bull team where Granville Street in Vancouver is today, and from there he moved on up the coast to Juneau, Alaska where he worked in the Treadwell Mine. In 1884, he came in over the Chilkoot Pass and bar-mined on the Stewart River until 1886 when the Fortymile was struck. He mined on the Fortymile for many years and in 1890 he, Frank Buteau and their other partners brought hydraulic mining to their claim on Franklin Gulch.

One night in the town of Fortymile, a man named Washburn began picking on Matlock's partner. Washburn was bigger than the partner and in his defence, George took up the fight. During the fight Washburn stabbed George in the back with a knife. Enraged, he broke loose and went outside. He returned with a gun and said, "I don't want to kill you, you s.o.b., but I am going to hurt you." He aimed for a flesh wound and shot Washburn in the thigh. Although he and Washburn later shook hands, the incident induced John J. Healy and Bishop Bompas to request the presence of the North West Mounted Police in the Fortymile area. In 1895 Inspector Constantine arrived with 19 constables in Fortymile—the first law enforcement group to arrive in the Territory. (Thus, the N.W.M.P., were already established by the time the Klondike was struck.)

Matlock went to Circle City, Alaska when gold was discovered there, and then moved on to the Klondike when it was struck. He left for Dawson City with a four-horse pack train, travelling on the river ice. He arrived in Dawson City just two days before the ice went out on the Yukon River. He freighted in the Klondike, and also borrowed money to buy in on some ground, but was not successful. He eventually left the Klondike, broke, and went to Fairbanks, Alaska. He then prospected the Seventymile and finally went back to the Fortymile. In the late 1900's (1917–1920) and early 1920's he worked for the July Creek Placer Company in Nation, Alaska. During this time, at 61, he was married.

Captain Al Mayo in his senior years at Rampart City, Alaska. *Richard Frank Collection, (Acc. #73-71-4) in the Archives, Alaska and Polar Regions Dept., University of Alaska, Fairbanks*

In 1932 he was at Eagle, Alaska but left Eagle with Fred Terwilliger and Al Stout to look for some hardrock he had found on the Fortymile years before. They prospected and trapped that winter, and in the summer of 1933 they moved on up to Gold Run. He died of a heart attack at Gold Run on October 8, 1933. Matlock is buried in the cemetery at Eagle, Alaska.[39]

MATSON, Charles

Charles Matson is listed as a miner at Keno Hill in Polk's 1923–24 directory.

MATSON, Mark

Mark Matson held claims on Dublin Gulch circa 1907.

MAWHINNEY, John

John Mawhinney staked on Steep Creek and Ledge Creek in the Mayo Lake area in 1903. He was employed by Elmer Middlecoff on Highet Creek in August 1910 and mined on Haggart Creek with Charles E. Merriman and Clarence E. Kinsey circa 1915.

MAY, Cecil G. & Esther

Cecil May was the Commanding Officer of the Royal Canadian Corp. of Signals staff in Mayo in the late 1930's. Roy Hallin and Art Hughes were the Signals staff at that time. He was married to Esther Broadfoot, daughter of Mrs. Sophie Broadfoot.

MAYNARD, John

John Maynard was born in Devon, England in 1867. He mined with Jahnke and Gill on Haggart Creek from 1908 to the 1920's. He worked for E. Middlecoff on Highet Creek for a short time in 1910[40] and lived in Keno City from the 1920's to the 1940's. John Maynard died in Mayo in 1952.

MAYO, Alfred Henry "Al" & Margaret

For Alfred Mayo's early history see the sections *McQuesten, Mayo and Harper* and *Early Bar Mining on the Stewart River*.

By 1893 Al Mayo was at the mouth of Minook Creek establishing Rampart City in

Alaska. He remained in Rampart and he and his wife Margaret, an Alaskan native, eventually retired to a small farm overlooking his beloved Yukon River. Al Mayo never left Alaska. He was always referred to as Captain Al Mayo because of all the years he had captained steamboats for the Alaska Commercial Company.

He made frequent trips to Fairbanks to visit his old friends from Circle City such as Dave Petrie, Al Morency, Dan Callahan, George Harrington, and Frank Buteau. Al Mayo was a humorist and a dispenser of neighbourly good will wherever he went.

Jack McQuesten used to say, "Al Mayo was the best trail mate a person could possibly wish for." Al Mayo died at age 76 at his home in Rampart on July 18, 1923, leaving eight children and twenty-three grandchildren.

McASKILL, Const. Donald Allen

Const. Donald McAskill was born in Stirling, Nova Scotia on November 4, 1896. He was a nephew of "McAskill the Giant" [Angus McAskill], Nova Scotia's famous 425 pound, 7' 9," strongman who toured the world from 1849–1853.

Donald McAskill served with the 25th Battalion Canadian Expeditionary Force in World War I in France and was awarded the Military Medal. When World War I ended he served with the Halifax police department and then joined the R.N.W.M.P., in St. John, New Brunswick in 1919 and served in Saskatchewan until 1921. He joined the Alberta police and then left to join the R.C.M.P., in Vancouver in 1923.

In 1931 he was stationed in Dawson City and in 1932 Inspector Dempster placed McAskill in charge of the Keno detachment. McAskill's great size and strength served as bait for local Mayo and Keno brawlers who could not resist provoking him; however, there was not a man in the district that ever bested McAskill in a fight. After a tumultuous posting in Keno he was transferred to Regina in June of 1934.

In 1938 McAskill was placed in charge of the Yellowknife detachment and then discharged with a pension on April 15, 1939. He enlisted in the Seaforth Highlanders, 1st Canadian Division C.A.S.F., and served overseas in World War II. He died at the Shaughnessy Military Hospital in Vancouver on January 2, 1953.

McCAFFERY, Joseph F.

Joe McCaffery owned and operated a hotel and restaurant from 1922 to 1931 on Centre Street in Mayo. Mrs. Alice Fisher (Bud Fisher's mother) came to Mayo to cook for his hotel and restaurant. He closed his business in 1931 but continued to live in the building during the 1930's and 1940's.

McCARTER, Alex & Daz

Alex McCarter was from Dawson, the son of Dawson pioneers. Alex first came to Mayo at the age of 14, in the early 1920's, to work for a survey crew. Through the 1930's he worked as a driver for the Transportation Department of Treadwell Yukon. Initially he was a tractor operator on the ore haul and later he drove trucks.

It was a three day round trip from Keno to Mayo for the Cat trains in winter. There were two Cat trains leaving Keno while two were returning from Mayo and they would pass each other midway. Each tractor pulled four sleighs and each sleigh carried 15 tons of ore. When trucks replaced Cats on the ore haul, travel time was reduced to one day for a return trip. In addition to his other duties, he drove the sick or injured from Keno to the hospital in Mayo. He recalls driving an injured miner, who had nearly every bone in his body broken in a mine accident, from Keno to Mayo. The trip took 13 tortuous hours and the only painkiller available was a bottle of brandy. The man died a few hours after he arrived at the Mayo hospital.

His truck was sometimes referred to as the "Maternity Special" because he drove so many expectant mothers from Keno to the Mayo hospital. Among the women he recalls driving down from the Hill are Mrs. Jeff Wynen (twice), Mrs. Peter Buyck (three times), Mrs. Jack Gerlitsky, and Mrs. Lovdahl.

He met his future wife, Daz Reighley, when she arrived in the Mayo district, after teaching school in Tagish, to work as a chambermaid for a hotel in Keno. Daz was from Leslieville, Alberta. She had seven sisters and one brother. She was an accomplished saxophone player.

They were married in Mayo on October 21, 1932 and Rose and Sam Mason-Wood were married the same day. Daz and Alex were Rose and Sam's bridesmaid and best man. The two couples left Mayo to drive out to Rupe and Mrs. Steeves for their wedding dinner but half way there, the car broke down. They walked back to Mayo and stopped for a visit at each cabin along the way. When they got back to town they had their wedding supper at the GN Cafe.

After Treadwell Yukon closed down, they moved to Whitehorse in January of 1941 where Alex went to work for Canadian Pacific Airlines as a mechanic. He then worked for Happy LePage driving Cat at the airport where the wages were better.

In 1945 the McCarters left Whitehorse for Kaslo in the Kootenays, British Columbia followed by a move to Vernon B.C. in 1947. Alex worked for G. M. Vernon Motors and later went into business for himself in 1954. He sold his business in 1972 to enjoy his retirement years in Vernon. Daz passed away on July 11, 1980. Alex celebrated his 84th birthday in 1989 and still lives in the same house that they bought in 1947.

McCARTHY, Edward E.

Edward McCarthy was the General Manager of Guggenheims Yukon Gold Company in Dawson at the time Louis Bouvette struck silver on Keno Hill (1919). McCarthy arranged the Keno Hill deal between Bouvette and the Yukon Gold Company and arranged deals with other early Keno Hill stakers as well. On an official visit to Keno Hill, McCarthy staked the Porcupine Fraction on October 3, 1919. McCarthy was promoted to the Guggenheim's New York office in 1920.

McCARTHY, J. J.

J. J. McCarthy came to Mayo from Dawson in about 1920. He was the chief accountant for Treadwell Yukon from the 1920's until 1946.

McCLOSKEY, James

He came to this district circa 1903–1909. He acquired the Grey Copper King claim from Robert Fisher in the 1920's. It is believed Mrs. McCloskey ran a small restaurant in Mayo sometime during the period 1903–1914.

McCOOMB, William "Bill" & Florence

Bill McCoomb came from the Eastern U.S.A. to Dawson City and when he learned his brother was ill that winter, he walked from Dawson to Telegraph Creek, B.C., in order to return to his home in the U.S. He journeyed north again in 1928 and on his way, Bill met his future wife, Florence, who was caring for special needs children in Billings, Montana. When the McCoombs arrived in the Yukon in 1946, Bill built a paddlewheel houseboat and the couple prospected and mined on the Teslin and Little Salmon Rivers.

They arrived in the Mayo district in 1948 and leased George Reynolds claims on Ledge Creek.

Mayo residents riding in Malcolm McCown's freight truck, 1923. Theos Whitney is third from right and Dorothy Hoggan is seventh from right. *F. & T. Whitney Collection*

Using only hand mining techniques, McCoomb located an old channel on Ledge Creek which resulted in a major placer gold discovery. The McCoombs lived on Ledge Creek until 1951 and then moved into Mayo and bought Joe Cantin's property on which the abandoned tennis court and clubhouse stood.

He was a talented boat builder. He built a barge at Mayo Lake using a sawmill with an 8 hp Briggs and Stratton engine that cranked the blade through the logs (rather than pushing the logs through the blade). It took two years for him to assemble the equipment and materials and build the barge. When it was completed, the barge floated the first heavy equipment to mining operations across the lake. His barge still sits on Ledge Creek today [1989].

Mrs. McCoomb worked at the school in the 1960's. By 1963, Bill was getting on in years and he hired Bert Liske to work the claim with him. McCoomb fell ill that summer and Liske had a plane fly him out from the claim. McCoomb died of cancer in a hospital in New Westminster, B.C. in 1965 at the age of approximately 74 years. Mrs. McCoomb continued to live in her home in Mayo until her death on December 29, 1989.

McCOWN, Malcolm Sunshine "Soapy"

Malcolm "Soapy" Sunshine McCown was nicknamed "Soapy" because he had driven a laundry wagon in Dawson before coming to Mayo circa 1920. He and Clem Sinyard were the co-owners and prospectors of the Hector claims on Galena Hill—some of the richest claims in the area.

Malcolm had a Model "T" gasoline taxi which he ran between Mayo and Minto Bridge in the 1920's. The fare was $1 a mile and if the taxi got stuck, the passengers had to get out and push! He left Mayo in 1939 to prospect his mineral property in Winnemucca, Nevada.

McCRIMMON, Harry

Harry McCrimmon came to Mayo from Dawson where he had been mining on Hunker Creek. He arrived in this district circa 1920 and prospected at Keno Hill.

McCUISH, Murdock Jr.

Murdock McCuish Jr., was born in Dawson in 1914. He was the son of Murdock McCuish Sr.

The McLaren home in Mayo, 1926. Front, from left: Lillian McLaren and Raoul Binet. Back row: Ellie Ballentine, Roy McDermid, and Tommy McKay. *Ione Christensen Collection*

The McCuish's were a pioneer Dawson family. Murdock Jr., died in Mayo in 1936 while waiting around for a job while the Elsa mill was being built.

McCULLUM, Reverend A. C. & Jean

Reverend McCullum was the Anglican minister in Mayo circa 1930. He and his wife Jean had two sons, Hugh and James. Reverend McCullum served as a padre in the Canadian Army during World War II with the rank of Major.

McDERMID, Roy

– Archie Gillespie, *The Mayo Miner*, October 23, 1936 and October 30, 1936

Roy McDermid was born in Buckingham, Quebec on July 10, 1899. He came to Dawson with his parents at the age of six and grew up in Dawson where he attended the Dawson public and high schools. His mother was Lily McDermid, (sister of George F. Potter of Mayo and daughter of George S. Potter of Vancouver). Mrs. K. A. Forbes, also of Mayo, was Roy's aunt.

During his school days Roy took an active interest in athletics. He played on many of the famous old high school hockey teams and was a keen devotee of baseball, boxing and other athletic sports in Dawson City before World War I. He also delivered newspapers for the *Dawson News* both in town and on Dominion Creek.

During Canada's involvement in World War I Roy tried to join up with the Black Contingent in Dawson but he was unable to because he was only 17. However, Roy worked his way to Victoria B.C. where he was able to sign up by putting his age as 18. He subsequently went to France where he saw active service with his Yukon pals then eventually went on to Berlin after the Armistice.

Roy returned to Dawson from overseas in 1919 and came to Mayo in 1920 as one of the first members of the government crew working on the new Mayo to Keno road. He later worked in the mines on Keno Hill, ran the hoist for the Treadwell Yukon Company, Ltd. on Keno Hill for several seasons, then became one of the company's most expert tractor operators.

In 1933 he entered the mining game on Haggart Creek in partnership with Ed Barker and Kit Watters. While working with Barker, Watters, and Norman Wightman on Haggart in the summer of 1936, Roy fell ill. He went to Mayo for medical treatment then was sent to a hospital in Vancouver B.C. Roy died in Vancouver on October 22, 1936, at age 37.

McDIARMID, Jack & Mary

Jack McDiarmid walked from Whitehorse to Mayo and worked at the Elsa in the early 1930's. He also mined on the Stewart River and was a ferry operator on the Stewart Crossing ferry.

In 1957 Jack and his wife Mary built the Stewart Crossing Lodge. The McDiarmids then bought the Tim-O-Lou Motel in Mayo and renamed it the North Star Motel. They operated this business until 1974. [See *Stewart Crossing* by Wendy Andre.] Their children are Jean, Roy (deceased), Allen, Buryl and Joe.

Jack's brother, Alister, mined with Jack on the Stewart in the late 1930's and early 1940's. After Alister and his wife Judy married in Dawson in 1948, they made their home in Elsa where he worked for the mine.

McDIARMID, Richard "Dick" & Hazel

In 1918 Dick McDiarmid and his wife Hazel were bar mining at Joe Gee Bar (also written Joe Jay Bar), three miles above Clear Creek. They had also started a farm there, and had two horses, farm machinery, and a few hens. They remained on the Stewart River though the 1920's and the 1930's. Records show that in 1922 he purchased a wooddealer's business license from the Town of Mayo. He was Jack and Alister McDiarmid's uncle.

McDONALD, Allan D.

Allan McDonald was a blacksmith for Keno Hill Ltd. at Keno Hill circa 1923–24.

McDONALD, Ambrose

Ambrose McDonald is listed as a miner on Keno Hill in Polk's 1923–24 directory.

McDONALD, Charles

Charles McDonald is listed as a miner on Keno Hill in Polk's 1923–24 directory.

McDONALD, Danny

Danny McDonald arrived in Dawson around 1900 with his brothers. He was a blacksmith by trade and moved to Elsa in the 1930's. He returned to Dawson and died there.

McDONALD, Edward "Spike"

– Archie Gillespie, *The Mayo Miner*, January 6, 1939

"Spike" McDonald was born in Nova Scotia, and worked for the Treadwell Yukon Company in the 1930's. He died of pneumonia at age 38 in the Mayo hospital. He was related to John A. McDonald, a Mayo resident living in Dawson at the time of Spike's death.

McDONALD, John

John McDonald is listed as a porter at J. E. Binet's hotel in Polk's 1923–24 directory.

McDONALD, John

John McDonald is listed as a miner at Keno Hill in Polk's 1923–24 directory.

McDONALD, John A.

John A. McDonald lived for a time in the Mayo district and by the late 1930's had become the proprietor of the New Westminster Hotel in Dawson. He ran for Territorial Council for the Dawson district in 1937 and was elected. He was related to Edward "Spike" McDonald.

McDONALD, Katherine

Katherine McDonald was a staff nurse at the

Mayo hospital in the late 1930's.

McDONALD, Kenneth Duncan

– Archie Gillespie, *The Mayo Miner*, April 29, 1938

Kenneth McDonald worked in the Mayo silver camp in 1922 and later in Dawson. In 1923 he moved to Stewart, B.C., where he worked at his trade as a blacksmith. He died at age 76 in the Stewart, B.C. hospital on April 3, 1938.

McDONALD, Rodney

Rodney McDonald is listed as a miner at Keno Hill in Polk's 1923–24 directory.

McDOUGALL, Elizabeth

See *Elizabeth McDougall Titus*.

McFADDEN, Hugh

Hugh McFadden was mining on Scroggie Creek in 1918 and into the 1920's.

McGEACHY, D. C. "Dunc" & Ruth

"Dunc" McGeachy managed the Chateau Mayo Hotel for Ed Barker and later ran a taxi business in Mayo. He was a Yukon Council member for the Mayo district from 1955 to 1958. The McGeachys operated a hotel in Keno City in the 1960's. They have two sons, Michael and Brian. The McGeachys retired to Penticton, B.C., where they continue to reside (1989).

McHENRY, Dutch

Dutch McHenry is responsible for the legend of the Lost McHenry Gold Mine supposedly located in the Lansing or Hess country. It is said he packed out 40 pounds of gold to Dease Lake in 1870.

McHUGH, John

John McHugh is listed at Keno Hill in Polk's 1923–24 directory.

McINNIS, Angus

Angus McInnis is listed at Keno Hill in Polk's 1923–24 directory.

McINTOSH, Allan

Allan McIntosh was one of the original four stampeders who staked the "Swedes" mine on Duncan Creek in 1901 and mined at Minto Lake in 1903. He was listed as a mail carrier in Mayo in Polk's 1915–16 directory and for a time was a dredgemaster in Dawson. He later ran a taxi service in Keno until the 1930's. He died in Keno City in the 1930's.

McINTOSH, Joseph S. "Joe"

Joe McIntosh, a member of the "McIntosh Apple" family, came to the Mayo district from Ontario.

He logged on the Stewart River with Arthur Oligny. He was appointed Liquor Vendor in Mayo after Frank Carscallen. McIntosh died in 1936 and is buried in the Mayo cemetery.

McINTYRE, Angus & Ida

Angus McIntyre was born at St Elmo Ontario, December 2, 1860. At the age of 19 he left home to work in Michigan logging camps then to Pat Portage where timber was being cut for the construction of the C.P.R. (Canadian Pacific Railway). He worked on this project across the prairies and by 1884 he and partners obtained a contract to build two miles of railway at Kay Falls. He was present at the driving of the last spike on November 7, 1885 at Craigellachie by Lord Strathcona and is one of the onlookers in the famous picture of that event. From 1887 to 1897 he drove horses and ran pack trains in the Kootenay country, where, at Nelson, B.C. he married Ida Johnson, the first marriage recorded in that town.

In October 1897 he boarded the steamer *Humboldt* at Vancouver and landed at Skagway, Alaska where he ran pack-trains for the Bartlett Bros from Skagway to the Summit. He then went on to Dawson where he was soon back in the pack-train business between Dawson and Glacier Creek establishing what is now known as the "Top of the World" Highway. From 1905 until 1908 he ran the pack-train supplying material for the construction of the Twelve Mile flume and ditch. Then he was a foreman on the ditch construction from 1908 to 1911 when he became city foreman for Dawson City.

In 1920 he went to Mayo for Greenfield and Pickering in charge of the ore haul from Keno Hill to Mayo. By 1925 he was the Dawson agent for Greenfield and Pickering who held the Whitehorse–Dawson mail contract.

It was during this time that he was given the nickname "Four Post" McIntyre, by Archie Gillespie, a reporter for the *Dawson News*.

Caterpillar "Cat train" outside the GN Cafe in Mayo. From left: Roy McDermid, Gordon McIntyre (in background), and Ed Barker. *F. & T. Whitney Collection*

Whenever Archie would ask about the mail service, Angus' stock answer was, "We made four posts today." (A "post" was the distance between roadhouses, approximately 25 miles.)

In his later years he worked as a teamster for Y.C.G.C. and in 1942, due to failing eyesight he retired at the age of 82 to Kelowna to live with his daughter, Hazel Cruickshank. He died there in 1950 at the age of 90. His wife Ida died in Vancouver the same year at the age of 80. The McIntyres had three children: Hazel, Edna and Gordon. [See Gordon McIntyre's story in the *Pioneer Accounts*.]

McINTYRE, Archie

Archie McIntyre found placer gold and staked the Discovery claim on Jarvis Creek on May 14, 1903. (Jarvis Creek is a tributary of Minto Creek).[41]

McINTYRE, Gordon A.

See *Pioneer Accounts*.

McINTYRE, Mabel Estelle

Mabel (née Huffman) McIntyre was born in Mayo on December 7, 1912. Her father was Grant Huffman and her mother was Emma Jimmy (daughter of Jenny Jimmy).

Grant Huffman named the "Mabel" claim for his daughter when he staked it on February 23, 1913.

She had one brother, Albert, and they received all of their schooling at Mooseheart School in Chicago, Illinois. They returned to Mayo in the mid 1920's to live with their father on his homestead at the foot of the Mayo Canyon (the Huffman homestead is now owned by Simon Mason-Wood).

Albert had studied sign painting at Mooseheart School, and while he was in Mayo he drew the Mickey Mouse on the GN Cafe sign. He stayed in Mayo only long enough to earn his fare back to Illinois. He never returned.

Mabel left home shortly after Albert and went to work for Mrs. O'Loane in Keno. She later worked as a chambermaid in one of the Keno hotels and then for Binet's Hotel in Mayo. She was appointed postmaster in Mayo in 1942 and remained in that position until she retired in 1972. She died while holidaying in Vancouver in 1986. She is remembered for her hearty sense of humour, her direct manner, and her kindness to those in need.

McINTYRE, William

William McIntyre is listed at Keno Hill in Polk's 1923–24 directory.

McIVOR, Murdoch

Murdoch McIvor was one of a party of eight Scottish prospectors who staked Thistle Creek in September 1898.[42]

McIVOR, Patrick

Patrick McIvor is listed as a miner at Keno Hill in Polk's 1923–24 directory.

McKAMEY, Raphael Lloyd "Ray" & Frances

Ray McKamey grew up listening to gold rush stories. His great grandfather, Granville McKamey from Marion County Missouri, stampeded to the California gold rush in 1848 and in 1852 took the first horses to the Cariboo gold rush in British Columbia. His father, John Blackwood McKamey, mined in the Yukon on the Sixtymile in 1894. His father's stories sparked a lifelong interest in the Yukon for his son and when he was offered the option of a job in Kitimat, B.C., or one in Mayo, Yukon, by

Northern Construction in 1950, he picked Mayo.

Northern Construction and J. W. Stewart Ltd., of Vancouver, B.C., were the prime contractors for the construction of the Mayo River hydro electric plant. McKamey recalls operating the crane that lowered the turbine and described the delicate and precision operation as a "nail biting" experience. He was in charge of clean-up when the project was completed and he was the last man to leave.

When he and his wife, Frances, first arrived in Mayo in 1950, staff quarters were not yet available so they lived in the building that had been Jim MacDonald's cafe (formerly Jim Ferrell's). The McKameys moved out to the hydro site when camp facilities were built in 1951. When the project was completed, he moved one of the houses from the site into Mayo for the family to live in. This house is today The Mayo Variety Store.

With the hydro work finished, he accepted a job with the Yukon Territorial Government. He was in charge of road maintenance for Pelly, Stewart Crossing, McQuesten River, Mayo, Duncan Creek and the Elsa road. The YTG was so cash poor in those days that he was equipped with only one grader, one truck, and one old pick-up truck. There was no money to hire help so Frances drove the truck while Ray shovelled gravel on glaciers. Glaciers on the Duncan Creek road

Installing the turbine, 1952. *J. Long - N.C.P.C. Collection*

were a major problem and fires were built in metal barrels and set on the glaciers to try and melt them. Old-timer Dan Ferguson sometimes worked for him on the glaciers. McKamey recalls Ferguson's reply whenever anyone asked him what he thought the weather outlook might be, "Only damn fools or damn liars predict the weather."

The wait between YTG pay cheques was three months and he finally quit his road maintenance job when he decided he "couldn't afford to work for the YTG any longer."

In the 1950's, he staked copper prospects in the Wernecke Mountains. He and Jim MacDonald found a crashed U.S. bomber and its six man crew while prospecting for copper in the McArthur Range.

In 1958 he was elected the Mayo district's representative on Territorial Council and served two consecutive terms.

Two of their three children, Brian and Tara (now Detaeye), were born in Mayo. Their third child, Linda (now Newman), while still a student at the Mayo School, attended the 1967 exposition in Montreal and was chosen to push the countdown button for the opening of Expo '67.

Frances, Ray and the McKamey children now live in British Columbia.

McKAY, Albert Edward

Albert McKay, Thomas William Johnston, and George Gordon staked claims on Empire Creek on April 11, 1904, and McKay continued to prospect and mine the creek for some years after.

McKay was accidentally shot by Henry McWhorter at Gordon Landing and died as a result of the accident on October 2, 1912.

McKAY, Phillip John

Phil McKay was a carpenter. He was employed by Treadwell Yukon at Wernecke and Elsa. Phil played the fiddle at dances.

McKAY, Thomas "Tommy" or "Tom"

Tommy McKay was born in Moose Jaw, Saskatchewan and was a cousin of Louis Bouvette. He came to the district to mine and prospect circa 1900. He prospected the eastern Ogilvie mountains and the Mayo and Keno areas in the 1920's. McKay and his partner Axel Erickson

Margaret Taylor & Tommy McKay. *T. McKay Collection*

were the owners and prospectors of the Nabob, Shamrock, and claims in the Calumet group. He remained in the Mayo district through the 1930's and the 1940's.

A long time friend and former Mayo resident, Nora McLaren Swenson recalls that, "Tommy was a quiet person, not boastful and kept his business to himself."

"Tommy" McKay was a Yukon Councillor for the Mayo district from 1932 to 1934.

McKAY, William F.

William McKay and "Snoose" Erickson found massive low grade Galena on McKay Hill in the Beaver district during the 1920's. McKay died in 1929 and is buried in the Mayo cemetery.

McKEE, Jack

Jack McKee was a well-known pioneer of the Yukon, Alaska, the Cariboo, and other mining camps in various parts of the world. He worked for Treadwell Yukon at Keno Hill in the 1930's and was injured when the wall of a building fell on him at Keno Hill in 1936.

McKENNA, Frank

Frank McKenna mined on Haggart Creek and Highet Creek circa 1910.

McKENZIE, Edward J. "Eddie"

"Eddie" McKenzie operated a general merchandise store in Mayo from 1923 to 1926, and also operated an ice cream parlour in 1926. Nora (McLaren) Swenson recalls that he was a thin man with an abundant supply of energy. At one time he even put on a minstrel show. "Eddie" McKenzie was a prominent figure in Mayo.

McKENZIE, W. D.

– Archie Gillespie, *The Mayo Miner*, March 12, 1937

W. D. McKenzie was a well-known Yukon pioneer and was famous for his recitations and readings. He worked as an orderly in the Mayo hospital. He died in Dawson City on March 9, 1937.

McKINNON, James

"Whispering" Jimmy McKinnon was one of the teamsters on the ore haul from Keno Hill to Mayo during the 1920's.

McKINNON, Malcolm J.

Malcolm McKinnon is listed as a miner in the Mayo area in Polk's 1915–16 directory.

McKINNON, Peter T.

Peter McKinnon is listed as a prospector and miner in the Mayo and Keno Hill area in Polk's 1923–24 directory.

McLAREN, Malcolm & Sarah

Malcolm McLaren with his wife Sarah (née Campbell) built the two-storey log house on the little hill adjoining the Anglican rectory. It later became "Ma" Fisher's boarding house. The McLarens had seven children: Earl, John, Mary, Dorothy, Lillian, Kathleen and Nora. See *Pioneer Accounts*.

McLAUGHLIN, James

James McLaughlin leased land on the Stewart River with partners Roden and Lindstrom from the 1910's to the early 1930's.

McLAUGHLIN, Patrick

Patrick McLaughlin was a member of the R.N.W.M.P., during the early days in Dawson. When a judge in Dawson found one of "Paddy's" arrests not guilty, he said to the judge, "Your Worship, let me know who your friends are and I won't arrest them." He mined on Keno Hill in the 1920's. He continued to live in Mayo until his death in the late 1940's or early 1950's.

Malcolm McLaren in Mayo, circa 1920. *McLaren-Cheatham Collection*

McLEAN, Daniel

Daniel McLean is listed at Keno Hill in Polk's 1923–24 directory.

McLEAN, John "Jack"

"Jack" McLean was one of a group who first prospected and found silver-lead ore on Silver Hill, part of Settlemier Ridge, in late 1923.[43] He lived in Keno during the 1930's and 1940's.

McLEAN, Murdock

Murdock McLean worked on Keno Hill during the 1930's.

McLEAN, Robert S.

Robert McLean is listed as a miner in the Mayo area in Polk's 1915–16 directory.

McLEAN, Shirley W.

Shirley McLean was a diesel engineer for Treadwell Yukon at the Elsa in the late 1930's.

McLEAN, Terry & Dora

Terry McLean was a mechanic for Treadwell Yukon at Elsa in the late 1930's. He and his wife, Dora, had two sons, Terry and Duncan. The family moved to Alaska.

McLELLAN, John

John McLellan is listed as an engineer for Keno Hill Ltd. in Polk's 1923–24 directory.

McLEOD, Angus A. "Cultus"

Angus McLeod found a vein of silver-lead ore on the banks of McLeod Creek (a tributary to Lightning Creek) in about 1915. He continued to mine and prospect in the Keno Hill area in the 1920's, and in 1921 staked the Sylvia claim. He owned the Bluebird, the Tin Can, and the Eagle claims. Angus lived in Keno until about 1950, then moved to Mayo where he eventually died.

He was also known as "Cultus" McLeod.

McLEOD, King

King McLeod is listed as a miner at Keno Hill in Polk's 1923–24 directory.

McMASTER, John D.

John McMaster is listed as a miner in the Mayo area in Polk's 1923–24 directory.

McMILLAN, John

John McMillan is listed at Keno Hill in Polk's 1923–24 directory.

McNAB, Bob

Bob McNab was a banker during his time in Mayo, and played the violin at dances. He was later transferred to Dawson and then to Vancouver.

McNEILL, Frank Edward

Frank McNeill was a prospector in the Mayo district during the 1920's. He staked the Bluebell claim on May 7, 1920 and the Mary claim on May 9, 1921. These claims were later acquired by Treadwell Yukon.

Frank McNeill. *D. Walloch McLaren Collection*

McNEILL, James H.

James McNeill was the Yukon Territorial Government Superintendent of Works and Buildings for the Dawson and Mayo districts in the 1920's. McNeill's home was in Dawson but he supervised many projects in the Mayo district; among them was the construction of a bridge over the Mayo River and the building of the first hospital in Mayo.

McPHERSON, Christopher Anderson "Sandy"

"Sandy" McPherson was born in Acton, Ontario on September 1, 1882. He left home at 19 and arrived in Alaska in 1904 where he had some success placer mining near Fairbanks. He prospected Groundhog Creek on Cook Inlet, Alaska then sold the property to a promoter. Hearing that "Swiftwater" Bill Gates of Klondike fame had struck it rich in Bolivia, he left for South America with two men named Hanning and Wilson to investigate.

In Lima, Peru they discovered that "Swiftwater's" strike was only a rumour, but they were in South America at a time that allowed them to witness the building of the Panama Canal in 1912. McPherson then went to the California gold fields and prospected for a time with a man named Sharpe before returning to Alaska.

When he heard that Louis Bouvette had discovered silver in the Mayo district, he came to Mayo in the fall of 1921 only to discover that everything on Keno and Galena Hills was staked. That winter, McPherson, Horace Glendenning, and three others cut logs on contract for Fred Lefebvre. In the spring he and Glendenning returned to Alaska to prospect and trap on Charlie River.

They returned to Mayo in 1925. In March of 1928, McPherson and partners Jack McLean and Jack McHugh got an option on the old Silver King mine from Thomas Aitken. They set to work but after a dispute over the possible location of an ore vein, the Silver King option was split between McPherson and McLean/McHugh. Glendenning returned to Mayo in 1929 to assist McPherson on his half of the Silver King option and in March that same year the two men struck a solid ore vein and sold the option to Treadwell Yukon for $125,000, receiving $12,500 down.

Later McPherson placer mined unsuccessfully in Barkerville, B.C. and returned to Keno in 1936 to work with Charlie Beck on Formo's property. He returned to mine the Silver King in 1942 with Ellef Bjonnes and Ellis Johnson, but the attempt to find new ore was unsuccessful.

McQUESTEN, Leroy Napoleon "Jack"

For "Jack" McQuesten's early history see the sections *McQuesten, Mayo and Harper* and *Early Bar Mining on the Stewart River*.

By 1893 McQuesten had helped to build Circle City, a time when that region on the headwaters of Birch Creek was producing $400,000 a season. He was still running his trading post at Circle City when the Klondike was struck, but he did grubstake others and purchased interests in claims on Eldorado and Bonanza Creeks. In 1898 he netted $10,000 from the sale of a small claim he held in Dawson.

McQuesten was made president of the Yukon Order of Pioneers at the first meeting of the lodge in Fortymile. In 1896 the Fortymile and the Circle City lodges combined to give him a $500 gold watch. The presentation was made at noon in front of his store in Circle City in the presence of practically every member of the community. McQuesten was held in high esteem by all who knew him, and although he was one of the shrewdest traders in the North, he was also very tolerant of the frailties of prospectors and miners.

McQuesten left the Yukon in late 1898 with over $200,000 and built a large home in Berkeley, California for his wife and children. He invested widely in mining interests in California and the west without losing his fortune. When the Alaska-Yukon Pacific Exposition was held in Seattle in 1909, he was appointed as the official representative of the Yukon.

While in Seattle he had to undergo an operation for bunions, but subsequently contracted blood poisoning. A few days later the "Father of the Yukon" was dead at the age of 73. After his death, Mrs. McQuesten managed his estate.

Jack McQuesten's name was a byword for integrity and honesty. He seemed to retain an optimism about his fellow man and he was known as an intelligent judge of character. Nowhere in the literature of the Yukon is it possible to find a critical or unkind word about him. It is rare anywhere to find a man as highly regarded during his own lifetime as was McQuesten.[44]

McRAE, J. E.

J. E. McRae found placer gold and staked the Discovery claim on McRae Creek on June 24, 1903, during the rush to the Minto Lake and Highet Creek finds. McRae Creek is a small tributary of Highet Creek. (His name has also been written J. D. McRay.)[45]

McRAE, Kenneth

Kenneth McRae worked as a cook for Treadwell Yukon at Wernecke and Elsa. He went to Dawson when Treadwell Yukon closed their operations and cooked at Yukon Consolidated Gold Corporation camps.

McVICAR, Archie & Barbara

Archie McVicar is listed in Mayo in Polk's 1923–24 directory. His wife's name was Barbara.

McWHORTER, Henry William

Henry McWhorter, originally from Pennsylvania, came into the Mayo district during the Duncan Creek stampede. On October 2, 1912, he accidentally shot and killed Albert E. McKay at Gordon Landing.

In February of 1913 he staked the Silver King on what had been Jake Davidson's Hell's Gate claim, then mined the Silver King in 1915 before selling it.

He left Mayo to return to the Alaskan gold fields where he married and built a fortune. In 1919 he bought a cattle ranch near Yreka, California and later financed a bank in Fortuna, California which he operated for many years. The bank was eventually bought out by The Bank of America.

Henry McWhorter died in February 1935, and was survived by three sons and one daughter.

MEGER, Adolph & Marion

Adolph Meger came from Biggar, Saskatchewan, and worked for Kimbel's sawmill in the 1930's. He married Marion Rachel Lindsay in Mayo on March 25, 1939, at a double ceremony with Hazel Moen and Roy Linn. Adolph began work with the Treadwell Yukon Company at the Silver King Mine in April of 1939. Three of the Meger children were born at the old hospital in Mayo; daughters Isabel (now Warnock), Arleen (now Kiff), and "Dee" Doreene (now Hassall). When the Treadwell Yukon mine closed, the Meger family moved to British Columbia where the children still reside. Marion Meger and Mike Murdock's wife, Isabel, were sisters.

MERCURE, Richard "Dick"

In 1920, Dick Mercure carried out the assessment work on Raoul Binet's Friendship claim and Binet gave him a half interest in the claim. The claim was sold to Keno Hill Ltd., and Binet and

Mercure shared the $40,000 proceeds

In the 1930's Dick Mercure worked with Matt Butijer (Butyer) on Butijer's Stone claim and then worked Jerry Drapeau's Silver Basin claims with Fred Guilbault. In 1947 Mercure and Fred Matthews leased the Chateau Mayo Hotel and beer parlour from Jim Mervyn. Mercure died in 1948 and is buried in the Mayo cemetery.

MEREDITH, Edward "Ed"

Ed Meredith, born in Wales, was one of the original members of the R.N.W.M.P., having joined that force in 1886. He came to the Yukon during the Klondike gold rush and remained in the Dawson district until World War I when he went overseas with the Yukon Infantry Company. He staked a claim on Sulpher Creek in the Klondike district in 1908.

He came to Mayo in the early 1920's and mined on the left fork of Duncan Creek. In May 1939 ill health forced him to leave Mayo for the Shaughnessy Military Hospital in Vancouver, British Columbia.

MERRIMAN, Charles E.

Charles Merriman mined on Haggart Creek with Mawhinney and Kinsey circa 1915.

MERVYN, James Hastings "Jim" & Julia

Jim Mervyn was born on a farm near Stratford, Ontario in 1866. As a young man he trained in Chicago as a carpenter and then worked his way across the United States and Canada. By 1897 he owned a sawmill near Walla Walla, Washington, but when he heard of the Klondike strike, he sold the sawmill and headed for Dawson, arriving too late to stake any of the better ground. He stampeded to Duncan Creek and in 1902 mined on Parent Creek with little success. Mervyn staked claim #8 on Steep Creek on August 21, 1903.

Following that he became a trapper with a partner named Billy Bowes. After two years in the bush they came out with 500 dark marten pelts which they took to Dawson to sell, but the market price there was so low that Mervyn took the furs to Vancouver where he rented a store and advertised his own auction. The venture was profitable and he decided to become a fur trader. To help finance his plan he worked as a miner for Elmer Middlecoff on Highet Creek during the summer of 1911.[46]

In 1912 he bought Lansing Post from J. E. Ferrell and operated it until 1937. He used his own coins, for trade complete with his name on them like the old traders had done. (See page 169.)

He married a Mackenzie River woman named Julia, and they had 11 children. Julia was a Lansing woman who had originally come from Fort Good Hope and her sister was Senoah (Niddery). After Lansing was flooded in 1936, most of the people living there moved away. Mervyn moved to Mayo and his son-in-law, Jimmy Wood, ran the Lansing Creek post for him during the winter of 1937, after which it closed.

In Mayo, Mervyn collected all of his trade coins and threw them into the Stewart River in front of Binet's Hotel. During the winter of 1937–38, he visited his old home in Stratford, Ontario, and also a brother who was living in Kaslo, B.C. He then returned to Mayo and bought Gene Binet's Chateau Mayo hotel and store, and began operating it on July 1, 1938.

He had two homes in Mayo. The first house he built is now owned by Margaret and Dick Wallingham. This house was the R.C.M.P. residence and lockup before Wallinghams purchased it. The Mervyn family lived in it for only a short time and then purchased a second house near the Whitehorse-Elsa road junction, where they lived while in Mayo.

Mervyn ran the hotel until about 1947 when he leased it to Fred Matthews and Dick Mercure, who ran the hotel and beer parlour until about 1950. After that he sold the hotel to Ed Barker and the stock in his store to the Northern Commercial Company. He returned to Ontario in 1954 and died there in 1957 at age 91. (See his daughter, Margaret Wood's story in *Pioneer Accounts*.)

MEYERS, Jack

Jack Meyers was foreman at the Calumet in the late 1940's.

MIDDLECOFF, Elmer Sr. & Nell

See *Pioneer Accounts*.

MIDDLECOFF, Elmer Land Jr.

– Archie Gillespie, *The Mayo Miner*, September 18, 1936

Elmer Middlecoff Jr., was the only son of Nell

and Elmer Middlecoff Sr. Elmer Junior was born in Seattle, Washington, on December 5, 1910, and came to the Mayo district when the Middlecoff family returned to Highet Creek in the summer of 1911. Elmer Jr., grew up in the Mayo district and attended the Mayo School. He made only one trip "outside," after his arrival as a small baby, and that was in the fall of 1929. His first adventure with the outside world was one that he enjoyed immensely and always remembered. During his teenage years, Elmer Jr., became an ardent hunter and was considered one of the best in the district. He never tired of hiking over the mountains and trails in pursuit of his favourite pastime.

In November, 1933, Elmer Jr., married Miss Mary Johnson of Mayo and the couple made their home on Highet Creek. In 1934 their daughter Jean was born and in the summer of 1936, their son Roy was born.

From the time that he was a growing young man, Elmer Jr., had worked with his father on their Highet Creek mining property. He proved himself of inestimable value to his father with his long practical experience and knowledge of the placer mining game. By the 1930's he had begun to assume more and more of his father's mining responsibilities and by 1936 Elmer Jr., was in charge of the Highet creek operations.

Tragically, Elmer Middlecoff Jr., was killed in a mining accident at Highet Creek on the morning of September 14, 1936. The accident occurred when a huge mass of blue clay caved in while Elmer Jr., was tending the hydraulic monitor. Although the accident happened so quickly that it was hard to judge just what happened, it was generally conceded that the force of the cave-in striking the monitor caused it to swing suddenly, dealing the young miner a terrific blow on the temple. As Elmer fell to the ground unconscious, the three men who were working in the cut at the time; C. R. Dickie, Tom Gregory and Walter Bleasdale; rushed to his aid. They immediately packed him to the mess house close by and made him as comfortable as possible on a cot. Dickie then started out for Minto Bridge to summon Dr. Nash. Nell and Elmer Middlecoff Sr., were in Mayo at the time of the accident.

Elmer Jr., died an hour and fifteen minutes from the time that he had been struck, never regaining consciousness.

Elmer Middlecoff Jr. with a bear cub. *Middlecoff-Bleiler Collection*

Upon receipt of the emergency call, Dr. Nash and Bud Fisher left in the latter's car at once, proceeding as far as the Highet Creek dredge where they left the car and hastened on foot to the scene of the accident, about a mile and a half further up the creek. Unable to do anything further the doctor and Bud returned to Mayo the next morning, bringing with them Mrs. Middlecoff Jr., and the children and Elmer Jr.'s sister Flora Middlecoff.

From Minto Bridge the sad, tragic news was phoned to Mayo. Middlecoff Sr., left at once by team to bring in the body of his son. Two members of the Highet Creek crew accompanied Middlecoff Sr., on his sad mission down the Highet Creek trail to Minto Bridge where they were met by Bud Fisher and his truck and brought into Mayo.

News of the accident came as a terrible shock not only to Elmer Jr.'s wife Mary and his parents, but to every resident of Mayo and the district. Elmer Middlecoff Jr., was well-known and well liked by older veterans of the camp and by the younger generation alike. Elmer Jr., was only 25 years old at the time of his death; his daughter Jean was just two years old and his son Roy was but a few months old.

Mary Middlecoff is now Mrs. Jack McDiarmid, daughter Jean is married and lives in Norway, son Roy was killed in a motor vehicle accident on the Dawson highway in 1987.

MIKESICH, Tom

Tom Mikesich staked the upper Discovery on Steep Creek on September 16, 1903. The claim lapsed and was restaked by Charles Augustus Anderson on January 1, 1906. Mikesich also staked claim No. 1 on Ledge Creek on August 18, 1903.

MILLAR, Mrs. M. E. (Mary B.)

Mary Millar (formerly of Vancouver, B.C., and Elkhorn, Manitoba) came to the Mayo district in the 1930's and lived in Elsa. Mary and her husband, Malcolm Edgar Millar (who never lived in the Yukon), had five children: Violet Millar (Mrs. H. W. Lacey) who lived in England and who visited her mother in Mayo in 1938; Reginald Thomas Millar who lived in both Elkhorn and Manson, Manitoba, before he came to Mayo; William Cecil Edgar Millar who lived in Vancouver and later Monte Creek and Revelstoke, B.C., but was never in the Yukon; and Rodney Malcolm Millar and Donald Millar who attended school in Vancouver and then came to Mayo. Millar brought Rodney and Donald to Mayo when they finished their education, Rodney arriving around 1935 and Donald a year later.

She moved the family from Elsa to Mayo in 1937. Her flower garden was one of the loveliest in Mayo.

Mary's brother was James Barr, who owned the large wholesale firm of "Barr and Sons" in Glasgow, Scotland, and was a Member of Parliament for Motherwell South, Lanarkshire, Scotland.

MILLAR, Reginald & Viola Annie

See *Pioneer Accounts*.

MILLAR, Rodney Malcolm & Elizabeth "Betty"

Rodney Millar was the third son of Mr. and Mrs. M. E. Millar. He was employed with Treadwell Yukon in the 1930's and was the owner and operator of the "Flying Husky Express" – the dog team that took the mail from Elsa to Keno. He married Elizabeth Evelyn Ida Kelly, youngest daughter of Mr. and Mrs. James Darton Kelly of Abbotsford, B.C., in Mayo on February 27, 1939. Elizabeth was a graduate of the Vancouver General Hospital nursing staff. She arrived in Mayo in December, 1938, and was engaged as Mrs. S. G. McLean's private nurse in Elsa, before her marriage. Betty was active in amateur theatre.

MILLER, George H. "Pop" & Ellen

George Miller was a partner in the "Little Guggs" Mining Co., (Rosmusen, Miller, Rockney, Lindquist and Godbout) circa 1910. George "Pop" Miller and his wife, Ellen, lived on Highet Creek. They had two children: a son, Henry who died at the age of 17 in 1926, and a daughter named Hilma.

Hilma was the postmistress in Mayo following Randy Maclennan's term. She married Roy Hallin of the Royal Canadian Corp. of Signals. "Pop" Miller became the school janitor in the 1940's and early 1950's.

"Pop" Miller was a keen gardener and had one of the finest gardens in Mayo with the loveliest display of flowers, chiefly flowering begonias, in his front yard.

MILLER, Herbert E. "Butch"

Herbert "Butch" Miller was a cook for the Department of Transport circa 1944 to 1950. He married Doris Fisher. Miller worked for Taylor and Drury in Mayo and Whitehorse. He was always in great demand for the softball team.

MILLER, Oscar

Oscar Miller had been a sailor before he came

to the Yukon where, in 1912, he mined on Duncan Creek. He found silver-lead ore at the head of Miller Creek (which flows from Galena Hill to Christal Creek) and staked the Dragon claim in 1921 and the Hobo in 1927 on a fraction which cut the Calumet. With the Calumet mine discovery, he sold the fraction for $30,000 to Treadwell Yukon receiving $10,000 down.

In the 1930's, he mined 15 miles from the mouth of Clear Creek with George Baudery and in 1937 he bought Joe Gergich's Clear Creek property. He was also one of Ed Barker's partners in the Haggart Creek Mining Co. The skills he had learned at sea came in handy in 1945 when he repaired the tramline cables for U.K.H.M.

Miller lived in Keno until his death in 1956.

MILOSEVICH, Nicolas "Little Nick"

Milosevich was a prospector and miner in the Keno Hill area in the 1920's and 1930's.

MILTON, John Joseph

John Milton, born in Brooklyn, New York, was a member of the Yukon Machine Gun Detachment during World War I. He is also listed as a miner in the Mayo area in Polk's 1915–16 directory.

MITCHELL, Mrs. Fanny

Fanny Mitchell came to Mayo from Black Hills. She had a bakery in Mayo in 1922 and 1923 on First Avenue near Centre Street.

MONNET, Father A.

Father Monnet was a native of Normandy, France and was the parish priest in the Mayo district in the 1930's. He left Mayo to be a military chaplain circa 1940. He acquired a Law Degree in Paris after World War II and returned to the Yukon to be the legal adviser for the Yukon Diocese. He died in Whitehorse, Yukon.

MOORE, Francis William

Francis Moore was one of the Duncan Creek stampeders, staking a claim on that creek in 1902. He staked on Edmonton Creek in June, 1904.

MOORE, Sam

Sam Moore lived in Keno in the 1920's.

MOORE, Walter Oliver

Walter Moore staked a claim on Edmonton Creek in June, 1904.

MORBERG, Raydor "Raider"

"Raider" Morberg and his brother, Roy, came from Saskatchewan during the Depression to work in the Kimbel Bros. sawmill. He married Maisie Profeit.

MOREAU, Samuel "Sam"

Samuel Moreau is listed at Keno Hill in Polk's 1923–24 directory. Sam Moreau and "Scotty" Gordon were mining on Sourdough Hill in 1926 and the Mayo *CHRONICLE* (March 13, 1926) reported that Moreau and Gordon's ore was "of a very high grade." During the 1920's and the 1930's he rented the liquor store premises to the Yukon Territorial Government in Keno and was also the liquor vendor. He owned the police quarters in Keno. Moreau and Henry Dubois did assessment work for Bermingham and Settlemier on the Artic Mastiff claims in the early 1930's. Moreau died in 1942 and is buried in the Mayo cemetery.

MORGAN, Joseph W.

Joseph Morgan is listed as living in Mayo and at Minto Bridge in the 1920's.

MORIARTY, Daniel

Daniel Moriarty is listed as a miner in the Mayo area in Polk's 1915–16 directory.

MORISHIGA, Frank

Frank Morishiga mined with Jimmy Sugiyama on the Happy claim in 1938.

MORNEAU, David

David Morneau is listed as a prospector in the Mayo area in Polk's 1915–16 directory and as a miner and prospector in Polk's 1923–24 directory.

MORRELL, May Gillespie Fielder

See *Pioneer Accounts*.

MORRIS, Charles

Charles Morris is listed as living in Mayo in Polk's 1923–24 directory.

MORRISON, Donald "Don"

Don Morrison and his partners Harry Colley and Tom McKay owned and prospected the Calumet group of claims. See *Pioneer Accounts*.

MORRISON, Hector

Hector Morrison was born in 1856 in Zorra,

Ontario, of Scottish parentage. He came to the Yukon in 1894 and in the fall of 1901 stampeded to Duncan Creek and remained in the district, prospecting the upper Stewart River, Hess River, and every creek that was struck in the district.

He was one of the original stakers on Highet Creek where he carried out a very successful drift mining operation. In 1907 he returned to Duncan Creek to mine.

In 1911 Morrison found a rich deposit of silver-lead ore in Gambler's Gulch on Keno Hill. He was also responsible for naming the Rogue River in Yukon and the Hungry Mountain south of Mayo.

He returned to Highet Creek where he worked for Elmer Middlecoff's placer mining operation from July to October 1917 and also for a short time in 1918.[47]

In September 1919, he went to work for the Guggenheims operation in Keno. On February 18, 1920 he staked the Lucky Queen claim. He also staked the Little Queen, and claims at Christal Creek and Hope Gulch. Keno Hill Limited, optioned the Lucky Queen and hired Morrison himself to work it for them. The Lucky Queen was returned to him when Keno Hill Limited ceased operating in the Mayo district. He was 75 years old when he sold his holdings to Treadwell Yukon in 1931.

On September 19, 1931, he married Grace Wier MacKay in Vancouver, B.C., and lived out the remainder of his life in comfort and style. He was a Mason and a devout Presbyterian. He could recite poetry by the hour and knew the names of all the wild flowers. He died in Vancouver on December 24, 1939.

MORWICK, Thomas

Thomas Morwick came to Mayo from Dawson and prospected in the Keno area. He died in 1933 and is buried in the Mayo cemetery.

MOSES, Chief Alfred "Albert""Chief Tom" & Alice

Chief Alfred Moses was known as "Chief Tom." He was leader of the Mayo people from 1925 to 1947. His daughter, Ella Moses, married Moses Lucas.

Ella and Moses' children are David "Little Dave" Moses, Alice Buyck, Flora Harper, Tommy Moses, Maggie Anderson, and Joan Ball. Chief Tom's son David Moses later served as Chief at the Old Village, as did his grandson Albert Moses.

MOSES, John

He was the Indian Special Constable who drove dog teams for the R.C.M.P., in their hunt for Albert Johnson, the "Mad Trapper," in 1932.[48]

MOSES, Mary

See *Pioneer Accounts*.

MOWATT, Alexander "Scotty"

Scotty Mowatt staked the Scotty claim on July 29, 1919. He was a teamster in Keno Hill and Mayo in the 1920's.

MUENZENBERGER, Horst

Horst Muenzenberger was born in Prussia on October 17, 1923, and spent his childhood in central Germany. He acquired the pipefitter trade during the years he spent in Germany.

He came to Canada in May 1955 and worked on construction in Edmonton, Alberta for a year. He came North in June 1956 to work for United Keno Hill Mines in Elsa and remained in their employ for 32 continuous years as boilerhouse operator, compressor operator, pipefitter, steamfitter and plumber. He holds the U.K.H.M. record for long service [to January, 1989].

MULDOON, Pat

Pat Muldoon was a truck driver for Kimbel Bros. in the 1930's. He came to this district from Saskatchewan with Walter Rude. In 1939 he went to work for Treadwell Yukon.

MURDOCK, Mike & Isabel

Mike Murdock was a Treadwell Yukon employee in the district in the 1930's. He was a member of the Treadwell Yukon bullgang in Mayo in 1937 and then returned to work at the Silver King. His wife, Isabel, and Marion Meger were sisters.

MURRAY, Mike

Mike Murray prospected in the Lansing and Rogue River country for many years in the early 1900's.[49]

NAGANO, George & Chiyo Ruth

George Nagano operated the GN Cafe in

Mayo from 1926 through to 1950. The congenial Nagano was famous for his banquets, children's Halloween parties, and retirement dinners.

Mrs. Nagano, Chiyo Ruth, was born in Japan in 1906, and came to Canada in 1914 and to Mayo in 1930. She died suddenly of pneumonia in the Mayo hospital on September 18, 1936. The Nagano children were all very young when their mother died; Marjory was 6, George Jr. was 4, and Jean was only 16 months. After her death, her sister Mary Burian helped to raise the Nagano children.

George Nagano died in 1951 and is buried in the Mayo cemetery near Mrs. Nagano.

NASH, C. W. "Charley"

Charley Nash was a Treadwell Yukon employee in the late 1930's. He was Dr. E. T. W. Nash's brother.

NASH, Dr. E. T. W.

Dr. Nash practiced medicine in Mayo from 1935 to 1941. He and his family had a recreational cabin at Five Mile Lake. He was the organizer of the Mayo Drama Club and directed many of its productions.

Dr. Nash and his family later settled in Victoria, B.C. where he established a medical practice.

NASH, Percival

Percival Nash was Frank Braine's partner. They travelled through the Lansing and Stewart River country from Edmonton on their way to the Klondike Gold Rush. After an unsuccessful search for gold in Dawson, they returned to the Northwest Territories to trap and trade. In 1902 they asked a number of Indian people from Fort Good Hope to guide them across the Mackenzie Mountains.

They built a trading post on the Stewart River at the mouth of the Lansing River and supplied it from Dawson. This post became known as

From left: George Nagano and children George Jr., Marjory and Jean at a Sunday School picnic, 1937. *G. A. McIntyre Collection*

Lansing Post. In 1905 Nash and an Indian boy from Lansing Creek assisted Charles Camsell in his Geological Survey of the Stewart River, Wind River, and Peel River to its junction with the Mackenzie River at the delta. Nash and the Indian boy guided them as far as the mountain's summit and then returned to Lansing Post.

NEGREAU, Mrs. Clara Frances

Mrs. Negreau homesteaded in the Mayo area in the 1920's. On November 25, 1921, she acquired two acres of land on the right limit of the Mayo River bounded on the east by the Cantin homestead and on the south by York and Douglas locations. She had a daughter, Maxwell. Negreau moved to Dawson and was hostess at the Principal Hotel in 1939.

NELSON (CLOSE), Annie

See *Pioneer Accounts*.

NELSON, "Big Steve" or "Little Steve"

Two prospectors known only as "Big Steve" and "Little Steve" found rich placer gold on Nelson Creek (a tributary of Mayo Lake) in 1916. One of their surnames was Nelson.[50]

NELSON, Thomas

Thomas Nelson was reported to have found gold in 1896 in the canyon four miles from the mouth of a creek now known as Haggart Creek and which he named Nelson Creek. That same year Thomas Haggart built two cabins on the creek and one on Dublin Gulch from which to prospect and mine, but he recorded no claims.

They were unsuccessful in the Klondike Gold Rush and returned to the creek in 1898. Thomas Haggart and his brother, Peter, Thomas Nelson and Warren Hiatt left Dawson but en route they quarrelled and split into two parties. Peter Haggart and Warren Hiatt reached Nelson Creek first. They staked Discovery and renamed the stream Haggart Creek after Peter Haggart.[51]

NEUGEBAUER, William

William Neugebauer worked at Elsa and later drove a taxi. He was killed in a car accident on July 23, 1964, on the Mayo road. Neugebauer is buried in the Mayo cemetery.

NICHOLS, Jack & Mary

Jack Nichols was a "catskinner" for T. C. Richards on the mail run before coming to Mayo. In this district he was a Treadwell Yukon employee at the Calumet but left in 1939 to join the Navy. He returned to Yukon after World War II and became a customs officer at Carcross. The Nichols had two daughters. His widow, Mary, lives in Whitehorse (1989).

NICHOLSON, Clint

Clint Nicholson was a Treadwell Yukon employee in the 1930's. He was a good boxer and was dubbed the "Silver Cyclone."

NICOL, Alexander J. "Alex"

See *Pioneer Accounts*.

NIDDERY, Norman & Senoah

Norman Niddery was a fiddler as well as a trapper and prospector in the Hess River country for many years. He worked on Highet Creek for Elmer Middlecoff from 1912 to 1918[52] and is listed in the Mayo-Keno Hill areas in Polk's 1923–24 directory. Niddery and George Ortell staked a Discovery claim on Johnson Creek on August 16, 1923.

He loved to play the fiddle and was still an accomplished fiddler in later life although his fingers were badly crippled from poling boats in his youth. His wife, Senoah, was Julia Mervyn's sister. Niddery left Mayo, and the Yukon, in the late 1960's to live with his nephew and died "outside."

NIXON, Charles "Charlie" & Mary

Charlie Nixon was a driver for Fisher Service Co., in the 1930's. He is Jean Boyle's brother. In 1935–36 Charlie and Alex Berry highgraded eight tons from the Twin Sisters claim and eight tons from the Lone Star claim. He and his Scottish-born wife, Mary, now reside in Richmond, British Columbia.

NOBLE, Hugh "Hughie"

Hugh Noble mined in Dawson City circa 1917 and, while there, bought land in the Municipality of South Vancouver. In the 1920's Hugh came to the Mayo district and prospected Keno Hill. Noble lived on Duncan Creek in his senior years.

There were no old age pensions at that time and it was difficult for elderly people who had no family and no income to make ends meet. Times were tough for senior bachelors who were no longer able to work continuously. To make matters worse, most of these old-timers were fiercely independent and would not ask for anything. Such was the case of Hughie Noble. In the dead of winter, when he had run out of money and future prospects looked bleak, he spread a canvas and blanket on the ground near his cabin (on Duncan Creek) and lay down on it. As the temperature was –40°F and he was clad only in light clothing, the life of this old-timer ended there.

NORDLING, Oscar

As a young boy Oscar Nordling arrived in Dawson with his family from Lulea, Sweden in 1914. He was the oldest child and he and his brothers Otto, and Tory were born in Sweden while their younger brother, Axel, and a sister were born in Dawson.

In the 1920's he mined at Keno Hill, then moved back to Dawson where he worked as a dredgemaster for Yukon Consolidated Gold Co. until the outbreak of World War II.

After the war he moved to Vancouver where he later died. He has a nephew in Dawson and a niece and nephew in Whitehorse (1989). His brother Tory still lives in Seattle.

NORMAN, Andrew

Andrew Norman is listed as a miner at Keno Hill in Polk's 1923–24 directory.

NORTH, Wilfred & Bertha

Wilfred North was Stan Barker's man in charge of telephones in Mayo during the early 1950's. The Norths had seven children and their oldest son, Nelson North, is a CBC announcer.

NOYD, Amos

Amos Noyd, born in Auburn, Ohio, was noted for being unusually tall. He mined on Henderson creek circa 1913–14. He joined the Yukon Machine Gun Detachment in World War I and upon his return, began a longtime employment with Treadwell Yukon.

OBESTER, Michael

Michael Obester is listed at Keno Hill in Polk's 1923–24 directory.

O'BRIEN, James Jonathan

Canadian-born James O'Brien went overseas with the Yukon Infantry Company during World War I. He is listed at Keno Hill in Polk's 1923–24 directory. O'Brien is buried in the Mayo cemetery.

O'BRIEN, John

John O'Brien is listed as quartz mining on Dublin Gulch in Polk's 1915–16 directory.

O'BRIEN, William

William O'Brien was born in Alberta. He went overseas with the Yukon Infantry Company during World War I and is listed as a miner at Keno Hill in Polk's 1923–24 directory.

OLIGNY, Arthur

– Archie Gillespie, *The Mayo Miner*, March 25, 1938 and April 1, 1938

Arthur Oligny was a native of St. Reamie, Quebec, and in his early years in the Yukon he mined in the Dawson district on Sulphur Creek. He spent about 30 years in the Mayo district engaged in mining, prospecting, logging, and trapping. It is understood that he was also a skilled millwright.

Oligny and Joe McIntosh had a logging operation up the Stewart River, and Arthur helped Joe build a cabin which he later bought following Joe McIntosh's death.

At one time Oligny and a partner became lost in the Troublesome Mountains behind Rampart. His partner died of exposure, but Oligny was rescued after 18 days in the mountainous wilds of that country.

Arthur Oligny died in his cabin in Mayo on March 24, 1938, at the age of 65.

O'LOANE, Richard M. "Dick"

"Dick" O'Loane was the manager of Taylor and Drury Ltd. in Mayo in the 1920's. He subsequently transferred to the Northern Commercial Company.

OLSON, Charles

Charles Olson is listed as a miner at Keno Hill in Polk's 1923–24 directory.

OLSON, Oscar

Oscar Olson is listed as mining on Johnson Creek in Polk's 1915–16 directory.

O'NEILL, John R.

Although John O'Neill never lived in the Mayo area, he had a great influence on the min-

ABOVE, Father Henk Huijbers and John R. O'Neill. BELOW, from left: Archie Martin, John R. O'Neill, Jim Cox, and Barry O'Neill, 1948. *Both photos by Herb Foster, O'Neill Collection*

ing game in this district during the late 1940's and early 1950's. In fact, he practically engineered the 1950's boom. He was from Almonte, Ontario, and had been a Federal Conservative Member of Parliament. He visited Keno shortly after United Keno Hill Mines Ltd., began mining on Galena Hill and optioned practically every property not controlled by U.K.H.M.

He was a godsend to the old-timers in Keno who had been holding on to their claims through the 1920's and 1930's with no returns. O'Neill optioned for big prices but typically paid a few hundred dollars a month to hold them. The first few companies formed out of these options were Silver Basin Yukon Mines, Yukeno Lead and Silver Mines Limited and Yukon Galena Hill Mines, promoted by Bryan W. Newkirk, a Toronto promoter.

Out of these companies, as they were refinanced and repromoted, evolved Yukeno Mines Limited, Mackeno Mines, and finally, Galkeno Mines Ltd. Most of the 30 odd companies taking part in the 1950's boom picked up O'Neill's options.

O'Neill's sons, Barry and Terry, were involved in his silver interests. Following the late 1940's and early 1950's boom, J. R. and Terry O'Neill went back to Ontario and did not return.

Barry O'Neill arrived in the district in 1948. He left in 1953 and returned in 1956 with his wife Mary-K. Barry O'Neill has continued his involvement with silver mining in this district to the present time.

ORTELL, George

See *Pioneer Accounts*.

OSATA, Fred

Fred Osata was a Treadwell Yukon employee at the Hector in the late 1930's. He partnered with Jimmy Sugiyama on Jimmy's Happy claim and they struck the rich Sugiyama vein in September 1936.

OTTAWAY, J.

J. Ottaway is listed as a miner at Keno Hill in Polk's 1923–24 directory.

PAGE, Veronica

Miss Veronica Page nursed overseas during World War I. Page, or "Pagey" as she was affectionately known, was a nurse and matron at the Mayo hospital in the 1920's and 1930's.

PALMER FAMILY

The Palmers originally came from Melfort, Saskatchewan. They had ten children, five boys and five girls: George, Wesley, Elmer, James, Lou (Mrs. L. B. Hartnell and later Mrs. Ed Kimbel), Hazel (Dalton), Voilet (Currie), Iva (Bilton), Maggie and Syd. George, Wesley and Elmer arrived in Mayo during the Depression to work for the Kimbel Bros. sawmill and their parents and the rest of the family followed, except for Maggie and Iva who did not come to Mayo. Iva Bilton, however, now lives in Whitehorse (1989).

Hazel Dalton and her brother George Palmer bought Mrs. Wootton's Club Cafe in 1939.

Wesley Palmer ran a daily car service between Mayo, the Silver King and Elsa in the 1930's. He and his wife, Florence, had two daughters, Norma and Leila. Their son, Wayne Palmer, was born after the couple left Mayo. Daughter Norma (now Mortimer) is the manager of Echo Answering in Whitehorse (1989). Wayne Palmer also lives in Whitehorse.

PALMER, R. H. "Bob"

"Bob" Palmer was also from Saskatchewan but he was not related to the Palmer family above. He owned and operated a general store and hotel in Keno Hill in the 1920's, but left the Yukon when the Wernecke mine closed. He went to Wells, B.C. but after mining started again on Keno Hill, he returned to Mayo and built the Silver Inn Hotel and Cafe.

In December, 1938, Palmer bought Binet's half of the Binet and Lefebvre sawmill and the name of the sawmill became Mayo Lumber Mills. After the mine closed in 1941 he left for Vancouver, B.C. selling both the Palmer Hotel in Keno and the Silver Inn in Mayo to Alex Arthur who reopened the Silver Inn in about 1947.

PARENT, Denis

Denis Parent prospected Parent Creek (a tributary to Duncan Creek) where he found placer gold and staked a Discovery in 1901. Parent staked claim No. 1 on Parent Creek on December 2, 1901, and when he transferred 50% of claim No. 1 to G. Beaudet on June 20, 1902, Denis named the Creek.

PARIS, Triffle

Triffle Paris was one of the Duncan Creek stampeders who came to the Mayo district in 1902. Paris acquired an interest in claims on Minto Creek in 1904 and was involved in the rush to Johnson Creek in 1915.

PARKER, Mr. & Mrs. John Frederick

The Parkers were well-known Mayo pioneers who lived in Mayo in the 1920's and part of the 1930's. Mrs. Parker was born in Australia in 1876. She was a baker in Mayo in 1922 and from 1928 to 1933 the Parker's operated a general store.

In 1932 she had a bakery in Sam Blackmore's rented cabin (later Cal Harris') on the alley. It is known that John Parker was ill at that time, probably suffering from silicosis. The Parkers later moved to Dawson and Parker worked for the Yukon Consolidated Gold Corporation. Mrs. Parker died in Dawson City in September 1939.

PATTERSON, John Melvyn

John Patterson was born on April 1, 1907, in Saint Joseph, Missouri. He was chief pilot for Rocky Mountain Airways when Clyde Wann of Yukon Airways and Exploration Company went to Colorado Springs, Colorado in 1928 to buy an Alexander Eaglerock airplane. Wann hired Patterson to take the plane to Whitehorse and then to stay on as Yukon Airways pilot. Patterson was also a qualified mechanic. Patterson went to St. Louis in September 1929 to take delivery of the *Queen of the Yukon II* when that plane was purchased by Yukon Airways. The *Queen of the Yukon II* was a five passenger Wright Whirlwind-powered Ryan Brougham. It had a cruising speed of 120 miles per hour, a high speed of 140 miles per hour, and was equipped for winter flying.

John Patterson was killed when the *Queen of the Yukon II* crashed into the ice of the Stewart River in front of Mayo on November 2, 1929. It was the first fatality in Yukon aviation history.

The Post Office Department had granted Yukon Airways permission to carry mail to Aklavik in the Northwest Territories, and 50 miles north of Aklavik if that proved feasible. Patterson was impatient to leave Mayo and return to Yukon Airways base in Whitehorse to prepare for his first flight. Bad weather, however, aborted the departure of [both the *Northern Light* and] the

Queen of the Yukon II from Mayo several times before it dawned clear and cold in Mayo on November 2.[53]

The *Northern Light* took off first and the *Queen* followed but failed to lift off whereupon Patterson taxied back to the upper end of the bar for another try. He got out of the plane and spent a few minutes apparently making an examination of the engine. Patterson then made a second attempt and rose into the air. (The *Queen* was bound for Dawson, she had no passengers and carried approximately 500 pounds of mail). Thirteen-year-old witness Dickson Steeves heard a noise like a pistol shot as soon as the plane was airborne. Rev. McCullum, Antone Besner and others also heard the noise and suspected it was a "backfire." Patterson no doubt realized the engine was malfunctioning and banked the *Queen* to return to Mayo. The plane was banked so steeply that the wings were almost vertical and it was in that position when the engine stalled. The *Queen* went down 150 yards from the air field and John Patterson was killed on impact.

The plane crashed on the ice on the opposite side of the river across from the hangar of Treadwell Yukon, the point from which it had taken off. It was about 80 feet off shore on the left limit. The plane went through the ice and was nose down in the water with the water level almost up to the rear end of the cabin. The left wing was entirely demolished and the left wheel was bent backwards and up so that it was laying parallel with the surface of the water. The right wing was intact. Several men who witnessed the crash arrived on the scene and began to take out some of the mail bags in an effort to free Patterson from the wreck. Very shortly, R.C.M.P. Sgt. W. J. D. Dempster arrived with more men, and planks were laid across the river over the unstable ice. Ropes and block and tackles were brought and the plane was pulled partly out of the water so that the body of the pilot could be recovered. The crash occurred shortly before 1:00 p.m. and the pilot's body was recovered around 3:00 p.m. Dr. D. R. Maclennan examined the body and stated that Patterson had been killed instantly. Mechanic Emil Forrest examined the engine after it was hauled out of the water and found the throttle wide open, and the spark and both magnetos on, indicating that the engine had been at full power when it crashed. Several people witnessed the entire incident, from the take off to the crash, from various points along the river bank in Mayo, about one third of a mile away. Dickson Steeves, Rev. Arthur McCullum, Antone Besner, George Besner, David W. Ballentine, and James Breaden were among the witnesses.[54]

Patterson's home had been in Denver, Colorado, and he had no family in the Yukon. The Mayo Yukon Order of Pioneers Lodge No. 3 called an emergency meeting and made Patterson an honorary member so that they could arrange for his funeral. An aircraft propeller marks his grave at the Mayo cemetery.

Forty years after the tragedy, Patterson's mother came to Mayo from Southern California to visit her son's grave. Mrs. Patterson was 84 years old when she journeyed to Mayo in 1969.

PATTON, James

James Patton was located on the Mayo-Minto Bridge Road 3.5 miles from Mayo in the 1920's. Patton owned 160 acres of land on the Yukon River below Independence Creek from approximately 1926 to 1950.

PAVISICH, Michael "Mike"

Michael Pavisich came from the Fortymile and staked Ledge Creek on September 18, 1903. He also staked claims on Keystone Creek, Minto Creek, Steep Creek and Cascade Creek. He continued prospecting in the Mayo-Keno Hill areas for a great many years.

PAVLOVICH, George & Audrey

George Pavlovich was born in 1902 and came to Canada from Hercegnovi, Yugoslavia in 1928.

He came North to work on the Alaska Highway and then walked from Whitehorse to Dawson where he worked at #4 Dredge on Bonanza Creek. He later ran the Occidental Hotel in Dawson and then worked on the steamboats.

He moved to Keno around 1950 and worked for Belle Keno, Mackeno, and then 14 years for United Keno Hill Mines as hoistman, timber boss, and boilerman.

He and his wife Audrey (née Anderson) have two sons, Kris and Danny. George Pavlovich died in 1977.

PAYNTER, Isaac N.

Isaac Paynter prospected Isaac and Idaho creeks circa 1916 in the Dawson district. He staked the Mowhawk claim on Keno Hill on May 16, 1928, and sold it to Treadwell Yukon. He also owned land near Mayo in the 1920's and 1930's.

PEILBECK, Lewis

Lewis Peilbeck is listed as a prospector in the Mayo area in Polk's 1915–16 directory.

PELLAND, Albert

Albert Pelland was Josie (Besner) Pelland's son. He worked as a bulldozer operator for Fisher Service Co. in the 1930's then later in the 1930's as a clerk in Jim Mervyn's store. His first wife Florence, was one of Julia and Jim Mervyn's daughters. After she died, he married Neoma Williams. They retired to Whitehorse and both passed away in the latter part of 1988.

PELLAND, Josephine "Josie"

"Josie" Pelland (née Besner) operated a candy, fruit and cigar confectionery in Mayo in 1923. She was the daughter of Mr. and Mrs. Ferdinand Besner and had two sons, Albert and Harvey.

PEPPARD, Ernest L.

Ernest Peppard was a miner with Treadwell Yukon at Wernecke. He is buried in the Mayo cemetery.

PETER, Charlie

– Archie Gillespie, *The Mayo Miner*, November 5, 1937

Charlie Peter was a long-time resident of the Mayo Indian Village and an excellent hunter and trapper. He died on November 1, 1937, survived by his wife, three children and one brother.

PETER, Lucy

See *Pioneer Accounts*.

PETER, Sam

See *Pioneer Accounts*.

PETERSON, Chris

Chris Peterson is listed as a miner in the Mayo area in Polk's 1923–24 directory.

PETIOT, Pete

Pete Petiot was a Treadwell Yukon employee at the Elsa mill in the 1930's. In September 1939 he moved to Mayo, opening a barber shop in room 1 of Mervyn's Hotel. He later moved to Whitehorse where he built on Main Street. He continued his barbering business and his wife had a beauty salon. He died quite young, of cancer, in Whitehorse.

A young Josephine "Josie" Besner. She later became Josie Pelland and ran a confectionery and cigar store in Mayo in 1923. *McLaren-Walloch Collection*

PHILIP, James C.

James was the license collector for the Town of Mayo in 1932. He died in 1933 and is buried in the Mayo cemetery.

PHILLIPS, James E.

Listed as manager of the Stewart Commercial Co., in Polk's 1915–16 directory and as mining in the Mayo area in Polk's 1923–24 directory.

PICKERING, Mr. & Mrs. J. E. "Jack"

"Jack" Pickering came from England where he had been in the transportation business. He started an express business in Dawson, then joined forces in 1915 or 1916 with Jim Greenfield, a livery stable owner.

Greenfield and Pickering obtained wood and freight contracts around Dawson and in this district hauled ore from the Silver King to Mayo. Their company was contracted by Keno Hill Limited to haul ore from Keno Hill to Mayo. The partners were two of the principals in Yukon Silver-Lead Mining company which operated a silver property on the side of Lookout Mountain in 1919. Jack Pickering staked the Pinochle claim on July 30, 1919.

The Pickerings made their home in Mayo.

PIKE, A. E. "Al"

Al Pike was Mine Manager at United Keno Hill Mines in 1953, and he was General Manager from 1957 until 1966. He was a 1933 University of British Columbia graduate in mining engineering and had worked in several British Columbia mines. D. R. Delaporte became General Manager following Pike's term.

PINDER, Beauchamp

Beauchamp Pinder is listed at Keno Hill in Polk's 1923–24 directory.

PLISSY, Charles

Charles Plissy is listed as a miner at Keno Hill in Polk's 1923–24 directory.

POLI, Cecil D.

Cecil Poli was Jack Alverson's prospecting partner. He was born in the Orkney Islands of Scotland, and at 18 he became a schooner sailor. He cruised the Caribbean and saw much of the world, then he did some prospecting in Quebec.

During the Depression he fished and lived off the sea near Estevan Point on the west coast of Vancouver Island.

He came to the Mayo district in 1935 and prospected Lookout Mountain and the McQuesten Valley. In 1936 and 1937 Poli, Jack Alverson, and George Potter prospected the south fork of the Klondike River for gold and they found and mined placer gold on Gem Creek.

During winter, he trapped for a living, and he trapped and prospected for a number of years around Mayo, on the Hess River, and at Henderson Creek.

In 1948, Poli restaked a former property of Alverson's near Haggart Creek calling it the Peso 1–12 claims. During the 1950's boom, Poli, Howard Colley and John Henry Gerlitzky staked the Leo claims and the Shanghai property. Poli and Alex Smith staked the twelve KPO claims in October, 1957, naming them after the San Francisco radio station they listened to on short wave during the long winters.

He continued to prospect in the Mayo district into the 1960's and then retired to Vancouver, B.C. He married Gladys Ellis, daughter of Jesse and Peter Ellis of Mayo.

POOLE, Don

Don Poole was a Treadwell Yukon employee at Keno Hill in the 1930's and was the brother of Ada (Mrs. Archie Gillespie).

PORTLOCK, William & Alice

William Portlock was a partner in Portlock and Abbot Bros., a partnership that mined on Haggart Creek from 1910 to the 1920's. His son, Tom Portlock, attended school in Mayo and later worked for White Pass and Yukon Route in Whitehorse. (See *Pioneer Accounts*)

POTTER, George Ferguson & Lena

George Potter came to the Yukon over the Chilkoot Pass. He was involved in the Duncan Creek stampede of 1901, promoted Haggart Creek and other ventures and was the first to drive a car overland from Dawson City to Whitehorse. He staked claim #10 above Discovery on Minto Creek in 1902.

George's daughter, Myrtle Mayo Potter, was the first non-native child to be born in the Town of Mayo. She was born in 1903 in the house that was later owned by Cora Rich. While George, his wife Lena, and daughter were living at Duncan Creek, Mrs. Potter became seriously ill, in December 1905, following an incomplete miscarriage. In a desperate attempt to save her life, Alec Nicol and George took Mrs. Potter to the hospital in Dawson with Nicol's dogteam. Mrs. Potter survived the arduous journey but died on January 2, 1906, a few hours after she arrived at the hospital.

The Potter's little daughter was only three years old when her mother died. Myrtle went to live with relatives in Vancouver, B.C., and later moved to Buckingham, Quebec. *The Mayo Miner* reported that Potter visited his daughter in Buckingham in 1938. Lena Potter was a relative (possibly a sister) of Mrs. George W. Saucerman (Clarence Saucerman's mother) of Dawson City.

In 1938, Potter purchased a new caterpillar in Chicago for use on his Haggart Creek property. The cat was an RD-4, hi-shovel lift, 5-roller track especially built for mining purposes. F. B. Towsley of Chicago joined Potter on his Haggart Creek venture at that time. Potter hauled 8 1/2 tons of equipment into Haggart, in March of 1938 including two heavy-service trailer sleds.

The Potters originally came from Buckingham in the Ottawa Valley of Quebec. His brothers were also involved in the mining game. Robert S. "Bob" Potter was one of Ontario's most widely-known mining operators and was one of the founders of the Potter-Dole Mines Ltd. His brother Ed arrived in Mayo in 1938 to work with him on the Haggart Creek property. His sister, Lily McDermid, was the mother of Roy McDermid and another sister, Hilda, was married to Kenneth Forbes.

George Ferguson Potter was shot on April 11, 1946, at 11:35 p.m., in Room 20 at the Chateau Mayo Hotel, by Peter Leonard Nord, over some hardrock claims on Johnson Creek. Mr. Potter died at 3:55 a.m. on April 12, 1946. George was in his seventies at the time of his death. (See *Shooting at the Chateau Mayo page 119*)

Myrtle Mayo Potter married Weldon Newton in Quebec in the late 1920's and the couple had three children: Robert "Bob" (deceased), Janet Ann Newton (now LeBreton), and George Stuart Newton.

POWERS, Thomas

Thomas Powers is listed at Mayo in Polk's 1923–24 directory.

PRATT, Lew

Lew Pratt is listed as a miner in the Mayo area in Polk's 1915–16 directory.

PROFEIT, Alex

Alex Profeit was the first school teacher in Mayo. He taught from 1913 to 1915. There were seven pupils in the 1913–14 school year. They were the three Kelly children, the three Ross children and Olive Kinsey. In the 1914–15 year there were six pupils. Profeit held the following creden-

Sixtymile River, spring 1914. From left: Garrison Costar, Chas. McPherson, George Potter, Chester Thomas. *A. K. Schellinger Collection*

tials: a first-class non-professional certificate from Ontario (a "non-professional" certificate was one granted for a written examination and later upgraded to a "professional" level following a satisfactory inspector's report); a second-class professional certificate from Toronto Normal School; and a second-class professional certificate issued by the Department of Education at Regina, Saskat-chewan. Profeit taught for five years before coming to the Yukon; one year in Ontario and four years in Saskatchewan. It is thought that he came to the Yukon in 1910 or 1911.

In 1911 he was night foreman on the construction of a new dredge at Grand Forks (Bonanza) near Dawson City.

From June to August 1918 Alex worked for Elmer Middlecoff on his placer property on Highet Creek, according to E. Middlecoff's 1910–1918 Time Book. Alex Profeit is listed in Polk's 1923–24 Alaska-Yukon Gazetteer and Business Directory as a fireman in Mayo. During the 1930's and 1940's he was a warehouseman.

Ella Profeit at Fraser Falls, circa 1920. *Dave Hager Collection*

Alex met his wife Ellen (Turgeon) in Mayo. They had a large family. Their children are Ella (Profeit) Rear, Isabel (Profeit) Tornai—deceased, Maisie (Profeit) Morberg, Bessie (Profeit) Friesen, Dorothy (Profeit) Genier, Charlie Profeit—deceased, James Profeit—deceased, William Profeit, Lawrence Profeit, and Mary Profeit—deceased.

PURDY, Charles

Charles Purdy was mining on O'Brien Creek circa 1913. He prospected in the Mayo area in the 1910's and owned property at Mayo Landing.

PUTNAM, Larry E. & Barbara

Larry Putnam was a University of Washington graduate in mining engineering. He was Wernecke's mining engineer in charge of the Silver King when the first shaft was sunk by the company in 1928, and he was in charge of Wernecke Camp while the Lucky Queen was being mined in 1932–33. He was also in charge of sinking the shaft on the Webfoot. Larry married Barbara Hargraves.

RADAKIVICH, Stephen

Stephen Radakivich is listed as a miner in the Mayo area in Polk's 1915–16 directory.

RAE, Frank

Frank Rae prospected the Wernecke Mountains with his partner Fred Hoffman. He was an old-time trapper and prospector in the Hart River region and had a cabin on Rae Lake. He and his partner worked there in the 1930's and the 1940's.

RAE, Samuel

Samuel Rae, born in England, went overseas with the Yukon Infantry Company during World War I and is listed as a miner in the Mayo area in Polk's 1915–16 directory.

RANKIN, Charlie

Charlie Rankin owned and operated the Northern Lights Hotel in Keno for a time in the 1920's. He also worked as road boss on the new roads that Treadwell Yukon was building in the 1920's. He died in November of 1930.

RANSOM, Percy & Elsie

Percy Ransom was the first bank manager in Mayo. He opened the Mayo Branch of the Bank of Montreal, located on First Avenue on April 2,

Larry Putnam with a Stanford Rod, May, 1930. *A.K. Schellinger Collection*

1928, and left the bank in July 1932. He was charged with embezzling in 1932, but a jury found him not guilty. His defence attorney at the trial was George Black.

His wife, Elsie, was a much-loved nurse at the Mayo Hospital. The Ransoms had two children, Ronald and Gwen.

RASMUSEN, Rudolph

See correct spelling Rosmusen, Rodolph

RAY, David A.

David Ray is listed as a miner in the Mayo area in Polk's 1915–16 directory. David, Harvey and Irvin Ray were brothers.

RAY, Harvey

Harvey Ray's father was a Nova Scotian who moved to California where Harvey was born.

Polk's directory shows that Harvey was mining in the Mayo area in 1915–16. He was also a trapper and prospector and had several cabins in the Mayo area, one was at Harvey Gulch (a short creek flowing into Highet Creek from Scheelite Dome) and another at the lower end of the canyon on Highet Creek where he mined on a bench for six or eight years.

He had two brothers who were also in the Yukon, Irvin and David. Ray made a trip to the U.S.A. in 1921 to see his mother but returned to the Yukon to spend the rest of his days prospecting.

RAY, Irvin & Elsy

Irvin Ray came to the Yukon when he was 17 and, while on the boat coming up the coast, he managed to impress an N.C. employee. He was told to contact Kippy Boerner, Manager of the Northern Commercial Store in Mayo, where he would be given a job.

Ray walked to Mayo from Whitehorse and records show that he was a clerk for the Northern Commercial Co. Ltd. in 1923–24. He left the store to become a mechanic at Wernecke Camp and returned to Washington State where he remained during the years 1929 to 1936.

In 1936 he married Elsy and the couple returned to Mayo to work for his brother, Harvey Ray, for two years on Highet Creek, followed by a partnership with Ed Barker who had the Haggart Creek Mining Co., mining on Haggart Creek. During World War II they took their equipment

Treadwell Yukon Cat, circa 1925–26. Charlie Rankin standing on Cat, Jack Alverson at left, Bill Hayden center front, with pipe. *Ralph Rogers Collection*

Harvey Ray, a pioneer prospector of both placer and hardrock deposits, circa 1925. He mined on lower Highet Creek. *Middlecoff-Bleiler Collection*

to Whitehorse and operated as Barker and Ray. Ray, Ed Barker, Jack Elliot and Wardy Forrest started Tourist Services in Whitehorse, one of the Yukon's first supermarkets. He mined on Shorty Creek in 1946.

Ray Lake west of Dezadeash Lake is named for Irvin Ray.

After he died, Elsy Ray married Robert J. Rowan.

RAYMOND, Horace "Frenchy"

"Frenchy" hopped a freight train in his home town of Quebec City in 1947 and worked his way across Canada. He heard that United Keno Hill Mines was hiring in Elsa, while he was in Edmonton, in 1949. He signed on and worked for United Keno until 1981. Since his retirement, he has lived in Mayo.

He is well-known for the many cars he has owned over the years. Numerous wedding parties in Mayo have enjoyed the loan of his Lincoln on their special day. By 1985, he had purchased his 25th new car since his arrival in Yukon.

REAR, Glen

Glen Rear came from Saskatchewan with the Palmer and Morberg brothers during the

Back from the hunt. From left: Irvin Ray and his wife, Elsy, and Ed Barker with caribou. *F. & T. Whitney Collection*

Depression. He was a logger, trapper and ice cutter, and he trapped in the Beaver district with Robert Greaves in 1937. He also had various carpentry contracts in Mayo during the 1930's, and built his own home on 4th Avenue between Wes Palmer's house and Jack Spence's place.

Rear played baseball and hockey, and wrestled in various competitions in Mayo and Dawson. He married Ella Profeit, daughter of and Mrs. Alex Profeit, on April 4, 1938, in Mayo.

RECKNESS, Ole

Ole Reckness is listed as a miner at Keno Hill in Polk's 1923–24 directory.

REDDY, Alfred

Alfred Reddy is listed in Mayo in Polk's 1923–24 directory.

REID, George

George Reid lived in Mayo in the 1930's. He ran the Mayo Library, occasionally did accounting work, and worked on the government road crew.

REINHARDT, Harry W.

Harry Reinhardt is listed as an accountant for Keno Hill Ltd. at Keno Hill in Polk's 1923–24 directory. The Reinhardts had two sons, one named Arthur.

RETALLACK, Thomas "Tom" & Athol

See *Pioneer Accounts*.

REYNOLDS, George & Phoebe Elizabeth (née Burns)

George Reynolds mined on Matson Creek circa 1913. In 1918, he and A. P. Krist were cutting hay at Swiftwater Slough on the Stewart River.

When he came to Mayo, he met and married his wife, Phoebe Burns who was nursing at the Mayo hospital.

In 1928, the Reynolds were mining with Hans R. Johnson on Ledge Creek. They bought Johnson out in 1932, and worked the ground until 1947. The Reynolds found a 10 1/4 ounce nugget on Ledge Creek in 1936. Reynolds mined on Hunker Creek in the Dawson area as well as in the Mayo area. He mined at Mayo Lake during the 1940's and lived in the cabin which is now owned by Shannon and Ken Cooper.

Phoebe was a British nurse during World War I and was decorated for her war service. She nursed at the Mayo hospital for several years in the late 1920's, and when Treadwell Yukon closed down and there was no doctor in the area (from 1942 to 1947) she worked as the district nurse.

Cora and son Gilbert Rich. *Rich Collection*

RICH, Mrs. Cora

Cora Rich came to Mayo around 1905. She was Gilbert Rich's mother, and Grant and Virgil Huffman's sister. She sewed the ore sacks for the first shipment of ore from the Silver King by Alverson and Huffman. She also had a marvellous garden that produced fresh fruits and vegetables which she sold to the steamboats docking in Mayo.

Her house was on the lot which is now owned by Don and Hazel Baker (1988). The Saskatoon bushes growing on that lot were originally planted by Mrs. Rich.

RICH, Mary

See *Pioneer Accounts*.

RICHARDS, Edward "Ted" & Dolly

Ted Richards conducted a laundry business at the Elsa beginning in 1937. In August of 1937 he secured a job in the Treadwell Yukon assay office at Elsa. Richards married Dolly Sullivan of Keno and Mayo. He joined the RCAF and was killed in action in Europe during World War II. A mountain east of Canyon Creek is named in his honour.

RICHARDS, Hickey C.

Hickey Richards drove Cat trains into Mayo in the late 1930's.

RICHARDS, T. C.

T. C. Richards worked for Pat Burns' company and made several overland cattle drives from Whitehorse to Mayo in the 1920's. He was manager for Burns and Co. Ltd. in the Yukon and carried mail into Mayo in the 1930's.

RIDDELL, Robert B.

Robert Riddell lived on the Riddell River (a tributary of the upper MacMillan River) and explored this great tract of wilderness by himself. In 1904 he helped Frederick C. Selous obtain big game specimens for various museums. Selous was a famous hunter and naturalist from Africa.[55]

RINGLE, Stephen

Stephen Ringle owned a 20-acre farm (Group 1004, Lot 88) known as Ringle Ranch in the 1920's.

RIVET, Father L. H.

– Archie Gillespie, *The Mayo Miner*, July 21. 1939

Father Rivet was a parish priest in the Mayo district in the 1930's. He later moved to Goldfields, B.C., where he died in 1939 after falling from a cliff.

ROBERTI, Fulvio

Fulvio Roberti grew up in Rimini, Italy. He came to Canada in 1953 and began work with United Keno Hill Mines in Elsa as a plumber on October 1, 1956, continuing with the company for more than 33 years. In addition to his duties as chief plumber, he is the mine's rescue specialist and Elsa's fire chief. Roberti met his wife Alla, a Belgian nurse, while she was working at the Mayo hospital and they were married in 1963. The Roberti's have three children. Alla Roberti passed away in Whitehorse on December 1, 1989.

ROBERTS, E. S.

E. S. Roberts is listed as Corporal in Charge of the R.C.M.P. at Keno Hill in Polk's 1923–24 directory.

ROBERTSON, Dale & Elizabeth "Betty"

Dale Robertson came to Mayo as an accountant for United Keno Hill Mines in 1946, then later left the company to become the Territorial Agent and Liquor Vendor in Mayo.

He subsequently worked for the Yukon Territorial Government in Dawson. At one point he also worked with Alex Wark at the Chateau Mayo, then moved to Whitehorse where he became Superintendent of Child Welfare.

His wife, Betty, was a nurse and often worked in the Mayo hospital. The Robertson's had three children, sons David, a teacher at Yukon College, and Neil and a daughter Phyllis.

Dale Robertson and his present wife Leona (née Woodcock) retired to Grand Forks, B.C.

ROBERTSON, David

David Robertson is listed as prospecting on Lightning Creek in Polk's 1915–16 directory.

ROBINSON, Harriet "Hallie"

"Hallie" Robinson was in her seventies when she came to Mayo from Dawson in the 1930's. She was the grandmother of Miss Desmond Robinson. She died in Mayo and is buried in the Mayo cemetery.

ROCHE, Laura

Laura Roche came to the Yukon in the 1930's to visit her old friend Mary Alice Fisher (of "Ma" Fisher's Boarding House) and ended up staying for several years. When Mrs. Fisher left the Yukon for good, Roche ran the boarding house for a while (circa 1936) and after that kept house (from 1936 to 1938) for Mrs. Fisher's son Bud, who was a widower with four children.

She left the Yukon in 1938 and died suddenly in Seattle while enroute to her home in Nebraska.

ROCKNEY, Charles

Charles Rockney was a partner in the Little Guggs Mining Co. from 1910 to the 1920's. He lived in a cabin on his Highet Creek claims.

RODEN, Charles A.

Charles Roden found and mined gold on Roden Creek (a tributary of the south McQuesten River) in 1907. (This creek was named for

Ralph Rogers standing in front of his tent office on the Mayo waterfront in 1922. Rogers was Treadwell Yukon's man in charge of loading the sacks of ore for shipping on the steamboats and unloading the sacks when they arrived in Mayo from the mine. *Ralph Rogers Collection*

Charles Roden but over time the spelling was altered and appears on maps today as Rodin Creek.) Rodolph Rosmusen staked the Discovery on Roden Creek on August 17, 1908.

Roden staked a claim on Paradise Creek on July 9, 1903. He continued to mine and prospect in the Stewart River area in the 1920's.

ROGERS, Ralph C. & Ruth

Ralph Rogers was a warehouseman for Keno Hill Ltd. in Mayo in 1921 and 1922. His office was a large tent placed beside the sacks of ore on the river bank.

In 1923 and 1924 he was warehouseman at Keno Hill. He was married to Ruth Rogers. He spent his retirement in British Columbia.

In 1983, Rogers generously donated his collection of Mayo photographs to the Mayo Historical Society and a number of those photographs appear in *Gold and Galena*.

ROSENBERG, Charles "Charley" & Laura

Charles Rosenberg was a member of the Taylor and Drury staff in Mayo in the 1930's, and was transferred to Taylor and Drury's store in Keno in February 1938.

He was from Laidlaw B.C. He married Laura Dale, daughter of Mr. and Mrs. W. Dale of Victoria, B.C., on August 2, 1938, in Mayo.

The couple had two sons, Donald and Murray.

ROSMUSEN, Rodolph

(The name has also been written "Rudolph Rosmussen," "Rudolph Rasmusson," or "Rudolph Rasmusen." A letter written in his own hand verifies the spelling of his name as Rodolph Rosmusen.)

He prospected the McQuesten River area with his partner Adolf Lindstrom in 1898 and mined placer gold near the mouth of Rodin Creek for a time. He staked a rich discovery at Minto Lake on June 15, 1903 with his partner Warren Elmer Hiatt, and was a partner in the Little Guggs Mining Company from 1910 to the 1920's.

He owned claims in the Silver Basin area which he sold to John R. O'Neill, who in turn sold them to Silver Basin Mines and the Yukeno Lead and Silver Mines Company Limited. Dan

Front row, left to right: Theos Whitney, Ruth Lesperance, and Eva Ladd. Back row: Dorothy Hoggan and Ruth Rogers, 1922. *Ralph Rogers Collection*

Steers worked his silver properties with him.

Rosmusen was also associated with Jack Hawthorne for a number of years.

ROSS, J. H.

J. H. Ross is listed as a miner at Keno Hill in Polk's 1923–24 directory.

ROSS, John Alexander "Jack" & Lena

See his daughter Mary Rich's story in *Pioneer Accounts*.

ROSS, Dr. C. C.

Dr. Ross was a physician in Mayo from 1925 to 1927. Dr. Ross's sister, Lillian Ross, nursed at the Mayo Hospital.

ROWAN, Mary

Mary Rowan lived in Mayo in the 1940's.

ROY, Joseph

Joseph Roy was a partner of Samuel Henry's on the Maisie Mae Ranch in 1918. He is also listed as a rancher at Scroggie Creek in Polk's 1923–24 directory.

ROYSTON, John Dobson "Jack"

Longtime Mayo resident Jack Royston was a trapper and market hunter. His area was on the Hess River, and until market hunting became illegal, he used to bring a load of moose meat to Mayo every fall.

He had a cabin on the Stewart River between Mayo and Fraser Falls as well as a cabin at Fraser Falls in the 1930's.

In Mayo, his place was the first log cabin upstream from the Anglican Rectory. Royston was also the president of the Mayo Lodge of the Yukon Order of Pioneers in the 1930's.

RUDE, Alf

Alf Rude lived in Mayo in the 1930's.

Ralph and Ruth Rogers making the trip from Mayo to Minto so that Mrs. Rogers could catch the stage from Dawson to Whitehorse in 1922. The temperature when they left 26-Mile that morning was –62°F. The temperature did not get above –50°F during Mrs. Rogers' 12-day trip from Mayo to Whitehorse. *Ralph Rogers Collection*

The Jack Ross family at Minto Lake. *L. Cheatham (McLaren) Collection*

RUDE, Walter

Walter Rude was a carpenter in Mayo in the late 1930's. He came from Melfort, Saskatchewan, with Pat Muldoon.

RUNACRES, Ed

Ed Runacres was a Treadwell Yukon employee in the 1930's.

RUNER, Pete

Pete Runer was a Treadwell Yukon employee on Galena Hill in the 1930's. The Runer vein on Mount Keno carries his name.

RUSSELL, Fenn

Fenn Russell operated a wood camp down the Stewart River in the 1930's.

RUVANE, M.

M. Ruvane is listed as a miner at Keno Hill in Polk's 1923–24 directory.

RYAN, Dave

Dave Ryan, an Australian, and his partner Ernest J. Corp prospected several properties in the Keno Hill area. The partners ran the post office at Keno City for several years during the 1930's. Corp and Ryan were still living in Keno in 1945.

RYDER, N.

N. Ryder operated a "jitney" in Mayo in 1927.

SAKATA, Massa

Massa Sakata came to Dawson in the early days, and it is believed he was a "houseboy" for E. O. Finlaison, who managed the Bank of Montreal in Dawson.

Sakata operated restaurants and barber shops in Dawson, Mayo, and Keno at various times. He was a cook for Treadwell in the 1930's and later owned a restaurant in Keno. He went back to Dawson and operated restaurants there until he returned to Keno during the silver boom in 1949 to reopen a cafe which he was still operating in 1966. Sakata retired to Whitehorse and died there.

SANDANGER, John

John Sandanger was born near Stavanger, Norway. He came to Canada when World War II ended and worked on the dredges in Dawson and then came to the Mayo district and rediscovered, developed, and mined on Johnson Creek. Klaus Djukastein became his partner in the 1950's, and when Hans Barchen joined the partnership the operation became known as Bardusan Placers Ltd. Sandanger retired to Vancouver, B.C., and died there.

SANDEEN, S. M. "Swan"

– Archie Gillespie, *The Mayo Miner*, September 30, 1938

"Swan" Sandeen cooked for the Yukon Gold Company at Keno Hill in the 1920's and later for Treadwell Yukon.

In the 1930's he was chef at the Hector Camp, and when he and Hector Morrison were both old men, they worked together on the Lucky Queen property. Swan helped Hector find the rich vein on the Lucky Queen which resulted in the option to Treadwell Yukon for $60,000. Swan Sandeen died in Vancouver in September 1938.

SANQUIST, Simon Peter "Pete"

– Archie Gillespie, *The Mayo Miner*, November 6, 1936

Pete Sandquist was born in Stockholm, Sweden, and in his younger days had served in the Swedish Navy before coming to North America. Pete was a pioneer of the Klondike and spent many years on Dominion Creek where he and his two brothers mined and ran a general store and post office. He moved to Dawson from Dominion around 1912, and his brothers left for their homes in California that same year.

Pete came to the Mayo district shortly after the silver camp opened up in 1920, and remained in the area until his death. He was working for Treadwell Yukon at the Silver King when he became ill. Pete Sanquist died following a lengthy illness, on November 5, 1936 in the Mayo hospital. It is believed that his mother was still living in Sweden at the time of his death.

A. K. "Shelly" Schellinger at the Elsa Mine Office, 1937. *A. Aho Collection*

SAUNDERS, John

John Saunders is listed as an assayer for Thomas Aitken in Mayo in Polk's 1915–16 directory.

SAVAGE, Victor

Victor Savage arrived in this district during the Duncan Creek stampede and staked claims on that creek in 1902.

He found and worked a placer gold deposit at Savage Gulch (a small stream flowing into Johnson Creek from Scheelite Dome) in the early 1900's (circa 1900–1905).[56]

SCHELLINGER, Alfred Kirk "Shelly"

Alfred "Shelly" Schellinger was born on July 12, 1887 at De Ruyter, New York. He obtained a degree in mining and geology from Stanford University in 1909.

He went to Dawson City in 1910 as a gold assayer and melter for the Guggenheims' Yukon Gold Company, and continued in that capacity until 1918. His wife Elsie, joined him in Dawson in 1913. The couple had two sons, Ken born in 1915 and Joe born in 1918. The family also lived in Mayo before returning to Palo Alto, California.

When the first ore from Louis Bouvette's Roulette claim was brought to Dawson, he assayed it and kept the silver beads from that assay as souvenirs for the rest of his life.

Realizing the value of the Bouvette discovery, the Yukon Gold Company sent him to the district to be the examining geologist-mining engineer. He was with Louis Bouvette when some of the original Keno Hill claims were staked in 1919. Schellinger staked the Keno claim on what became known as Keno Hill, on July 29, 1919. He remained in the Keno Hill district for more than 20 years, first with Keno Hill Limited as Production Superintendent, and later with Treadwell Yukon Company, Limited and its General Manager, Livingston Wernecke.

He played a leading role in the finding and production of more than 50,000,000 ounces of silver in galena concentrates and hand-sorted ores that also produced more than 100,000,000 pounds of lead at the smelter.

Much of his time and energy during his Yukon years was spent in geological and mining exploration activities to help his many untutored friends and prospectors. During the lean 1930's, his geological know-how and advice guided several hard-working but penniless prospectors to relatively rich discoveries, particularly in the Galena Hill area. Among these were the Hector and Calumet strikes made by Clem Sinyard and the Colley-McKay-Morrison partnership respectively.

He also had a variety of other interests. He introduced the first home radio receiver to the district in the early 1920's, and his photography hobby produced a collection of excellent photographs of Yukon mining activity and life in the early days.

In 1941, he retired to Palo Alto, California, where he spent close to another thirty productive years. During that time he was a World War II plant machinist and an early helicopter plant machinist. While in California, genealogy became another of his hobbies. Alfred Schellinger died in early 1969 in his 82nd year.[57]

SCHOGREN, Gustav R. "Gus"

Gus Schogren mined and prospected in the Mayo district from approximately 1905 to 1915. He operated a roadhouse with his partner Frank Chasni at Minto Bridge in the 1920's. Schogren died in 1929 and is buried in the Mayo cemetery.

SCHOLEFIELD, Gordon

Gordon Scholefield was a Bank of Montreal teller in Mayo in the 1930's, and later transferred to the Bank of Montreal in Dawson. He was also a sprint star and badminton ace.

SCHOLES, A. T. & Lena

A. T. Scholes worked on Duncan Creek in the 1930's. He married Lena Ross, daughter of Lena and John "Jack" Ross. The couple later moved to Vancouver, B.C.

SCHOLLY, John

John Scholly is listed in the Stewart River area in Polk's 1923–24 directory.

SCHROYEN, Frank & "Lil"

Frank and "Lil" Schroyen arrived in Elsa in 1949 where he worked as a carpenter and foreman for United Keno Hill Mines. The couple have two children. The Schroyens left Elsa in 1975.

SCOTT, John D. & Dorothy

John Scott was a mining graduate from the University of Washington. Livingston Wernecke gave him his first job in the mines in 1928–29 when Treadwell Yukon took over the Silver King mine and sank the first shaft into the vein. The

From left: Warren McFarland, A. K. Schellinger, Ruth MacFarland, Gladys MacKeen (Flemming), and Elsie Schellinger, circa 1922. *A. K. Schellinger Collection*

young Scott was on hand when Wernecke inspected the Calumet, Hector and No Cash in 1935. Scott wrote that at the time, "I was very low on the totem pole, just a 'Joe Boy', tagging along and saying, 'Yes, Mr. Wernecke'."

He went on to work as an engineer for Treadwell Yukon in the Mayo area for seven years, and worked under both Wernecke and Alfred Schellinger.

Scott's wife, Dorothy née Phelps, was a registered nurse. She sometimes came down from her home in Elsa to assist at the Mayo hospital. The Scotts have two sons, Michael and Bill.

John Scott lives in Whitehorse.

SCOTT, R. J. "Jack"

R. J. Scott was an R.C.M.P. constable in Mayo in 1923.

Dorothy (née Phelps) Scott with her firstborn son, Michael, circa 1940. *C. A. Boerner Collection*

SCOUGALE, James A.

James Scougale was a prosperous Dawson merchant who bought the discovery at Minto Lake from Makela and Scrivener and mined it with moderate success for a time. He was a pioneer prospector in the Mayo district circa 1901–1905 and mined on Scougale Creek (a tributary of the Beaver River).

Scougale died in 1927 and is buried in the Mayo cemetery.

SCRIVENER, James Thomas "Jimmy"

Jimmy Scrivener staked a Discovery claim on Paradise Creek (a tributary to Minto Lake) on May 19, 1903, and staked the old river bed at Minto Lake on the same day with his partner Peter Findlay Haggart. "Jimmy" Scrivener was an Irishman.

SCROGGIE, Ernest R.

Ernest Scroggie came from Rawdon, Quebec, leading a party of nine prospectors into the Stewart River area in 1898. On August 27, two of the men (J. G. Stephens of Deadwood, North Dakota and H. LeDuke of Quebec) found gold on Scroggie Creek. They staked the Discovery claim and the remainder of the party staked claims as well. The creek was named by Ernest's brother, Wealthy T. Scroggie.[58]

SEAHOLM, Hugo & Alberta

Hugo Seaholm worked for Ed Barker on Haggart Creek in the 1930's and early 1940's. His wife, Alberta "Berta" Neale, was a nurse at the Mayo Hospital. He was known to be an expert skier.

SEGSWORTH, Bob

Bob Segsworth was a mine superintendent for Keno Hill Mining in the late 1940's.

SETTLEMIER, Charles Reed

Charles Settlemier was a reporter for the *Dawson Daily News* in the 1910's and later became its editor.

He owned property at Mayo Landing and prospected widely in the Mayo area in the 1920's and 1930's and found silver-lead ore near Settlemier Creek (a tributary to the Beaver River) in 1923.[59]

He and his partner, Claude H. Bermingham, acquired the Arctic and Mastiff claims which

From left: Jack Stewart, Frank Short, and J. H. Farrell. *A. K. Schellinger Collection*

were later sold to United Keno Hill Mines.

He eventually became editor and manager of the *Fairbanks News-Miner*, and in 1937 he supervised a special edition of the paper entitled "*Goldfields Edition*." Charles was half-time editor of the *Alaska Weekly* in Seattle when he died.

SHAND, David C. "Davy" & Margaret "Peggy"

The Shands, who were both from Scotland, owned and operated the Hotel Stewart at Stewart River (also known as Stewart City or Stewart Island which had a population of 75 in 1903) from 1900 until Mrs. Shand sold the hotel in 1929 after her husband died.

Margaret Clark Shand wrote a book entitled *The Summit and Beyond* about the Shands' experiences in Stewart City. The book was published in 1959 by Caxton in Idaho.

SHANDRO, John & Winnifred "Winnie"

"Winnie" Shandro followed Mrs. Reynolds as the district nurse in the early 1940's. It was she who cared for George Ortell (see *Pioneer Accounts*) in a room at Mervyn's Hotel until a plane could get to Mayo to fly him to the hospital in Dawson City. She also administered care and what medicine was available to all those afflicted in the epidemics of whooping cough, chicken pox and measles that struck Mayo during the years 1940 to 1943.

SHARKE, Jack

Jack Sharke lived in Mayo in the 1930's.

SHAVER, Alex & Margaret Jane

Alex and his wife, Margaret Jane Shaver, had a homestead on the Pelly River and land 26 miles past the Mayo-Minto road in the 1920's. Shaver was an associate of Ira Van Bibber. He died in 1928 and is buried in the Mayo cemetery.

SHEARDOWN, Robert "Bob" & Margaret

Bob Sheardown was a Treadwell Yukon employee at the Silver King in the 1930's. He married Margaret Gillespie, daughter of R. L. Gillespie, and the couple moved to Whitehorse in the 1940's where he established a dry-cleaning business. He played the violin.

SHILLETO, Emery & Eileen

Emery Shilleto arrived in the Yukon in 1944 and remained employed by the U.S. Army, Air Force and Canadian Army until 1946. He came to Mayo during the winter of 1946-47. He worked for United Keno Hill at the time of the reopening of the Elsa-Calumet Mines. In the years that followed he was employed by a number of small mines in the area i.e. Klondike Keno, Comstock Keno, Belle Keno, and Mackeno. With the exception of Mackeno, these mines were short-lived.

In the late 1940's and early 1950's, he worked for Fisher Service in Mayo. Bud Fisher and Shilleto moved the first truck-transported load of concentrate from the waterfront in Mayo overland to Whitehorse using the newly constructed road from Mayo to Carmacks. They then continued from Carmacks to Whitehorse over the old winter road.

He married Eileen Mason-Wood, eldest daughter of Rose and Sam Mason-Wood of Mayo, in 1952. They had three children; Rod, deceased, son Blair lives in Whitehorse, and daughter Blaise in Whitehorse.

Shilleto became Mayo's Liquor Vendor and Territorial Agent in 1965 and continues in that position today (1989).

SHORT, Edward

Edward Short is listed as a miner in the Mayo area in Polk's 1915–16 directory.

SHORT, Frank R.

Frank Short, son of a Reno, Nevada, sheriff,

William C. "Billy" Sime was in charge of the Territorial Assay Office in Keno for 30 years. *G. A. McIntyre Collection*

graduated from Stanford University in 1898 as a mining engineer. He had managed various operations in the U.S.A. for the Guggenheims before they sent him from San Francisco to manage the exploration at Keno Hill.

After he completed his work at Keno Hill he returned to San Francisco. He later went to Pacific Tin's dredging operations in Malaya, and pioneered the flying of dredging equipment to the Bulolo gold fields in New Guinea.

SHUARD, A. R.

A. R. Shuard was a Treadwell Yukon employee at both Wernecke and Elsa in the 1920's and 1930's. He lived at Reid Lakes in the late 1930's.

SIME, Wm. C. "Billy" & Viola "Dolly"

Billy Sime was born in Dundee, Scotland and spent part of his childhood in Chile where his father was a mining engineer. He graduated from the University of Glasgow, Scotland, following in his father's footsteps as a mining engineer.

He came to the Yukon via the Stikine-Teslin route and arrived in Dawson in December of 1898. "Dolly" Sime, his wife, arrived in Dawson circa 1900 and worked as a dance-hall girl.

He worked as an assayer for the bank in Dawson and became well acquainted with the writer Robert Service. In 1901 Sime and three other men prospected the Stewart and Hess Rivers and trapped at Fairweather Lake.

He continued working for the bank for ten years, and was then put in charge of the Territorial Assay Office in Whitehorse.

Ten years later, the Territorial Assay Office was moved to Keno during the silver boom in the 1920's, and Sime, accompanied by Mrs. Sime, moved with it.

He operated the assay office in Keno for 30 years, and also served as Justice of the Peace while in Keno. He staked several claims in the Keno Hill area, grubstaked many prospectors, and was owner and prospector of the Rio claims on Galena Hill. He was involved with several of J. R. O'Neill's companies during the 1950's boom.

In 1952 he made a trip to Chile but became ill upon his return to Vancouver, B.C., where he subsequently died.

While he was away from the Yukon, his assay office burned down, destroying all of the priceless records contained inside.

SIMLEY, Steven

Steven Simley is listed at Stewart River in Polk's 1923–24 directory.

SINYARD, Clement "Clem"

Clem Sinyard was originally from Newfoundland, and came to the Yukon in 1899 to work with a construction crew on the White Pass Railway.

He staked the Hector claim on Galena Hill on July 21, 1920, and acquired the Jock claim. His financial backer was "Soapy" McCown, and the two men co-owned and prospected the Hector claims. In the 1930's, he struck a high-grade vein on the Hector and shipped ore from the claim.

After many years of hard work, he sold his Hector claim to Treadwell Yukon in the late 1930's when the price of silver was high. He received $200,000 for it, and after that bought an airplane drill to prospect the first 10 miles of the McQuesten River for placer gold with partners Tom and Charlie Rendall. Clem Sinyard died in 1949. He had a brother living in Baltimore, Maryland, in 1937.

From left: Ken Schellinger and Clem Sinyard at their hunting lodge. *A. K. Schellinger Collection*

SKONSENG, Thor M. "Ted"

Ted Skonseng came from Norway and worked in logging camps on Vancouver Island before coming to Mayo in 1935 to trap and log. He trapped in the Hess River country and worked as a woodcutter during the 1930's and 1940's. He worked as a carpenter for United Keno Hill in the 1950's.

He worked with Dr. Aaro Aho on various properties in Canada and Chile from 1963 until he died. He was involved in the discovery of the Faro mine.

In the late 1930's, he was taken hostage by a mentally disturbed trapper on the Hess River and forced at gunpoint to pull a sled from the Hess to Mayo. The trapper was later taken to an institution "outside" by the R.C.M.P.

Skonseng died an accidental death in Whitehorse.

SMITH, Alexander A. "Alex" & Alice "Babe"

Alex Smith was a Treadwell Yukon employee at Elsa from 1935 to 1940. He married Alice "Babe" Maclennan, daughter of Alexander Maclennan, in 1937 and they had three sons, Malcolm, Dwight and Warren.

He served on the Territorial Council for the Whitehorse area from 1945 to 1948. In 1949 he was employed as liquor vendor and territorial agent in Mayo, then bought and operated Dick Major's store at Elsa circa 1949 and later sold it to United Keno Hill.

SMITH, Clarendon Clark

Clarendon Smith was born in the U.S.A., possibly North Dakota. He went overseas with the Yukon Infantry Company during World War I. C. C. Smith is listed as a miner at Keno Hill in Polk's 1923–24 directory.

SMITH, George H.

George Smith is listed at Stewart River in Polk's 1923–24 directory.

SMITH, Harold

Harold Smith was a mining engineer and assayer at Wernecke Camp. Following that he was in charge at the Elsa during the first mill construction in 1935.

SMITH, John Valentine "Jack"

Jack Smith mined in the Mayo area from 1900 to the 1920's and made Mayo his home until the late 1940's. He lived in a cabin on the lot next to where the North Star Motel is today (1988).

Smith returned to his former home in Newfoundland and died there.

SMITH, Jonas

Jonas Smith owned property on the Mayo-Minto Bridge road in the 1920's, and is listed in the Polk's 1923–24 directory as a miner. Smith is buried in the Mayo cemetery.

John "Jack" Valentine Smith in Mayo. *G. J. Gordon Collection*

SMITH, T. C.

T. C. Smith is listed as a cook for Thomas Aitken in Polk's 1915–16 directory.

SMYTHE, Mrs. Emma

Emma Smythe owned and operated the Smythe Hotel and was postmistress at Stewart City circa 1916 to the 1920's.

SNELL, Coleman L.

Coleman Snell was a partner in the Cox, Snell and Cox Sawmill which operated at Mayo Landing from approximately 1903 to the late 1920's. The sawmill was located on the property which is today owned by the Mehaffey family. He also homesteaded on the 160 acres. Snell died in 1928 and is buried in the Mayo cemetery.

SOWALA, Samuel

Samuel Sowala is listed as a miner at Keno Hill in Polk's 1923–24 directory.

SPAN, Tony

Tony Span is listed as a miner at Keno Hill in Polk's 1923–24 directory.

SPARKS, David "Davie"

Davie Sparks was an Irishman who mined on Duncan Creek from 1903 to approximately 1920.

Sam Rae and R. S. McLean worked with him on his Duncan Creek property circa 1915. He held the ground on Duncan Creek from No. 8 up to No. 17 above Discovery. In the 1920's he prospected on Keno Hill.

SPENCE, John "Jock" "Jack"

This John Spence was originally from Scotland, and came to Mayo in the 1930's after working in Australia on the Sydney Harbour Bridge.

He did odd jobs, and trapped and prospected the area south of Mayo. His main trapping grounds were at the head of Sheep Creek about 80 miles from Mayo and he had a fishing cabin near the outlet of Ethel Lake.

In Mayo, he lived in a small frame cabin behind the cabin known as Rose Zenuik's. In 1938 he renovated the Pioneer Hall. Spence died in Mayo.

SPENCE, John Neilson

John Spence staked bar claims on the Stewart River in 1908. He is listed as a miner in the Mayo area in Polk's 1923–24 directory.

SPENCE, Robert

Robert Spence is listed in Mayo in Polk's 1923–24 directory.

SPRAGUE, Bowles Colgate

B. C. Sprague was a miner and prospector who found gold on Sprague Creek (which flows into the North McQuesten River) and lived there for many years. He staked on Clear Creek circa 1906, staked claims on Dublin Gulch in 1910 and in 1920, he made application to purchase 1.5 acres of land on the Mayo River. One half an acre of Sprague's land was absorbed when the old hospital site was established.

STEEVES, Dickson & Margaret

"Dick" Steeves was born in Mayo and lived in this area in the 1930's. He worked as a Cat skinner and master mechanic. He also had 8-mm movies taken of the workings at Keno Hill and the Elsa. He was the son of Harriet and Rupert Steeves and married Margaret Roll, a nurse at the Mayo Hospital.

STEEVES, Rupert S. "Rupe" & Harriet

"Rupe" Steeves and his wife, Harriet, came from Nova Scotia. They went to Dawson when they first arrived in the Yukon and then he obtained employment as a winchman on the Highet Creek Dredge in the Mayo area circa 1920.

He is listed as a carpenter in Mayo in Polk's 1923–24 directory. He worked for Treadwell Yukon off and on over the years.

The Steeves operated the roadhouse at Minto Bridge during the late 1920's and the 1930's. This roadhouse was owned by Treadwell Yukon.

STENBRATEN, John D. "Stampede John"

John Stenbraten was born in Sweden and was a noted Yukon prospector from 1904 to 1960. He prospected the Stewart River region at one point in his career, and mined at Keno Hill in the 1920's.

Stenbraten was well known throughout Yukon, Alaska, and northern British Columbia. He lived in Whitehorse and Atlin, B.C. for the remainder of his life. He married high school teacher Doris McMurphy in 1961. She still lives in Whitehorse (1989).

STEPHEN, John O.

Listed at Stewart River in Polk's 1923–24 directory.

STEWART, Alexander

Alexander Stewart is listed as mining in the Mayo area in Polk's 1923–24 directory.

STEWART, Allen

Allen Stewart is listed as a foreman for the Yukon Gold Co., at Keno Hill in Polk's 1923–24 directory.

STEWART, Arthur

Arthur Stewart is listed as a miner in the Mayo area in Polk's 1915–16 directory. Stewart died in 1930 and is buried in the Mayo cemetery.

STEWART, Jessie

Jessie Stewart was born in 1884 in Ryegiate, Vermont. She was a telephone operator for Mayo Utilities in Mayo and Keno in the 1920's and 1930's. Stewart died in 1945 and is buried in the Mayo cemetery.

STEWART, John A. "Jack"

Jack Stewart was a Nova Scotia hard-rock miner who had mined in Arizona for a time. In the early 1920's he was foreman of the Guggenheim's operation on Keno Hill where he supervised over twenty men.

STEWART, John Scott

John Stewart and William Catto mined an underground gold prospect on Dublin Gulch from 1915 to 1917 and called their operation Stewart and Catto. He was also a successful drift miner on Highet Creek, and worked for Elmer Middlecoff on Highet Creek in 1918.

STEWART, H. A. "Ray" & Elizabeth

Ray Stewart was born in Brighton, Maine on December 26, 1866. While he was working in the Coeur D'Alene silver district of Idaho, George Boswell told him about the Yukon gold prospects. He decided to go there himself in 1892.

When he reached Juneau he worked for a time at the Treadwell Mine to earn enough money for an outfit, then left for Dyea and the Chilkoot Pass. When he arrived at Fortymile he found that all the ground had been staked, but he managed to stake some of the deep ground on Miller Creek. Stewart worked his ground until 1894, then joined the Birch Creek stampede in Alaska.

He built the first large building in Circle for his horses and played a role in building about two-thirds of the cabins in Circle. He returned to Michigan just weeks before the big strike on Bonanza Creek in the Klondike and missed the initial staking.

He returned in 1897 with his brothers Arthur and Fifield and built a log store at Dominion Creek. Stewart, Arthur, and their partner Napoleon Huot bought claim 30 below Upper Discovery and several claims joining it from the Brennan brothers. The partners worked the ground for eight years and made a handsome stake.

On May 31, 1901 he married Elizabeth Sandison in Michigan, and the couple came to the Yukon on their honeymoon. In 1909 the partners sold their claims on Dominion Creek and Stewart returned to Michigan to start a fur farm. In 1915 he spent a winter in South America investigating the diamond mines in the Matto Grosso.

In 1918 he sold the farm in Michigan, moved his family to Seattle, and made plans to return to the Yukon. By this time the Stewarts had four children: Chester "Chet," Don, Gordon, and Evelyn.

In 1919 he prospected the head of the White River for native copper, and paddled down the White River to the trading post at the mouth of the Stewart River. There, he met Alfred Schellinger and crew who told him about the rich new silver prospects on Keno Hill. He returned the following season with a well-equipped outfit and teamed up with David A. Cunningham in Dawson. Cunningham had been his partner in a gold mining venture at Fortymile.

The two arrived in the Mayo district and began to prospect the Keno Hill and Galena Hill areas. In May 1920, Stewart staked the Poca Plata claim and by fall they had staked several other claims in the same area. The partners also staked the rich Ladue fraction and claims in the Lake Group. They sold the Ladue Fraction to Livingston Wernecke of Treadwell Yukon in 1921.

Stewart continued to prospect widely, and as his sons grew up, they joined him. He prospected Alaska, the Western States, Mexico and Brazil but always returned to the Yukon. He prospected from Keno to the Beaver and Wind River country with companions such as Emil Anderson and Jack Carpenter.

His son Don Stewart has written a book entitled *Sourdough Ray* detailing the adventures of Ray Stewart and his sons. Stewart lived to the fine old age of 98.

STEWART, William

William Stewart is listed at Stewart River in Polk's 1923–24 directory.

STRANBERG, Alex

Alex Stranberg was a carpenter in Mayo during the 1930's, and went to work for Treadwell Yukon in 1938 on the new tramway. He was an

From left: Ken Schellinger and Jimmy Sugiyama, 1938. Schellinger is using a precursor of the magnetometer. *A. K. Schellinger Collection*

expert skier.

STROM, Evert & Gladys

Evert Strom was born in Finland and raised in Sweden. He went to the Dawson area, from central British Columbia, in the 1930's and came to Mayo from Dawson circa 1937. He was employed by the Yukon government road crew as an excavator operator (shovel operator was the term used at the time) for many years. In his senior years he assisted Emery Shilleto in the Mayo liquor store (circa 1966–67).

Strom lived in a small log cabin near where Barry MacDonald's home is today. He numbered Norman Wightman and Wilf Gordon among his friends and he was married to Jack McDiarmid's cousin, Gladys. Strom died in 1972 and is buried in the Mayo cemetery.

SUGIYAMA, James "Jimmy"

Sugiyama mined in the Dawson district on Ten Mile Creek and Brewer Creek circa 1914–15.

He came to the Mayo district, bought one of the hotels in Keno from Harry Yamasaki and ran it for many years. He had a reputation for never turning a hungry man away.

He staked the Happy claim on Galena Hill on July 5, 1936 and closed the Galena Hotel in Keno to work on his silver property.

In September 1936 he and partner Fred Osata struck a rich vein on the property. It was called the Sugiyama vein and is still referred to by that name today.

SULLIVAN, John & Margaret

John Sullivan prospected in the Mayo district from 1920 to 1950. He was among the first men to reach the area during the rush to the silver-lead finds on Carpenter Creek in the early 1920's, and prospected in the Keno Hill area in the 1920's.

He and his wife Margaret ran a roadhouse at Half Way Lakes (also known as Sullivan Lake) during the late 1930's. They had several children: Jack, Glen, Mable, Dolly, and Laura. In the late 1930's he was a truck driver for Kimbel Bros. He died in Victoria around 1980. His son, Glen Sullivan, was killed in a mine accident at Stewart, B.C.

SUMMERS, Henry "Hank"

Hank Summers came from California to the Yukon in 1885. He bar mined on the Stewart River, joined the rush to the Fortymile in 1887, and later went to the Sixtymile. He made a fortune in the Klondike.

In 1913 he found gold on Miller Creek (a tributary to the Teslin River below Mason's Landing). He died in Whitehorse on December 6, 1913.[59]

SUNDERLAND, Floyd E. "Teddy"

Floyd "Teddy" Sunderland is listed as a waiter in Mayo in Polk's 1923–24 directory. "Teddy" and "Joe" were brothers and their mother was Mae Sunderland.

SUNDERLAND, Joel R. "Joe"

Joel Sunderland worked in the kitchen at Wernecke camp. He staked the Union claim and the Maiden's Hope on Keno Hill in 1919, and in May of 1920 he and his partner Frank McNeill staked the Friendship claim.

He also staked the Bluestone and other claims. In order to cover his fees at the Mining Recorder's office, he gave the Friendship claim to Raoul Binet in exchange for the fee money. The Friendship later proved to be an extremely rich claim.

After selling his Bluestone claim to Treadwell Yukon, he left for Seattle and San Francisco. It is said he spent his money on new cars and bought a race horse in Tijuana, Mexico. He eventually returned to work in the kitchen at Wernecke camp.

SUNDERLAND, Mrs. Mae

Mae Sunderland operated a roadhouse in Mayo from circa 1900 to the early 1920's. She was the mother of Floyd and Joel Sunderland.

Fred Kilbride (left) and Glen Sullivan listening to the radio, circa 1926. *Middlecoff-Bleiler Collection*

SUTHERLAND, Hugh

Hugh Sutherland originally came from Scotland. He is listed as a miner at Keno Hill in Polk's 1923–24 directory. He worked for Treadwell Yukon at the Elsa and died in a flu epidemic in 1936. His daughter Meg McCall was M.L.A., for Dawson from 1978–1982. His son, John, lives in California.

Sutherland is buried in the Mayo cemetery.

SUTHERLAND, William W.

William Sutherland was a Corporal in the R.C.M.P., stationed in Mayo in 1941. He married Antonette Hobbis, who worked at the Mayo Hospital.

SUTOLOVIC, Mr. & Mrs. Nick

Nick Sutolovic worked at Elsa during 1930's. He and his wife were a very warm hearted and friendly couple, their door was always open to children in the area. When the camp closed they moved to Whitehorse where they owned and operated the Capital Theatre for many years.

SUTTLES, John Jackson

John Suttles was born in Silver Creek, Kentucky on April 28, 1872. He was a member of the U.S. Army during the Spanish-American War (April 24, 1898 to December 10, 1898).

He came to the Mayo area and mined on Dublin Gulch from approximately 1899 to 1916. It was one of the larger operations in the district and employed several men. In October of 1914, a legal dispute occurred between Suttles and Frank Cantin (Cantin was mining on Haggart Creek) regarding Suttles tailings at Dublin Gulch. On February 16, 1915, the court ordered Suttles to pay $1,974.60 plus interest (from October 23, 1914) as well as court costs, poundage and incidental expenses. On February 24, 1915, Dawson Sheriff, George Brimston, seized Suttles' interests in and assets on claim numbers 11, 13, 15, 17 and 19, and Cantin acquired the claims. Local legend has it that Suttles placed a curse on Dublin Gulch following the loss of his claims.

On July 1, 1916, Suttles joined the World War I Canadian Expeditionary Force Yukon Infantry Company in Sidney, British Columbia. The company embarked on the SS *Canada* from Halifax, Nova Scotia, to go overseas on January 24, 1917. Suttles received the British War Medal for his service. When the war was over, he received an honourable discharge in Kingston, Ontario, on July 6,

From left: Dorothy McLaren, Mrs. Mae Sunderland and Lenore Walsh. *McLaren-Walloch Collection*

Back row, from left: Albert Taylor and Mrs. Isaac Taylor. Front row, from left: "Babe" Maclennan, "Betty" Maclennan, "Mickey" Maclennan, and Charlie Taylor, 1935. *Charles Taylor Collection, donated by Rick Cocks*

1919, and then returned to his home in Kentucky.

Suttles was not highly educated but he was musically gifted and could play almost any instrument. He enjoyed entertaining and played the guitar exceptionally well.

SWANSON, Fred

Fred Swanson is listed as a miner in the Mayo area in Polk's 1915–16 directory. In the 1930's he was a trapper, and then took over Bob Levac's trading post at Fraser Falls circa 1937. He owned a cabin upstream from the Anglican Church in Mayo.

SWANSON, Swan

Swan Swanson is listed as a teamster in Mayo in Polk's 1915–16 directory.

SWINEHART, Guy C.

Guy Swinehart is listed at Stewart River in Polk's 1923–24 directory.

SWIZINSKI, Martin

Martin Swizinski came from Alberta to the Mayo district in 1951 to work for United Keno Hill Mines. He left in 1951 and then returned to work for United Keno in 1955 and remained in their employ until 1964 when he moved to British Columbia to work for other mines. During his years with United Keno, he was an underground miner, shift boss and mine captain.

He came North again in 1971 to work for Canada Tungsten in Tungsten, N.W.T., and then returned to the Mayo district to boss Canada Tungsten's placer gold operation on Dublin Gulch.

In 1982 he optioned silver property on Mount Keno and formed Springmount mining company. He continues to mine the property today [1989].

TABER, Major Harold

Major Taber was one of the original members of the first Royal Canadian Corp of Signals detachments to come to the Yukon in 1923. He was later transferred to Mayo and was stationed here for several years (teaching school in Mayo part-time during 1925-26) before moving to eastern Canada.

TAYLOR, Albert D.

Albert Taylor, son of Isaac Taylor, worked as an accountant for Taylor and Drury's Mayo store from August 1935 to September 1936.

TAYLOR, C. D. N. "Red" & Margot "Margo"

"Red" Taylor, a professional engineer, was the "surface" engineer for United Keno Hill Mines and Assistant Manager of the company in the 1950's.

He supervised the building of the new hospital in Mayo in the early 1950's.

His wife, Margot Taylor, painted the famous *Wildflowers of Yukon and Alaska* while the couple was living in Elsa. Margot signed her paintings simply "Margo." She never had a show and painted mostly for family and friends but her wildflower painting is very popular and widely known throughout the Yukon. The Yukon Gallery in Whitehorse produces art cards and posters of *Wildflowers of Yukon and Alaska* and the original painting hangs in the Taylor's home in Victoria, B.C.

TAYLOR, Charles D. & Elizabeth "Betty"

See *Pioneer Accounts*.

TAYLOR, Fred

Fred Taylor ran the diesel plant for Treadwell's new mill at Elsa in the 1930's. When the price of gold went up in 1935, he left Treadwell's employ to placer mine gold at Dublin Gulch.

From left: Ann Taylor, Cliff Greig, and Fred Taylor. *Fred Taylor Collection*

When he first started mining he stayed at Ed Barker's camp and then moved to mine on lower Dublin with Ted Bleiler. Bleiler left the creek in 1936 and Taylor continued mining in 1937 and 1938 with two employees.

He marketed several tons of tungsten to the Canadian government during World War II and then again during the Korean War. He leased his ground from 1942 to 1946, mined again from 1946 to 1948 and then shut his operation down.

He began mining again with his two sons Frank and Jim in 1953. In 1955 he sold the upper half of his Dublin Gulch claim to Cliff Greig.

Frank Taylor, his wife Bonnie and their three children operate a placer mine (Duncan Creek Gold Dusters) on Duncan Creek today (1989). (See *Pioneer Accounts*)

TAYLOR, Isaac

Isaac Taylor was General Manager of Taylor and Drury in Whitehorse. See *Taylor and Drury Ltd. Goes to Mayo in 1919.*

TAYLOR, William

William Taylor was an Australian who mined on Barker Creek. He owned and operated the hotel at Stewart City in the 1930's, then sold the hotel to Bill Bamford in 1937. He later returned to New Zealand where he died in 1939. Taylor was Mrs. J. J. Van Bibber's uncle.

TAYLOR, William D. "Bill"

William "Bill" Taylor, eldest son of Isaac Taylor, managed Taylor and Drury's Mayo store from 1930 to 1932.

TEARE, Walter

Walter Teare was a blacksmith from the Isle of Man in the Irish Sea. He worked at Wernecke and Elsa and one season for Barker and Ray. Teare died in Mayo (circa 1950's or 1960's) and is buried in the Mayo cemetery.

TELFORD, Edward

Edward Telford was the R.N.W.M.P.'s Officer-in-Charge in Mayo in 1917. Telford was later appointed Yukon's Acting Gold Commissioner and Sheriff for the period 1922–1926. The Telfords had a daughter, Margery.

TESSIER, Arsene

Arsene Tessier staked claims on Duncan Creek in 1902 and mined there until 1907.

THOMAS, Roy & Edna

Roy Thomas was a truck driver for Treadwell Yukon in the 1930's. He was once in the

R.C.M.P., and took his discharge in Mayo. He married Edna Wharton and later settled in Victoria, B.C.

THOMEY, Bill & Doris

Bill Thomey was a Treadwell Yukon employee at the Silver King in the 1930's. He was also a guitar player. His wife's name was Doris.

THOMPSON, Ogden Pickett

Ogden Thompson came to the Yukon in 1898. He spent a number of years in the Thompson Creek area which flows north into the South McQuesten River. He was known as a good, hard-working prospector.[60]

Thompson staked a Discovery on Johnson Creek on October 26, 1914, where he mined profitably for some time.

THOMSON, Alba Root "Deep Hole"

– Archie Gillespie, *The Mayo Miner*, July 3, 1936

Alba Thomson was born in the Thousand Islands, on the St. Lawrence River just below Lake Ontario. He moved to Colorado at an early age where he lived through the gold rush days of that camp before coming north. He came to the Yukon in 1898 and mined for many years on Eldorado Creek and in other areas of the Dawson district.

He and his partner "Doc" Hayden prospected at Grand Forks. Alba gained the nickname "Deep Hole" Thomson as a result of a mining incident that happened in the early days of the Klondike. On 3-A Eldorado he sank an unusually deep shaft. During the winter the shaft filled with water and began to overflow and glacier for a distance of over five miles down Bonanza Creek. It is said to have cost the government in the neighbourhood of $50,000 to cap the hole.

In the Mayo district Alba made his home in Keno where he mined and prospected. In 1916, at age 70, he took a wrong turn on his way from Haggart Creek to Dawson and found himself on the Hart River. He built a raft and floated down to Fort McPherson on the Mackenzie. He returned to Dawson City in the summer via the Rat and the Porcupine Rivers, having travelled a distance of approximately 1,200 miles.

During the last months of his life Alba was in the habit of referring a great deal to his Eldorado shaft, claiming that there was still a secret channel of rich gravel there. He was a member of the Royal Arch and Knights Templar Lodges of Dawson and a member of the Shrine in Victoria.

He died in the Mayo hospital on June 30, 1936 at the age of 91. At the time of his death he had a son and daughter living in New York State.

THORNTON, Alfred B. "Cap"

– Archie Gillespie, *The Mayo Miner*, October 7, 1938

"Cap" Thornton was born in Liverpool, England in 1856. He was a colourful character of the Mayo district for many years, and spent most of his time in this part of the country in the vicinity of Cache Creek and Hansen Lake, near the McQuesten River.

In the early 1930's "Cap" left Mayo for the Stewart River district where he was located until he became ill and was taken to the hospital in Dawson. He died at St. Mary's hospital in Dawson on September 28, 1938, at the age of 82.

THURBER, Samuel J. "Sam"

Sam Thurber mined for a time on Duncan Creek in 1907 with Dan Matheson. In the 1920's, the partners acquired claims on Keno Hill that included the Malcolm, Butte, Pippin No. 1 and No. 2, Mary, Ora and Anaconda. Thurber was also associated with the Eagle claims and he owned the Solomon Fraction.

TIDD, Claude Britoff & Mary

Tidd was born in England and came to the Yukon circa 1914. He was an R.C.M.P. sergeant, and worked with the force in Dawson during World War I. He later worked with the R.C.M.P. in Old Crow, Ross River and Mayo (1930's). He married a nurse, Mary, from Fort Yukon.

In 1937 he retired from the R.C.M.P., and returned to Mayo to work for Treadwell Yukon's Mayo office. He stayed only one year and then went to Vancouver, B.C. where he served in security for the duration of the war.

He left Vancouver and went to live in King's Lynn, Norfolk, England, where he died. His wife, Mary, returned to her family home in Pennsylvania. Tidd was a great photographer of both still photographs and movies. His photograph collection is in the Yukon Archives. He was also a talented musician and played the piano and trombone.

TITUS, Elizabeth McDougall

Elizabeth McDougall was a nurse at the Mayo hospital when it first opened in 1922. She was promoted to Matron in 1923. She met and married Lou Titus while she was nursing at the hospital in Mayo. She was born in 1887 and died in Mayo in 1929. She is buried in the Mayo cemetery.

After she died, her niece Catherine McDougall arrived from Calgary, Alberta, to keep house for "Uncle Lou" Titus. Catherine worked for a time in Treadwell Yukon's Mayo office and then married Bud Fisher in 1938.

TITUS, Lewis Henry "Lou"

When Lou Titus first came into the district he was in charge of a dredging operation on Highet Creek in 1919. There is a large canvas poke imprinted with the name "Titus Dredge Co." on display in the Gold Room of the old Bank of Commerce building in Dawson City.

For many years in the 1930's he was head of transportation, warehousing and supplies for Treadwell Yukon and also ran the Mayo power plant in the 1930's in partnership with Alex Nicol. Titus married Elizabeth McDougall, matron of the Mayo hospital.

He continued as warehouseman for Treadwell Yukon until the mine closed in the 1940's, then worked in Mervyn's store until his death. Titus died in 1945 and is buried in the Mayo cemetery.

TOLMIE, Dan

Dan Tolmie, born in Ontario of Scottish parents, owned a saloon in Dawson City at one time.

He became Charlie Brefalt's partner in the development of the Elsa and No Cash claims and after the claims were optioned, he conserved his money and lived frugally in his old log cabin in Keno. Dan Tolmie died "outside."

TOMOFF, Mike

Mike Tomoff, of Russian background, owned the Northern Lights Hotel in Keno before Bobby Greaves did, and acquired the Bluebird claim on Galena Hill.

TORMEY, William

William "Bill" Tormey is listed as a prospector in the Mayo area in Polk's 1923–24 directory. He was originally from the eastern U.S.A.

He kept to himself and did not mix with the townspeople. He would come into Mayo to get his mail and promptly return to his neat green cottage near Joe "Rabbit's" farm. He seemed to have an affection for children and would stop to chat with them.

They found his knowledge of nature fascinating and would accompany him on the walk to his cabin while he explained and showed them the behaviour of various birds and pointed out the varieties of trees and flowers along the way.

He would invite the children in for "tea" which would consist of canned pineapple and "store" cookies. He had a huge collection of books and magazines which the children were allowed to read as well as a wonderful record collection which they were allowed to play.

Bill Tormey never spoke about himself but it was obvious to them that he was a well-educated man. He may have been a teacher or professor. It was not known if he had a family in the States.

TOURPIN, Sid

Sid Tourpin was a Mayo old-timer. He died in 1926 and is buried in the Mayo cemetery.

TOWNSEND, Norton Turner "Nort"

"Nort" Townsend went overseas with the Yukon Infantry Company during World War I. He is listed as operating a general store at Keno Hill in Polk's 1923–24 directory. He was the son of Turner Townsend, of the pioneer Dawson family.

TRAPAIRISH, George

George Trapairish lived in Mayo in the 1930's.

TREACY, Tom

Tom Treacy, an Irishman from County Mayo, came to Canada with his sister in the late 1920's. They settled in Edmonton where he worked as a labourer in the construction industry. His sister returned to Ireland circa 1930 and he came to the Yukon. For a number of years he worked in the Dawson area, where he acquired about 600 acres of land, with a building in the Indian River area. After his death, his heirs sold the property to Paul White whose company is mining on the Indian River [1989].

Tom came to the Mayo area in the late 1940's and worked at various labouring jobs. He was employed by Ray Stewart on the Galena Hill claim and did such a good job that Ray gave him a half interest. He also worked for United Keno Hill Mines, Ltd. as a miner.

Treacy died in the Mayo hospital circa 1962. He was good to his friends and a delightful Irish character with the gift of blarney. His nephews, John and Mike Mitchell, live in Carmacks.

TRESIDDER, Charles Martin

Charles Tresidder worked on Highet Creek for Elmer Middlecoff in 1910 and 1911[61] and is listed as a miner in the Mayo area in Polk's 1915–16 directory.

He trapped on Tresidder Creek (which enters the Stewart River about ten miles above Stewart Crossing) for some years and continued to prospect in the Mayo area until the 1940's. Tresidder Creek was named for him in 1941.

He owned an acre of land on the left limit of the Mayo River bounded on the east by Cantins homestead, on the south by the R.N.W.M.P. Reserve and on the west by the Mayo River.

TROMBLEY, Lloyd A.

Lloyd Trombley is listed as quartz mining on Dublin Gulch in Polk's 1915–16 directory.

TUNNICLIFFE, Samuel

Samuel Tunnicliffe was formerly a member of the R.C.M.P., in Carmacks. He operated the Broadway Hotel in Mayo for a short time. The hotel was not successful and some accounts say it was only open for a few weeks.

The Broadway was a two-storey log building located just beyond Churchward's Sheet Metal Shop. After the operation was abandoned, the building continued to supply windows and doors to Mayo homebuilders for years. It is believed Tunnicliffe originally came from England.

TURGEON, Charles Blaise

See *Pioneer Accounts*

TURNER, John "Jack"

Jack Turner was one of the Duncan Creek stampeders. He staked claim No. 4 above Discovery on Duncan Creek on December 3, 1901. On April 20, 1903, Turner et al acquired 75% of the upper end of Jake Davidson's upper Discovery on Duncan Creek. Turner mined on Duncan, just below the Discovery, for many years. Thomas Williamson worked with him on his claims in the 1920's. Turner died at approximately 70 years of age in 1934 and is buried in the Mayo cemetery.

VALENTINE, Rev. W. & Mrs.

Rev. Valentine was the Anglican minister in Mayo in 1937 and 1938 and was an Army Chaplain during World War II.

VANACORE, Giovanni "Johnny Vinegar"

Known by all as "Johnny Vinegar" he is listed in Polk's 1923–24 directory as a gardener for Frank Doyea, who had a greenhouse in Mayo.

Jack Alverson and Jack Turner skinning a wolf at Turner's McQuesten Lake cabin, March 1919. *A. K. Schellinger Collection*

Vanacore was born in Italy and lived in Mayo for quite some time. He was a miner at the Wernecke Camp before farming the property he owned near the Mayo airport.

He grew raspberries which he made into wine, produced wonderful blueberry and dandelion wine and some of the best homebrewed beer. He supplied Mayo with carrots, potatoes, turnips and cabbages in the 1930's and 1940's.

VAN BIBBER, Ira & Eliza

Ira Van Bibber lived for more than 60 years in the Yukon. The Van Bibbers had 14 children and many of their descendants live in Mayo: their sons George (and his wife Maria) and John James "J. J." (and wife Clara) lived in Mayo for many years. His son Pat (and his wife Ada) and his daughter May Fairclough continue to live in Mayo today (1989).

VAN BIBBER, Pat

Son of Eliza and Ira Van Bibber. See *Pioneer Accounts*.

VAN CLEAVE, C. M. "Charley" & Henrietta

"Charley" Van Cleave originally came from Minburn, Iowa. He and his wife, Henrietta, had previously operated a roadhouse in the Dawson City area before coming to Mayo in the 1920's to operate the Van Cleave Roadhouse on the Duncan Creek Road near the Mayo Lake Road junction. The Van Cleave's two sons always referred to the two little lakes just past the Van Cleave Roadhouse as Henrietta and Charley Lakes. "Charley" was also a prospector and trapper.

They later bought "Ma" Fisher's house in Mayo and lived there for a few years before moving to Vancouver.

VEALE, Alex

Alex Veale was mining on Isaac Creek in 1915–16 and he is listed at Thistle Creek in Polk's 1923–24 directory.

VEILHUR, Mathias

Mathias Veilhur is listed as a miner in the Mayo area in Polk's 1915–16 directory. He staked on Johnson Creek in 1915.

VERREAU, Eli

In the 1900–1920 period Eli Verreau was the mail carrier between Eagle and Dawson before Percy DeWolfe.

He came to Mayo in the early 1920's and cut wood for mining companies in the Duncan Creek-Mayo River area. He lived in a cabin on Duncan Creek and had another cabin at the foot of Mayo Lake. Verreau died in 1954 and is buried in the Mayo cemetery.

VOLKMAN, B. M.

B. M. Volkman was an early Dawson pioneer who operated a sawmill in Dawson. He came to Mayo during the 1920's boom and operated the Yukon Saw Mill Co., but left after just a short stay in the district.

VORTHIN, Henry "Dutch Henry"

"Dutch Henry" operated a roadhouse at Minto Bridge sometime during the period 1900 to 1919 and was a well-known character in the Mayo silver district where he was employed as a cook at the silver camps. He lived his last years in Mayo in a house behind Jack Smith's cabin and he died at St. Mary's hospital in Dawson on March 30, 1938.

VUCCER, Antone

Antone Vuccer is listed as a miner in the Mayo area in Polk's 1915–16 directory.

VUCER, John

John Vucer, born in Bosnia [Yugoslavia], is listed in Polk's 1915–16 directory as a miner in the Mayo area. He left the Yukon in 1916 to go overseas with the Yukon Infantry Company during World War I.

WADE, Fred H.

Fred Wade was an Australian miner who arrived in the district during the Duncan Creek stampede in 1902. He mined at Minto Lake in 1903.

WAECHTER Bros.

Waechter Bros. operated a retail butcher's business in the Town of Mayo in 1922 and 1923. The company had a floating refrigeration plant on the waterfront.

WALBY, William

William Walby was listed as quartz mining on Dublin Gulch in Polk's 1915–16 directory and as prospecting in the Mayo area in Polk's 1923–24 directory. He held placer claims on Dublin Gulch

Joe Walsh's home on Duncan Creek with Lillian McLaren by the doorway. *L. (McLaren) Cheatham Collection*

and lived for quite a number of years in a cabin at Minto Bridge circa 1915. Walby died in 1930 and is buried in the Mayo cemetery.

WALL, W. A. "Art" & Dorothy

Art Wall was United Keno Hill's Mill Superintendent from approximately 1952 until 1965. It is believed he had previously worked for Treadwell Yukon. H. E. Pawson became Mill Superintendent when he retired. The Walls had one daughter, Betty.

WALLI, Jack

Jack Walli was involved with the development of various silver properties in the Mayo district during the 1950's boom. He formed the company Klondike Keno and he and Jack Hogan mined Mount Keno on an independent basis. He helped to build the potash mines of Saskatchewan, Rio Blanco in Chile, as well as Great Lakes Nickel.

WALLINGHAM, Richard "Dick" & Margaret "Maggie"

Dick Wallingham came to the Yukon in 1947 to work for Yukon Consolidated Gold Company in the Klondike. He landed a job stripping at Hunker Creek running the monitor with Harry Pogson as his foreman. He arrived in Dawson in April [1947], left in October of the same year, and went to Whitehorse where he worked for three years. During the years he was based in Whitehorse, he drove a truck moving pipeline on the Canol Road and cut and hauled wood for the Army bases with partner Andy Hooper.

In 1950 he came to the Mayo district to work for Mackeno Mines as a truck driver and remained with Mackeno until 1954 when he joined the Chateau Mayo staff. He worked as a bartender at the Chateau Mayo until 1956 and then he and Roxy Carriere became partners and leased the hotel from Alex Wark. When Roxy retired in 1962, Wallingham and his wife, Maggie (née Mason-Wood), leased the hotel until 1967 and then bought the Chateau Mayo from Alex Wark.

They continued to own and operate the hotel until it burned down on March 23, 1986.

The Wallinghams have three children, Elaine (McLeod), Sam, and Walter.

WALSH, Joseph A. "Joe" & Lenore

Joe Walsh came to the Yukon in 1904. In 1912 he walked a Keystone drill under its own power all the way from Dawson City to Duncan Creek where he and his wife had a cabin. He drilled Duncan Creek with partners William Bramley and John "Jack" Adair and later drilled Rodin Creek.

In the 1920's he mined and prospected in the Keno Hill area. He was also Livingston Wernecke's chief prospector during the period when airborne prospecting parties were being sent out all over the territory by Treadwell Yukon. Walsh and pilot Everett Wasson were involved in the first B.C. air search and rescue. They found Paddy Burke's downed plane on the Liard River in 1930 and rescued the survivors.

In 1938, he and Sid Barrington bought three miles of placer leases on Boulder Creek in the McDames area of Cassiar district for $85,000, and took 75 tons of mining equipment into the property.

WARD, F. C.

F. C. Ward operated a roadhouse at the mouth of Scroggie Creek in 1918.

WARNER, George

– Archie Gillespie, *The Mayo Miner*, November 5, 1937 and November 12, 1937

George Warner was born in Virginia, U.S.A. He came North during the Klondike gold rush and was widely known in the Yukon, particularly in the Dawson district.

He was also well-known in the Mayo district where he worked for Treadwell Yukon for many years. George also worked as a warehouseman at Elsa camp. He died at the Mayo hospital on November 4, 1937 at age 68.

WASSON, Everett & Florence (née Jones)

Everett Wasson was a pioneer bush pilot in the Yukon and northern British Columbia. He flew a Fairchild FC2-W2 for Treadwell Yukon from 1928 to 1935. From 1935 until 1940, he was General Manager and Chief Pilot of the White Pass and Yukon Route's Aviation Division.

Wasson was involved in the first British Columbia aerial search and rescue while he was a pilot for Treadwell Yukon.

The downed plane, piloted by Paddy Burke with his engineer, Emil Kading, had flown from Atlin, B.C., to Liard Post to pick up prospector Robert Martin. They left Liard Post for Atlin with Martin aboard on October 11, 1930. They encountered a snow storm east of the Pelly Range and Burke landed the Junkers on the Liard River to wait out the storm. They took off the following day, but the weather in the mountains had worsened and when they returned to the river, the plane's pontoons were ripped open by rocks when they landed. The three men were stranded without winter clothing and with very little food.

Everett Wasson, pioneer bush pilot who lived with his wife and family in Mayo in the 1930's.
Middlecoff-Bleiler Collection

They had plenty of matches but only one axe and one rifle with twelve shells. The men decided to trek north to a food cache at Junkers Lake, at the headwaters of the Liard River. They estimated the distance to be 20 miles, when it was actually nearer to 60. They struggled for seven days through eight-foot-deep snow, packing their gear. When the party halted on October 24 they had no food left and Pilot Burke was suffering from exhaustion. Finally, on November 15, Emil Kading managed to drop a caribou with a single shot. Kading and Martin quickly prepared a soup for Paddy, but it was too late and Pilot Burke died. Kading and Martin struggled on for another five miles before they also succumbed to exhaustion.

Although word had gone out from Atlin that Burke's plane was missing, a series of aircraft accidents occurred which delayed search efforts until November. It was at this point that Everett Wasson, flying a Treadwell Yukon plane, became involved. Everett flew to Liard where he was incorrectly informed that Burke had intended to fly over Teslin, Yukon, on his way to Atlin. This misinformation caused Wasson to lose time flying hundreds of dangerous miles. Wasson then enlisted the aid of Mayo trapper and prospector Joe Walsh and the pair flew back over the Liard where they spotted the Junkers on December 4, 1930. Wasson and Walsh landed at the only available spot, 16 miles from the Junkers, and hiked into the site only to find the plane deserted. Wasson and Walsh returned to the skies and, while searching for a spot to fly across the Pelly Range on December 6, they spotted smoke and then two men waving from the forest below. Walsh kicked out two boxes of food, one of them Walsh and Wasson's own supply, with a note saying Wasson would land the plane some ten miles away on a lake and they would hike to where the men were. Wasson and Walsh were on the trail overnight with no food. They continued on at daylight, overshot their mark and were backtracking when they heard two rifle shots nearby and soon reached Martin and Kading. The four men struggled back to Wasson's plane on December 10. They had a great deal of difficulty starting the plane's engine but finally it fired and they were soon winging their way back to Whitehorse. Wasson, Walsh and Sgt. Leopold returned to the Liard on December 16 to claim Paddy Burke's body.[62]

Everett Wasson and Joe Walsh were honoured by the Federal and British Columbia governments for their rescue efforts.

The Wassons were well-known in Mayo where they lived during the 1930's. They had a daughter, Barbara.

Wasson died in Alexander Valley, California, in December of 1958.[63]

WATERS, W. C.

W. C. Waters was an R.C.M.P. Corporal in Mayo from January 1919 to January 1921. In 1923 he was Constable in Charge at Fortymile and also ran the Post Office.

WATSON, Robert

Robert Watson was a graduate in mathematics of the University of Toronto and a fellow alumnus of Joseph Keele who was a geologist with the Geological Survey of Canada working in the Yukon.

After being cheated of a claim on Bonanza Creek in 1896, he became so disgusted that he cut himself off from his fellow men. He withdrew to Watson Creek (a tributary of the Stewart River) where he lived in almost complete isolation. He trapped for a living, and made a close search of Shakespeare's works for the origins of modern slang terms as a pastime. He died and was buried on Watson Creek in the latter part of 1905.[64]

WATT, Garnet A.

Garnet Watt was a timekeeper for the Yukon Gold Company in the 1920's. He staked the Mathole claim, naming it after his sister. He later became a representative of Mutual Life (Vancouver, B.C.) and travelled to the Yukon every summer. He was married and had a family in Vancouver.

Watt died on the road to Haggart Creek circa early 1940's.

WATTERS, Jesse "Kit"

– Archie Gillespie, *The Mayo Miner*, January 20, 1939 and January 27, 1939

Kit Watters was born in Fredericton, N.B. He came to the Yukon during the Klondike gold rush in 1898 with Captain George Black and several other friends from Eastern Canada. After arriving in the Yukon, he and his partner George Black mined on Livingstone Creek in the Whitehorse district.

In the early days Kit worked for the White Pass on the overland mail route. He later went to Dawson where for many years he engaged in various pursuits then arrived in the Mayo district in about 1920. In Mayo he worked for the freighting firm of Greenfield and Pickering. From 1923 to 1938 Kit worked for Treadwell Yukon and then retired to Mayo.

Kit Watters was Ed Barker's first partner on Haggart Creek and helped Barker during his preliminary prospecting and mining explorations.

He was a rollicking, jovial pioneer character, droll of speech and generous in action. His word was law unto himself; never to be broken. His friendship was something once made, never forgotten. He was a most colourful and magnetic character. He died in a room at the Mervyn Hotel on January 16, 1939 at age 63.

WAUGH, Elizabeth (née Black)

Elizabeth Black was born on March 9, 1880, in Newcastle, Tyne, England. She came to Canada in 1909 to marry Aaron Waugh in Dawson City. Their daughters Frances and Mildred Waugh were later born in Dawson.

After her husband's death in 1929, she ran a roadhouse in Carmacks until she became ill in 1938 and was confined to St. Mary's hospital in Dawson. Her daughter, Mildred, cared for her in Dawson for a time and then she came to Mayo to live with daughter Frances Portlock. Mrs. Waugh died in Mayo on July 22, 1940, and is buried in the Mayo cemetery.

WAUGH, Frances

Frances Waugh, daughter of Elizabeth and Aaron Waugh of Dawson City, was a staff nurse at the Mayo General Hospital in the late 1930's. She married Thomas Portlock, son of Alice and William Portlock, on October 30, 1937.

She lives in Victoria, B.C. near her son Franklin and his wife and her grandchildren (1989).

WAUGH, Mildred "Millie"

Mildred Waugh visited her mother for a short time in Carmacks in 1937 and then came to Mayo in August of that year, to visit her sister, Frances Waugh. She relieved staff nurses at the Mayo hospital on several occasions during the year she lived in Mayo. She left Mayo in August, 1938, to care for her ailing mother in Dawson.

She married Chuck Harrison, a teller at the Dawson Branch of the Bank of Montreal, in Dawson. The couple met in Mayo while Harrison was at the Mayo Branch of the Bank of Montreal.

WERNECKE, Livingston & Mabel

Livingston Wernecke was born on January 16, 1883. His father was a fur trader in Livingston, Montana. In 1898, at the age of 15, he went to Dawson with botanist Robert S. Williams. He graduated with honours in mining and geology from the University of Washington in 1905. Shortly after graduation, he married Mabel Rushton. They had two children: a son Theodore and a daughter Claire.

Wernecke was draftsman and construction engineer on the Nevada Consolidated Copper Company plant at Ely, Nevada. He also did some surveying in Nome, Alaska; geologic reconnaissance of Alaska's Kenai Peninsula and Graham Island; engineering on the Copper River Northwestern Railway at Cordova, Alaska; construction with the Gastineau Company; and engineering and geology for Alaska Treadwell and Alaska Juneau Gold Mining Companies from 1914 to 1921. During that time, he scouted mineral properties for the Bradley interests throughout Alaska (particularly in the Kuskokwim area) and the Yukon.

In June of 1921, Fred Bradley, President of Treadwell, sent Wernecke to Mayo to examine silver prospects in the district. He was given complete authority to make deals on behalf of the company and he became the central figure in the Mayo silver district for many years.

He established Wernecke Camp and built a home there, but, because there were no schools near, his children attended schools in Victoria, B.C., and Seattle and Tacoma, Washington.

Wernecke was General Superintendent of Treadwell Yukon from 1922 to 1941.

He conducted extensive air reconnaissance prospecting throughout the Yukon from 1928 to 1931. The Depression brought an end to the air prospecting program. He kept the mines going during the hungry thirties with silver prices ranging from 25 cents to 42 cents an ounce, lead at 4 cents a pound and a penalty was charged if there was much zinc in the concentrates. There was no return on the zinc at all.

Wernecke had a cottage at Mayo Lake in the 1920's. The cottage was later moved to Mayo and the lake site was flooded out in 1952 by the Mayo River hydro project.

In 1935, he bought a large house in which he had a well-equipped scientific laboratory, in the Claremont area of Berkeley, California where he and his wife lived until his death in 1941. The Werneckes' daughter also lived there while she attended the University of California.

After moving to Berkeley, Wernecke worked

Livingston Wernecke. *Yukon Archives Collection*

out of San Francisco. By this time he was General Manager and Consulting Engineer for Treadwell Yukon Company, a geologist for Alaska Juneau Gold Mines, and President and General Manager of three other mining companies in the Western States.

In order to save time, all of his trips north to the Yukon were by air. His Bellanca airplane, which was formerly owned by the famous Hollywood screen actor Wallace Beery, was equipped with the latest in flying instruments and devices, including 2-way radio equipment.

In October of 1941, Wernecke and his pilot, Chuck Gropstis, flew into Mayo and he gave orders to the Treadwell Yukon staff to shut the mine down until World War II was over. He left Mayo and was to stop at Juneau and Hyder, Alaska on his way back to Seattle. After examining the Hyder property, Wernecke and his pilot left to fly back to Seattle on October 21, 1941. When they had not arrived in Seattle by October 23, Treadwell President, Phil Bradley sent out an alarm.

Wernecke and Gropstis crashed the Bellanca trying to rescue two survivors from another plane crash. They went down near Clam passage behind Salal Island near Millbank Sound (125 miles north of Alert Bay). Wernecke and Gropstis both died.

The survivors of the other crash, Gerald Bodding of Seattle and Harry Sherman of Dayton, Oregon, survived by salvaging rations and sleeping bags from Wernecke's plane. With the death of Livingston Wernecke all Treadwell Yukon activity stopped.

Wernecke's activities played a very important part in the economic history of the Yukon. This brief summary cannot do justice to the man or his work.

WESTMAN, Robert Morris "Bob"

Bob Westman came from Dewberry, Alberta. He was a Treadwell Yukon miner at the Calumet circa 1936–1940. He left the district in June of 1940, went to Toronto to join the Royal Canadian Air Force, and was killed in action April 29, 1943. A mountain east of the Beaver and Rackla rivers is named in his honour.

WHARTON, Edna

Edna Wharton ran Sullivan's roadhouse at Half Way Lake for a time in the late 1930's while Mrs. Sullivan was away. Wharton later worked at the Silver Inn, and in 1938 ran a dry-cleaning and pressing business in Mayo. Her children were Gwen and Stephen. She married Roy Thomas, an ex-R.C.M.P. officer and truck driver for Treadwell Yukon.

WHITE, Clare E.

Clare White graduated from the University of Alberta in 1927 as a mining engineer, and spent some time as superintendent of the north central district of Cominco. He took over as general manager of United Keno Hill Mines Ltd. in 1953. In 1957 he left for Labrador.

WHITE, John

– Archie Gillespie, *The Mayo Miner*, January 28, 1938

John White was born in Ontario and moved

Evelyn (left) and Ethel Middlecoff phototgraphed in Mayo, circa 1919. Ethel later married Stanley White. *G. & E. Lee Collection*

with his family to British Columbia at a very early age. He was a veteran of World War I. John and his partner Carl Miller came to the Yukon from Alberta with a string of pack horses via the old Liard Trail.

From the Kluane district in southern Yukon, he and Carl proceeded to Carmacks where they began a pack train operation between Carmacks and Mount Freegold.

After selling his interest in that enterprise, John came to Mayo in the spring of 1936 where he worked for Treadwell Yukon on Galena Hill. John White died in January of 1938 in the Mayo hospital at age 39.

WHITE, Mr. & Mrs. McLeod

McLeod White was a mining engineer who lived at Elsa. He had also lived in Atlin, B.C. The Whites had a son, Alan. Mrs. White, predeceased by her husband, died in Whitehorse circa 1987.

WHITE, Richard & Ivy

Richard White was a barber in Mayo in 1935. His wife's name was Ivy.

WHITE, Stanley J.

Stanley White, from London, England, was a bookkeeper in Mayo in the 1920's.

He worked on Highet Creek for Elmer Middlecoff and he married Ethel Middlecoff in the 1920's. They had two daughters: Lucille, who died in Vancouver, B.C. in 1987, and Bernice of Nanaimo, B.C. The family left Mayo for Dawson where White operated a dry goods store from the 1930's until his death circa 1950.

WHITNEY, Frank A.

Frank Whitney was the principal of the Dawson public school in 1917–18. He moved to Mayo in the early 1920's and became a clerk for Taylor and Drury Ltd. During the late 1920's and the 1930's he was a bookkeeper for Treadwell Yukon at the Wernecke Camp and then was timekeeper for Treadwell Yukon in their Mayo office from 1938 to the 1940's. In the 1940's, he worked for White Pass.

John and Nancy Whitney at Wernecke Camp in March, 1929. *Ralph Rogers Collection.*

Whitney and his wife, Theos, had two children, Nancy and John. They moved to Vancouver B. C. where he died when he was in his 80's. In 1986, Theos Whitney also died in Vancouver at the age of 92.

John Whitney joined the Bank of Montreal after graduating from St. Georges School in Vancouver. He began his career in Dawson City under Arnold T. Hall, followed by a posting to the main branch in Vancouver. He retired from Head Office as Vice-President in Charge of Credit for all of Canada and currently lives at Qualicum Beach, B.C. His sister Nancy Moulton lives in Vancouver, B.C.

WIGHTMAN, Alexander James "Sandy"

Sandy Wightman was the owner of the Betty claim on Galena Hill. He lived either on the claim or in Keno.

WIGHTMAN, Norman Buchanan

Norman was a ships engineer from Glasgow, Scotland, where he served his apprenticeship. During the Depression he was hired by White Pass as a fireman on one of the river boats.

Treadwell Yukon needed a machinist and he was hired. He worked at Wernecke until 1930 or 1931, when he quit along with Bon Kunze, Irvin Ray, and some others over Ed Barker's dismissal as master mechanic. He moved to Mayo and lived there until 1939 when he left to join the Air Force in World War II.

He returned after the war and remained in Mayo until his death circa 1970. He lived in the cabin just beyond the first Harry Ewing residence and south of the old hospital road.

WILLIAMS, Frank

Frank Williams came to the Yukon in 1898. He prospected and trapped in the Mayo area and east to the Mackenzie Mountains.

WILLIAMSON, Thomas

Thomas Williamson is listed as a miner in the Mayo area in Polk's 1915–16 directory. In Polk's 1923–24 directory he is listed as a trapper in the Mayo area and a miner at Keno Hill. Williamson died in 1935 and is buried in the Mayo cemetery.

WILLIAMSON, William Hodge "Billy"

Billy Williamson, born in Motherwell, Lanark, Scotland, came to the Mayo district during the Duncan Creek stampede in 1902. On October 7, 1903 he staked a claim on Cascade Creek.

He went overseas with the Yukon Infantry Company in 1917. When the war was over, he returned and worked as a miner for the Treadwell Yukon Company in the 1920's and 1930's.

He later leased the Lucky Queen from Treadwell Yukon and had it diamond drilled but found no additional ore. He did some high-grading during the late 1940's and made a few ore shipments.

He had a lovely tenor singing voice.

From left: Frank Whitney, Doreen Lee on his knee, Evelyn Lee and Theos Whitney, Mayo, circa 1937. *G. & E. Lee Collection*

WILSON, Charles

Charles Wilson was a Texan who came in to Mervyn's trading post at Lansing one season, bought some supplies then left with only a pack on his back and one pack dog. He did not appear again for 11 years.

Wilson reportedly staked the first prospects at the Britannia Copper Mine near Vancouver. He prospected for Dutch McHenry's lost gold find for several years, and prospected up the Hess River. He is reported to have brought out 50 to 60 pounds of quartz from the Hess region. These quartz samples contained free gold and were the richest samples ever brought into Mayo. Wilson was a loner who only occasionally went to trade at Lansing, always paying for his supplies in gold. He lived for a time near the sawmill in Mayo but returned to Lansing in 1939 where he later died.

WILSON, Herbert G.

Herbert Wilson is listed as mining on Highet Creek in Polk's 1915–16 directory.

WILSON, Yorke

Yorke Wilson was a cashier in the Mayo office of White Pass in the 1930's. When J. A. Fairborn retired in 1938, Wilson succeeded him as Mayo's White Pass agent. He was the son of Yukon Councillor George Wilson of Whitehorse. Wilson married Dorothy Rogers and they had two girls, Dorothy and Carol.

WINSLOW, E. T.

E. T. Winslow came to Mayo from Victoria, B.C. in October of 1938 to work at the Bank of Montreal.

WOLFE, Emil

Emil Wolfe is listed as a miner at Keno Hill in Polk's 1923–24 directory. Wolfe is buried in the Mayo cemetery.

WOOD, James Edward & Margaret "Maggie"

James Wood was a member of the Royal Canadian Corp. of Signals and manned the Mayo Radio Station in the 1930's. He went to Whitehorse during World War II and was a diesel operator for Yukon Electric. He married Jim Mervyn's daughter, Margaret May Mervyn. They have a son, Douglas, and a daughter, Bobby. (See *Pioneer Accounts*)

WOOD, Simon Stanley M. "Sam"

(See Mason-Wood, Simon Stanley "Sam")

Sam was the Mayo Mining Recorder in the

Party at the Wernecke Camp. Back row, from left: H. B. Smith, Norman B. Wightman, Jack Nichols, and Ed Kunze. Front row, from left: Margretta Gaundroue (Mrs.), Larry Putnam, Barbara Putnam (Mrs.), and Elmer Gaundroue. *Ed Kunze Collection*

1930's. His wife Rose (née Turgeon) was from Mayo. Their children are Mrs. Eileen Shilleto, Mrs. Margaret "Maggie" Wallingham, and Simon "Si" Mason-Wood. (Also see *Pioneer Accounts*)

WOOTTON, Mr. & Mrs. A. W. "Art"

– Archie Gillespie, *The Mayo Miner*

Art Wootton and his partner Fred Hale mined the Stewart River bars in the vicinity of 26-Mile in the 1930's. His wife operated a restaurant in Mayo initially called Buenos Aires Place, then re-named the New Cafe, and later named the Club Cafe.

Mrs. Wootton sold the restaurant to Mrs. Hazel Dalton and her brother George Palmer in March 1939. Art Wootton was mining on Barker Creek at the time.

WOZNIAK, Leo & Louise

Leo Wozniak was born in Posnan, Prussia [Poland] and his wife, Louise, was born in Mannheim, Germany. He came to Canada in 1956 and worked in Uranium City, Saskatchewan and Elliot Lake, Ontario. He arrived in the Mayo district in 1958 to work for United Keno Hill Mines at Calumet. His family joined him in 1960. When the Calumet mine closed down theirs was one of the last families to leave the town. Wozniak injured his back shortly thereafter and stopped underground mining to become the blacksmith at U.K.H.M.'s machine shop in Elsa.

In the early 1960's he partnered with Herman Moutner and Alouise "Louis" Hofer to prospect Duncan Creek. The trio gradually made their way on foot to Steep Creek, prospecting as they went. In 1962–63 they staked on Steep Creek and began mining. He bought his partners out in 1964–65. He was still working full time for United Keno Hill Mines and ran his placer gold operation during evenings and weekends, and spent all of his vacation time sluicing and working the claim.

In 1974–75 he acquired a claim on Anderson Creek and sold his Steep Creek property in 1977. Sons Hans and Manfred became his partners in 1977 and the claim has remained in the family since that time.

Mary and Aya Yamasaki in Keno during the 1920's. *T. McKay Collection*

During U.K.H.M.'s major layoff in 1982, the position of blacksmith was discontinued and Wozniak went to work on the bull gang.

Leo Wozniak died in December of 1983. Louise Wozniak has since moved outside of the Yukon. Their sons continue to mine on Anderson Creek.

WYNEN, Mr. & Mrs. Jeff

Jeff Wynen operated a woodsaw business in Mayo in the early 1920's. He moved to Dawson City where he spent most of his life before retiring to Victoria, B.C.

YAMASAKI, Harry & Tome

The Yamasakis worked in Dawson as cooks for the Guggenheim's Yukon Gold Company's operations there. When the Yukon Gold Company set up a camp in the Mayo district, the company brought the Yamasakis to Keno to cook.

Harry Yamasaki built a roadhouse in Keno which he called the Galena Hotel and the family operated it through the 1920's. After his wife's death, he sold the hotel and moved his family, daughters Mary and Aya and baby son, back to Japan.

YAMASHITA, Thomas

Thomas Yamashita is listed as a cook in Mayo in Polk's 1923–24 directory.

YARMAN, Frank

Frank Yarman is listed as a miner in the Mayo area in Polk's 1915–16 directory.

YELLAND, Harry

Harry Yelland was the N.C.O. in charge of the Royal Canadian Corp. of Signals station in Mayo during the 1930's. He moved to a station in West Edmonton in the late 1930's.

ZACCARELLI, Ralph & Marion

Ralph Zaccarelli was born in Dawson City. He came to Mayo from Dawson in the 1950's and was the liquor vendor and territorial agent for several years.

He later worked as an electrician for United Keno Hill Mines Ltd., then moved to Whitehorse where he worked for the Department of Public Works as an electrician. The Zaccarellis had two children, Rilla and Ralph, who both live in Whitehorse at present (1988).

ZAHN, Joe

Joe Zahn mined on Dublin Gulch in the early days. His wife died in childbirth at Dublin Gulch. He staked claims on Stand-To Hill in 1920. In his senior years pioneer Zahn lived in Keno.

ZUCCO, Antonio

Antonio Zucco is listed as a miner in the Mayo area in Polk's 1923–24 directory.

Endnotes

1 R. Coutts, *Yukon: Places and Names*. (Sidney, B.C.: Gray's Publishing, 1980), p. 1.
2 R. Coutts, pp. 9-178 respectively.
3 R. Coutts, p. 15.
4 R. Coutts, p. 16.
5 R. Coutts, p. 19.
6 R. Coutts, p. 20.
7 W. R. Hamilton, *The Yukon Story*. (Vancouver, B.C.: Mitchell Press Ltd. Second Ed, 1967).
8 R. Coutts, pp. 30-193 respectively.
9 R. Coutts, p. 85.
10 R. Coutts, p. 49.
11 Information taken from Elmer Middlecoff's Time Book 1910–1918.
12 R. Coutts, p. 53.
13 R. Coutts, p. 54.
14 R. Coutts, pp. 54-55.
15 R. Coutts, p. 55.
16 Information taken from Elmer Middlecoff's Time Book 1910–1918.
17 R. Coutts, p. 136.
18 R. Coutts, p. 91.
19 R. Coutts, p. 92.
20 R. Coutts, p. 96.
21 R. Coutts, p. 29.
22 Information supplied by Barry O'Neill.
23 R. Coutts, p. 87.
24 R. Coutts, p. 120.
25 H. S. Bostock, *Pack Horse Tracks*. (Canada Geological Survey, 1979), pp. 135-136.
26 R. Coutts, pp. 121-122.
27 R. Coutts, p. 177.
28 R. Coutts, p. 128.
29 R. Coutts, p. 131.
30 Information given by Paul Germaine.
31 R. Coutts, pp. 140-141.

[32] R. Coutts, p. 150.
[33] R. Coutts, p. 152.
[34] R. Coutts, p. 154.
[35] R. Coutts, p. 277.
[36] H.S. Bostock, p. 158.
[37] R. Coutts, p. 166.
[38] R. Coutts, pp. 181-226 respectively.
[39] Information by Fred Terwilliger & Linda MacDonald
[40] Information taken from Elmer Middlecoff's Time Book 1910–1918.
[41] R. Coutts, p. 139.
[42] R. Coutts, p. 265.
[43] R. Coutts, p. 173.
[44] R. Coutts, pp. 174-175.
[45] R. Coutts, p. 175.
[46] Information taken from Elmer Middlecoff's Time Book 1910–1918.
[47] Information taken from Elmer Middlecoff's Time Book 1910–1918.
[48] R. Coutts, p. 191.
[49] R. Coutts, p. 192.
[50] R. Coutts, p. 195.
[51] R. Coutts, p. 119.
[52] Information taken from Elmer Middlecoff's Time Book 1910–1918.
[53] Jeanne Harbottle, "Clyde Wann, Father of Yukon Aviation." *Alaska Journal*, Vol. 3 No. 4, Autumn 1973.
[54] R.C.M.P. Report on First Yukon Air Fatality, Mayo Staff Sgt. W. J. D. Dempster and Dawson Superintendent R. Field, November 4th 1929. Yukon Archives.
[55] R. Coutts, p. 223.
[56] R. Coutts, p. 232.
[57] Written by his son Ken Schellinger who lives in Perth, Australia [1989].
[58] R. Coutts, p. 235.
[59] R. Coutts, p. 184.
[60] R. Coutts, p. 266.
[61] Information taken from Elmer Middlecoff's Time Book 1910–1918.
[62] Frank H. Ellis, "First Air Rescue was Epic of Valor" (*The Vancouver Sun*, January 9, 1954).
[63] R. Coutts, p. 279.
[64] R. Coutts, p. 279.

Mount Haldane (a.k.a.: Lookout Mountain) a Mayo district landmark. *L. Bleiler*

PART FOUR ADDENDUM

ADDENDUM

Since *Gold & Galena* was first printed in 1990 new information has been provided which expands the book's historic data or corrects material already appearing in the first printing. Written information that was acquired by the Mayo Historical Society has been edited and included in this section. The addendum might have been much larger had all of the verbal information pertaining to the book's contents been written down. Alas, this task is beyond the capacity of the very small number of people who do the work of the Society.

The Society will continue to collect the written family histories of the Mayo region, and perhaps at some point in the future, another update may be written.

CLARK, J.V. (also see p. 356)

A thespian, an opera enthusiast, and photo buff, James Vincent Clark was born in the small town of St. George, New Brunswick, on September 16th, 1924. As a young man, he attended St. George Superior School in NB, where he was awarded the Lord Beaverbrook Scholarship in his senior year. After graduating, he attended the University of New Brunswick, and later McGill Medical School, where he graduated in 1952 with honours after serving a two year stint as a gunner in the RCAF.

Dr. James V. Clark, Mayo's longest serving Doctor. *(1976) R. Aitkins*

A matter of months after completing his internship at St. Mary's Hospital in Montreal, Clark was Yukon bound. Serving first as the resident doctor for United Keno Hill, and then later for the surrounding communities, Clark fell under the "Spell of the Yukon", and remained in the area until his death some 42 years later.

Clark was known best for his love of Wagner's operas and his "bedside manner". In 1976, after the Mayo Composite School was burned to the ground, the community honoured the normally publicity shy doctor by naming the newly built school after him.

Clark remained a fixture in the community, and served as the local medical professional until his retirement in 1993. It was only a short year afterwards that he passed away, on September 27, 1994 a week and a half past his 70th birthday. (Ryan MacGregor)

COOPER, Ken K. (see S. Cooper p. 264)

Shanon and Ken Cooper met at a curling bonspiel in Mayo in 1973 and were married by Si Mason Wood in the "old blue house" of Besner's on June 30th, 1974.

Ken came to Mayo in 1966 working for CN Telecommunications (now NorthwesTel) after having been in the Royal Canadian Airforce.

Together with his three young children of a previous marriage, Ken and Shanon raised a family of seven; Valerie, Gregory, Ryan, Lana, Kenny, Ross and Timothy. All the children were born at the Mayo General Hospital and delivered by Dr. Clark. (Shanon Cooper)

FELL, Fred

According to family history, Fred Fell travelled to Skagway, Alaska from Vancouver, B.C. In Vancouver he had been working on the docks. He arrived in Skagway between 1898 and 1900. He hiked up the Chilkoot Pass and rode a bicycle from Lake Bennett up the frozen rivers to Dawson City. His brother, Herbert Fell, was either with him or had followed him. On searching archival records, the family has not found anything of these early years. Placer claim certificates from 1903 to 1905 for claims staked on Minto, McIntyre, Jarvis, Paradise and Highet Creeks were located.

Fred sold his Highet Creek mining interests to his brother Herbert. It is believed that Fred left the area at that time while Herbert stayed in the area, teaming up with Andrew Johnson in 1907.

Herbert married Lilly Chapman in Prince Rupert. Their daughter, Ethel Hiatt Fell, was probably born in Dawson City in 1912. ("Hiatt" was named for the creek on which the Fells had mined.) Herbert is thought to have left the north in September 1912. Apparently Herbert helped to keep the peace in the Dawson City or Mayo area. (by Fell's grand daughter, Shirley Quiring)

Fred Fell, wedding day 1909 *Quiring Collection*

KLIPPERT, Conrad & Ina-Mae

"Con" and Ina-Mae Klippert arrived in Mayo in 1963 with their three children, Charmaine, Dan and Kim. Con and his older brother Bert were partners in construction, mining and Klippert Bros. Lumber supplying round and square timber to United Keno Hill Mines for a number of years. In the late 70's with the boys in the respective families being young men, Con and Bert dissolved their partnership and continued business in their own companies. Conrad died on October 11, 1998.

Conrad Klippert with Dan, Kim and Charmaine *(1957) Ina-Mae Klippert*

Charmaine, the eldest child of Con and Ina-Mae has a son Joshua. She manages Gateway Housing and has a small business in Whitehorse where she and her son live.

Dan Klippert operates a placer mine on Seattle Creek with his sons Justin and Kasey. Dan's wife Dee manages Yukon Housing in Mayo. They also have a daughter Koral.

Kim, the youngest son, placer mines on McNeil Creek and has an interest in a local garage and fuel distribution business. Kim is married to Cheryl and they have two children, Stephanie and Byron. Cheryl works at the Mayo school as a teaching assistant and at the gas station.

Con and Ina-Mae also have a foster-son Carl Blanchard. He and his wife Rose have three sons. They live in High Level, Alberta where Carl works for Nova and Rose drives the school bus.

McDONALD, Dan K. (p. 418)

"Danny" McDonald apprenticed his trade at the turn of the century in many of the huge machine shops that operated in Acadia back East. One of his brothers was Rod McDonald, a dredgemaster for YCGC. The other brother met with a fatal accident in Fairbanks, Alaska. His sister, Maggie, married Dick Diment of Dawson City.

Danny McDonald worked as a blacksmith and mechanic for dredging companies at 40-mile, 60-mile and in the Klondike. He was much in demand because he could weld with or without oxygen and acetylene. In about 1922 he moved to Keno to contract the construction of the wagon road between Keno and the Wernecke mill and camp using corduroy construction (spruce trees were cut and placed side by side crossway to the road direction and then covered with earth to make a firm roadbase) nearly all of the way because of the muddy sidehill. After helping to install the power plant on Duncan Creek he drove one of the first ten-ton Holtz tractors on the ore haul. In 1936 he helped Barker move the aerial tramway and mill from the Wernecke Camp to the Calumet Mines and Elsa.

In 1940 he returned to Dawson to join Black's Infantry Company. He returned to Dawson after the War to resume his mining. In 1949 he hired on as blacksmith at Calumet for UKHM and worked there until 1959 when he retired to his cabin in Dawson City.

He was the YOOP's representative to attend the 100th Anniversary of the opening of the Rideau Canal in Ottawa. In 1965 he retired to an apartment in Victoria and after a few years died of cancer of the gullet in the Veterans Hospital. His grave is under an evergreen tree at the foot of the Veteran's graveyard in Victoria, B. C. He did *not* die in Dawson as stated on p. 418. ("Red" Taylor)

RUNER, Harry

In the spring of 1940, as an 18 year old, fresh out of school, Harry was given his fare to the Yukon by his father Pete Runer (p.447). Harry and Pete worked the summer of 1940 underground at Calumet. When Calumet closed for the winter, Harry partnered with Ellis Johnson staking a claim on Duncan Creek about 4 miles south of Keno. They spent the winter there sinking a hole that was too far away from the pay dirt.

When Calumet opened again in the spring in 1941, they both went back to work to pay off the $500 debt that they owed Dick Major for grub. They worked there until the camp closed in October when Harry and his father Pete left for Vancouver. There, Harry joined the RCAF and never returned to the Yukon. Harry's father and brother did return to the Yukon to work for several more years. (Harry P. Runer)

TOWSLEY, Lyall H.

Towsley made a visit to the Yukon at age 14 spending much of the summer of 1941 on Haggart Creek with his Uncle Bob Potter. (see Potter p. 438).

WOZNIAK, Leo & Louise (p. 473)

Leo Wozniak was born in Annahuette, Germany, *not* Posnan.

He acquired a placer claim on Anderson Creek in 1974-75 and sold his Steep Creek property in 1977. Leo's sons Hans and Manfred joined him in the gold mining venture on Anderson Creek in the late 70's and early 80's. Leo Wozniak died in 1983. His widow Louise moved to Calgary, Alberta, and son Hans moved to Victoria, B.C. in 1987.

Manfred and his family remained and continue to operate the placer mine on Anderson Creek (Margrit Wozniak)

ZIMMERLEE, Art & Lily

Art Zimmerlee, Meda Alcock's father, had Trading Posts at Selkirk and Russell Creek, being partners with Mr. Schofield. The smaller Russell Creek Post in the MacMillan River region, was managed by Zimmerlee, while the main one at Ft. Selkirk on the Yukon River at the mouth of the Pelly River, was managed by Schofield. Schofield was the brains of the partnership. He was no "bush wrangler", and be-

Art Zimmerlee at Russell Creek Trading Post with furs at the end of the trapping season. *Meda Alcock Collection*

cause of the difficult travel by boat, he never saw Russell Post.

Art Zimmerlee was the "woods man". He trapped, hunted, built log cabins, and in his early life had been cook and guide for American trophy hunters. He came to the Yukon with his lifelong friend Alex Coward in 1911, stopping in Whitehorse where they built their first boat. That was the beginning of a long treacherous and lonely life in the Yukon. Thirty years for Zimmerlee and many more for Coward. With his own boat he was able to transport supplies from Selkirk to Russell Creek.

Major Armstrong who had the gold mine that we called the "Concession" some miles up the creek, was a neighbour. A great deal of mining equipment was taken in by horsepack train, possibly in 1924 by Mr. J. O. Erickson, later owner and Manager of the Regina Hotel in Whitehorse. Mr. Erickson had a busy season travelling through the wilderness of the Pelly and MacMillan Rivers ending up at the "Concession" lugging heavy equipment for Major Armstrong. Some of that mining equipment is still there.

Upon leaving the Yukon, Zimmerlee had the heartache of parting with his new boat called the *Dolphin*. He and Alex put together this "dream boat" to travel the Pelly and MacMillan Rivers. To the best of anybody's knowledge it was included in the sale of Russell Post. To this day the Zimmerlees wonder about its fate.

To all concerned, Art Zimmerlee was a fur man and cared little about mining.

Schofield and Zimmerlee sold the Selkirk Store to the Hudson's Bay Co. in about 1938. The following year Russell Post was sold to Oscar Erickson.

Zimmerlee, retired to B.C. settling in Richmond with his family. His wife Lily (Horsfall), was daughter of Julia and Joe Horsfall. The Horsfall's lived out their lives in Whitehorse, where both are buried. (Meda Alcock)

ERRATA

p.41 Photo, left to right Phillip, Frank, Louis and Joe Cantin (Doreen Lee)

p.50 Top photo date incorrect, probably circa 1916 (Doreen Lee)

p.116 Photo bottom picture. Man in apron likely "Dutch Henry" Vorthin. (Doreen Lee)

p.131 Top photo should be dated 1937/38. Unknown girl is Evelyn Wilms. (Doreen Lee)

p.132 Photo top picture. Back row - Alice Mervyn not Ruth MacDonald. Unknowns are Gwen and Stephen Wharton. (Doreen Lee)

p.142 List of Doctors: Drs. Homer and Baxter practised after Dr. Nash. (Doreen Lee)

p.146 Bottom photo: Boy in front is Bill Jeffrey. Front row, unknown next to Doris Fisher is likely Audrey Breaden (or Bernice Kazinsky). (Doreen Lee)

p. 147 Middle photo: name should be Thelma Jeffrey, not Nancy. (Doreen Lee)

p.154 Top photo: Standing next to Yorke Wilson is Jack Nichols. (Doreen Lee)

p.335 Photo: Person on left is Jack Smith. (Doreen Lee)

p.388 Joseph Horsfall (not Hosfall) died in Whitehorse December 31, 1943 and is buried in the Old Pioneer Cemetery. Julia Kirby Sim Horsfall, whose father was a Scot, is thought to be related to Jack McQuesten. Joseph and Julia had 9 children, 8 girls and 1 boy. The boy died and is buried in Ft. Selkirk. As of Feb. 2, 1999, all of the Horsfall children had died. (Rosalie Brown, grand daughter)

p.429 Father Monnet was from Brittany. (Doreen Lee)

p.454 Rupe and Harriet Steeves were both from Moncton, New Brunswick, *not* from Nova Scotia as stated in the *Pioneer List*. Harriet's father, Mr. Dickson was a MLA and Speaker of the House there for many years. (Doreen Lee)

p.457 Sullivan: son Jack died in Victoria in 1980's. (Doreen Lee)

p.471 Photo: Date should be 1934 . (Doreen Lee)

THE MAYO HISTORICAL SOCIETY

The Mayo Historical society was born at a meeting on March 23, 1981 when a group of interested townspeople met to discuss the possibility of preserving historic buildings. A Society was soon formed to research sites and seek funding for restoration. By 1982, however, the focus of the group expanded to include the possibility of writing a history of the whole Mayo district including not only sites but also family stories.

A form letter was sent out in the spring of 1982 to 200 individuals, businesses and government departments requesting the history of their activities in the area. Replies were enthusiastic and many included photos as well as stories. Between 1982 and 1990 additional stories were solicited from local and former residents. Address books were combed to establish a data base and eventually 485 letters went out. Pioneer lists were drawn up and circulated amongst "old timers" for proofing and additions.

With a donated copy stand, and a borrowed camera, the Society set up shop in the community hall to copy people's photo albums. A tape recorder was purchased to record the stories that were told. Slowly but surely the little community booklet that was envisioned gave way to a 500 page hard cover coffee table book that was published in 1990. Printing of *Gold & Galena* was made possible through raffles, pre-sales, a grant from The Yukon Foundation, a loan from the Yukon Economic Development Department and a grant from Yukon Lotteries that was repaid. The majority of the book was compiled by Society members, Linda MacDonald and Lynette Bleiler. The maps and the conversion from typewriter

Silver mining display at Binet House (1999) *L. Bleiler*

Renovated Binet House (right) and Annex serve as an Interpretive Centre for the Mayo Region. *L. Bleiler Collection*

to computer files were done by Lowell Bleiler. Finally on March 26, 1991 two thousand copies of the book arrived in Mayo.

With some of the proceeds from the first printing, the Society established a Fund with the Yukon Foundation for projects associated with the Mayo District. Sales of the book con-

Official opening of the Binet House Annex (1993), Corp. Rick Noack, Joyce Ronaghan (Pres., Silver Trail Tourism Assoc.) Lyn Bleiler (Pres., Mayo Hist. Soc.), Mike McGinnis (Mayor) and Piers McDonald (MLA for Mayo). *Anne Meaney-Leckie*

tinue to provide a major source of funds for the Society's objective of preserving Mayo's history. With the second printing of 2000 volumes, the Society will be able to continue its undertakings.

Historic preservation of buildings in Mayo was still of interest and indeed work along these lines paralleled the book's development. By June 1993 work was completed on the renovation of Gene Binet's house. Binet was one of the founding fathers of Mayo and built one of its first two-story homes.

The annex to the house was opened to the public that summer and displayed a selection of the Society's collection of photos. In the summer of 1994 the house itself was opened as an Interpretive Centre, showcasing the local flora, fauna, geology, mining and outback. The displays were financed and overseen by the Historical Society with a grant from the Canada/Yukon Mineral Development Agreement.

In 1999 the display was renovated again under the auspices of the Society. Also in 1999 the Executive of the Mayo Historical Society made the decision to reprint it's successful community history book. Corrections were made and addendum were added. Future upgrades to the Binet House displays will be funded in part, by sales of the second printing of *Gold & Galena*.

Over the last 18 years the Society has been blessed with innovative, creative, and enthusiastic members on their executive board. These have included: Don Baker, Harry Ewing, Linda MacDonald, Joyce Ronaghan, Lyn Bleiler, Verna Klippert, Ed Sveinson, Elizabeth Sveinson, Annie Caswell, Sherry Prichard, Ina-Mae Klippert, Bev Mason-Wood and Anne Meaney-Leckie.

The Society's work with historical projects in the new millennium will begin with planning the Centenary of Mayo in 2003.

Photographic and written information is still being sought by the Society, however large artifacts that come into its possession will continue to be put on permanent loan with the Keno City Mining Museum 38 km. north of Mayo on the Silver Trail.

Keno City Mining Museum, Keno. *L. Bleiler*

REFERENCE LIST

INTRODUCTION

Linda E. T. MacDonald

GEOGRAPHY

Geological Survey of Canada. *Memoir 284, Yukon Territory Selected Field Reports of the Geological Survey of Canada 1898–1933*, compiled and annotated by H. S. Bostock, (Pages 26, 128, 129, 130, and 161). Reprinted with permission of the Minister of Supply and Services, 1989.

THE NA-CHO NYÄK DUN

Cruikshank, Julie. "Through the Eyes of Strangers: A Preliminary Survey of Land Use History in the Yukon During the Late Nineteenth Century." February, 1974. Report to the Yukon Territorial Government and the Yukon Archives.

Gillespie, Archie–*The Mayo Miner*, Mayo, Yukon, 1935–1940. Various editions and articles.

Hager, Mary. Original story as told by Mary Hager to her granddaughter Nancy Sinnott.

Interviews with Mayo Elders: Buyck, Helen– Conducted by Linda MacDonald

Peter, Lucy & Sam–Conducted by Doris Ewing and Linda MacDonald

Hager, Mary–Conducted by Nancy Sinnott

Germaine, Paul–Conducted by Catherine Germaine

Mayne, Richard Charles. *Four Years in British Columbia and Vancouver Island*. London: J. Murray, 1862.

McClellan, Catharine. "The Yearly Round on the Stewart River" page 156. Reprinted from *Part of the Land, Part of the Water: a History of the Yukon Indians*. Vancouver: Douglas and McIntyre, 1987. By permission of Douglas & McIntyre.

McIntyre, Gordon. Written information from Gordon McIntyre.

Morissett, Father E. OMI. Paper prepared by E. Morissett, OMI, 1987, from Father Lechevallier's "La Croix dans les Roucheuses." Copy donated to the Mayo Historical Society by Father Henk Huijbers.

Yukon Archives. Anglican Church Records. Identified only as Chapter 4–Mayo, Mayo Indian Village, Elsa, and Keno City.

Map showing McQuesten Village Location. Mayo Mining Recorder's Office.

Map of the "Old Village" on the Stewart River. Mayo Mining Recorder's Office. Information for insert of house locations and residents names supplied by Mary Hager, Helen Buyck, and Lillian Popadynec.

MCQUESTEN, MAYO & HARPER and EARLY BAR MINING

Ogilvie, William. *Early Days on the Yukon & the Story of its Gold Finds*. Ottawa: Thornburn & Abbott, 1913.

Snow, George T. Manuscript in Snow Family Papers, MS38. Historical Library Section, Alaska State Library, Juneau, Alaska.

Wright, Allen A. *Prelude to Bonanza: The Discovery of Exploration of the Yukon*. Whitehorse: Arctic Star Printing, 1980.

MINING CONTINUES

Coutts, R. *Yukon Places & Names*. Sidney, B.C.: Gray's Publishing, 1980.

Gillespie, Archie–*The Mayo Miner*, Mayo, Yukon, 1935–1940. Various editions and articles.

National Archives of Canada, R.N.W.M.P. Records, RG 18, Volume 549, File MOS 106-111.

STEWART RIVER POST

Coutts, R. *Yukon Places & Names*. Sidney, B.C.: Gray's Publishing, 1980.

Gillespie, Archie–*The Mayo Miner*, Mayo, Yukon, 1935–1940. Various editions and articles.

Hudson's Bay Company Archives RG7/RBL E/97 "Stewart River" Winnipeg, Manitoba.

Sawatsky, Don. *Ghost Town Trails of the Yukon*. Langley, B.C.: Stagecoach Publishing Co. Ltd., 1975.

Shand, Margaret Clark. *The Summit and Beyond*. Idaho: Caxton, 1959.

Snow, George T. Manuscript in Snow Family Papers, MS38. Historical Library Section, Alaska State Library, Juneau, Alaska.

THE PLACER GOLD CREEKS

Bostock, H. S. *Pack Horse Tracks – recollections of a geologist's life in British Columbia and the Yukon 1924–1954*. Ottawa, Geological Survey of Canada, 1979. Open File 650.

Coutts, R. *Yukon Places & Names*. Sidney, B.C.: Gray's Publishing, 1980.

Debicki, R. L. Ed. *Yukon Mineral Industry 1941 to 1959*. Whitehorse: Indian and Northern Affairs, 1982.

Geological Survey of Canada. *Memoir 284, Yukon Territory. Selected Field Reports of the Geological Survey of Canada 1898–1933*, compiled and annotated by H. S. Bostock.

Interview with Ed Bleiler, who has placer mined on Dublin Gulch and Highet Creek from 1933 to the present [1989]. Has been a long-time resident of the Mayo area.

Mayo Mining Recorder's Office. Letter to Mayo Mining Recorder, R. L. Gillespie, from Livingston Wernecke, September 2, 1928.

Mayo Mining Recorder's Office. Mayo Mining Recorder's reports for 1934 and 1935.

Yukon Archives. Aaro Aho Collection, 82/161, *Keno Hill: An Era of Individualism in Yukon's Great Silver District*, ca. 1972.

Yukon Archives. Letter from White Pass & Yukon Route to Mr. G. P. McKenzie, Gold Commissioner dated October 30, 1919.

Yukon Archives. Mining Recorders Records (Placer) Series 10 1-7 Vol 267.

Yukon Archives. *Yukon Basin Gold Dredging Co., (Ltd.)* A prospectus of the Yukon Basin Gold Dredging Company with property on Stewart River. President William Ogilvie. circa 1908.

Mayo Landing

Geological Survey of Canada. *Memoir 284, Yukon Territory. Selected Field Reports of the Geological Survey of Canada 1898–1933*, compiled and annotated by H. S. Bostock.

R.C.M.P. Ottawa. Correspondence between R.C.M.P. Historian and the Mayo Historical Society, May 1982.

SILVER MINING

Geological Survey of Canada. *Memoir 284, Yukon Territory Selected Field Reports of the Geological Survey of Canada 1898–1933*, compiled and annotated by H. S. Bostock.

Mayo Mining Recorder's Office. Photocopies of various records.

Ronaghan, Roland, Mayo Mining Recorder. List of Some of the Early Silver Claims Staked in the Mayo District.

United Keno Hill Mines Limited, Elsa Yukon. Letter and documents from UKHM to Mayo Historical Society, dated March 26, 1982, October 3, 1988, June 6, 1989.

Yukon Archives. Aaro Aho Collection, 82/161, "Keno Hill: An Era of Individualism in Yukon's Great Silver District", ca. 1972.

Yukon Archives. Johnston, Diane. "Carrying the Territory: Silver-Lead Mining in the Mayo District 1919–1942." 81/100 F-166 Yukon Archives.Whitehorse, Yukon, 1981.

Yukon Archives. Mayo Board of Trade–General Vol. 53, File 32252 (Volume II, Series l).

MAYO

Conversations with pioneers.

Dawson Daily News. Yukon and Gold Fields Edition, 1922.

Gillespie, Archie–*The Mayo Miner*, Mayo, Yukon, 1935–1940. Various editions and articles.

Taylor & Drury Ltd.

Taylor, Charles. "Taylor & Drury Ltd Goes to Mayo in 1919" written specifically for Gold & Galena, 1988.

Yukon Archives. Johnston, Diane. "Carrying the Territory: Silver-Lead Mining in the Mayo District 1919–1942." 81/100 F-166 Yukon Archives. Whitehorse, Yukon, 1981.

N.W.M.P./R.N.W.M.P./R.C.M.P.

Gillespie, Archie–*The Mayo Miner*, Mayo, Yukon, 1935–1940. Various editions and articles.

Public Archives of Canada, B Division Monthly State Reports, RG 18,Volumes 2492 to 2495.

R.C.M.P. Ottawa. Information from 1912 Annual Report.

R.C.M.P. Ottawa. Letters and information from the R.C.M.P. Historian, Ottawa. GH 1510-2, GH 1520-4, CMR-88-11-70. May 1982, November 1988.

Webb, Melody. *The Last Frontier*. Albuquerque: University of New Mexico Press, 1985.

Yukon Archives. Aaro Aho Collection, 82/161, "Keno Hill: An Era of Individualism in Yukon's Great Silver District", ca. 1972.

District Mining Recorder's Offices

Coutts, R. *Yukon Places & Names*. Sidney, B.C.: Gray's Publishing, 1980.

McIntyre, Gordon. Mayo Mining Recorder 1947–1965 provided early information in conversations with Linda MacDonald.

McIntyre, Mabel. Information on the first Mayo Mining Recorder's cabin in Mayo.

Ronaghan, Roland, Mayo Mining Recorder. List of District Mining Recorders (by name) as well as information on location of various Recording Offices around the district.

Anglican Church

Yukon Archives. Anglican Church Records

The Catholic Church

Huijbers, Father Henk, OMI Compiled the list of priests who served in the Mayo District.

Lechevallier, OMI Paper donated by Father Huijbers, Mayo: Extract from "Cross in the Rockies" by Lechevallier, OMI.

Mayo School

Almstrom, Marjorie. Research paper (and attachments) titled "A History of the Mayo School, 1913–1949".

Conversations with Mary (Ross) Rich, Lillian (McLaren) Cheatham and Nora (McLaren) Swenson.

McIntyre, Gordon. "Mayo School 1932–1941" written specifically for Gold & Galena by Mr. McIntyre. November 1983.

Walloch, Dorothy (McLaren). Handwritten notes on the Mayo school, 1989.

Yukon Archives YRG l, Series 4, Vol. 13, File 311, Parts A and B. Researched and summarized by Suzanne Laberge.

Mayo General Hospital

Clark, J. V. Dr. Supplied information on the Mayo General Hospital.

Duncan, Allan M.D. Letter from Dr. Duncan to Mayo Historical Society dated 6th May, 1989.

Gillespie, Archie–*The Mayo Miner*, Mayo, Yukon, 1935–1940. Various editions and articles.

Mayo Historical Society. Original Town of Mayo Business License Register 1922–1936.

McIntyre, Gordon. Supplied information on the Mayo General Hospital.

Yukon Archives. Mayo General Hospital Volume II, Series I, Vol. 68, File No. 9; Volume V, Series 4; Vol. 25, File No. 403-3.

Recreation

Conversations with area pioneers.

Gillespie, Archie–*The Mayo Miner*, Mayo, Yukon, 1935–1940. Various editions and articles.

McIntyre, Gordon. Supplied written notes and photographs.

The Mayo Fire Department

Gillespie, Archie–*The Mayo Miner*, Mayo, Yukon, 1935–1940. Various editions and articles.

Mayo Volunteer Fire Department, Mayo, Yukon.

Taylor, Charles and Betty. Supplied information on Alexander Maclennan, Mrs. Taylor's father, along with other information.

Yukon Archives. Mayo Board of Trade–General Volume II, Series I, Vol. 53, File 32252.

Buildings & Artifacts

Bobillier, Father M., OMI, Dawson City, Yukon. Letter to Mayo Historical Society dated April 16, 1986, regarding the "Old Cemetery" at the Mayo-Elsa road junction.

Dawson Daily News, Yukon and Gold Fields Edition 1922.

Denman, Brian. *Yukon Tokens & Medals* Collections and Research, New Westminster, B.C.

Gillespie, Archie–*The Mayo Miner*, Mayo, Yukon, 1935–1940. Various editions and articles.

Letters from, and conversations with, area pioneers.

MacDonald, Linda. Representative drawing of old cabins along the riverbank near the Anglican Church with cabin locations and owner's names supplied by Gordon McIntyre.

Simpson, Scott and Stewart, D. N. "The Tokens of James Mervyn" Canadian Numismatic Journal Vol. 18, No. 10, November, 1973.

Yukon Government Land Titles Records Whitehorse, Yukon.

Veterans

Baker, Don. History of The Royal Canadian Legion Keno Hill Branch No. 3.

McIntyre, Gordon and Buyck, Wes. List of the Men who enlisted from the Mayo District during World War II.

National Archives of Canada, Personnel Records Centre. Copies of the sailing lists of the 2nd Cdn. Mounted Rifles and Yukon Machine Gun Detachment for SS GRAMPIAN sailing from Halifax on July l, 1915.

Parks Canada, Dawson City. Copy of Yukon Infantry Company Nominal Roll of Officers, Non-Commissioned Officers and Men for SS CANADA sailing from Halifax on January 24, 1917.

The Daily Province, Vancouver, May 20, 1919. Research by Lesley Buchan, Accessions Archivist, Yukon Archives.

Yukon Archives *Joe Boyle, King of the Klondike* by William Rodney; biography of Joe Boyle, Yukon Archives Search File.

Yukon Archives. *Martha Louise Black's Scrapbook: World War I*; Microfilm #62, 82/234 Yukon Archives Manuscript Collection.

Legislators

Conant, Cheryl. Paper titled "History of Administration in the Yukon Territory." Yukon Archives, 1981.

McIntyre, Gordon. Provided written notes.

Yukon Government Records. List of Territorial Councillors and M.L.A.'s for Mayo District. List provided by Piers McDonald, Mayo MLA.

TRANSPORTATION, COMMUNICATION & ELECTRIC POWER

Sternwheelers

Coates, Ken. Letter to Mayo Historical Society dated 9th September, 1988.

Cohen, Stan. *Yukon River Steamboats*. Missoula, Montana, Pictorial Histories Publishing Company, Second Printing 1983.

Mayo Historical Society. Taylor & Drury Store Records for the Mayo area.

Taylor, Charles, Whitehorse. Former Mayo resident and businessman provided additional information in conversations with Linda MacDonald.

Wilson, William H. "The Mayo Ore Deal" The Alaska Journal, Winter, 1979.

Trails and Roads

du Plessis, Raymond L., Q.C.Letter to Mayo Historical Society dated June 23, 1987.

Federal Department of Consumer and Corporate Affairs, Ottawa.

Hamilton, W. R. *The Yukon Story*. Vancouver, B.C.: Mitchell Press Ltd., 2nd Ed., 1967.

Nielsen, Erik. Letter to Mayo Historical Society dated November 12, 1985.

Trotter, J. B. "Angus McIntyre – Frontiersman". In Kelowna Courier, 1950.

Woodall, Robert G. *The Postal History of Yukon Territory, Canada*. Lawrence, Mass.: Quarterman Publications Inc. 1964, revised edition 1976.

Yukon Archives. Central Registry FilesVolume II, Series I, Vol. 4, File 182.Volume II, Series I, Vol. 13, File 2759.Volume II, Series I, Vol. 15, File 3262.Volume V, Series 4, Vol. 24, File 401.Volume V, Series 4, Vol. 27, File 410 (710).

Yukon Archives. Mayo Valley Railway Limited, Volume II, Series I, Vol. 30, File 32649.

Mayo Utilities Co. Ltd.

McIntyre, Gordon. Provided written notes.

Northwestel. Letter, and attachments, to Mayo Historical Society dated December 30, 1982.

Yukon Archives. Central Registry Files Volume III, Series l, Vol. 36, File 25201. Volume III, Series 2, Vol. 31, File 33771

Aviation

Cameron, R. B. "Bob", Yukon Aviation Historian, Whitehorse, Y.T. Notes 1983 and 1989.

Gillespie, Archie–*The Mayo Miner*, Mayo, Yukon, 1935–1940. Various editions and articles.

Harbottle, Jeanne. "Clyde Wann, Father of Aviation", The Alaska Journal, Autumn 1973, Vol. 3, No. 4.

Harbottle, Jeanne. "White Pass Aviation and its Rivals", The Alaska Journal, Autumn 1974, Vol. 4, No. 4.

Roxborough, Bill. Letter to Mayo Historical Society dated January 28, 1990.

Woodall, Robert G. *The Postal History of Yukon Territory, Canada*. Lawrence, Mass.: Quarterman Publications Inc. 1964, revised edition 1976.

Communications

Bennett, Gordon. *Yukon Transportation: A History*. Canadian Historic Sites: Occasional Papers in Archeology and History, No. 19. Ottawa: National Historic Parks, 1978.

Coates, K. and Morrison, W. *Land of the Midnight Sun: A History of the Yukon*. Edmonton: Hurtig Publishers, 1988.

Hamilton, W. R. *The Yukon Story*. Vancouver, B.C.: Mitchell Press Ltd., 2nd Ed., 1967.

MacLean, H. D. *Yukon Lady*. Surrey: Hancock House, 1985.

Woodall, Robert G. *The Postal History of Yukon Territory, Canada*. Lawrence, Mass.: Quarterman Publications Inc. 1964, revised edition 1976.

Yukon Archives. *The Yukon Territory 1926*. Ottawa: Department of the Interior, 1926. 971.21CA

Hydroelectric Power

Humphreys, E. W. "Ted." Paper titled "The N.C.P.C. in the Yukon."

Long, Joe. Paper written specifically for *Gold & Galena* titled "Northern Canada Power Commission." June 3, 1982.

Northwest Territories Power Commission. Map of Mayo River Hydro Development.

Yukon Archives. Central Registry Files Volume II, Series I, Vol. 10, File No. 2090.

OTHER SETTLEMENTS

Andre, Wendy. "Stewart Crossing."

Coutts, R. *Yukon Places & Names*. Sidney, B.C.: Gray's Publishing, 1980.

Yukon Archives. *Polk's 1915–16 Alaska-Yukon Gazetteer and Business Directory*. Seattle: R. L. Polk & Co., 1915. Vol. XI.

Yukon Archives. *Polk's 1923–24 Alaska-Yukon Gazetteer and Business Directory*. Seattle: R. L. Polk & Co., 1923. Vol. XII.

Mayo Mining Recorder's Office.Various records.

Mayo Historical Society.

THE PIONEER LIST

Anglican Church Records.

Bleiler, Edwin

Bolotin, Norm. *Klondike Lost: A Decade of Photographs by Kinsey & Kinsey*. Anchorage: Alaska Northwest Publishing Co., 1980.

Bostock, H. S. *Pack Horse Tracks – recollections of a geologists life in British Columbia and the Yukon 1924–1954*. Ottawa, Geological Survey of Canada, 1979. Open File 650.

Coutts, R. *Yukon Places & Names*. Sidney, B.C.: Gray's Publishing, 1980.

Gillespie, Archie–*The Mayo Miner*, Mayo, Yukon, 1935–1940. Various editions and articles. Mayo Historical Society and Yukon Archives. (Yukon Archives copies of *The Mayo Miner* researched by Peg Crook.)

Mayo Mining Recorder's office. Various records.

McIntyre, Gordon

Middlecoff, Elmer. Time Book for 1910–1918. The Time Book lists Middlecoff's employees on his Highet Creek placer mine.

Recollections of numerous district elders, pioneers and relatives of pioneers.

Research and interviews by Linda MacDonald and Lynette Bleiler.

Rich, Mary

Taylor, Charles and Betty

Terwilliger, Fred

Yukon Archives. Aaro Aho Collection, 82/161, "Keno Hill: An Era of Individualism in Yukon's Great Silver District", ca. 1972.

Yukon Archives. *Polk's 1915–16 Alaska-Yukon Gazetteer and Business Directory*. Seattle: R. L. Polk & Co., 1915. Vol. XI.

Yukon Archives. *Polk's 1923–24 Alaska-Yukon Gazetteer and Business Directory*. Seattle: R. L. Polk & Co., 1923. Vol. XII.

[Note: Polk's directories were published intermittently with issues being printed for the years 1903–04, 1907–08, 1909–10, 1910–12, 1915–16 and 1923–24. Mayo listings appeared only in the 1915–16 directory and 1923–24 directory which was the last combined Alaska-Yukon Directory and Gazetteer published.]

GLOSSARY

Adit A horizontal or nearly horizontal passage driven from surface into a hillside for exploration, working or de-watering a mine.

Assay Analysis of a sample of ores or minerals to determine the amount of valuable metals contained.

Assessment Work Annual work required to be done to retain mineral leases granted by government.

Bar This term was given to sandbars that formed in a creek or river. It increased in size as dirt, sand, gravel, black sands and gold were deposited.

Bedrock Rock of the earth's crust, in place.

Bench A flat area above a stream or river.

Breakup Thaw of snow and breakup of ice on rivers and lakes in the spring.

Cache A small log building, built off the ground on three or four poles, used to store meat and food supplies out of reach of animals.

Cat A tracked type of tractor, usually equipped with a blade for pushing earth.

Cirque A rounded basin gouged from a mountain by a glacier.

Claim A mining property granted by government lease.

Claim Jumper Someone who seizes, or illegally restakes a claim which has already been filed by another prospector.

Clean-up A term used to describe the removal of gold concentrates from the riffles of rockers, sluices, dredges, etc., after the gold-bearing gravel has been washed.

"Colours" Small specks or flakes of gold less than 1 mm in diameter.

Concentrator A milling plant that produces a concentrate of the valuable minerals or metals. Further treatment is required to recover the pure metal.

Discovery The first mineral claim granted in an area – generally a larger piece of land than any other for that region.

Dredge A barge with a rotating bucket line feeding sluice boxes, which digs and recovers gold from gravels below water level.

Drift A horizontal tunnel opening in or parallel to a vein, structure, or long dimension of an orebody.

Fault A fracture along which the walls of rock have moved relative to each other.

Float Loose pieces of rock (mineralized or not) dislodged and moved from their source by gravity, water or ice.

Fraction Smaller than full size mining claim staked between other claims.

Galena Lead sulphide crystals, always containing lead and often containing silver.

"Glacier" A local term used for buildup of ice from water seepage during winter months, often substituted for the term *aufies*.

Gold Pan A pan, usually 10 to 18 inches in diameter, with sloping sides for washing away gravel and sand, leaving heavy minerals in its bottom.

Greenstone Dark greenish rock, usually hard and dense.

Grey Copper Steel-grey copper antimony sulphur mineral often containing silver (tetrahedrite); also called freibergite if high in silver.

Grizzly A grating, constructed of rails, placed over the top of a sluice box or ore pass for the purpose of stopping large pieces of rock or ore that may hang up in the sluice box or pass.

Ground Sluicing The use of water to wash away overburden or gravel in an inclined ditch.

Grub Food.

Grubstake Financing or supplies of food etc., furnished to a prospector in return for an interest in any discoveries made.

High-grading Mining only the highest grade (most valuable) ore.

Hydraulicking Moving of gravel or overburden with high pressure water from a swivelling steel cannon or "monitor", the direction being controlled by a handle which deflects the nozzle.

"Jitney" Taxi-cab.

Lay Local term for a sub-lease on a mining property.

Level The horizontal openings on a working horizon in a mine. It is customary to work mines from a shaft, establishing levels at regular intervals (125 feet on average at United Keno).

Lode A mineral deposit in solid rock.

Low Grade Deposits of gold, silver, or other metals, which exist in insufficient quanities to be worked economically, except through large-scale methods.

Mill (concentrator) Combination of crushers, grinders, flotation cells, filters, etc., to separate valuable minerals from rock.

Monitor An apparatus fitted with a nozzle used in hydraulic mining. Water is forced through the nozzle under great pressure, then directed against gold-bearing gravel. The material is then washed into sluices where the gold can be separated.

Option An agreement to purchase a property reached between the property vendor and some other party that wishes to explore the property further.

Ore Shoot A concentration or lens of ore along a vein or structure.

Ore, Orebody A mineral deposit that can be mined at a profit under existing economic conditions.

"Outside" A term used by Yukoners for any place outside of the Territory's boundaries, particularly southern areas.

Overburden Soil or glacial material overlying bedrock or paydirt horizon.

Panning Using a gold pan to test for heavy minerals.

Pay, Paydirt Placer gravels that are economically mineable.

Permafrost Ground permanently frozen below seasonal thawing.

Placer Claim A property staked for placer minerals, generally gold.

Placer Valuable mineral or minerals (gold, platinum, tin, tungsten, diamonds, etc.) concentrated in stream gravels by water action over tens of thousands of years.

Pocket An unusually large concentration of gold in a small area.

Prospect A mining property, the value of which has not been proven by exploration.

Prospecting The act of searching for gold, silver, copper, lead, or any other valuable metal.

Pup Short creek, tributary to another.

Recording Claims Filing a legal record of staking at a mining recorder's office.

Retort A cast iron pot with a tight-fitting lid, connected to a condenser by a tube, which separates gold from mercury through heating.

Riffles Cross-bars or other devices in the bottom of a sluice box for catching gold or other heavy minerals.

Right or Left Limit (of a Stream) Right or left side when looking downstream.

Rim In placer mining, a bedrock surface of an older stream bed above deeper bedrock surface in present stream bed.

Rocker, Rocking A rocker is a box, usually made of planks, 3 1/2 to 4 feet long, about 2 feet wide, and 1 1/2 feet deep. The top and one end of the box is open, and the lower end and sides slope gradually until they reach the bottom. At its head is attached a closely-jointed box with a sheet-iron bottom, pierced with holes sufficiently large to allow pebbles to pass through. The machine is provided with rockers like a child's cradle, with cleats placed on the inside to arrest the gold in its passage. The rocker was placed by the water's edge and one miner fed it with dirt, while the other rocked the box and supplied it with water. The dirt to be washed was thrown into the upper iron box and a continual stream of water was poured in, causing the gold and pebbles to pass down to the bottom where the water carried the stones away while the cleats held the gold. If the gold was very fine, a piece of cloth was laid along the bottom box and covered with mercury (8 to 10 pounds) to arrest the gold. When a party of miners worked with rockers, they divided the labour of rocking, carrying water and digging equally among themselves. The rocker could also successfully be worked by one man.

Rough-locked Usually, ropes tied around the bottom of a sleigh for braking while going down hills.

Ruby Silver Silver antimony sulphide.

Scheelite Calcium tungstate.

Shaft Vertical or inclined opening extending downward for mine access.

Showing Occurrence of mineralization of possible economic value.

Single Jack Short handled four pound sledge hammer for drilling with hand-held steel.

Sluice Box An inclined trough with riffles at its bottom to catch gold or other heavy materials as gravel and sand are carried through it by water.

Sniping Mining on a small scale, usually by hand.

Sorting Picking out ore from waste, usually by hand.

Staking Claims Placing squared posts bearing staker's name and area being claimed as a mineral property.

Steampoint A sharp-tipped steel pipe which is driven into frozen ground, through which steam can be forced to thaw several feet of prospect pit or shaft.

Stope An excavation in a mine from which ore is being or has been extracted.

Stripping Removing overburden, gravel or valueless rock until the economic mineral zone is reached.

Tailings Waste material left after the valuable minerals have been removed from ore or gravel.

Tramline An aerial cable carrying buckets of ore.

Trench A long, narrow excavation dug through overburden, or blasted out of rock, to expose a vein or ore structure.

Tunnel A horizontal or nearly horizontal underground passage open to air at both ends, often loosely applied to adit.

Vein A more or less continuous body of minerals occupying a fissure or zone, differing in nature, and abruptly separated from the enclosing rock; often steeply inclined, usually a fracture filled with introduced minerals. Also referred to as a lode.

"Walking" To move a piece of heavy equipment from place to place under its own power.

INDEX

(Photograph captions are not indexed)